Thin-Film Diamond I

SEMICONDUCTORS
AND SEMIMETALS
Volume 76

Semiconductors and Semimetals

A Treatise

Edited by R.K. Willardson
CONSULTING PHYSICIST
12722 EAST 23RD AVENUE
SPOKANE, WA 99216-0327
USA

Eicke R. Weber
DEPARTMENT OF MATERIALS
SCIENCE AND MINERAL
ENGINEERING
UNIVERSITY OF CALIFORNIA
AT BERKELEY
BERKELEY, CA 94720
USA

Thin-Film Diamond I

SEMICONDUCTORS
AND SEMIMETALS

Volume 76

CHRISTOPH E. NEBEL

Physics Department/Walter Schottky Institute
Technische Universität München
Garching, Germany

JUERGEN RISTEIN

Institute for Technical Physics
University of Erlangen-Nuernberg
Erlangen, Germany

ELSEVIER
ACADEMIC
PRESS

Amsterdam • Boston • Heidelberg • London • New York • Oxford
Paris • San Diego • San Francisco • Singapore • Sydney • Tokyo

ELSEVIER B.V.
Sara Burgerhartstraat 25
P.O. 211, 1000 AE Amsterdam
The Netherlands

ELSEVIER Inc.
525 B Street, Suite 1900
San Diego, CA 92101-4495
USA

ELSEVIER Ltd
The Boulevard, Langford Lane
Kidlington, Oxford OX5 1GB
UK

ELSEVIER Ltd
84 Theobalds Road
London WC1X 8RR
UK

1st. edition 2003

Library of Congress Cataloging in Publication Data
A catalog record is available from the Library of Congress.

British Library Cataloguing in Publication Data
A catalogue record is available from the British Library.

ISBN-13: 978-0-12-752185-5
ISBN-10: 0-12-752185-2
ISSN:0080-8784

⊗ The paper used in this publication meets the requirements of ANSI/NISO Z39.48-1992 (Permanence of Paper). Printed in The Netherlands.

Contents

This page is intentionally left blank

List of Contributors

X. JIANG (1), *Institut für Werkstoffechnik, Universität Siegen, Paul-Bonatz-Str. 9-11, 57076 Siegen, Germany*

EBERHARD BLANK (49), *Ecole Polytechnique Fédérale de Lausanne (EPFL), Materials Science, CH-1015 Lausanne, Switzerland*

R. KALISH (145), *Department of Physics, Solid State Institute, Technion-Israel Institute of Technology, Haifa 32000, Israel*

A. DENEUVILLE (183), *Laboratoire d'Etudes des Propriétés Electroniques des Solides, Centre National de la Recherche Scientifique, University Joseph Fourier - Grenoble I, BP 166, 38 042, Grenoble Cedex 9, France*

S. KOIZUMI (239), *Advanced Materials Laboratory, National Institute for Materials Science, 1-1 Namiki, Tsukuba 305-0044, Japan*

C. E. NEBEL (261), *Walter Schottky Institut, TU-München, Am Coulombwall, D-85748 Garching, Germany*

MILOŠ NESLÁDEK (325), *Limburgs Universitair Centrum, Institute for Materials Research (IMO), Wetenschapspark 1, B-3590 Diepenbeek, Belgium*

KEN HAENEN (325), *Limburgs Universitair Centrum, Institute for Materials Research (IMO), Wetenschapspark 1, B-3590 Diepenbeek, Belgium*

MILAN VANĚČCK (325), *Academy of Science of the Czech Republic, Institute of Physics, Cukrovarnicka 10, CZ-16200 Prague 6, Czech Republic.*

ROLF SAUER (379), *Abteilung Halbleiterphysik (Department of Semiconductor Physics), Universität Ulm, D-89069 Ulm, Germany*

This page is intentionally left blank

Preface

Diamond: From gemstone to semiconductor

Diamond is a famous gemstone but also a very promising 'wide band gap' semiconductor that may one day rival silicon as the material of choice for high power and high frequency electronics. Its mechanical, optical and electronic properties are in many respects outstanding. Diamond has an indirect band gap of 5.47 eV leading to excitonic light emission at a wavelength of 235 nm, i.e. deep in the ultraviolet regime. Its thermal conductivity of 20 W cm^{-1} K^{-1} at room temperature is the highest known for any material, and many of its electronic properties (electric breakdown field, dielectric constant and saturation drift velocity) make it an ideal semiconductor. Drift mobilities for electrons and holes measured at room temperature in highly pure monocrystalline diamond have been reported to be 4.500 and 3.800 cm^2/Vs, respectively. Diamond is resistant to radiation damage, chemically inert and biocompatible. The surfaces of diamond, their terminal to the outside world in all applications, exhibits a number of features that are unique among all semiconductors. They can be engineered between hydrophilic, with a large work function and positive electron affinity in the case of oxygen termination and hydrophobic, with a rather small work function and a negative electron affinity in the case of hydrogen termination. The latter conditions lead to a two-dimensional p-type surface conductivity that is not fully understood yet but already exploited in a number of surface field effect devices.

The rarity of natural diamond largely prevented its exploitation as a technical material. With the invention of synthetic growth techniques at high pressure and temperature in the 1950s, diamond became available in large quantities and was used mainly for mechanical applications. But it was the discovery of a chemical vapor deposition (CVD) process in 1981 that laid the basis for diamond as a material for electronics. Polycrystalline diamond, grown on non-diamond substrates, became available in the form of extended thin films and free-standing plates. Many properties of high-quality polycrystalline diamond approach those of the best natural diamonds, but the presence of grain boundaries limits electronic performance. Increasing activities are therefore dedicated to grow monocrystalline diamond by microwave plasma-assisted homoepitaxial CVD techniques.

Whereas p-type doping by boron is well established for long, the pursuit for an appropriate donor remained unsuccessful for decades. Back in 1997, however, the first successful n-type doping of CVD diamond using phosphorus has been reported. This marked a breakthrough for diamond bipolar electronics. Meanwhile the first ultraviolet light-emitting p-n junctions are realized.

The basic requirements for diamond electronics are now available and progress in growth, doping and characterization of diamond has provided the tools for an efficient further development as a semiconductor.

This volume reviews the state of the art of thin film diamond. It covers growth issues, structural properties, doping and defect characterization, hydrogen in and on diamond as well as surface properties in general, and finally turns towards applications of diamond in electrochemistry, as detectors, as cold cathode emitters and in surface acoustic wave device applications. The eight chapters, written by recognized experts who represent academic and industrial institutions from Europe, Japan and the U.S., cover thoroughly the most recent results. Because the chapters may be read independently, the editors retained some overlapping among the paragraphs. The authors were encouraged to adopt a tutorial format to their contributions to make them as useful as possible to both experts and non-experts in the field of semiconductor research and development. The editors want to thank the invaluable help of Vijala Kiru and the entire Editorial staff of Elsevier for their excellent and patient work.

Christoph E. Nebel
Jürgen Ristein

CHAPTER 1

TEXTURED AND HETEROEPITAXIAL CVD DIAMOND FILMS

X. Jiang

INSTITUT FÜR WERKSTOFFECHNIK, UNIVERSITÄT SIEGEN, PAUL-BONATZ-STR. 9-11, 57076 SIEGEN, GERMANY

1. Introduction

The achievement of single crystal growth of large-area semiconductor films on different kinds of substrate materials is of technological importance in many solid state electronic devices. In addition, the heteroepitaxial structure is also of scientific interest because it is dependent on the relationship between the atomic arrangements in the substrate and the film. The crystalline perfection of the film further depends on the physicochemical nature of the substrate surfaces.

Combining a wide variety of outstanding material properties (Field, 1979; Angus and Hayman, 1988), diamond can be regarded as one of the most important materials for applications ranging from tribological, optical, thermal and electrochemical to electronic applications. With the development of various chemical vapor deposition (CVD) methods during the past years economic preparation of synthetic diamond films has become possible. For a series of fascinating and technological relevant applications, high-quality epitaxial monocrystalline films are required. By homoepitaxy diamond layers can be grown on natural or synthesized diamond crystals (Derjaguin et al., 1975; Spitsyn et al., 1981; Nakazawa et al., 1987; Kobashi et al., 1988; Yoshikawa et al., 1990). But natural diamond and synthesized diamond crystals are expensive and small (<0.5 mm). To overcome this problem the only way is to grow diamond films on appropriate substrates heteroepitaxially.

After the successful heteroepitaxial nucleation of diamond on β-SiC epitaxial films and on bare silicon wafers, obtained respectively, by research groups in the USA and Germany in 1992 (Jiang and Klages, 1993), considerable technical progress and scientific understanding have been achieved in the area of heteroepitaxy of CVD diamond. The decisive role of ion bombardment for the bias-enhanced epitaxial diamond nucleation has been recognized. It has been shown that a narrow parameter window exists for the growth of well-oriented diamond films, especially with respect to the choice of nucleation time and of kinetic conditions during deposition. A precise control of the nucleation was found necessary for the epitaxy. The epitaxial crystal growth is of Volmer–Weber type

ISBN: 0-12-752185-2
ISSN: 0080-8784

leading to polycrystalline film deposition with epitaxial orientation of individual crystallites with slight misorientation. Direct diamond epitaxy on silicon and on some metallic substrates has been demonstrated at the nucleation sites by cross-sectional high-resolution electron microscopic investigations. Large-area, single crystalline diamond films could not yet be prepared, in spite of the significant improvement of the crystallographic perfection.

In this chapter the advances in diamond heteroepitaxy and in textured diamond growth will be reviewed and the problems, which remain to be resolved, will be discussed. These advances raise hopes that single crystalline diamond films are not far beyond reach.

2. Heteroepitaxy of Diamond on Different Substrate Materials

2.1. c-BN

Because of its close lattice parameter with diamond ($a_D = 3.567$ Å; $a_{c\text{-BN}} = 3.612$ Å; lattice misfit = 1.3%) and its relatively high surface energy ($\cong 4.8$ J m^{-2} for the (111) surface of c-BN (De Vries, 1972), 5.3–9.2 J m^{-2} for low-index plane of diamond (Field, 1979)), c-BN is the most suitable substrate for diamond heteroepitaxy. Indeed, heteroepitaxial diamond was first and reproducibly grown by CVD on c-BN (Jones and Gunnarsson, 1989; Koizumi et al., 1990; Argoitia et al., 1993a,b; Koizumi and Inuzuka, 1993; Tomizuka and Shikata, 1993; Argoitia et al., 1994). The nucleation and growth of CVD diamond layers on both N- and B-terminated {111} faces have been investigated and show different behaviors, mainly caused by the different strengths of B–C and N–C bonds. While boron-terminated faces show high diamond nucleation density, the diamond crystals are hardly formed on the nitrogen-terminated faces (Koizumi and Inuzuka, 1993). The diamond crystals grow epitaxially on c-BN faces with a parallel, cubic-to-cubic orientation relationship. Due to the high surface free energy of c-BN a high nucleation density of diamond crystals of about 10^{10} cm^{-2} can be achieved without any pretreatment of the substrate surface.

Unfortunately, large c-BN single crystal cannot be grown at present. Cubic BN films have been grown in a polycrystalline form with an extremely small grain size (several nanometers) by PVD processes. Such films are, however, useless as substrates for the diamond epitaxy.

2.2. β-SiC and α-SiC

The second substrate candidate chosen for the diamond heteroepitaxy is β-SiC. In spite of the large lattice misfit of about 22% ($a_{\beta\text{-SiC}} = 4.359$ Å) Stoner and Glass (1992) obtained highly oriented diamond crystallites using the bias-enhanced nucleation method. Kawarada et al. (1997) further improved the nucleation and growth process and obtained continuous [001]-oriented diamond films with a high orientation perfection. The X-ray pole figure of the film also shows a cubic-to-cubic orientation between diamond and substrate, with the full width at half maximum (FWHM) value of the intensity peak being smaller than 1°.

In the case of α-SiC oriented diamond crystals were grown in the following orientation relationships which is often observed in the cubic-hexagonal systems: $(0001)_{\alpha\text{-SiC}}//(111)_{\text{diamond}}$; $[11\bar{2}0]_{\alpha\text{-SiC}}//[1\bar{1}0]_{\text{diamond}}$ (Suzuki and Argoita, 1996). Possibly due to the different lattice structure only poorly oriented, mostly isolated crystallites were obtained.

2.3. Si

Silicon, as the basic material for today's microelectronics, is strongly favored as a substrate material for thin film diamond devices. The most crucial point to overcome for a successful heteroepitaxial growth is the large lattice mismatch of 52% of diamond to silicon ($a_{\text{Si}} = 5.431$ Å) and the much lower surface energy of silicon (1.5 J m^{-2} for Si (111) plane). After the first successes of CVD diamond growth in the early 1980s many research groups attempted to achieve diamond heteroepitaxy on Si. Most of the experiments, however, yielded only randomly oriented island growth of diamond, although there was some evidence of local epitaxy of diamond on (001) Si substrates (Narayan et al., 1988; Jeng et al., 1990).

In 1992, Jiang and Klages reported that, in a MWCVD process, [001]-oriented diamond films can be epitaxially grown on the (001) plane of single crystal silicon, also by applying a negative electrical potential to the substrate without the intentional formation of an intermediate layer (Fig. 1). They found by high-resolution electron microscopy that diamond (001) grew directly on silicon (001) (Jiang and Jia, 1995; Jia et al., 1995). They stated that β-SiC is not a necessary interfacial layer required for diamond heteroepitaxy on Si. Later, Yang et al. (1994) and Stubhan et al. (1995) also realized heteroepitaxial growth of diamond (001) on silicon (001) in a HF-CVD system.

The structural quality of films has been gradually improved during the recent years. It has been possible to reduce the misorientation on Si (001) substrates from a best value of about 9° in 1992 (Jiang et al., 1992), as determined from the FWHM of X-ray rocking curves, to a best value of about 2° in 1998 (Jiang et al., 1998a–c). By a strictly controlled deposition process thick films with average lateral grain sizes >30 μm were deposited which show a high crystal quality as confirmed by the similar electronic properties as diamond single crystal (Janischowsky et al., 1999; Jiang et al., 2000). Figure 2 shows the scanning electron microscopic morphology of a 140 μm-thick film. The film surface is fully covered by (001) faces of diamond grains.

2.4. Graphite and BeO

Heteroepitaxial growth of diamond was also attempted on other kinds of substrates. (111)-oriented diamond particles have been deposited on the (0001)

FIG. 1. Scanning electron microscopic image obtained from a heteroepitaxial diamond film on Si with a thickness of about 6.5 μm (Jiang and Klages, 1993).

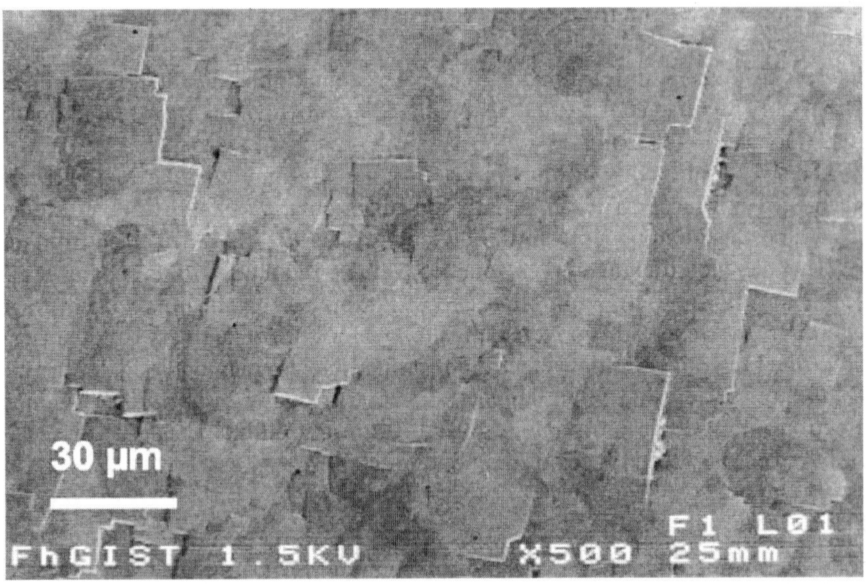

FIG. 2. Scanning electron microscopic image of a [001]-oriented, 140-μm-thick diamond film on Si (Jiang et al., 2000).

faces of graphite. Due to the poor orientation of diamond to graphite only isolated grains are reported to form and no continuous films have been obtained (Suzuki et al., 1994).

BeO has the hexagonal wurzite structure with Be–O bonding lengths only 7% greater than the C–C bond in the diamond crystal. Using HF-CVD, poorly oriented diamond crystallites were formed on a BeO single crystal (Argoitia et al., 1993a,b). Due to the toxic substrate material it is questionable whether the technique will have practical applications.

2.5. Metallic Substrates

Due to the close lattice match of some metals to diamond many experiments have been performed aiming at achieving better oriented diamond films than for depositions on the silicon and silicon carbide substrates. The first attempt on Cu failed due to a poor nucleation density. Earlier reports about the epitaxial nucleation by ion implantation could not be reproduced (Lifshitz et al., 1989). On Ni and Pt substrates, in contrast, [111]-oriented diamond films could be grown (Zhu et al., 1993; Tachibana et al., 1997) by seeding diamond particles and subsequently by plasma annealing. A minimum value of the grain misorientation of about 1° tilt misorientation has been achieved by the same group on Pt/Ir/Pt/sapphire substrates.

Recently, superior-quality heteroepitaxial growth of diamond on iridium layers has been reported (Ohtsuka et al., 1996; Ohtsuka et al., 1997; Schreck et al., 1999). The iridium layers have been deposited on the (001) planes of MgO or on mechanically polished (001) planes of $SrTiO_3$. Iridium has a lattice constant of 3.840 Å, which is close to that of diamond and it does not form carbide under the conditions of diamond deposition. The catalytic effect which results in the formation of graphitic carbon and happens on substrates containing Fe, Co and Ni does not occur on Ir. Thus Ir would be a good candidate as substrate for diamond heteroepitaxy. By using bias-enhanced nucleation and subsequent growth by a microwave plasma CVD process, highly oriented diamond films have been grown on Ir (001) faces. The size of single crystal domains can be significantly increased with increasing film thickness, while the XRD polar and azimuthal spread for the crystal orientations are less than 1° (Schreck et al., 2001). Although being expensive this approach may offer a promising route for realizing large area diamond epitaxy.

In terms of practicality, the realization of diamond heteroepitaxy directly on silicon is particularly attractive, because Si wafers are easily available and extensively used in electronic industry. The epitaxial growth of diamond on Si is highly desirable, particularly in view of its convenient integration of diamond electronics with Si device technology. Therefore the realization of heteroepitaxy on Si has become a great challenge of immense technological and scientific interest. In the following I will concentrate mainly on this issue.

3. Nucleation of CVD Diamond Films

3.1. Practical Significance of Nucleation

Nucleation is the first and critical step of CVD diamond growth. The control of nucleation is essential for optimizing the diamond film properties necessary for targeted applications such as grain size, orientation, transparency, adhesion, and roughness.

The investigation of diamond nucleation not only leads to insights necessary for the controlled growth of diamond films suitable for various applications, but it can also provide insight into the mechanism of diamond growth. To date, the understanding of diamond nucleation is very limited. Carbon atoms can form different types of chemical bonds via sp^1, sp^2, and sp^3 hybridization. Under the experimental conditions used in CVD, diamond consists solely of sp^3 bonds and is thermodynamically metastable compared to graphite, which is composed of sp^2 bonds. It is an interesting and yet intriguing problem why metastable diamond can be grown on diamond or non-diamond substrates under CVD conditions.

3.2. Methods for Diamond Nucleation

During the early development of CVD diamond deposition, diamond single crystals were used as substrates (Eversole, 1962; Lander and Morrison, 1964; Spitsyn et al., 1981). Later, diamond seeds were used (Matsumoto et al., 1982). The earliest efforts were limited to homogeneous growth or homoepitaxy of diamond. In 1982 Matsumoto achieved a breakthrough in growing diamond on non-diamond substrates without using diamond seeds, resulting in a low nucleation density of diamond nuclei. A continuous film could not be formed. Mitsuda et al. (1987) found that scratching of the substrate surface with diamond powder could greatly enhance the nucleation density. Since then, substrate scratching has become the most common and powerful method for achieving a high nucleation density and fine grains of uniform size. For silicon substrates, which have been studied intensively, a nucleation density of 10^7-10^8 cm^{-2} can be routinely obtained after scratching with diamond powder. In contrast, with non-scratched substrates, the nucleation density only reaches 10^4 cm^{-2}. Besides diamond powder other abrasive powders such as c-BN, TaC, SiC, and even iron can be used for scratching the substrate surface in order to enhance nucleation density. Nevertheless, diamond powder is regarded as most effective among the hard materials.

Later, investigations have revealed that coating the substrate surface with graphite (Feng et al., 1993), amorphous carbon (Morrish and Pehesson, 1991), C_{60}, and even mechanical oil (Morrish and Pehesson, 1991) can enhance greatly the nucleation density. However, these methods including the scratching method mentioned above, cannot lead to oriented heterogeneous nucleation or epitaxial growth on non-diamond substrates.

Yugo et al. (1991) reported the bias-enhanced nucleation method by which they obtained a high density of nucleation on a mirror-polished substrate (without scratching) using the MWCVD system. They applied a negative bias to the substrate during nucleation and obtained diamond nucleation with a density as high as $10^9-10^{10}\,\text{cm}^{-2}$ on Si. Subsequent developments of bias-enhanced nucleation by Jiang et al. and Stoner et al. have led to the heteroepitaxial growth of diamond on silicon (Jiang and Klages, 1993; Jiang et al., 1993a–c) and on silicon carbide (Stoner and Glass, 1992) substrates, respectively.

For the popular HFCVD method, a negative substrate bias cannot induce enhanced nucleation since the gas reactants consisting of atomic hydrogen and hydrocarbon radicals are neutral species. However, when a plasma is generated by the proper choice of bias voltage, an enhancement of diamond nucleation similar to MWCVD can be achieved also for the HFCVD process (Stubhan et al., 1995; Chen and Lin, 1996).

Other methods for enhancing diamond nucleation have also been advocated. One method is the nucleation by straining the substrate lattice, while the other one is by ion implantation into the substrate prior to the introduction of methane into deposition chamber.

For the heteroepitaxial diamond deposition the bias-enhanced nucleation has been proved most successful and in the following the nucleation schemes for deposition on Si and β-SiC will be discussed.

3.3. Heterogeneous Nucleation of Oriented Diamond

Generally, diamond heteroepitaxy on silicon and β-SiC proceeds in two stages, namely, nucleation and growth. Because substrate scratching causes serious damage to the arrangement and periodic structures of the surface atoms, epitaxial growth cannot be achieved via this technique. To avoid the substrate scratching with diamond powder bias-enhanced nucleation was applied to a mirror-polished Si substrate in a MWCVD system. Figure 3 shows a schematic diagram of the process parameters. Prior to the bias-enhanced nucleation, in situ hydrogen plasma etching was performed in order to remove the native surface oxide layer. By carefully controlling the nucleation process, diamond nucleation with more than 90% [001]-oriented nuclei was achieved (Jiang and Klages, 1996).

The investigations of the nucleation process reveal a narrow parameter window for epitaxial nucleation. The crucial parameters are the substrate temperature, the methane concentration, the applied bias voltage during nucleation and the nucleation time. A critical bias voltage exists, beyond which the energy barrier for the formation of stable nuclei can be overcome (Jiang et al., 1993a–c). The experimental results showed different values of the critical bias voltage for various reaction chambers. Typically, the critical bias voltage is approximately 80–100 V, depending on other parameters, above all the process pressure (Fig. 4).

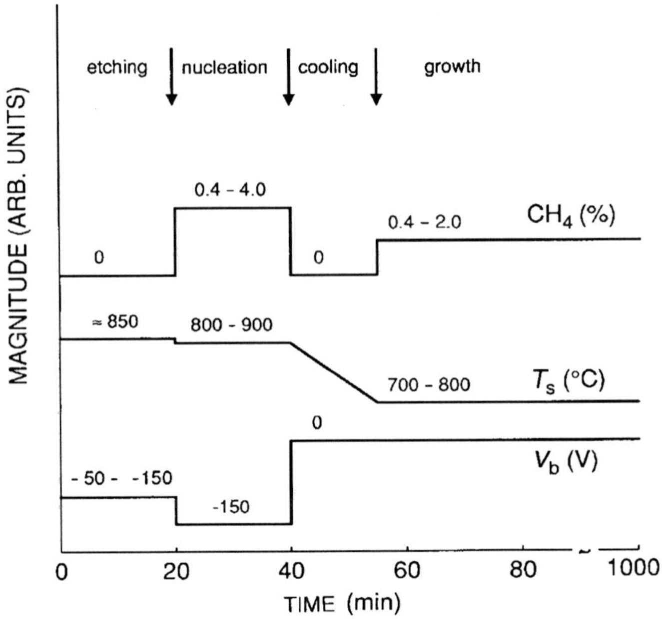

FIG. 3. Schematic diagram of the MWCVD process for preparing heteroepitaxial diamond films.

Atomic force microscopy (AFM) and reflection-high-energy electron diffraction (RHEED) were employed to investigate the early stage of diamond nucleation (Jiang et al., 1993a–c, 1994). The RHEED patterns obtained for different nucleation times suggested that a certain period of time is necessary for

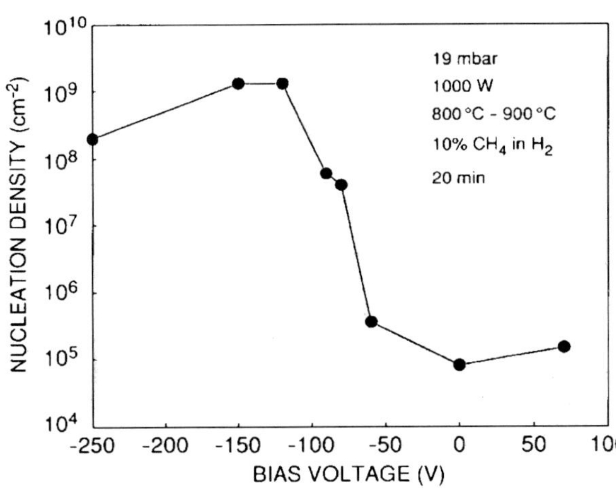

FIG. 4. Diamond nucleation density as a function of substrate bias voltage (Jiang et al., 1993a–c).

the initiation of nucleation. It has been demonstrated that the diamond nuclei formed at the start of the nucleation are oriented. As the nucleation time increases, the nucleation density increases dramatically to $5 \times 10^{10}\ \text{cm}^{-2}$ (Jiang et al., 1994a,b) and the nuclei are randomly oriented. These results confirm the importance of controlling the nucleation process during epitaxial film growth (Jiang and Klages, 1996). Schreck et al. (1997) studied this processing window for heteroepitaxial nucleation of diamond on Si(001) using the bias process by X-ray diffraction texture measurements. The temporal development of the azimuth pole density distribution is shown in Figure 5(a). It can be found that the bias time lies within a distinct time interval. As the bias voltage $|V_b|$ decreases, the biasing time and the time window for an optimal crystal alignment increase sharply (Fig. 5(b)).

A possible mechanism of the loss of diamond epitaxy beyond the parameter window for optimized diamond growth is suggested by Jiang et al. (1996) and Schreck et al. (1997) based on investigating diamond growth under bias condition. The authors reported on slightly misoriented crystallites grown homoepitaxially on {100} facets under bias conditions. According to their results, the crystal misorientation can be traced back to either the formation of defects during the homoepitaxial growth of the crystallites induced by ion bombardment or the re-nucleation of strongly misoriented grains. A valuable hint provided by these investigations is that, even for homoepitaxial diamond growth which has the perfect lattice match, a misorientation of the subsequently grown crystallites occurs, when the underlying crystal lattice is damaged. The lattice distortion-induced misorientation of the heteroepitaxial diamond films on silicon and on other kinds of substrates explain the narrow parameter window for a successful bias-enhanced epitaxial nucleation.

3.4. Mechanisms of the Bias-enhanced Nucleation

A complete explanation of the mechanism of bias-enhanced epitaxial nucleation is still a object of controversial discussion and being studied intensively. This is due to the difficulties to distinguish the aspects of plasma-chemical vapor deposition (presence of reactive gas species of thermal energy at the substrate) and physical vapor deposition (PVD, hyperthermal ionic species bombarding the substrate). After numerous investigations different models have been proposed. Yugo et al. (1993), Gerber et al. (1995) and Sattel et al. (1996) suggested a shallow ion implantation model in which the sp^3 bonded carbon clusters, formed by low energy ion implantation, act as nucleation precursors. The negative bias causes the positively charged ions in the growth chamber to accelerate towards and bombard the substrate surface, thereby removing the contamination and facilitating cluster formation at the surface. These events in turn advance diamond nucleation. Stoner et al. (1993a,b) on the other hand suggested the critical process should be the change in plasma chemistry, such as

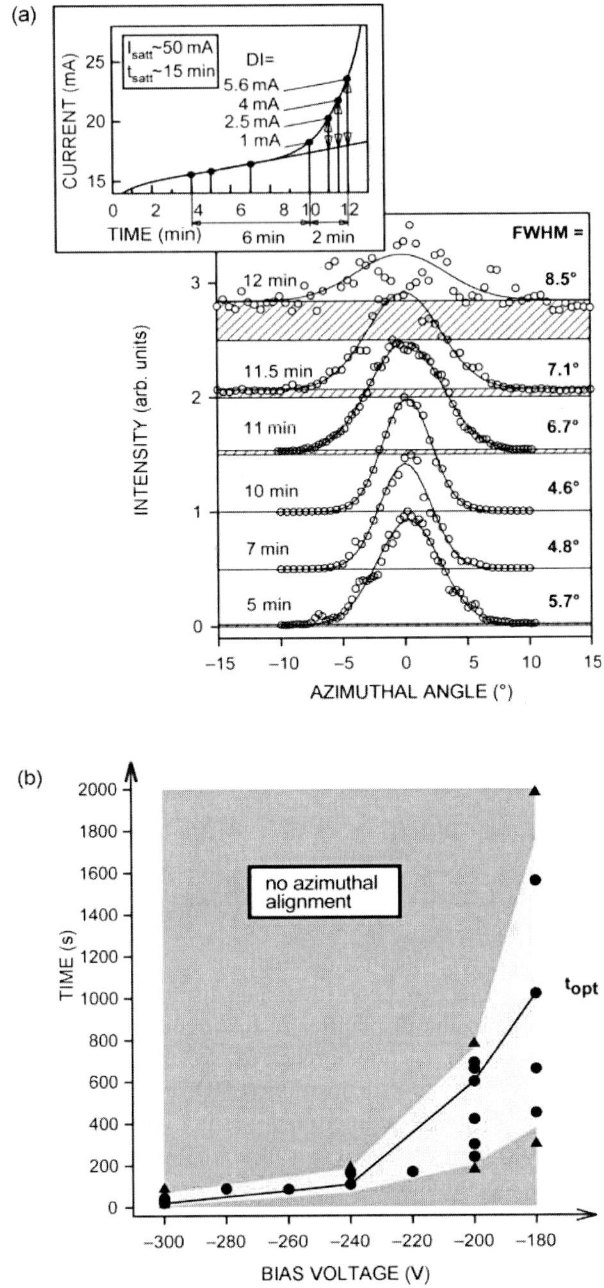

FIG. 5. (a) Azimuthal {220} pole X-ray intensity for diamond films obtained for a variation of the biasing time; (b) variation of the process time window for oriented nucleation with the bias voltage (Schreck et al., 1997).

the increase in the concentration of atomic hydrogen caused by substrate biasing and the formation of a carbide surface layer. Jiang et al. (1994) found that the temporal evolution of the overall nucleation density corresponds well with a surface kinetic model established by Tomellini et al. (1991) involving immobile active nucleation sites, germs and nuclei. They also suggested that, in addition to surface defects (point defects, steps and sp^3 bonded carbon clusters) serving as the nucleation sites, the enhanced surface diffusion and sticking probability of carbon on silicon due to ion bombardment should be the decisive factors. The enhancement of the surface diffusion of carbon species has been identified by investigation of the distribution of the first nearest-neighbor distances (Jiang et al., 1994a,b).

Generally, the nucleation of a new phase requires formation of a nucleus exceeding a critical size and becoming thermodynamically stable. According to the kinetic theory the nucleation may be described by the following sequence as schematically shown in Figure 6: (a) vapor species arrive on the substrate surface (atoms, molecules, and/or radicals) and are adsorbed; (b) the species diffuse over the substrate surface and part of the adatoms desorb from the surface into the vacuum; (c) the adatoms combine and form clusters due to fluctuations of the local species concentration; (d) nuclei grow by capturing of the adatoms or by direct impingement of the atoms from vapor phase. The critical size of the nuclei depends on the total free enthalpy of formation related to the free energies of surfaces of nuclei and substrate. Because of the strong bonding and an energetic advantage, the nuclei form preferentially on surface defects, such as steps, dislocations or surface impurities. Such surface defects are often called 'active' sites for nucleation.

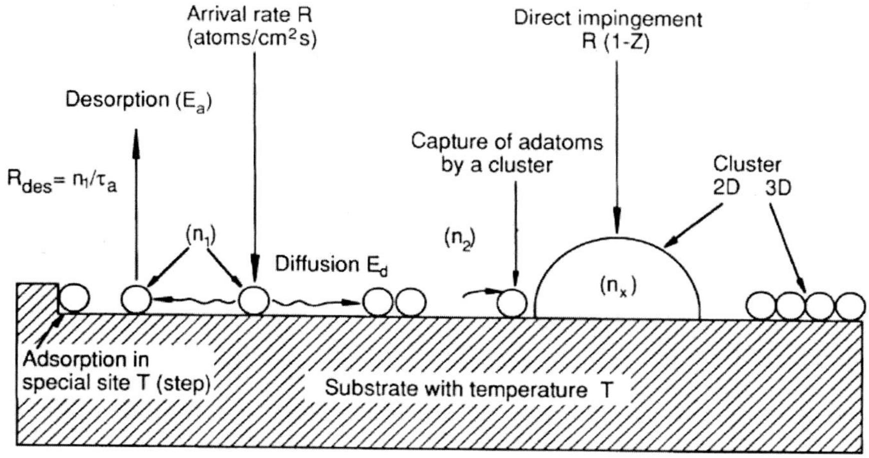

FIG. 6. Elementary processes during island nucleation.

The diamond nucleation in a BEN process should follow this sequence. In comparison with this sequence the above-mentioned different models have only considered the possible key factors of the processes critically influencing the nucleation. For example, the subplantation of hyperthermal ions leads to the formation of sp^3-bonded amorphous carbon cluster. This initial foreign phase (with respect to the diamond phase) can transform itself into a diamond nucleus by atom reorganization. In addition to a possible variation of plasma chemistry, this leads to a strongly reduced formation enthalpy of diamond nucleus. The mobilization of surface species enhanced by the ion bombardment can accelerate the nucleation. In the following some important results demonstrating the important role of substrate biasing for diamond nucleation will be summarized.

3.4.1. Role of Ion Bombardment for the Bias-enhanced Diamond Nucleation

To identify the effect of substrate biasing, different experiments have been carried out, in which BEN was performed on topographically structured silicon wafers containing grooves with vertical side walls obtained using reactive ion etching (Fig. 7) (Flöter et al., 1995; Jiang et al., 1995) as well as SiO_2-masked silicon substrates (Jiang et al., 1994a,b; Yugo et al., 1994). The height of the vertical walls is only about 2 μm. When a negative electrical potential was applied to the substrate, the positively charged ions in the plasma were accelerated towards the wafer. An ion flux with a preferred direction perpendicular to the substrate will be formed. Due to the mean free path of the ions of approximately 5 μm under gas pressures around 25 mbar, the ions flying towards the substrate cannot 'feel' the groove surface structure. As a result, the faces which are parallel to the wafer surface will be bombarded whilst the side walls will not (please see the schematic diagram of Figure 7). Therefore diamond was only nucleated on surfaces parallel to the wafer, while the side walls hardly experienced any diamond deposition, demonstrating that the directional flux of energetic species towards the substrate surface is essential for the observed phenomena.

3.4.2. Role of Kinetics During Bias-enhanced Nucleation

It was demonstrated (Jiang et al., 1998a–c) that the temporal evolution of overall nucleation density is not in conflict with a surface kinetic model involving immobile 'active sites', 'germs', and 'nuclei' as has been suggested by Tomellini et al. (1991) to describe the formation of diamond crystallites on Si. Their modified three-step model introduces an extra step for the generation of nucleation sites and has the kinetic scheme $R \rightarrow N^+ \rightarrow m \rightarrow N$ where R, N^+, m and N are densities of the surface centers that can be transformed into a nucleation site, the

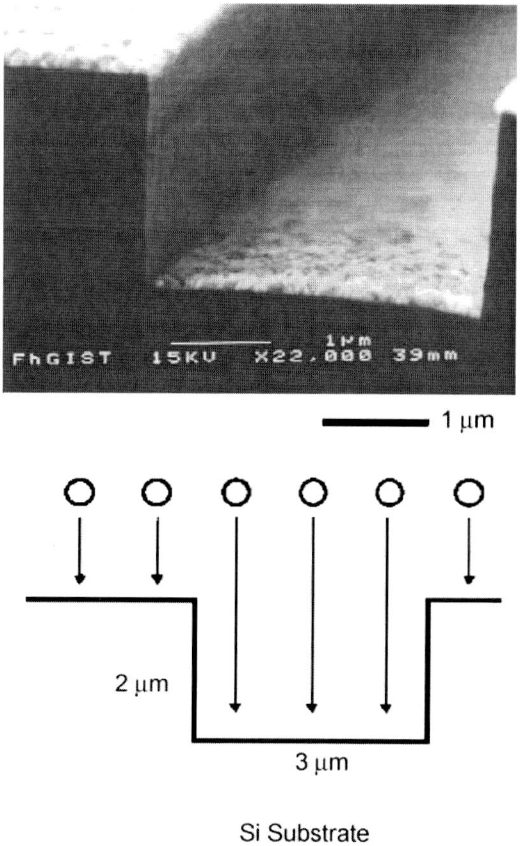

FIG. 7. Selective growth of diamond crystals on a grooved silicon wafer containing both faces perpendicular and parallel to the substrate. The different nucleation densities on the faces confirm the decisive role of ion bombardment for the diamond nucleation. Lower part: sketch of substrate surface and incident ions (Jiang et al., 1995).

nucleation sites, the unstable clusters, and the stable nuclei (= nucleation density).

The nucleation density N (islands per cm^2) is plotted in Figure 8 as a function of nucleation time t, which is counted from the induction time t_{ind} is defined as a time that the aggregates formed initially have been detectable. t_{ind} can also be understood as the period for production of the nucleation sites. The time-dependence of nucleation density was obtained not only by directly counting nuclei from the SEM images, but also by establishing a relationship between the particle radius distribution function and the nucleation rate through the knowledge of the nucleus growth law. From Figure 8 it is evident that, once the nucleation begins, the nucleation density increases rapidly to $4.3 \times 10^{10}\,cm^{-2}$ within 8.5 min. The main part of the substrate surface is then covered by the nuclei and the

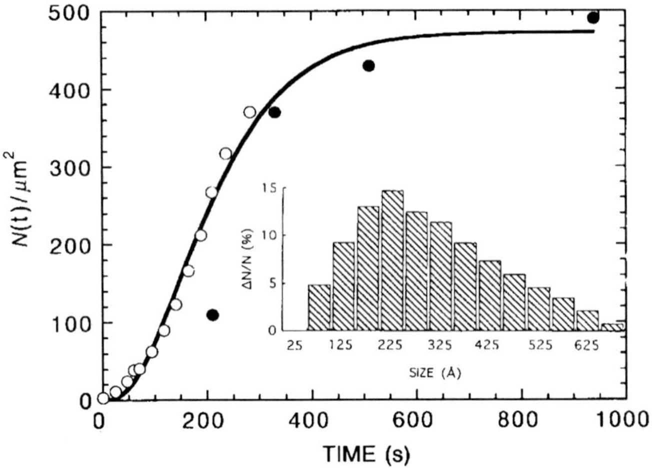

FIG. 8. Nucleation density $N(t)$ (islands per cm^2) vs. t deposition time. In the abscissa an induction time of 6.5 min is subtracted. The curves are obtained by computer modeling using the three-step kinetic models proposed by Polini and Tomelini (1995). The open and full circles in the plots represent data calculated from crystal size distribution (inset) and are obtained by direct particle counting, respectively (Jiang et al., 1998a–c).

increase of the density, i.e., the nucleation rate, is lowered with increasing coverage.

From Figure 8 it can be noticed that the nucleation kinetics derived from the size distribution function (open circles in the plots, deposition time = 12 min) is close to the result directly measured by particle counting at different times (full circles) and the diamond nucleation agrees very well with kinetics of the established surface process.

3.4.3. Role of Surface Diffusion During Bias-enhanced Nucleation: Measurements of First-nearest-neighbor Distance Distribution of the Nuclei

Evidence of surface diffusion of growth species is obtained by measurement of first-nearest-neighbor distance distribution of the nuclei (Jiang et al., 1994a,b). If surface diffusion of mobile species (adatoms) towards immobile nuclei is involved in the nucleation process, the presence of nuclei, acting as sinks for the mobile species, should reduce the probability of formation of new nuclei in their immediate vicinity. The first-nearest-neighbor distance distribution of the nuclei should therefore deviate from a random distribution. This was actually observed in SEM and AFM investigations showing a depletion zone around each nucleus.

Figure 9 compares the measured distributions with calculated distributions (solid lines) of the first-nearest-neighbors of a random distribution of diamond

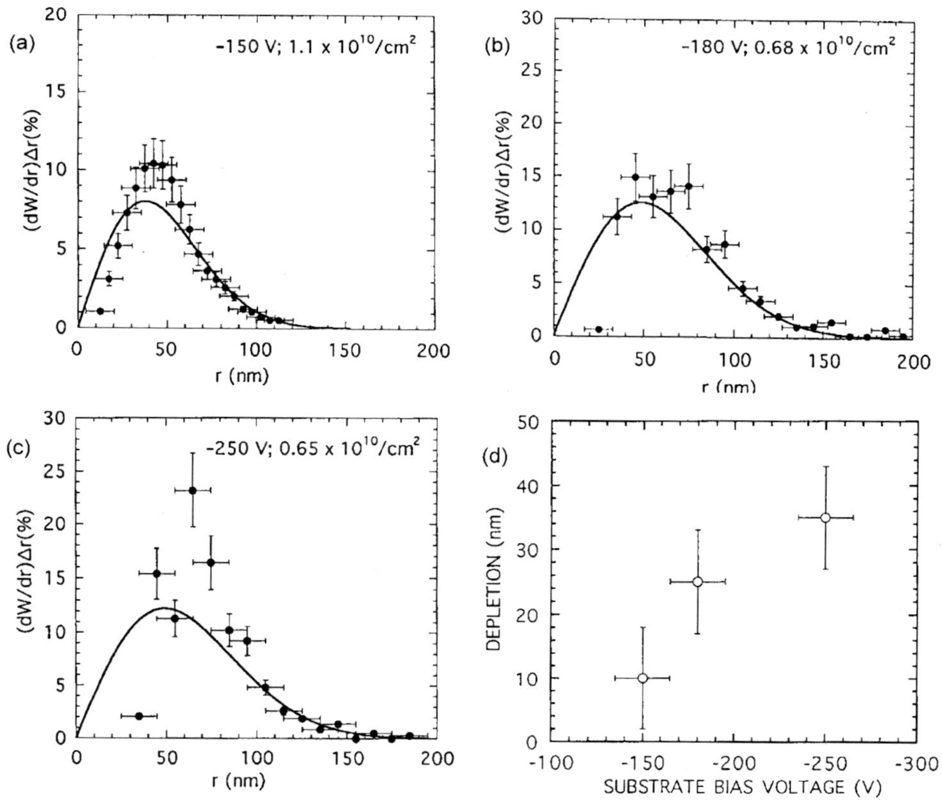

FIG. 9. Comparisons of measured nearest-neighbor distance distributions and the distributions predicted from a random nucleation model for samples prepared under different substrate bias voltages. (a) $V_b = -150$ V; (b) $V_b = -180$ V; (c) $V_b = -250$ V; (d) depletion distance vs. V_b (Jiang et al., 1998a–c).

nuclei deposited under − 150, − 180, and − 250 V, respectively. The nucleation densities of different samples were kept at a similar range and were 1.1×10^{10}, 0.68×10^{10} and $0.65 \times 10^{10}\ \mathrm{cm}^{-2}$, respectively. All the measured distributions are shifted to larger distances in comparison to the random distribution, and a depletion zone around each cluster was observed. The depletion zone of the nuclei increases from 12 nm for $V_b = -150$ V to 35 nm for $V_b = -250$ V (Jiang et al., 1998a–c).

These results imply that the diamond nucleation is strongly influenced by the existing nuclei—an effective 'repulsive interaction' among the clusters exists and the clusters tend to separate. The surface diffusion of the adsorbed species leads to a decrease of the precursor concentration and the nucleation frequency drops down to zero on the border of one island. The fact that the depletion is a function of substrate bias voltage confirms the contribution of surface diffusion to the depletion. Due to the bias induced ion bombardment the mobility and the diffusion of adsorbed species on the substrate surface will increase and the spatial extension of the surface diffusion-caused depletion zone will increase. The depletion zone size should correspond to the minimum distance from a nucleus up to which the precursor concentration is influenced by the existing nucleus. The role of the surface diffusion of the species during CVD diamond growth has been investigated also by Flanklach and Skokov (1995).

To summarize the results obtained for the nucleation stage, the decisive role of ion bombardment for the diamond nucleation and the surface diffusion of the species forming diamond has been confirmed. The diamond nucleation rate agrees with the value calculated by a kinetic model of surface processes. Based on these results the nucleation sequence can be understood as follows:

(i) *Formation of nucleation sites.* The nucleation sites can be considered as sites having high local surface energy, such as defects and surface impurities. Due to an increased ion bombardment, a shallow implantation of hyperthermal ions can occur if the ion energy is large enough (>20 eV) (Sattel et al., 1996). A growing amount of C and Si interstitials result in densification and localized stress and increase the local surface energy. One important factor of applying a substrate bias for accelerating the nucleation is the abstraction of the adsorbed hydrogen of the substrate surface. This leads to an increase in the sticking coefficient of hydrocarbon radicals to form high energy nucleation sites.

(ii) *Formation of carbon clusters due to the enhanced surface diffusion.* Applying a negative substrate potential will increase the bombardment by positively charged energetic ions and thus increase the mobility of surface precursors and hence the probability of forming clusters. The enlargement of the first-nearest-neighbor distance provides a strong evidence for the adatom diffusion.

(iii) *Formation of stable diamond nuclei.* Once the clusters reach the critical size, which depends on the experimental conditions and is still unknown, the growth stage begins. The attraction of the positively charged hydrocarbon species, which leads to a carbon supersaturation at the substrate surface and the increasing generation of atomic hydrogen probably assists in increasing nucleation rate.

4. Interface and Grain Misorientation

4.1. Cross-sectional TEM Investigation of Interfaces

So far, the epitaxial crystallites of diamond grown on Si have the following crystallographic relationship to the substrate:

Diamond{hkl}//Silicon{hkl}

and

Diamond⟨hkl⟩//Silicon⟨hkl⟩.

The interface structure between a heteroepitaxial diamond and silicon with a perfectly epitaxial orientation was revealed by HRTEM in 1995 (Jiang and Jia, 1995; Jia et al., 1995) A high-resolution lattice image of the interface is shown in Figure 10. The cubic-to-cubic orientation relationship between diamond and silicon, i.e., $(001)_{\text{diamond}}//(001)_{\text{silicon}}$ and $[110]_{\text{diamond}}//[110]_{\text{silicon}}$, is clearly demonstrated. No secondary phases like β-SiC, graphite and amorphous carbon can be recognized in the interface area. Inspecting the image at a glancing angle along the two sets of {111} planes, it becomes obvious that the large lattice mismatch between silicon ($a_{\text{Si}} = 0.543$ nm) and diamond ($a_{\text{D}} = 0.357$ nm) is accommodated by the introduction of a 3 : 2 registry (with 1.5% mismatch) for the diamond lattice with respect to that of silicon, e.g., every three Si (111) fringes are matched well with four diamond (111) fringes (the multiple mismatch is about 1.5%). Following the lattice fringes from Si, across the interface, to the diamond we find that every third lattice plane of diamond terminates at the interface as denoted by arrows, forming individual 60° interface misfit dislocations, with a spacing of about 7 Å between two such dislocations.

The lattice image shown in Figure 10 provides us with strong evidence that diamond crystals can be epitaxially grown directly on silicon in spite of the large lattice misfit. A thin epitaxial intermediate layer is unnecessary for the epitaxy. This important conclusion has, however, been questioned in recent years due to the fact that different interface configurations have been observed in different laboratories. In early studies, the diamond films with random orientation were nucleated by scratching the substrates with diamond seed crystals and it was found that diamond was grown on Si through a β-SiC intermediate layer

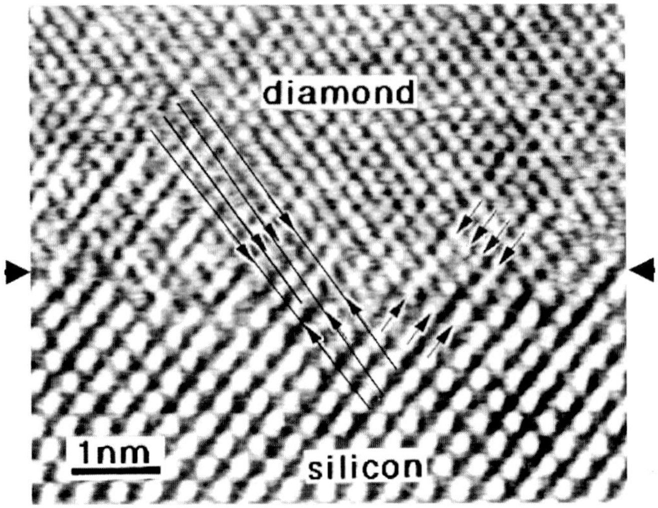

FIG. 10. A lattice image of the interface between diamond and silicon. The micrograph is taken in the [110]-zone axis of diamond and silicon crystals (Jiang and Jia, 1995).

(Williams and Glass, 1989). This β-SiC layer was also frequently found after the diamond nucleation stage by means of BEN (Fig. 11; Wurzinger et al., 1997; Wittorf et al., 1997a,b, 2000). Considering the large lattice mismatch the formation of a β-SiC intermediate layer seems to be reasonable since the misfit between diamond and β-SiC, $\delta = (a_{\beta\text{-SiC}} - a_D)/a_D = 22\%$, is much smaller than that between diamond and silicon, $\delta = (a_{Si} - a_D)/a_D = 52\%$. From this point of view and the available experimental findings it is even believed that the diamond nuclei were only formed on a β-SiC layer. The observed direct contact of epitaxial diamond to silicon would be considered as a result of an extended growth of diamond nuclei on a β-SiC layer to the surrounding area where the β-SiC layer was etched away. The interface structure formed at the nucleation stage might be different from that formed during the extended growth process.

In order to investigate the interfaces located both at the nucleus sites and in the surrounding area cross-sectional specimens for HRTEM investigation were prepared from the diamond film wafers after a very short growth process of only a few minutes (Jiang and Jia, 2000). Figure 12(a) shows a cross-sectional image of a diamond grain with a lateral grain size of about 120 nm. At the central region of the interface a hillock on the silicon substrate is clearly seen, as denoted by an arrow, with a lateral size of only 10 nm. {111} twinning occurs in this grain and the majority of the lamellae start on the sides of the hillock. Considering the widely accepted phenomenon that the silicon substrate can be etched during the nucleation

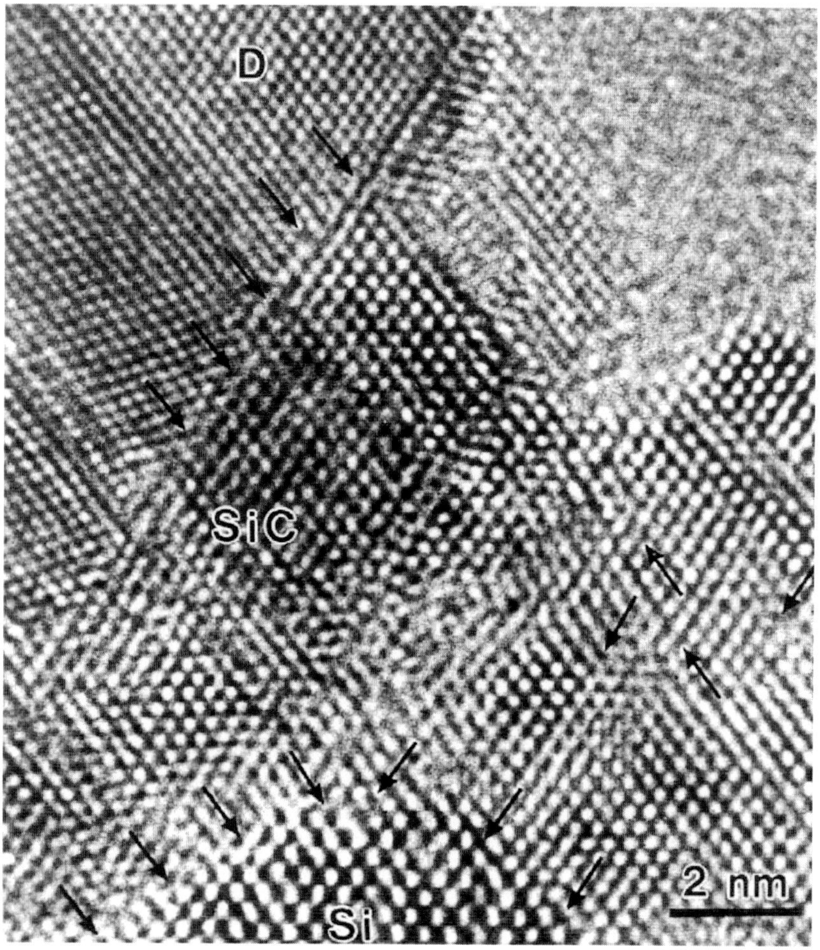

FIG. 11. Epitaxially oriented diamond crystal on top of the SiC interlayer. Terminating {111}-lattice planes at this interface and at the SiC/Si interface are indicated by arrows (Wittorf et al., 1997a,b).

and growth processes this hillock should be the place under the protection of the early formed diamond. Therefore the top of this hillock must be the nucleation site of diamond.

Figure 12(b) shows the enlarged lattice image of the hillock area in Figure 12(a). A direct bonding of the diamond lattice with silicon substrate is evident at top of the hillock. No β-SiC can be recognized. It is additionally seen that the top of the hillock is faceted. The dotted line traces the morphology of the hillock. A contrast irregularity of the hillock region can be seen which is due to a local lattice strain or a local thickness difference of the TEM sample.

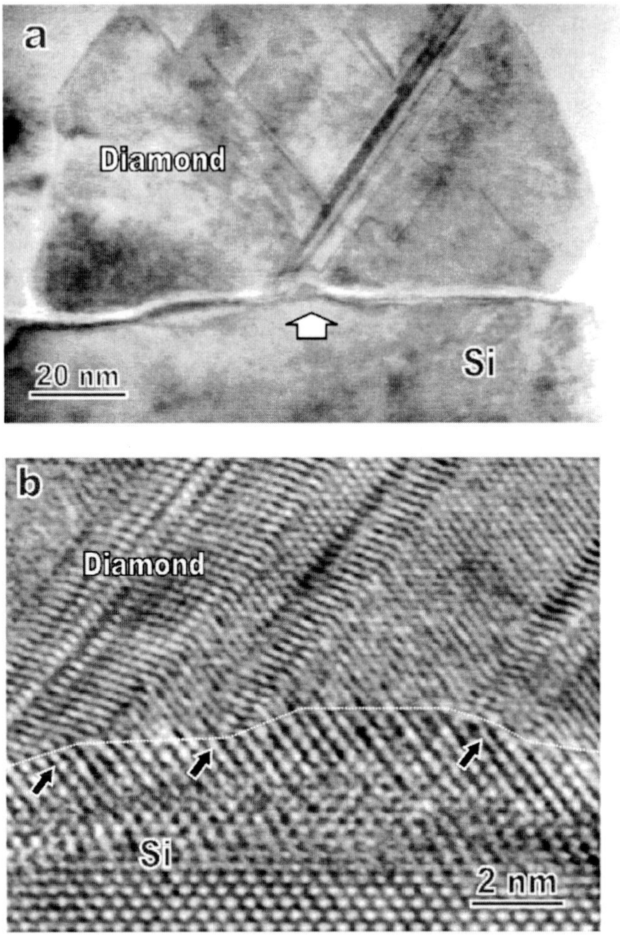

FIG. 12. (a) A low magnification image of a diamond grain. An arrow shows a hillock at the central part of the interface. The majority of twins occur near the hillock. (b) Enlargement of the hillock area in (a). A dotted line traces the facet morphology of the hillock. Three arrows mark the facet edges on the sides of the hillock, where the {111} lamellae start (Jiang and Jia, 2000).

An interesting point is that twin lamellae clearly start at the facet edges on the sides of the hillock denoted by arrows and run into the grain. From this lattice image a clockwise tilt of the diamond lattice by 1° can be measured with respect to the silicon substrate. Numerous experiments on the diamond nucleation showed that, if the BEN process is performed at moderate bias voltage and growth temperature and the incubation time is short, diamond can nucleate directly on Si substrate, instead of the formation of a β-SiC or an amorphous carbon layer.

4.2. Experimental Investigations of Crystal Misorientation

As mentioned in the previous sections the crucial point preventing single crystal diamond growth is the so far observed crystallographic orientation deviation of individual diamond grains with respect to the silicon substrate and thus small-angle grain boundaries. In the investigations of the crystallographic orientation of diamond films to reveal grain misorientation, X-ray rocking curves with line widths of up to 5° demonstrate the slight orientation deviation as indicated in Figure 13. AFM analysis of the [001]-oriented crystallites confirm the orientation deviation and a computer fitting of the measured crystal tilting of the (001) facets around the [110] crystal axis reveals excellently the statistical Gaussian distribution. The Gaussian distribution was found to be independent of the crystallographic direction of the grains along which the crystal tilting was measured (Schiffmann and Jiang, 1994).

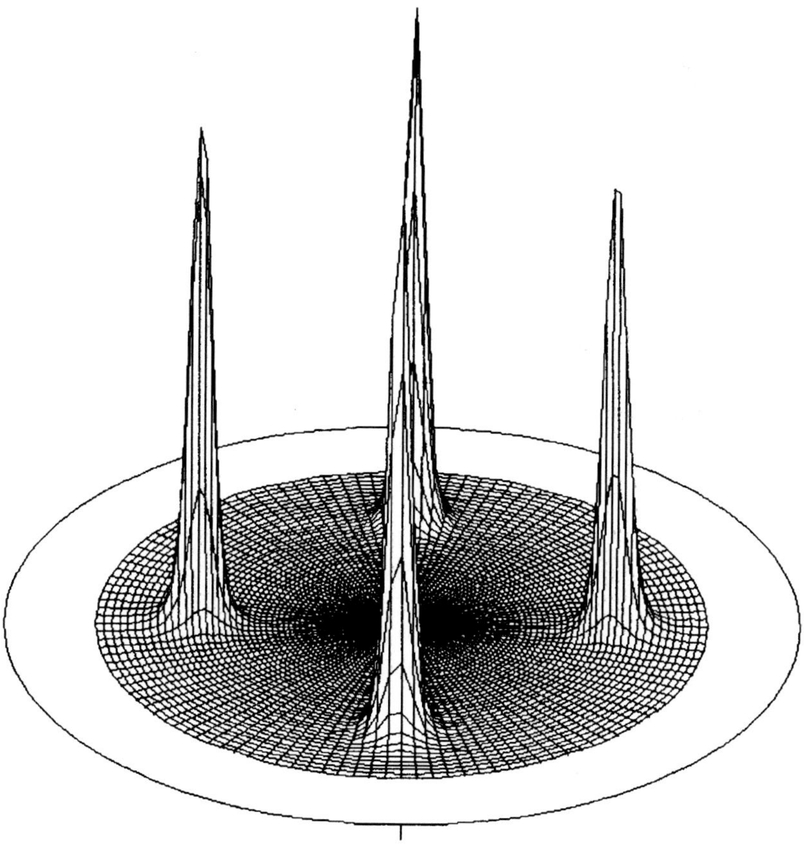

FIG. 13. {111} X-ray pole figure of a 15 μm thick [001]-oriented diamond film (Jiang and Jia, 1995).

In fact, crystal misorientation has also been observed for diamond crystallites grown on cubic silicon carbide (001) surface (Stoner and Glass, 1992). Different assumptions about the observed crystal tilt have been proposed, while the popular one is that the tilt is caused by the lattice mismatch between diamond and silicon (Zhu et al., 1992). According to such a model, all of the grown crystallites will tilt at a certain angle to accommodate the lattice matching. Obviously, this model would face difficulty in interpreting the statistical tilt distribution and the location of the maximum of the tilting angle distribution at 0° (Fig. 13). Furthermore, the recent observation of ion bombardment-induced misorientation (Jiang et al., 1996) in bias-enhanced chemical vapor deposition (BECVD) has shown that ion bombardment could also influence the crystal orientation. Even for diamond growth on an ion-bombarded diamond crystal where the lattice mismatch is zero, misorientation was clearly observed.

HREM also revealed the crystal tilting by showing the {111} lattice fringes with direction deviating from the direction of Si{111} fringes (Jiang and Jia, 1995; Jia et al., 1995). In many cases, the orientation of the diamond grains deviates slightly, i.e., by a few degrees, from the ideal one. Nevertheless, these grains also nucleate directly on the substrate and exhibit a well-defined structural relation to the substrate lattice. Figure 14(a) shows an example. The diamond grain in this image has a deviation of 4° from the ideal orientation by a rotation around the common $[1\bar{1}0]$ − zone axis. This small deviation is compensated for by varying the number of the two types of {111} planes which terminate at the interface. Figure 14(b) shows the Fourier-filtered image. About 2/7 of the (111) planes terminate at the interface in the upper picture, while a termination of about 2/5 of the $(\bar{1}\bar{1}1)$ planes is visible in the lower picture of the figure. The variation of the numbers of the terminating planes occurs in such a way that their total number is almost unchanged with respect to the case of the ideal orientation. The total number of the 60° interface dislocations was found to remain constant for a misorientation up to 14°.

In Figure 14(a) a certain extent of roughness of silicon surface such as surface steps is evident. In Jiang and Jia (2000) it was found that one of the most important factors influencing the orientation of the crystalline nuclei is surface roughness of the silicon substrate. Figure 15 shows a lattice image of a diamond grain misoriented with respect to the cubic-to-cubic relation. The HRTEM lattice image also shows a hillock at the central region of the grain/substrate interface. The top (001) facet of the hillock has a dimension of several $(1\bar{1}0)$ atomic planes only. The diamond grain in this image has a deviation of 6° from the ideal orientation by a rotation around the common (silicon and diamond) $[1\bar{1}0]$ zone axis. Note that a twin boundary is seen once again to start at a facet edge marked by an arrow. With respect to the grain orientation, a better situation is evident for the diamond grain in Figure 12(b) than in Figure 15. The good orientation of the diamond grain can be attributed to a relatively flat morphology of the hillock on the substrate. In addition, we observe a flat interface between the well-oriented grain and silicon substrate displayed in Figure 10. From the above results it is evident that the misorientation of the diamond grains on (001) silicon is closely

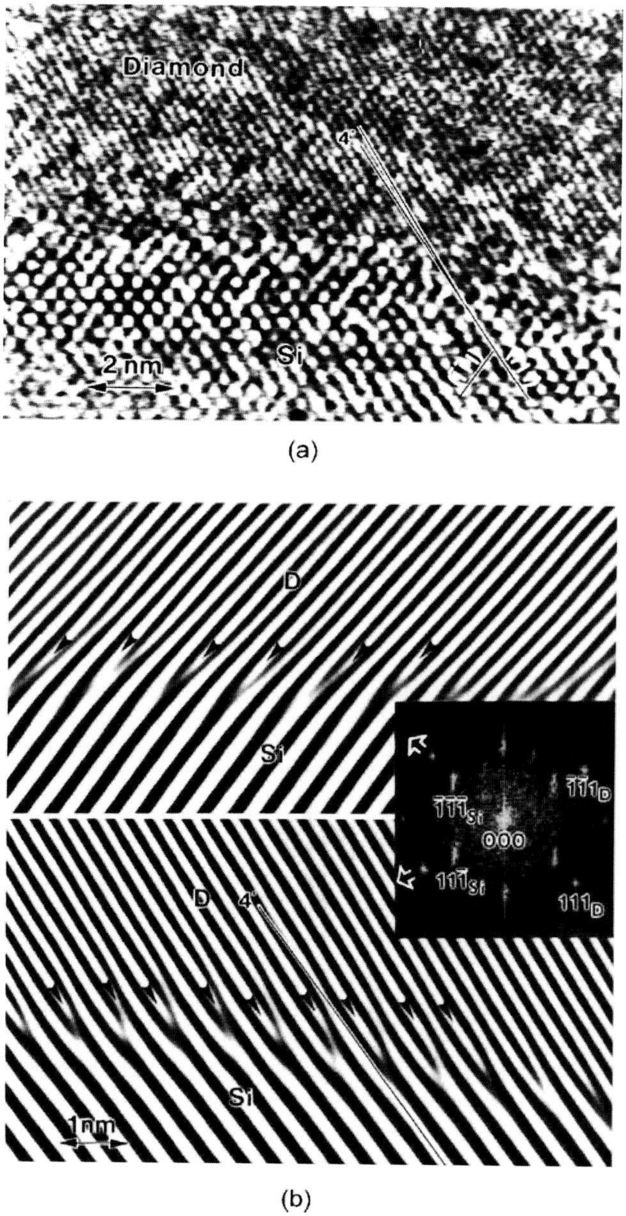

FIG. 14. (a) A $[1\bar{1}0]$ lattice-fringe image of the interface showing a diamond crystal which exhibits a 4° deviation from the ideal orientation; (b) Fourier-filtered images obtained from (a) by using the (111) spot row (upper picture) and the $(\bar{1}\bar{1}0)$ spot row (lower picture) of the corresponding power spectrum shown as an inset (Jia et al., 1995).

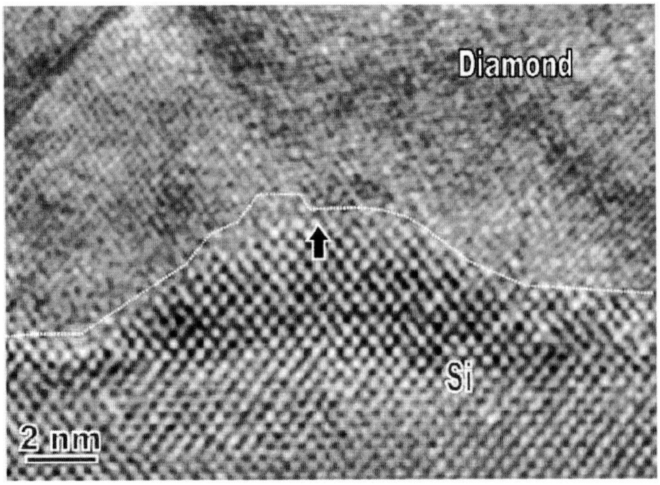

FIG. 15. High-resolution lattice image of a diamond grain with a tilt angle of about 6° around the [110] axis, showing a small (001) facet on top of the hillock. An arrow denotes a facet edge inducing a twin boundary (Jiang and Jia, 2000).

related to the appearance of the hillocks on substrate where diamond grains directly contact the silicon.

4.3. INVESTIGATION OF THE CAUSES OF THE CRYSTAL MISORIENTATION

Above observations provide us with strong evidence that crystal tilting is related to structural distortion of the substrate surface and thus to the concomitant local strain. The tilting induced by the surface roughening is obviously very complicated with a range of possibilities. For heteroepitaxial diamond growth on Si(001), however, the local strain is not only caused by surface roughening, but also by the lattice mismatch. The mismatch dislocation-induced local strain may result in crystal tilting especially when ion bombardment causes surface etching.

In order to understand the origin and nature of the experimentally observed diamond grain tilting, the three-dimensional nucleation and growth processes were studied in a step-by-step manner by means of the molecular orbital PM3 calculation (Zhang et al., 1996; Jiang et al., 1998), which is based on the MNDO semiempirical Hamiltonian of Hartree–Fock theory. A cluster model composed of more than 100 silicon atoms with a hydrogen-saturated boundary was selected to simulate a rough silicon (001) substrate involving a silicon island on the surface. At the initial stage of diamond nucleation, an adsorbed CH_2 may be bonded with and bridge two surface Si atoms, while two hydrocarbon or a single C–C species may form a heptagon with the neighboring surface Si atoms, as schematically shown in Figure 16(a). According to the PM3 calculation, the

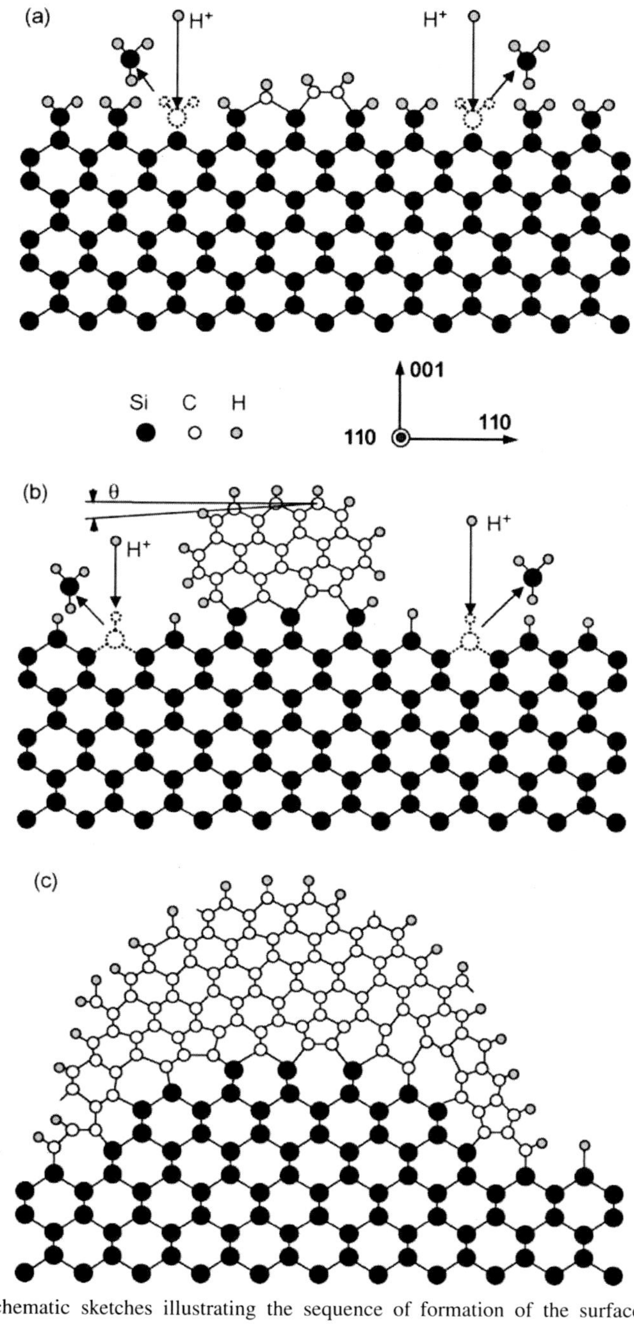

FIG. 16. Schematic sketches illustrating the sequence of formation of the surface hillock and diamond nucleus: (a) the initially deposited carbon configuration and H^+ ion etching on the silicon (001) surface, (b) further etching of the silicon surface and formation of tilted diamond nucleus, and (c) further growth of the nucleus and formation of the surface hillock due to etching.

Si–C and C–C bond lengths tend to be close to those in β-SiC and in diamond, respectively, due to structural relaxation. However, the carbon-related bond angles may deviate appreciably from the tetrahedral value (109.47°) because of internal strain. As a result, the carbon atoms are not located on the same plane parallel to the silicon surface. By overcoming the deviation in the tetrahedral bonding orientation, deposition may continue towards formation of a diamond embryo. At the interface between the diamond embryo and silicon, an interfacial dislocation will appear accompanying the formation of a pentagon/heptagon bonding configuration, as shown in Figure 16(b). The formation of such a diamond embryo actually partially compensates the bond tilt created during the deposition of the first carbon layer. Due to the limited contact area between the diamond embryo and the Si(001) terrace, on the other hand, the crystal tilt will remain and the structure mismatch will be accommodated during crystal growth by varying the number of the terminating {111} planes at the interface (pentagon/heptagon bonding configurations) and a lattice tilt of several degrees will remain. This is schematically demonstrated in Figure 16(c).

The model shows that the crystal tilt is already introduced at the initial clustering stage and can interpret the HRTEM result in Figure 15 quite well. However, if the pentagon/heptagon unit is positioned symmetrically in the diamond embryo, the local lattice strain which is presently considered to be the driving force of the tilt will be balanced in a symmetrical way. On the other hand, if such a diamond embryo is formed in a sufficiently large area of (001) terrace, the initial tilt should be significantly reduced or eliminated by a homogeneous distribution of the pentagon/heptagon unit. The fact that a relatively flat top of the hillock in Figure 12 leads to a better orientation ($\theta = 1°$) of the diamond grain corroborates the theoretical model. In practice, the deposition conditions in the nucleation stage should be chosen in such a way that the nucleation rate is as high as possible and the etching effect of the plasma is as small as possible in order to obtain a well-oriented diamond film on silicon substrate.

5. Microstructure and Growth Phenomena of [001] Diamond Films

5.1. Microstructure of [001] Diamond Films

The epitaxially [001]-oriented films which can be prepared to date in many laboratories (Hessmer et al., 1994; John et al., 1994; Maeda et al., 1995; Tachibana et al., 1996) via a bias-enhanced nucleation process consist of columnar grains with a lateral size of several micrometers.

Figure 17 shows a typical cross-sectional morphology of a 10 μm thick [001] diamond film in SEM. From this picture it can be clearly seen that the individual islands grow together and the columnar film structure is formed. Investigations of

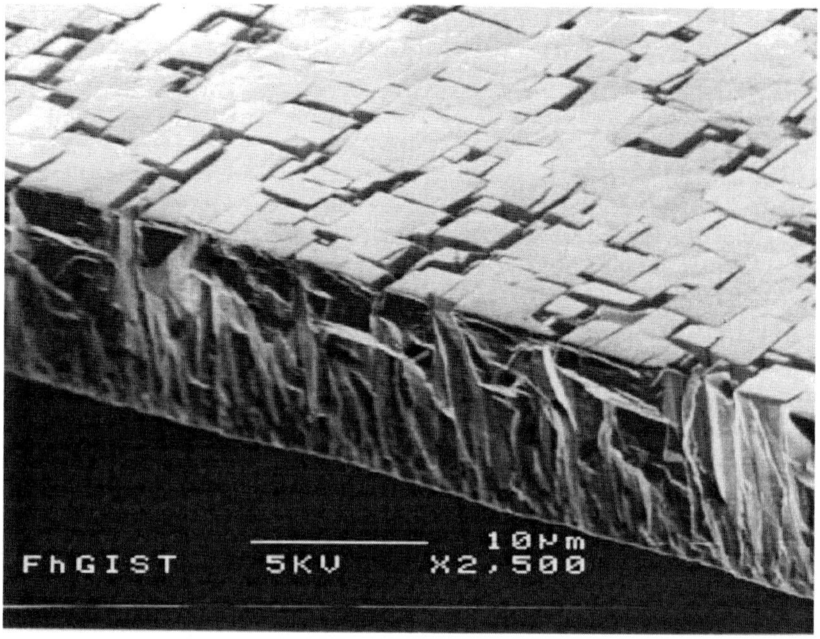

FIG. 17. SEM micrograph of a [001] diamond film showing the film surface and the columnar film structure (Jiang and Klages, 1996).

the orientation relationships of diamond grains near the interface in similarly grown CVD diamond films by TEM show that the randomly oriented diamond grains were overgrown in the near-interface region and columnar growth already begins at a film thickness of about a few hundred nanometers leading to the observed film morphology. The reason is that the applied textured growth conditions utilizes an 'evolutionary selection' (Van der Drift growth) (Van der Drift, 1967) which will be discussed in detail in Section 6. Since most of the individual grains grow approximately in the [001]-direction, they have similar growth rate leading to grain boundaries nearly parallel to each other and perpendicular to the silicon substrate. Figure 18 is a cross-sectional TEM micrograph showing the morphology of the columnar diamond grains. The grain coarsening at the top region is caused by boron doping which will be discussed later. The arrangement of the (001)-oriented columnar diamond grains with an average lateral dimension of 1 μm and the grain boundaries are clearly visible. The grain boundary planes are in general parallel to the (110)- and the $(1\bar{1}0)$-lattice planes. Many defects, mainly in the form of dislocations, microtwins and stacking faults, appear along the grain boundaries. These defects are considered to arise from the local mechanical stresses and/or the {111} facet growth of the crystallites. In the grain interior the density of defects is relatively low.

FIG. 18. Cross-sectional TEM micrograph recorded in the [110]-direction showing the columnar film structure and the change of the grain boundary orientations in the boron-doped top layer, for example, the grain boundary between grains A and B. Grain C is overgrown by grain E and the grain boundary between the two grains is switched (Jiang et al., 1998a–c).

Between the diamond grains the orientation deviation is accommodated by the introduction of small angle grain boundaries of symmetric and asymmetric geometry (Wittorf et al., 1997a,b). It is found that the structures of such small-angle grain boundaries can be described by a dislocation model. These grain boundaries are on average parallel to the {110}-plane and contain in many cases also micro-facets with {111} faces. Figure 19 shows such a grain boundary section, taken in a [100]-zone axis orientation. Both grains are seen in the [001]-zone axis orientation, and the {220}-lattice fringes with distances of 1.26 Å are clearly visible. The misorientation between the two grains is about 3–4° and is compensated by the introduction of dislocations (arrows). The average distance between the dislocations is 1.9 nm. Similar to the observations reported for [110]-zone axes lattice images, no inclusions or additional second phases can be observed at these grain boundaries. The small-angle misorientation between grains is observed to be connected with arrays of dislocations in all cases.

5.2. Crystal Coalescence During Deposition

In order to prepare films with large lateral grain size and to reduce crystal defects two other growth modes were applied (Jiang and Jia, 1996). In the first

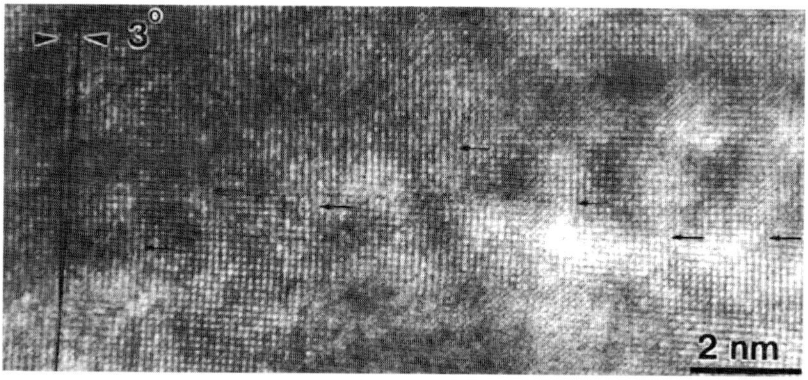

FIG. 19. Enlarged area of small-angle grain boundary. Terminating (220)-planes are indicated by small arrows. Neighboring grains show both sets of the {220}-lattice fringes with distances of 0.126 nm. HRTEM micrograph taken in [001]-zone axis (Wittorf et al., 1997a).

mode the misorientation of grains should first be reduced to such a small value that the coalescence of grains can occur, thus leading to a single crystalline top layer (Jiang and Jia, 1996). In the second mode, a process favoring lateral growth is applied, allowing the grains to overgrow adjacent ones and resulting likewise in crystals with large lateral dimensions (Section 5.3). Figure 20 schematically shows these two growth phenomena.

From the cross-section SEM micrograph in Figure 17 an interesting feature concerning the microstructure change can be seen. Some grain boundaries disappear in the vicinity of interface resulting in coarsening of the columnar structure. In order to investigate this phenomenon in atomic details HREM investigations were performed on cross-section specimens (Jiang and Jia, 1996). Figure 21(a) shows an HREM lattice image of a small-angle grain boundary between two epitaxially grown diamond grains. The image is recorded with the electron beam parallel to a ⟨110⟩ zone axis of the right-hand side grain (DII),

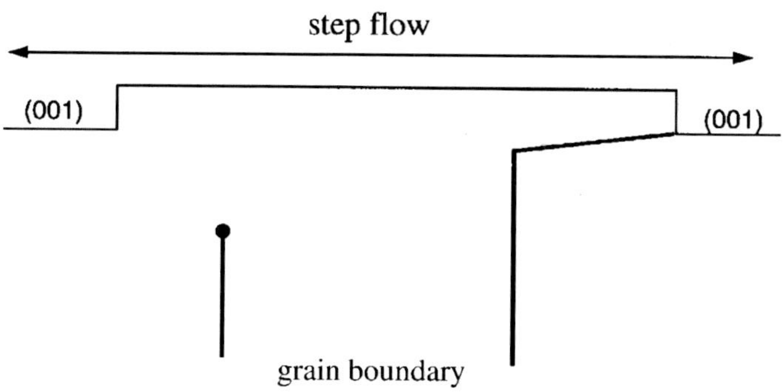

FIG. 20. Schematic representations of two [001] grain growth phenomena—coherent coalescence of grains and overgrowth of grain by lateral step flow growth of neighboring grain.

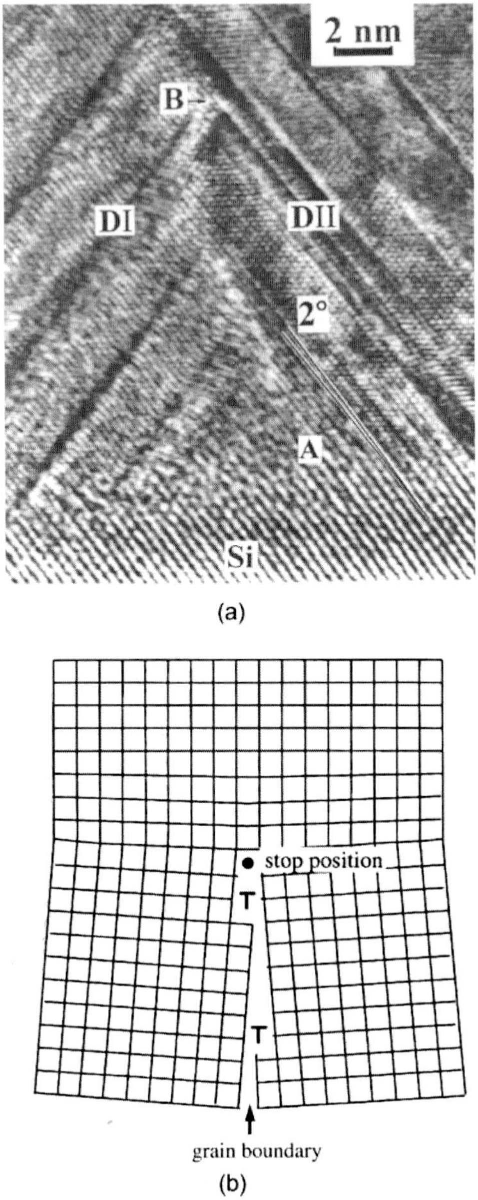

FIG. 21. (a) HREM image of a grain boundary showing disappearance of a 2° small-angle grain boundary at B; (b) schematic representation of stopping a low-angle grain boundary by introduction of a disclination (Jiang and Jia, 1996).

which has a small angle deviation from the orientation of the left grain and the silicon substrate. The left-hand side grain (DI) was grown in the ideal [001] parallel epitaxial orientation to the silicon. This can easily be seen by inspecting the image at a glancing angle along the direction of the marked line. The letter 'A' denotes the starting position of the boundary at the interface between the film and the substrate. Both diamond grains exhibit a multiple-twin structure near the grain boundary. From the image a 2° tilt angle around the viewing direction with respect to the substrate and the left-hand side grain DI can be measured near the interface for the right-hand side grain DII. The boundary separating the grains is therefore a 2° low angle tilt grain boundary. The grain boundary looks relatively sharp, indicating that it is nearly in an edge-on position and nearly parallel to the (110) plane. The most important information obtained from Fig. 21(a) is the disappearance of the grain boundary at the position indicated by the letter 'B'. The two initial individual grains have coalesced and formed a single crystal. This result confirms the SEM observations and is another evidence that individual [001] diamond grains can coalesce forming a single crystal, even in the case where they have a slight orientation deviation.

Due to the high elastic modulus of diamond (about 1020 GPa) the energy for straining the diamond lattice and forming dislocation arrays should be extremely high. It is therefore expected that two contacting grains with a slight orientation deviation will have the tendency to form a single crystal thereby terminating the grain boundary. At this terminating position a disclination with a corresponding rotation angle is necessary to compensate for the misorientation (Fig. 21(b)). To determine the highest value of misorientation angles which can be compensated by such disclinations further investigations are necessary. Nevertheless, this observation provides evidence for the fact that large-area single crystalline grains can be grown based on the coalescence of the individual grains.

5.3. 'Change Mode' Growth for Preparing Films with Large Grains

Heteroepitaxial [001]-oriented diamond films with improved lateral grain size and surface smoothness could also be prepared using a [001]-textured growth process followed by a lateral growth step by varying the fastest growth direction (Kawarada et al., 1997; Jiang et al., 1998a–c). This variation can be achieved by changing the process parameters, first of all the substrate temperature and the methane concentration.

The best lateral growth could be realized by a [110] step-flow growth process of the (001) surface. The results indicate that the diamond crystals increase their lateral dimensions at the (001) film surface by changing the grain boundary plane orientations from preferentially vertical to preferentially parallel directions with respect to the (001) growth faces. The grains with relatively large angle deviation from the ideal epitaxial orientation are overgrown by those with relatively small angle deviation. As a result, the degree of orientation perfection of the films improves in comparison to that of films prepared by the established process of [001]-textured growth. The presence of boron in the gas phase was found to encourage the step-flow lateral growth. It was possible to achieve deposition of a thin boron-doped diamond film characterized by a FWHM value of the measured tilt angle distribution of only 2.1°.

5.4. Boron-doping Induced Lateral Growth

Lateral growth is enhanced by small additions of boron to the gas phase. This has been demonstrated by an experiment in which only 10 ppm boron was incorporated into the gas phase in a subsequent second growth step on a diamond film which was grown heteroepitaxially on (001) silicon substrate (Jiang et al., 1998a–c).

The morphology change due to the lateral growth was clearly demonstrated by the cross-sectional SEM image shown in Jiang et al. (1998a–c). The cross-sectional TEM micrograph of Figure 18 also shows the grain structure and the behavior of the grain boundaries resulting from the lateral growth with boron incorporation. It is most interesting to notice that the traces of the grain boundaries between the columnar diamond grains are preferentially oriented parallel to the [001] growth direction. At the top region of the cross-section, the grain boundary planes disappear or change their orientation from a vertical to a direction nearly parallel to the (001) growth faces. A coarsening of the columnar structure occurs resulting in a top layer with relatively large areas. Diamond crystallites with increased average lateral dimensions of the (001) growing face resulted from this deposition.

Step-flow-induced lateral growth can occur if the growing face of the film is completely covered by the (001) facets. Such step-flow-induced growth has been observed previously on the (001) face of a single-crystal diamond substrate during homoepitaxy (Lee and Badzian, 1995; Hayashi et al., 1996). It was found that multi-atomic steps belong to terraces of approximately 8 nm in lateral dimensions which propagate along the [110] direction of the surface resulting in net film growth in the perpendicular [001]-direction. SEM and TEM observations (Jiang et al., 1998a–c) indicate that the grain boundary plane orientation is related to the existence of the {111} facets leading to the formation of pits on the growth surface. If the (001) growth faces are separated by the {111} faces, surface pits are formed. In this case the step flow cannot occur and the grain boundary propagates in a direction vertical to the substrate surfaces. The step flow stops if a step is encountered on the other grain which moves in the reverse direction.

6. Textured Diamond Films

The formation of a texture, i.e., a preferred crystallographic orientation of crystalline grains, is a well-known phenomenon which is not restricted to vapor growth of thin films. The solidification from a melt or gas phase usually starts from random nuclei on the substrate. In general, a textured film growth can, according to the established mechanisms, be realized either by an evolutionary competition process of the statistically oriented grains or by particle flux-determined selective growth process. In the former case, the so-called van de Drift mechanism, only grains with their fastest crystal growth direction parallel to the film growth direction can survive, if the growth process proceeds to a large thickness, while other grains will be overgrown (Wild et al., 1993). After the oriented diamond grains have grown together their size remains nearly constant during further film growth. The columnar structure is a typical result of the Volmer–Weber growth mode resulting in grains of textured crystal orientation and in grain boundaries parallel to the film growth direction. In the latter case the grains with certain special orientation will be allowed/supported to grow by the ion impact while growth of other grains will be suppressed, due to a selective ion etching or re-sputtering process.

6.1. Textured Film Growth by Evolutionary Selection

A cubic diamond crystal tends to take the shape of a cubo-octahedron consisting of the $\{001\}$ and $\{111\}$ faces. The growth of diamond grains occurs by propagation of $\{001\}$ and $\{111\}$ planes. The crystal degenerates into a cube or an octahedron as limiting habits for vanishing portions of $\{111\}$ and $\{001\}$ faces, respectively. The relative fraction of the two types of faces depends on the ratio of growth rates in the $\langle 100\rangle$ and $\langle 111\rangle$ directions, V_{100}/V_{111}, which represents the ratio of the vertical distances of the $\{001\}$ and $\{111\}$ faces from the nucleation point (origin) and can be definitively controlled by the deposition parameters. For computer simulations of the diamond film growth Wild et al. (1993) defined a 'growth parameter' $\alpha = \sqrt{3}(V_{100}/V_{111})$. Figure 22 shows schematically crystal shapes 'grown' for different α. The arrows indicate the largest diameter which is the direction of fastest growth of an individual crystal. If $\alpha \geq 3$ the $\{100\}$ planes disappear and the contour of a diamond grain is formed by four $\{111\}$ faces and a pure octahedron is realized. In order to grow a cubic diamond crystallite, the growth parameter $\alpha \leq 1$ is required. Clausing et al. (1992) demonstrated that all forms of cubo-octahedrons, including cubes and octahedrons, can be produced as single isolated crystallites by adjusting the growth parameters α between 1 and 3.

For the growth of an oriented film the evolutionary selection proceeds accompanied by a competition process between the crystallites. As growth proceeds, more and more crystals are overgrown by adjacent crystals and the

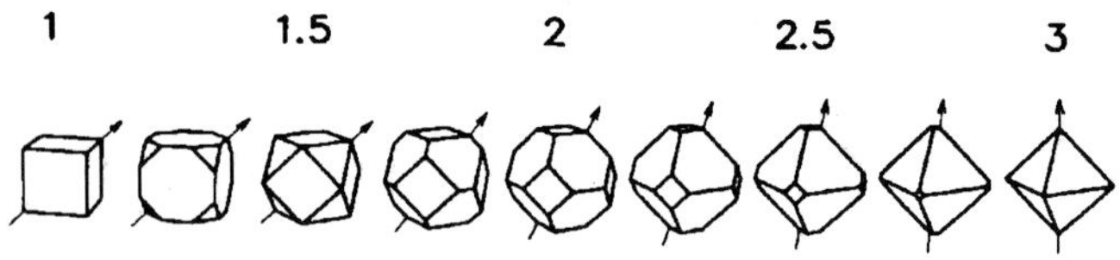

FIG. 22. Crystal shapes for different values of the growth parameter. Arrows indicate the fastest growth directions.

number of crystals extending to the surface decreases progressively. The survival probability of individual crystals depends on their vertical growth rates. Only those crystals with a fastest growth direction (arrows in Figure 22) parallel with the film growth direction will survive, whereas other orientations are gradually buried.

To study the growth of diamond films during prolonged deposition, i.e., for the case that the film thickness is much larger than the lateral grain size, an analytical method for simulating three-dimensional (3D) gas phase film growth has recently been developed (Jiang et al., 2002). Using a model system composed of 4×10^4 grains growth of diamond films by $\{001\}$ and $\{111\}$ deposition under different initial conditions and with varying growth parameters can be simulated. It is demonstrated that the rate of structure evolution is sensitively influenced by the growth parameters. A nearly linear proportionality between the average grain size and the thickness of films is obtained for large film thickness. The evolution of size and orientation (texture) distribution for different film thicknesses are shown in Figure 23. The proportionality constant varies by changing α. Furthermore, the orientational distribution is drastically narrowed down.

Under conditions, which are probably representative for most diamond growth experiments reported in the literature, a preferential orientation of $\{110\}$ planes parallel to the substrate surface was characterized due to the high growth rate of $\{110\}$. Thick films grown under the extreme conditions with $\alpha \approx 1$ exhibited a $\langle 111 \rangle$ texture, whereas $\langle 100 \rangle$-texured films were obtained with $\alpha \approx 3$. For $2.5 < \alpha < 3$ the top face of the growing grain is always an exposed $\{100\}$ plane and the coverage of the film surface by $\{100\}$ plane becomes more and more dominant as the film thickness increases. Figure 24 is a SEM micrograph of a thick film sample with a small misorientation of the crystallites (α is close to 3) as measured by a X-ray rocking curve with FWHM $\approx 4°$.

The growth parameters for achieving a certain film texture were reported by Wild et al. (1994). A $\langle 100 \rangle$ film texture is favored by high methane concentrations and low deposition temperature (Fig. 25). A third parameter determining the $\langle 100 \rangle$ film texture is the concentration of molecular nitrogen in the gas phase. A small amount (several tens of ppm) of nitrogen leads to the formation of nearly perfect $\langle 100 \rangle$ texture (Locher et al., 1994).

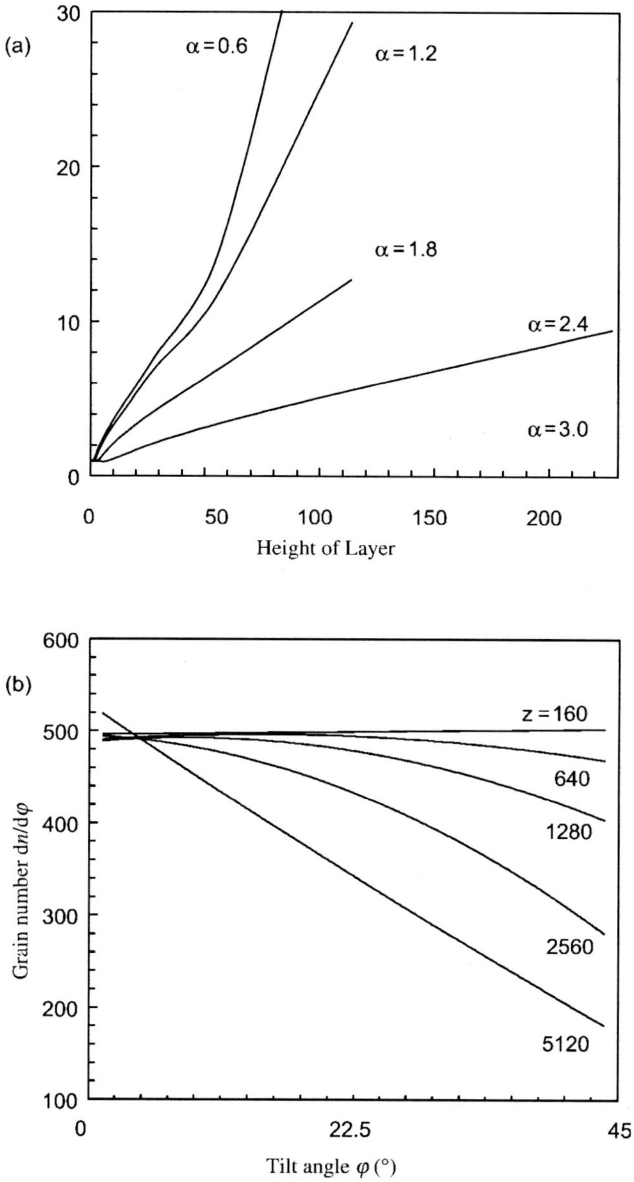

FIG. 23. (a) Average grain size vs. the height (thickness) of layer for different growth parameters $\alpha = \sqrt{3}(V_{100}/V_{111})$; (b) distribution of the tilt angle of grains at different film thicknesses. The orientation distribution density is defined as $dn/d\varphi$ with dn being the number of the grains in the range from φ to $\varphi + d\varphi$ (Jiang et al., 2002).

FIG. 24. SEM micrograph of a ⟨100⟩-textured diamond film.

6.2. TEXTURED FILM GROWTH BY ION BOMBARDMENT

A new mechanism leading to the growth of textured diamond films was discovered recently (Jiang et al., 1996; Jiang et al., 1998a–c). It was found that the application of a proper bias potential to the substrate leads to a strong

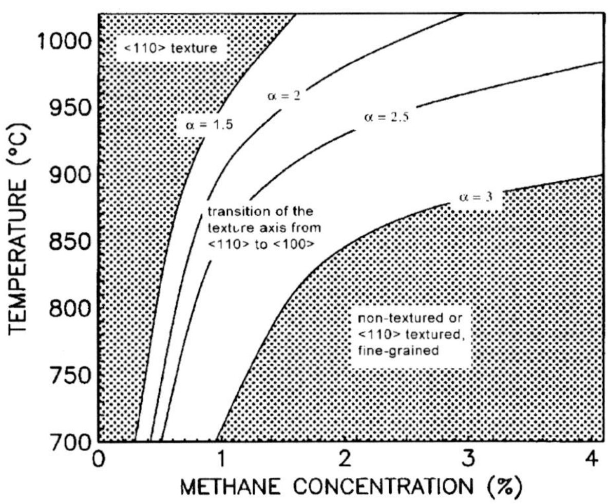

FIG. 25. Experimental parameter domains for preparing textured diamond films (Wild et al., 1994).

selection of the favored growth direction. The selection effect is believed to be due to the direction-dependent H^+ ion etching of the diamond crystal and therefore a film texture can be achieved for small thickness (Zhang et al., 1997). Both of the above-mentioned approaches were used to prepare epitaxially oriented films (Jiang et al., 1998a–c).

Diamond depositions were performed onto large-grained diamond film substrates of various orientation by applying a negative bias voltage of up to − 150 V to the substrate. Other process parameters are similar as those for a conventional growth. The electrical potential was kept throughout the experiments so that the substrate surface was continuously bombarded by energetic ions. According to the axially symmetric experimental geometry, the ions flow in a direction perpendicular to the substrate surface as confirmed by the selective nucleation via biasing the substrate (Section 3.4).

Figure 26 shows the SEM images of the diamond films prepared under different bias voltages. Under bias-induced ion bombardment (substrate bias voltage $V_b = -125$ V), a significant variation of the film morphology was achieved: the film morphology was dominated by diamond crystals of [001] orientation. The epitaxial [001] grains protrude from the film surface and the non-[001] top grains keep lower positions. With a bias voltage of − 150 V, the surface is covered by smaller crystallites with a random orientation.

Further experiments showed that the bias-induced textured growth depends mainly on the parameters which influence the particle energy (e.g., substrate bias voltage and process pressure) and occurs within a very small parameter window. If one works at a higher pressure the ion energy will decrease due to the increasing probability of collisions between ions and uncharged atoms. Correspondingly, the bias voltage for the textured growth should be increased. Figure 27 shows a SEM surface image of a film after 20 h growth under bias-induced ion bombardment on a 10 μm-thick random diamond film substrate prepared without bias. The top layer of the film changes from a randomly oriented film to a completely [001] textured film. Due to the slightly increased gas pressure the bias for a optimal texture was increased to − 150 V in comparison to the films shown in Figure 26(b). The [001]-oriented crystallites shown in Figure 26 have a lateral size of about 0.5 μm, smaller than the underlying crystallites. Domains containing crystallites of parallel grain orientation are clearly seen.

To study the development of the orientation and the orientation relationship between crystallites of the top layer and the underlying grains, diamond films were grown, under bias condition, onto a specially prepared ⟨111⟩ textured polycrystalline diamond film. Figure 28 shows the SEM surface images of films prior to and after bias-assisted deposition at different deposition times. Prior to bias-assisted deposition the triangular (111) face of the large diamond grain is essentially smooth except for some triangular growth hillocks with edges along the ⟨110⟩ direction (Fig. 28(a)). After growth for 3 h (Fig. 28(b)) the hillocks became steep and a few small triangular islands were formed on the substrate

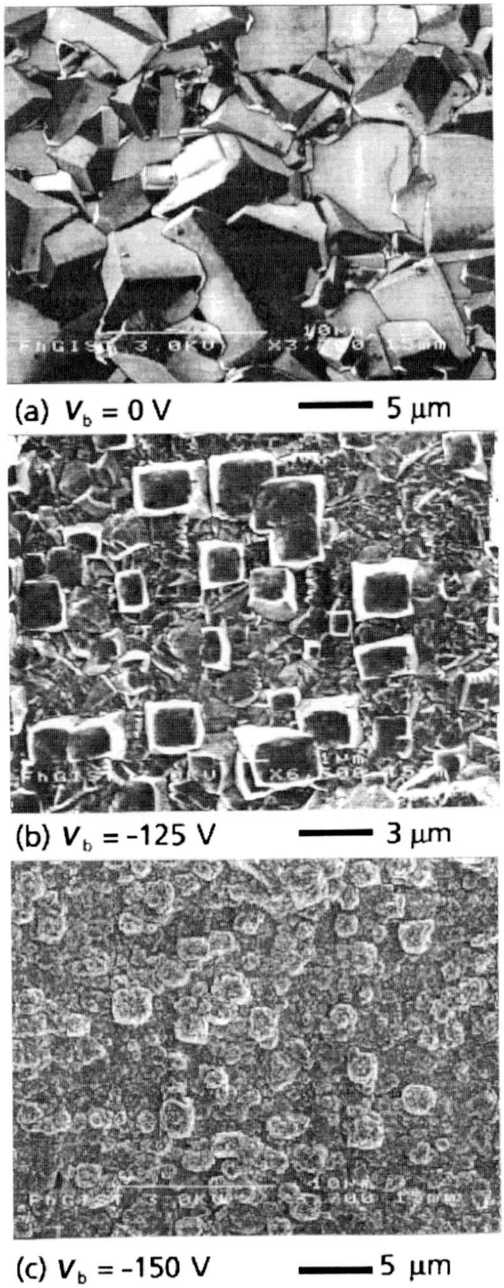

FIG. 26. SEM micrographs of diamond films grown on Si under various bias voltages. V_b = (a) 0 V; (b) − 125 V; (c) − 150 V (Jiang et al., 1998a–c).

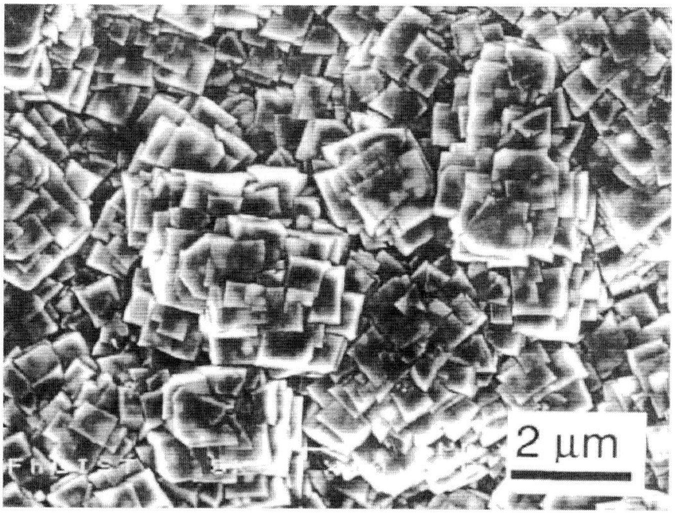

FIG. 27. SEM micrograph of [001]-oriented diamond film grown on Si under bias voltage of $V_b = -150$ V for 20 h (Jiang et al., 1996).

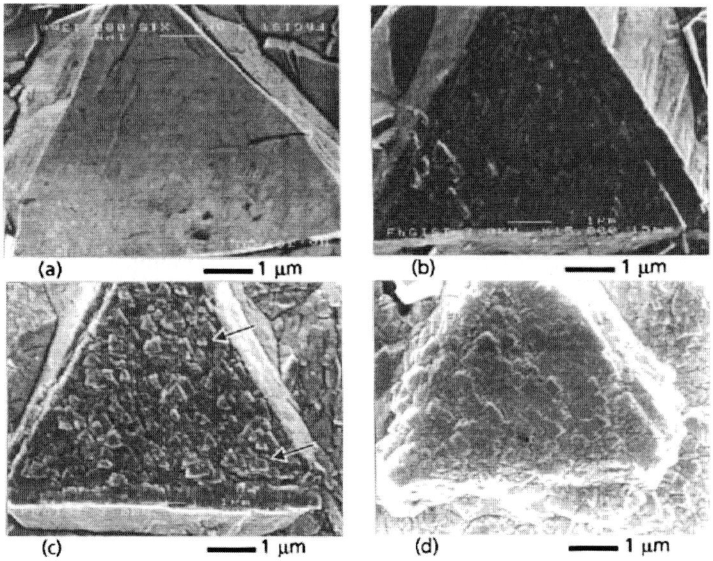

FIG. 28. Scanning electron microscopic images of the (111) diamond facets for deposition at − 150 V bias voltage. (a) Prior to bias-assisted deposition; (b), (c), (d) after bias-assisted deposition for 3, 5, and 10 h, respectively (Jiang et al., 1998a–c).

(111) surface. These top islands have their (111) crystal facets parallel to the (111) substrate surface and their ⟨110⟩ edges aligned parallel to each other and to the corresponding substrate ⟨110⟩ directions. The (111) face of the large underlying crystal is clearly rougher than that before bias-assisted deposition. As the growth time increases, the amount of the top islands increases and a few grains of [001] direction are formed among these triangular islands after bias-assisted deposition for 5 h, as indicated by the arrows in Figure 28(c).

A surprising phenomenon was observed after bias-assisted deposition for 10 h. As shown in Figure 28(d) the substrate (111) face is now mainly covered by a top layer of [001]-textured crystallites, i.e., their (001) facets are parallel to the substrate (111) facet. These top [001] crystallites are even in-plane oriented with their ⟨110⟩ edges aligned parallel to each other and parallel to one of the ⟨110⟩ axes of the underlying crystals. The observed phenomenon is obviously a result of crystal re-nucleation producing crystallites with a certain orientation relationship, but is not in identical crystal direction to the substrate. The growth orientation of the top crystallites is clearly determined not only by the substrate orientation ($\langle 110\rangle_{\text{top crystallites}}//\langle 110\rangle_{\text{underlying crystal}}$) but also by the ion flux direction, i.e., the [001] axis is parallel to the ion flux. If only the effect of ion bombardment were considered, an ⟨001⟩-textured top layer would grow, i.e., the in-plane orientation of the top layer should be random. In contrast, if only the effect of substrate crystal direction were considered, the homoepitaxial growth should continue.

6.3. Mechanism of the Ion-induced Textured Diamond Growth

It is a complicated task to interpret the effects of ion bombardment on diamond film growth. The observed [001] textured film growth is, obviously, not a result of evolutionary selection. Under the used experimental conditions of the film growth in Jiang et al. (1996, 1998a–c) the {100} surfaces grow more slowly than other surfaces if a negative substrate bias is not applied. From the cross-sectional SEM observations, it was found that for the deposition on (111) diamond surface there exists a sharp transition of the preferred crystal orientation from [111] to [001]. The ion-flow-direction dependent film growth can therefore be only a result of the ion impact, a physical factor in diamond CVD.

It was found for PVD thin film deposition assisted by ion bombardment that the degree of orientation order was a function of the intensity of the ion/atom flux ratio and of the incident angle of the ions (Yu et al., 1985). The orientation of the films was influenced by the sputtering yield which is dependent on the relative orientation of the film crystal to the ion flux. To explain the phenomenon that a (001) textured layer was deposited on a diamond film of random crystal direction, the role of the substrate bias voltage must be considered. The diamond growth is generally a combined process of deposition and re-etching taking place concurrently. A film grows if the deposition rate is larger than the etching rate. During the deposition process of diamond films by CVD, atomic H and H^+ ions in the plasma are known to cause etching, H^+ ions etching much faster than atomic H (Yamada and Sone, 1983). For the deposition under bias condition, H^+ ions are accelerated towards the substrate, and the reactive ion etching caused by H^+ ions is expected to play a dominant role. The etching yield of H^+ is probably dependent on the relative orientations between the ion flow and the growing grain. It can therefore be argued that the selection mechanism for grain orientation is due to the difference in etching yields between grains which are oriented (in our case the [001] textured grains) and those which are not. The different etching yields on different faces, which can be as high as a factor of 5 in some PVD processes, combined with re-nucleation, leads to growth for aligned grains, and hence to an orientation order. The selective etching by H^+ ions and its effects on the oriented growth of diamond films were studied by using pure hydrogen as reactant gas and are discussed in Section 6.4.

The ion impact damages the surface of the substrate and introduces a large amount of active sites for secondary nucleation (re-nucleation) interrupting the homoepitaxial growth. This leads to a decrease in the efficient free path (diffusion length) of the adsorbed surface precursors and furthermore the average lateral size of the top [001]-textured crystallites decreased to only about 0.5 μm, much smaller than that of the underlying substrate crystals.

6.4. Selective Etching of H^+ Ions and its Use for Textured Film Growth

In order to test the reported selection mechanism due to the orientation-dependent etching yields, the selective etching effect of H^+ ions on the differently oriented diamond faces was investigated using pure hydrogen as reactant gas. Furthermore, this selective etching was used for the preparation of oriented diamond films (Jiang et al., 1998a–c).

The partially [001]-oriented diamond films deposited via a bias-enhanced nucleation on mirror-polished (001) silicon wafers were etched in situ in hydrogen plasma by setting an electrical potential of −150 V to the substrate. After etching the samples grew further without bias. The surface morphologies and the cross-sections of the film were observed by SEM before and after the bias-assisted hydrogen etching process. The average etching rate of the films was about 0.05 $\mu m\ h^{-1}$. By H^+ etching under optimized conditions the re-nucleation can be avoided and the [001]-oriented crystals become more dominant in the image. The (001) faces are rougher. The average size of the (001) facets of the large [001] grains increases significantly.

The contribution of the etching by atomic hydrogen to the selectivity can be ruled out because it causes only an improvement of the film quality, but not of the film orientation (Cline et al., 1991; Olsen et al., 1994). Due to the lower atom density of the {100} face relative to the {111} face more H^+ ions can channel through the lattice of oriented (001) faces. In contrast, the {111} faces have the highest atom density and the impact probability of hydrogen ions with atoms on the {111} plane should be high. The etching efficiency of H^+ ions is therefore lower for (001) faces than for {111} faces. To explain the phenomenon that the (001) faces which are parallel with the substrate surface become larger after H^+ ions etching, a re-deposition should be taken into account during the H^+ ions etching process. The H^+ ions will etch off the carbon atoms from the diamond face forming hydrocarbon species (such as CH, CH_2 and CH_3) which could be neutral or positively charged. The positively charged hydrocarbon clusters will be accelerated towards the substrate due to the negative bias potential and then neutralized by capturing electrons from the substrate; so the carbon partial pressure near the substrate surface is higher than that for the etching process without substrate bias. The re-deposition process can occur during the etching procedure.

The critical dependence of the etching rate and the re-nucleation of diamond crystallites by hydrogen ions on the bias voltage applied enables a control of not only the textured, but also the epitaxial diamond growth by the variation of the ion energy. As a logical consequence the selective etching effect was utilized to improve the epitaxial growth of [001]-oriented films (Zhang et al., 1997).

The selectivity of H^+ ion etching on differently oriented diamond grains and its assistance to improve the orientation of diamond film are clearly different from the method including a textured growth via evolutionary selection. The orientation grade of films prepared by evolutionary selection increases with the film thickness while the selective etching effect is helpful for obtaining an [001]-oriented diamond film at small film thickness.

7. Summary and Conclusions

Significant progress in heteroepitaxial diamond growth has been achieved during the past few years. The synthesis of single crystalline diamond films and their applications as high temperature electronic materials still remain a great challenge for materials scientists. The crucial point preventing single crystal diamond growth is the observed crystallographic orientation deviation of individual diamond grains with respect to substrate and the resulting small-angle grain boundaries.

Reduction of the average misorientation angle between grains in (001)-oriented CVD diamond films has been experimentally achieved by employing special modes of crystal growth. It was found that diamond grains with a very small misorientation angle of about 2° can coalesce to form a larger diamond grain by terminating the small-angle grain boundary between the grains by a disclination. It was furthermore demonstrated that the presence of boron in the gas phase during film deposition favors a step-flow lateral growth mode. This leads to overgrowth of the diamond grains of relatively large misorientation by the highly oriented neighboring grains. Changes of the grain boundary orientation from an inclined, nearly perpendicular to a more parallel direction with respect to the substrate surface were observed by TEM. The parallel direction indicates the overgrowth of neighboring diamond grains. The films prepared in such a way show an improved smoothness of the film surface and an increased lateral size of diamond grains.

In order to realize large area, high quality epitaxy, further efforts are needed. For the heteroepitaxy of diamond on silicon approaches for a much-improved surface treatment and an exact control of nucleation must be taken into consideration. The surface treatment is the first and essential requirement. The substrate surface must be clean, free from surface contamination and oxidation and the dangling bonds of the Si surface atoms must be saturated by hydrogen. During the process of bias-enhanced nucleation the surface roughening due to the ion etching and the formation of amorphous carbon must be avoided.

The bias voltage, biasing time, and pressure strongly influence crystal orientations. The bombardment of energetic ions above a critical energy is necessary for the formation of nuclei. A negative influence by ion bombardment on the alignment of diamond grains has also been demonstrated. An improved epitaxy requires a compromise of the positive and negative influences. On the other hand, at present the base pressure of deposition chamber during CVD diamond nucleation and growth is low and the residual gas in the chamber would contaminate or even oxidize the Si surface and change the surface status. Therefore improvement of the base vacuum in the growth chamber and the purity of gas source would be helpful in achieving better epitaxy.

Acknowledgements

The critical reading of and suggestions relating to this chapter by W. Jäger are especially appreciated.

References

J. C. Angus and C. C. Hayman, Low-pressure, metastable growth of diamond and diamond-like phases, *Science* 214, 913–921 (1988).

Argoitia, et al., Heteroepitaxially grown diamond on a c-BN {111} surface, *Appl. Phys. Lett.* 63, 1336–1338 (1993a).

Argoitia, et al., Diamond grown on single-crystal beryllium oxide, *J. Appl. Phys.* 73, 4305–4312 (1993b).

Argoitia, et al., Heteroepitaxy of diamond on c-BN: growth mechanisms and defect characterization, *J. Mater. Res.* 9, 1849–1865 (1994).

Q. Chen and Z. D. Lin, Electron-emission-enhanced diamond nucleation on Si by hot filament chemical vapor deposition, *Appl. Phys. Lett.* 68, 2450–2452 (1996).

Clausing, et al., Textures and morphologies of chemical vapor deposited (CVD) diamond, *Diamond Relat. Mater.* 1, 411–415 (1992).

Cline, et al., Cyclic deposition of diamond: experimental testing of model predictions, *J. Appl. Phys.* 72, 5926–5940 (1991).

Derjaguin, et al., Structure of autoepitaxial diamond films, *J. Crystal. Growth* 31, 44–48 (1975).

R. C. De Vries, "Cubic Boron Nitride: Handbook of Properties". G.E. Representative, Vol. 72. CRD, New Jersey, 1972, pp. 178.

W. G. Eversole (1962). US Patent No. 3, 030, 187; 3, 030, 188.

Feng, et al., Effect of graphitic carbon films on diamond nucleation by microwave plasma-enhanced chemical-vapor deposition, *J. Appl. Phys.* 74, 2841–2849 (1993).

J. E. Field, "The Properties of Diamond". Academic Press, London, 1979.

M. Flanklach and S. Skokov, On the role of surface diffusion in diamond growth in "Diamond Materials IV" Electrochemical Society Proceedings (K. V. Ravi and J. P. Dismukes, Eds.), Vol. 95-4, pp. 1–12. Electrochemical Society, New Jersey, 1995.

Flöter, et al., Deposition of diamond on patterned silicon substrates, *Diamond Relat. Mater.* 4, 930–935 (1995).

Gerber, et al., Experimental characterisation of bias-enhanced nucleation of diamond on Si, *Diamond Relat. Mater.* 4, 559–562 (1995).

Hayashi, et al., Stepped growth and etching of (001) diamond, *Diamond Relat. Mater.* 5, 1002–1005 (1996).

Hessmer, et al., Correlation between breakdown voltage and structural properties of polycrystalline and heteroepitaxial CVD diamond films, *Diamond Relat. Mater.* 3, 951–956 (1994).

Janischowsky, et al., Growth of high quality, large grain size, highly oriented diamond on Si (100), *Appl. Phys. Lett.* 75, 2094–2096 (1999).

Jeng, et al., Oriented cubic nucleations and local epitaxy during diamond growth on silicon {100} substrates, *Appl. Phys. Lett.* 56, 1968–1970 (1990).

Jia, et al., Heteroepitaxial diamond films on silicon (001): interface structure and crystallographic relations between film and substrate, *Phys. Rev. B* 52, 5164–5171 (1995).

X. Jiang and C. L. Jia, Diamond epitaxy on (001) silicon: an interface investigation, *Appl. Phys. Lett.* 67, 1197–1199 (1995).

X. Jiang and C. L. Jia, The coalescence of [001] diamond grains heteroepitaxially grown on (001) silicon, *Appl. Phys. Lett.* 69, 3902–3904 (1996).

X. Jiang and C. L. Jia, Direct epitaxy of diamond on Si(001) and surface-roughening-induced crystal misorientation, *Phys. Rev. Lett.* 84, 3658–3661 (2000).

X. Jiang and C.-P. Klages, Heteroepitaxial diamond growth on (100) silicon, *Diamond Relat. Mater.* 2, 1112–1113 (1993).

X. Jiang and C.-P. Klages, Recent development in heteroepitaxial nucleation and growth of diamond on silicon, *Phys. Status Solidi (a)* 154, 175–183 (1996).

Jiang, et al., The effect of substrate bias voltage on nucleation of diamond crystals in a microwave plasma assisted CVD process, *Diamond Relat. Mater.* 2, 407–412 (1992).

Jiang, et al., Epitaxial diamond thin films on (001) silicon substrates, *Appl. Phys. Lett.* 62, 3438–3440 (1993a).

Jiang, et al., Deposition and characterization of diamond epitaxial thin films on silicon substrates, *Appl. Phys. A* 57, 483–489 (1993b).

Jiang, et al., Atomic-force-microscopic study of heteroepitaxial diamond nucleation on (100) silicon, *Appl. Phys. Lett.* 63, 1203–1205 (1993c).

Jiang, et al., Nucleation and initial growth phase of diamond thin films on (100) silicon, *Phys. Rev. B* 50, 8402–8410 (1994a).

Jiang, et al., Approach of selective nucleation and epitaxy of diamond films on Si(100), *Appl. Phys. Lett.* 65, 1519–1521 (1994b).

Jiang, et al., "Studies of heteroepitaxial nucleation and growth of diamond on silicon" Society Proceedings, Vol. 95-4, pp. 50–55. Electrochemical Society, New Jersey, 1995.

Jiang, et al., Diamond film orientation by ion bombardment during deposition, *Appl. Phys. Lett.* 68, 1927–1929 (1996).

Jiang, et al., Coalescence and overgrowth of diamond grains for improved heteroepitaxy on silicon (001), *J. Appl. Phys.* 83, 2511–2518 (1998a).

Jiang, et al., Effects of ion bombardment on the nucleation and growth of diamond films, *Phys. Rev. B* 58, 7064–7075 (1998b).

Jiang, et al., Local strain in interface: origin of grain tilting in diamond (001)/silicon (001) heteroepitaxy, *Phys. Rev. B* 58, 15351–15354 (1998c).

Jiang, et al., High quality chemical-vapour-deposited diamond films: recent progresses, *Diamond Relat. Mater.* 9, 1640–1645 (2000).

Jiang, et al., Simulation of thin film growth: an investigation of the evolution of grain size and texture in CVD diamond films, *Appl. Phys. A* 74, 217–224 (2002); Jiang, et al., *Relat. Mater.* 2, 1112–1114 (2002).

John, et al., Localized epitaxy of diamond on (100) silicon, *J. Mater Res.* 9, 3083–3087 (1994).

R. O. Jones and O. Gunnarsson, The density functional formalism, its applications and prospects, *Rev. Mod. Phys.* 61, 689–746 (1989).

Kawarada, et al., Heteroepitaxial growth of highly oriented diamond on cubic silicon carbide, *J. Appl. Phys.* 81, 3490–3493 (1997).

Kobashi, et al., Synthesis of diamonds by use of microwave plasma chemical-vapor deposition: morphology and growth of diamond films, *Phys. Rev. B* 38, 4067–4084 (1988).

S. Koizumi and T. Inuzuka, Initial growth process of epitaxial diamond thin films on cBN single crystals, *Jpn. J. Appl. Phys.* 32, 3920–3927 (1993).

Koizumi, et al., Epitaxial growth of diamond thin films on cubic boron nitride {111} surfaces by dc plasma chemical vapor deposition, *Appl. Phys. Lett.* 57, 563–565 (1990).

J. J. Lander and J. Morrison, Surface reactions of silicon with aluminium and with indium, *Surf. Sci.* 2, 553–565 (1964).

N. Lee and A. Badzian, Effect of methane concentrations on surface morphologies and surface structures of (001) homoepitaxial diamond thin films, *Appl. Phys. Lett.* 67, 2011–2013 (1995).

Lifshitz, et al., Subplantation model for film growth from hyperthermal species: application to diamond, *Phys. Rev. Lett.* 62, 1290–1293 (1989).

Locher, et al., Nitrogen stabilized (100) texture in chemical vapor deposited diamond films, *Appl. Phys. Lett.* 65, 34–36 (1994).

Maeda, et al., Formation of highly oriented diamond film on carburized (100) Si substrate, *J. Mater. Res.* 10, 158–164 (1995).

Matsumoto, et al., "Proceedings of the Seventh International Conference on Vacuum Metallurgy". Iron and Steel Institute of Japan, Tokyo, Japan, 1982, pp. 386–391.

Mitsuda, et al., The growth of diamond in microwave plasma under low-pressure, *J. Mater. Sci.* 22, 1557–1562 (1987).

A. A. Morrish and P. E. Pehesson, Effects of surface pretreatments on nucleation and growth of diamond films on a variety of substrates, *Appl. Phys. Lett.* 59, 417–419 (1991).

Nakazawa, et al., X-ray section topographs of a vapour-grown diamond film on a diamond substrate, *Thin Solid Films* 151, 199–206 (1987).

Narayan, et al., On epitataxial growth of diamond films on (100) silicon substrates, *Appl. Phys. Lett.* 53, 1823–1825 (1988).

Ohtsuka, et al., Epitaxial growth of diamond on iridium, *Jpn. J. Appl. Phys.* 35, L1072–L1074 (1996).

Ohtsuka, et al., Fabrication of epitaxial diamond thin film on iridium, *Jpn. J. Appl. Phys.* 36, L1214–L1216 (1997).

Olsen, et al., A mechanism of CVD diamond film growth deduced from the sequential deposition from sputtered carbon and atomic hydrogen, *J. Mater. Res.* 9, 1546–1551 (1994).

R. Polini and M. Tomelini, Analysis of size distribution functions of diamond crystallites formed in the early stages of chemical vapour deposition, *Diamond Relat. Mater.* 4, 1311–1316 (1995).

Sattel, et al., Ion induced nucleation of diamond, *Phys. Status Solidi (a)* 154, 141–153 (1996).

K. Schiffmann and X. Jiang, Investigation of heteroepitaxial diamond films by atomic force and scanning tunneling microscopy, *Appl. Phys. A* 59, 17–22 (1994).

Schreck, et al., Limitations of the process window for the bias enhanced nucleation of heteroepitaxial diamond films on silicon in the time domain, *J. Phys. Phys.* 81, 3092–3095 (1997).

Schreck, et al., Diamond/Ir/$SrTiO_3$: a material combination for improved heteroepitaxial diamond films, *Appl. Phys. Lett.* 74, 650–652 (1999).

Schreck, et al., Diamond nucleation on iridium buffer layers and subsequent textured growth: a route for the realization of single-crystal diamond films, *Appl. Phys. Lett.* 78, 192–194 (2001).

Spitsyn, et al., Vapor growth of diamond on diamond and other surfaces, *J. Crystal. Growth* 42, 219–226 (1981).

B. R. Stoner and J. T. Glass, Textured diamond growth on (100) beta-SiC via microwave plasma chemical vapor deposition, *Appl. Phys. Lett.* 60, 698–700 (1992).

Stoner, et al., Epitaxial nucleation of diamond on β-SiC via bias-enhanced microwave plasma chemical-vapor-deposition, *Diamond Relat. Mater.* 2, 142–146 (1993a).

Stoner, et al., Highly oriented, textured diamond films on silicon via bias-enhanced nucleation and textured growth, *J. Mater. Res.* 8, 1334–1340 (1993b).

T. Suzuki and A. Argoita, Current status of heteroepitaxy of CVD diamond, *Phys. Status Solidi (a)* 154, 239–254 (1996).

Suzuki, et al., Oriented diamond on graphite flakes, *Appl. Phys. Lett.* 65, 540–542 (1994).

Stubhan, et al., Heteroepitaxial nucleation of diamond on Si(001) in hot filament chemical vapor deposition, *Appl. Phys. Lett.* 66, 1900–1902 (1995).

Tachibana, et al., Azimuthal rotation of diamond crystals epitaxially nucleated on silicon {001}, *Appl. Phys. Lett.* 68, 1491–1492 (1996).

Tachibana, et al., X-ray diffraction pole figure measurements of diamond films grown on platinum (111), *J. Appl. Phys.* 82, 4327–4330 (1997).

Tomellini, et al., A model kinetics of nucleation at a solid surface with application to diamond deposition from the gas phase, *J. Appl. Phys.* 70, 7573–7578 (1991).

T. Tomizuka and S. Shikata, Cross-sectional transmission electron microscopy observations and selected-area electron diffractions of interfaces of epitaxially grown diamond thin films on cubic boron nitride substrates, *Jpn. J. Appl. Phys.* 32, 3938–3942 (1993).

A. Van der Drift, Evolutionary selection, a principle governing growth orientation in vapour-deposited layers, *Philips Res. Rep.* 22, 267–288 (1967).

Wild, et al., Chemical vapour deposition and characterization of smooth {111}-faceted diamond films, *Diamond Relat. Mater.* 2, 158–168 (1993).

Wild, et al., Oriented CVD diamond diamond films – twin formation, structure and morphology, *Diamond Relat. Mater.* 3, 373–381 (1994).

B. E. Williams and J. T. Glass, Characterization of diamond thin films: diamond phase identification, surface morphology, and defect structures, *J. Mater. Res.* 4, 373–384 (1989).

Wittorf, et al., Grain boundaries in diamond films on Si(001), *Mater. Res. Soc. Symp. Proc.* 466, 27–32 (1997a).

Wittorf, et al., Microstructure and growth of MWCVD diamond on $Si_{1-x}C_x$ bufferlayers, *Diamond Relat. Mater.* 6, 649–653 (1997b).

Wittorf, et al., Electron microscopy of interfaces in chemical vapour deposition diamond films on silicon, *Diamond Relat. Mater.* 9, 1696–1702 (2000).

Wurzinger, et al., TEM investigations on the heteroepitaxial nucleation of CVD diamond on (001) silicon substrates, *Diamond Relat. Mater.* 6, 752–757 (1997).

R. Yamada and K. Sone, On the influence of incident energy of protons on chemical erosion of graphite, *J. Nucl. Mater.* 116, 200–205 (1983).

Yang, et al., Structural study of diamond film formed on silicon wafer by hot-filament chemical vapor deposition method, *Appl. Phys. Lett.* 65, 3203–3205 (1994).

Yoshikawa, et al., Study of crystallographic orientations in the diamond film on cubic boron nitride using Raman microprobe, *Appl. Phys. Lett.* 57, 428–430 (1990).

Yu, et al., Alignment of thin films by glancing angle ion bombardment during deposition, *Appl. Phys. Lett.* 47, 932–933 (1985).

Yugo, et al., Generation of diamond nuclei by electric field in plasma chemical vapor deposition, *Appl. Phys. Lett.* 58, 1036–1038 (1991).

Yugo, et al., Nucleation mechanism of diamond in plasma chemical vapor deposition, *Diamond Relat. Mater.* 2, 328–332 (1993).

Yugo, et al., in "Kimura, Advances in New Diamond Science and Technology" (S. Saito, N. Fujimori, O. Fukunaga, M. Kamo, K. Kobashi, and M. Yoshikawa, Eds.), p. 175. MYU, Tokyo, 1994.

Zhang, et al., Structural modeling of the possible growth of oriented textured single-crystal diamond film on a silicon (111) surface, *Appl. Phys. Lett.* 69, 1086–1088 (1996).

Zhang, et al., The selective etching of H^+ ions and ist effects on the oriented growth of diamond films, *J. Appl. Phys.* 82, 1896–1899 (1997).

Zhu, et al., Diamond and beta-SiC heteroepitaxial interfaces: a theoretical and experimental study, *Phys. Rev. B* 47, 6529–6542 (1992).

Zhu, et al., Oriented diamond films grown on nickel substrates, *Appl. Phys. Lett.* 63, 1640–1642 (1993).

This page is intentionally left blank

CHAPTER 2

STRUCTURAL IMPERFECTIONS IN CVD DIAMOND FILMS

Eberhard Blank

ECOLE POLYTECHNIQUE FÉDÉRALE DE LAUSANNE (EPFL), MATERIALS SCIENCE, CH-1015 LAUSANNE, SWITZERLAND

1. Introduction

The unrivalled physical, chemical and mechanical properties of diamond have stimulated the search for new ways to synthesise and shape this outstanding material for more than a century. With the arrival of the chemical-vapour-deposition (CVD) process, a new dimension has been added to the potential use of diamond as an engineering material, which has activated world-wide efforts aiming at its integration in modern fabrication technology. An impressive body of knowledge has been accumulated during the last two decades in the fields of CVD processing, property evaluation, and product development. Much of this work is concerned with defects at all size scales, which always play a key role in materials science. While most chapters of this book deal with defect engineering, which is fundamental for taking advantage of the intrinsic properties of diamond, this chapter focuses on imperfections accompanying crystallisation from the gas phase. The state-of-the-art in the control, analysis and properties of structural imperfections, mainly dislocations, grain boundaries and substrate interfaces, is reviewed. Point defects and impurities are discussed in less detail since they are the subjects of other chapters in this book. In many respects, this approach is complementary to the approach taken by recent review articles where emphasis has been placed on point defects (Zhu, 1994; Nazaré and Neves, 2001).

The chapter begins with some morphological aspects of diamond thin film growth and then introduces the phenomenology of twin formation. After a short review on impurities in single- and polycrystalline films, the present understanding of dislocations and grain boundaries is discussed in detail. Attention is given to the role of imperfections during film growth and to their implications in determining properties. Much room is imparted to the results of atomistic and molecular dynamics modelling. These techniques have become not only reliable and complementary to high-resolution transmission microscopy and spectroscopy, they also represent unique tools in the investigation of the electronic structures associated with dislocations and grain boundaries. The final section on

ISBN: 0-12-752185-2
ISSN: 0080-8784

epitaxy focuses on diamond films with low defect densities. The role of interface structures in heteroepitaxial growth is emphasised.

Only growth-related imperfections are dealt with in this chapter. The defect structures of nano-crystalline diamond films are not taken into account systematically. Defects resulting from post-treatments, such as ion implantation or thermal treatment, have not been included. Nor have surface related structures and properties been treated, as these defect structures are fully discussed elsewhere in this book.

2. Structure and Morphology at Different Size Scales

2.1. Diamond Crystals and Film Morphology

Diamond is the cubic polymorph of carbon and belongs to the space group O_h^7 ($Fd3m$). The carbon atoms are tied together by covalent bonds oriented along $\langle 111 \rangle$ directions. One s orbital and three p orbitals blend into four sp^3 hybrid orbitals (σ bond) which point to the corners of a tetrahedron occupied by the four nearest neighbour atoms (Fig. 1). The C–C bond length is 1.54 Å. The Bravais

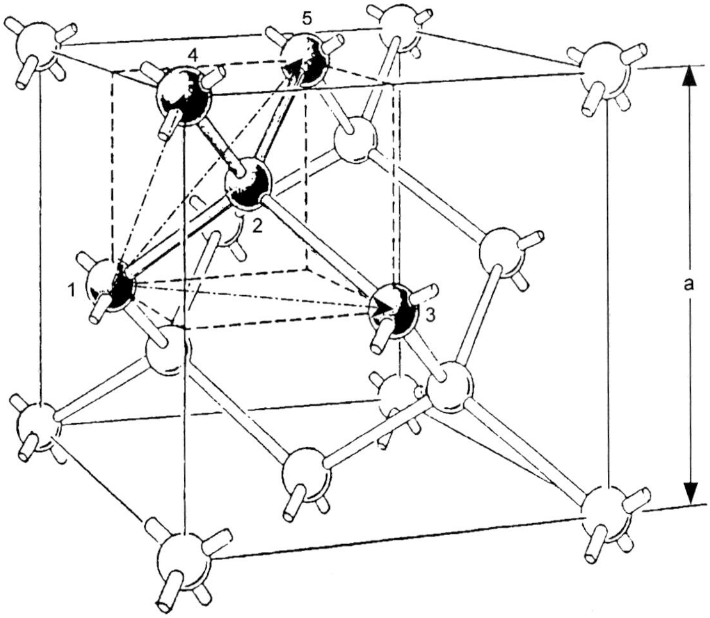

FIG. 1. Structure of diamond. The carbon atoms have tetrahedral coordination and occupy the lattice sites of two interpenetrating face-centred-cubic Bravais lattices of dimension a. The smallest unit cell is indicated by dashed lines and is occupied by atoms 1 and 2. The vectors joining atoms with atoms 3, 4, 5 represent the translation basis (after W. Shockley, "Electrons and Holes in Semiconductors", Van Nostrand, Princeton, 1950).

lattice of diamond is face-centred cubic (fcc) with a lattice constant of 3.567 Å (Lang and Pang, 1998). The unit cell is composed of two simple fcc point lattices which are shifted with respect to each other along the cube diagonal by a quarter of the diagonal length. In other words, the primitive basis consists of two atoms at (000) and $\left(\frac{1}{4}\frac{1}{4}\frac{1}{4}\right)$ associated with each lattice point. In a perfect crystal, each atom is surrounded by four nearest neighbours in tetrahedral coordination and 12 next-nearest neighbours. Diamond does not have a closest-packed structure. In terms of a hard sphere model, only 34% of the lattice volume is filled by the carbon atoms. Nevertheless, there is no other material with a higher molar density than diamond (Angus, Buck, Sunkara, Groth, Hayman and Gat, 1989).

CVD grown diamond crystals are bounded by {100} and {111} facets. In a few cases, {110} facets have been observed (Kuang, Badzian, Tsong, Lee, Badzian and Chen, 1996; Sun, Yu, Jiang, Li, Liu, Jin and Zou, 1997). The facets oriented to each other according to the symmetry properties of the crystal lattice determine the shape of single crystalline particles. Depending on the relative growth rate of the {100} and {111} facets which is measured by the growth parameter $\alpha = \sqrt{3}v_{100}/v_{111}$ (Wild, Koidl, Müller-Sebert, Walcher, Kohl, Herres, Locher, Samlenski and Brenn, 1993), the particle shape varies from octahedral to cubo-octahedral and to cube shape, as illustrated in Figure 2 (Clausing, Heatherly, Horton, Specht, Begun and Wang, 1992; Wild et al., 1993). The growth parameter has become a convenient means for the morphological characterisation of CVD diamond particles and films. The actual value achieved in a deposition

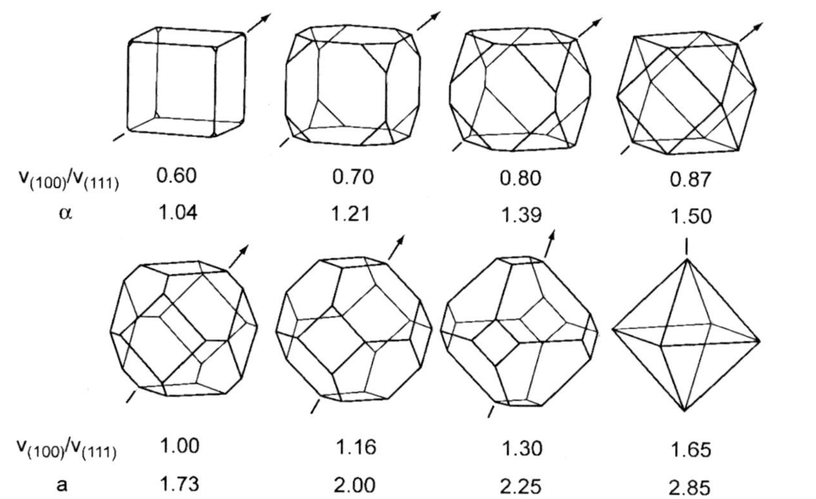

FIG. 2. The shape of defect-free diamond crystallites is determined by the ratio of the growth velocities in the ⟨100⟩ and ⟨111⟩ directions or by the α value. The arrows indicate the direction of fastest particle growth which corresponds to the largest diameter of the polygonised particle (after Menon, P. M., Clausing, R. E., Heatherly, L., and Feigerle, C. S. (1998) *Diamond Relat. Mater.* 7, 1201–1206).

process is easily evaluated from the geometrical relationship between the facets of individual particles (Maeda, Ohtsubo, Irie, Ohya, Kusakabe and Morooka, 1995; Rigby, Cherian, Lister, Rudolph and White, 2000). Higher carbon contents in the gas phase tend to increase the α value, whereas higher substrate temperatures and lower pressure decrease it (Wild, Kohl, Herres, Müller-Sebert and Koidl, 1994; Seehofer, 2002). This dependence is not well understood, however. While it is generally accepted that methyl radicals and acetylene, both of which are present in the gas atmosphere, act as precursors for diamond formation, the mechanisms for $\{100\}$ and $\{111\}$ growth are still under discussion. Growth experiments with homoepitaxial films have revealed very early that cube and octahedral facets exhibit different activation energies, which depend differently on methane flow and substrate temperature (Chu, Hauge, Margrave and D'Evelyn, 1992; Maeda et al., 1995). Growth of $\{111\}$ facets at low partial pressure of acetylene should be controlled by the nucleation of new monolayers on the flat surface, according to a three-dimensional, atomic scale Monte Carlo simulation by Bataille, Srolovitz and Butler (2001). The reality appears to be more complicated, however. In a recent experiment conceived to create, at the diamond growth surface, an atmosphere consisting exclusively of methyl radicals in hydrogen, or exclusively acetylene in hydrogen, it has been demonstrated that the absence of acetylene does not impede the nucleation of new $\{111\}$ monolayers (D'Evelyn, Graham and Martin, 2001). Moreover, gas atmospheres with only acetylene precursors lead to the formation of octahedra ($\alpha = 3$), indicating that $\langle 100\rangle$ growth is much faster than $\langle 111\rangle$ growth, whereas the deposition from atmospheres containing only methyl radicals leads to cubo-octahedral crystals ($\alpha \approx 1.8$). It appears that defects on $\{111\}$ facets, mainly microtwins, screw dislocations, provide much higher nucleation rates for new $\{111\}$ layers than homoepitaxial nucleation considered in the Monte Carlo simulation.

The relative growth velocities of $\{100\}$ and $\{111\}$ facets also play a major role in the formation of crystallographically textured, polycrystalline films. Texture has been shown (Wild, Herres and Koidl, 1990; Clausing et al., 1992) to develop conforming to the model of competitive grain growth proposed by van der Drift (1967). If growth begins with a random orientation of nuclei that have the same α value, only those crystallites for which the fastest growth direction is perpendicular to the substrate plane will survive. Less favourably oriented grains are overgrown because the tips of the fast growing grains are always ahead by a nose and thus have an opportunity to expand laterally. For a given value of α, the fastest growing direction is equivalent to the longest diagonal which may be drawn through the body of the particles, as shown in Figure 2, where these directions are marked by arrows. The α value thus determines the shape of individual particles and the crystallographic texture of diamond films once the slowly growing grains have been overgrown by the rapid ones. Values of 1, 1.5, and 3 correspond to $\langle 111\rangle$, $\langle 110\rangle$, and $\langle 100\rangle$ textures, respectively. Advantage of the effect of grain selection by competitive grain growth has been taken in depositing diamond films with sharp ⟨100⟩ fibre-texture where the surviving grains share a common ⟨100⟩ zone axis in growth direction about which they are randomly oriented (Wild et al., 1993). The tilt component of the grain misalignment with respect to the substrate can be strongly reduced by this mechanism, to below 0.5°. The twist component experiences almost no changes. It should be noticed, that sharp fibre textures are obtained for ⟨001⟩ but not the other growth directions because frequent twinning induces local deviations from the preferred orientation.

2.2. Twinning and Stacking Faults

2.2.1. *Crystallographic Description*

Twinning is a growth defect describing a symmetrical orientation relationship between two adjacent crystals. In diamond, the two crystals share a common ⟨111⟩ axis about which they are rotated by 180° or, equivalently, 60° with respect to each other. There are several ways to visualise this symmetry operation. Referring to the tetrahedral atomic arrangement shown in Figure 3, twinning by rotation about the vertical $[1\bar{1}1]$ bond axis brings the atoms from a staggered (Fig. 3a) into an eclipsed stacking conformation (Fig. 3b). The $(1\bar{1}1)$ plane perpendicular to the rotation axis will be called the *twin plane*. The eclipsed conformation would also be obtained by a rotation of the upper tetrahedron (Fig. 3a) by 109.5° about the $[\bar{1}\bar{1}0]$ axis located on atom D, in which case atom A flips to B and atom B to the upper right corner of the reference (110) plane, while atom C rests in place. Stacking of structural units in staggered conformation leads to the regular crystalline arrangement of cubic diamond exhibiting folded, six-membered carbon rings in chair conformation (Fig. 3c). In case of twinning, the six-membered rings crossing the twin plane are in the boat conformation (Fig. 3d). The $(1\bar{1}1)$ twin plane, not depicted in Figure 3d, is located between the two B layers and represents a mirror plane for the structural units on both sides. Using the coincidence site lattice (CSL) notation, the orientation relationship which designates two lattices originating from each other by first-order twinning is $\Sigma = 3$ (referring to the ⟨110⟩ projection of Figure 4a, the planar translation unit cell common to the lattices on both sides of the boundary is three times larger than the unit cell necessary to generate the structures on either side of the boundary).

Two-dimensional representations of the diamond lattice generally use ⟨110⟩ projections where the six-membered rings are aligned behind each other and where the symmetrical orientation relationship between the parent and twin lattices shows up clearly. In Figure 4a, the twin plane is located between the double layers bC and Cb which are in the boat conformation. The twin plane coincides with the *twin boundary* separating the twinned crystal units. It will be shown later that twin boundaries can adopt arbitrary shapes and

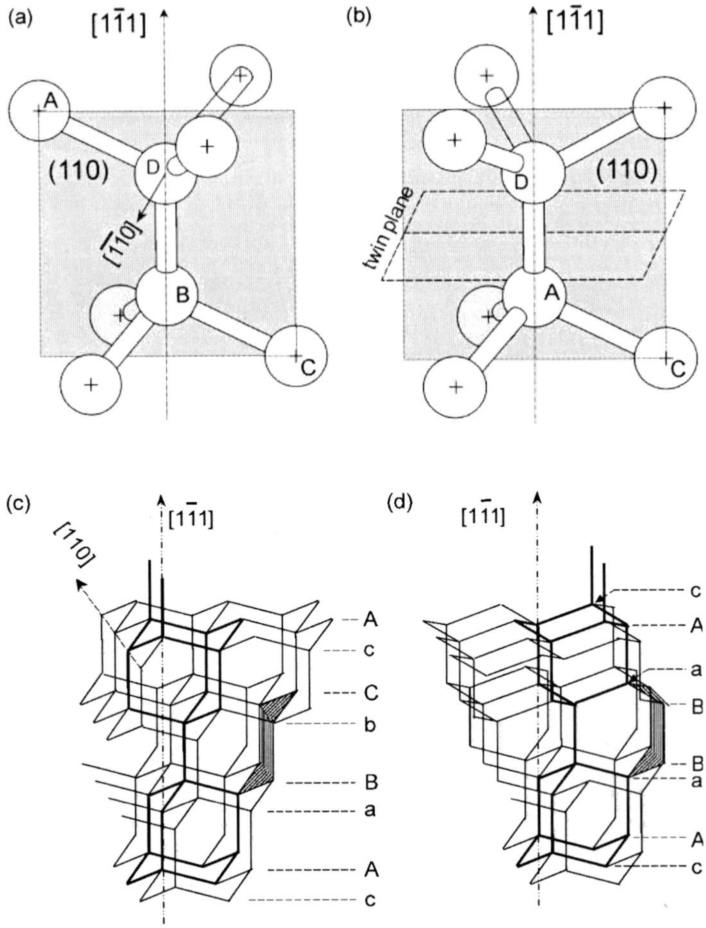

FIG. 3. Twinning at the atomic scale brings the atoms of the regular (a) staggered into the (b) eclipsed position (looking in the [1$\bar{1}$1] direction). This operation corresponds to a rotation by 60° or 180° about the [1$\bar{1}$1] bond direction between atoms B and D, or to a rotation about the [1$\bar{1}$0] axis passing through atom D. (c) If many tetrahedral units as shown in (a) are stacked in a regular way, the diamond crystal lattice is obtained with its typical six-membered rings in chair conformation. (d) Similar to (c) but a twin plane is introduced between the double layers aB and Ba. The six-membered rings crossing the twin plane are in boat conformation.

orientations which depend on how they are formed. This does not change the $\Sigma = 3$ relationship of the twinned lattices. Two different notations are used in Figure 4 for indexing the stacking sequence. In the first column, the planes with atoms in mirror position with respect to the twin plane are designated by identical upper or lower-case letters, whereas the standard notation in the second column uses systematically upper-case letters for the three conjugate carbon double layers linked by bonds in the $\langle 111 \rangle$ stacking direction. If in Figure 4a another double layer bC in boat conformation with respect to Cb is added, a defect called *intrinsic stacking fault* is formed (Figure 4b). The stacking sequence bC, Cb, bC thus contains two twin boundaries on both sides of the double layer Cb which itself can be considered as the thinnest possible twin plate in a regular diamond lattice. In the second notation used in Figure 4, the intrinsic stacking fault signifies a missing A layer in the regular stacking sequence ABCA... An *extrinsic stacking fault* (Fig. 4c) is generated if another double layer (Ba) that is in chair conformation with respect to Cb is inserted between the two twin boundaries of Figure 4b. The lattices on both sides of the stacking faults are out of phase by (1/6) $\langle 112 \rangle$, equivalent to the Burgers vectors of Shockley partial dislocations. Only if a third double-layer is added the phase shift of the parent lattice is brought back to zero.

Diamond polytypes are formed when the stacking sequence of the diamond lattice is disturbed at regular distances in the stacking direction. The incorporation of a twin plane between all double layers produces the 2H polytype representing a hexagonal diamond phase called lonsdaleite. The stacking pattern is analogous to that of an intrinsic stacking fault continued over long distances (/CB/CB/...). Lonsdaleite has been identified in diamond produced by shock wave loading of graphite powder (Knight and White, 1989) or in diamond grown by the high-pressure, high-temperature technique (HPHT) (Ergun and Alexander, 1962; Bundy and Kasper, 1967; Frondel and Marvin, 1967). Reports confirming the formation of 2H diamond in CVD processing are scarce. Chen, Chang, Lin and Chen (1996) have observed this phase in heteroepitaxial growth on (110) silicon wafers. Silva, Amaratunga, Salje and Knowles (1994) report to have found hexagonal diamond grains of about 30 nm size together with graphite and silicon carbide in carbon films deposited on silicon substrates. There is also evidence for the existence of 6H hexagonal diamond (Frenklach, Kematick, Huang, Howard, Spear, Phelps and Koba, 1989; Howard, Huang, Yuan, Frenklach, Spear, Koba and Phelps, 1990) with the stacking sequence /ABCACB/...). These and other polytypes (4H, 8H, 15R, 21R; H hexagonal, R rhombohedral) have been postulated by analogy with the seven most common silicon carbide polytypes (Spear, Phelps and White, 1990)

2.2.2. *The Formation of Twins*

2.2.2.1. Energy of Stacking Faults and Coherent Twin Boundaries. Twins and stacking faults are abundant in diamond. The most common boundary type in diamond are Σ 3 {111} boundaries which are coherent (Figs. 3b and 4a). Their energy is very low because the distances between nearest and next-nearest neighbours do not change across the boundary. Only the positions of the next-nearest bonds vary, involving a change of the distances between third-nearest neighbours. Energy values are not available explicitly but can be estimated as about half of the energy of intrinsic stacking faults. From the

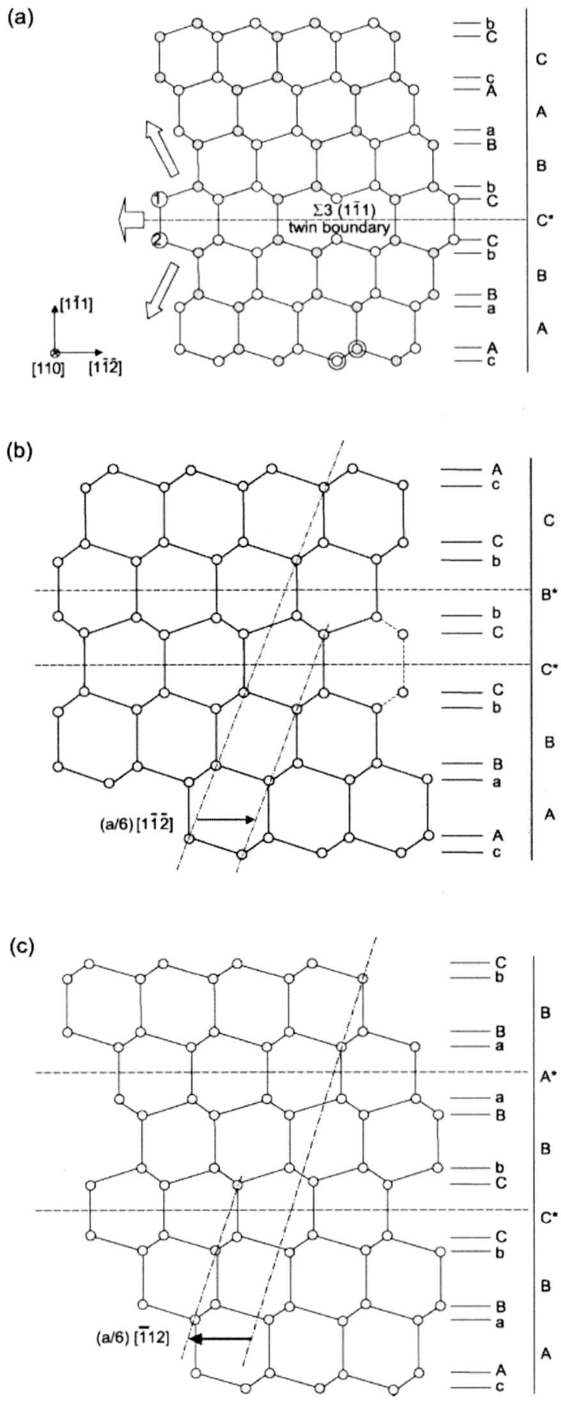
(a)
Σ3 (1$\bar{1}$1) twin boundary
[1$\bar{1}$1]
[110]
[1$\bar{1}\bar{2}$]
C
A
B
C*
B
A
(b)
(a/6) [1$\bar{1}\bar{2}$]
C
B*
C*
B
A
(c)
(a/6) [$\bar{1}$12]
B
A*
B
C*
B
A

separation of partial dislocations observed by the weak beam technique in transmission electron microscopy (TEM), Pirouz, Cockayne, Sumida, Hirsch and Lang (1983) have obtained a stacking fault energy of 0.279 ± 0.041 J m^{-2} for type IIa diamond, and Suzuki, Ichihara, Takeuchi, Ohtake, Yoshikawa, Hirabayashi and Kurihara (1992) $0{,}290 \pm 0{,}041$ J m^{-2} for CVD diamond. Dorignac, Serin, Delclos, Philipp, Ats and Vandenbulcke (1997), using a microscope with ultra-high resolution (1250 kV acceleration voltage), report 0.276 ± 0.003 J m^{-2}, based on the separation of 8.3 ± 0.1 nm between the 30° partials of screw dislocations in CVD diamond. All author groups apply the elastic constants given by Grimsditch and Ramdas (1975) for their estimations. Early first-principle theoretical calculations arrive at stacking fault energies rather close to experiment (Persson, 1983) but the results of atomistic modelling based on either first-principle calculations (Shenderova, Brenner, Nazarov, Romanov and Yang, 1998) or the assumption of semi-empirical interatomic potentials (Altmann, Lodge and Lapiccirella, 1983) sometimes reveal substantial discrepancies with experiment. Nevertheless, the reliability of theoretical modelling has been considerably improved during the last few years (for a brief discussion of various theoretical models used in simulation, see Section 3.2.1.1). Applying density-functional tight-binding (TB) simulation techniques, the theoretical stacking fault energy recently has been calculated as 293 J m^{-2} which lies well within the range of the reported experimental values (Blumenau, Fall, Jones, Heggie, Briddon, Frauenheim and Öberg, 2002b). Also the elastic constants deviate by less than 10% from their experimental counterparts. In conclusion, the energy of a coherent first-order twin boundary is estimated at about 0.14 J m^{-2}, approximately one half of the stacking fault energy. The twin nucleation energy calculated below, using an independent approach, agrees satisfyingly well with this value.

2.2.2.2. *Nucleation Energy of Twins.* It is well established that diamond grows by the nucleation and propagation of steps on otherwise flat {100} and {111} facets (Janssen, von Enckefort, Vollenberg and Giling, 1995; von Enckefort, Janssen, Schermer and Giling, 1995a; von Enckefort, Janssen, Vollenberg and Giling, 1995b). Since homogeneous, two-dimensional nucleation with three ad-atoms is required to form a stable nucleus on a smooth {111} diamond facet, as for germanium and silicon, the probability for the nucleation of a new growth step by

---

FIG. 4. ⟨110⟩ projections of defective diamond lattices. White and grey atoms are shifted with respect to each other by $a/42$ in [110] direction (a is the lattice parameter). Two notations are given for the stacking sequences (see text). (a) Diamond lattice containing a coherent twin boundary (mirror plane); on the left side, a re-entrant corner is formed. The projected distance of the encircled atoms in double-layer cA amounts to 0.089 nm and is not resolved in HRTEM. (b) Stacking sequence in case of an intrinsic stacking fault; the lattices on both sides of the stacking fault are shifted by $a/6\ [\bar{1}1\bar{2}]$. (c) Stacking sequence in case of an extrinsic stacking fault; the lattices on both sides of the stacking fault are shifted by $a/6\ [1\bar{1}2]$.

this mechanism is very small compared to the formation of growth steps at re-entrant corners. (Tiller, 1991; Angus, Sunkara, Sahaida and Glass, 1992) Nevertheless, twinning must precede the formation of re-entrant corners. In diamond deposition at low substrate temperatures or from carbon-rich gas atmospheres, the situation changes because two-dimensional nucleation may become the predominant nucleation mechanism (Michler, Stiegler, von Kaenel, Moeckli, Dorsch, Stenkamp and Blank, 1997). In order to explain the homogeneous nucleation of growth steps on smooth {111} layers, it has been argued recently that a new nucleus, which consists of a 6-membered ring with three ad-atoms standing out from the {111} facet and the three other atoms being embedded in the {111} facet can oscillate between the chair and boat conformation (Seehofer, 2002). The energy difference is expected to only slightly exceed that for free cyclohexane rings which also oscillate between these two conformations. The enthalpy difference for cyclohexane in chair and boat conformation is 23 kJ mol^{-1} and the activation energy for this transformation 46 kJ mol^{-1} (Christen and Vögtle, 1992). It is interesting to compare this energy difference with the energy of a twin boundary by assuming that the {111} twin boundary is composed of cyclohexane rings in boat conformation (in a $\langle 110 \rangle$ projection just one carbon double layer is added). As each boat conformation occupies a projected area of 4.53×10^{-21} m^2 in the twin boundary, a boundary energy of 0.12 J m^{-2} is calculated which is close to half of the stacking fault energy, as one would expect from the above discussion. As third-nearest neighbours are not involved in this estimation, the energy of a real twin boundary is supposed to be slightly greater. Concerning the oscillation of the free-standing part of the cyclohexane ring between chair and boat conformations, the activation energy of 0.24 J m^{-2} obtained in an analogous manner probably underestimates the real value because constraints exerted by the diamond lattice on the oscillation are not taken into account. Obviously, these energy considerations refer only to coherent $\Sigma = 3$ {111} type twin boundaries terminating at free surfaces. If the twinned sections are included in a parent crystal, the overall excess energy for twin formation becomes considerably larger owing to lattice distortion, misfit dislocations or grain boundaries conciliating matrix and twin at the boundaries which are not twin planes.

2.2.2.3. Re-entrant Corner Growth. In faceted growth of materials with fcc lattices, in particular diamond-type lattices, re-entrant corners are formed at the outcrops of twin boundaries and stacking faults which become favourite sites for the nucleation of new growth layers because two ad-atoms are sufficient to create a nucleus with a high chance of survival (Angus et al., 1992). The re-entrant corner which is formed by the joining {111} facets of a twinned crystal is shown on the left side of Figure 4a. Similar re-entrant corner exists at the outcrop of intrinsic stacking faults and extrinsic. Penetration twins, which will be explained below, also provide re-entrant corners at the intersection of {100} and {111} facets. Such intersection have been identified as active sites for the emission of growth steps both in monocrystalline (von Enckefort et al., 1995b) and polycrystalline diamond films (Sun, Yu, Jiang, Li, Liu, Jin and Zou, 1996). In Figure 6b an isolated particle with growth terraces emanating from the edges at the intersection between cube facets and the {111} facets of penetration twins is shown. Re-entrant corner growth has a strong impact on the morphology of polycrystalline films (Shechtman, 1994). An illustration of the importance of twinning for polycrystalline diamond film growth is given in Figure 5 where the numerous facets of large diamond grains are related to each other via twin boundaries in re-entrant corner configuration. Re-entrant corners provide self-perpetuating sources for the nucleation of new steps and there is evidence that they control the growth rate in the regime of surface controlled diamond growth if the propagation of growth steps is fast (Angus et al., 1992; Bataille et al., 2001; D'Evelyn et al., 2001).

2.2.3. *Contact and Penetration Twins*

Twinning thoroughly affects the shape of individual particles and the morphology of diamond films. In order to better distinguish between the various morphological features of twin-related growth, twins in diamond have been classified into contact and penetration twins. The reader should keep in mind, however, that twinning at the atomic scale always signifies a rotation of the parent and twin lattices by 60° or, equivalently, 180° about a ⟨111⟩ bond direction, i.e., perpendicular to the twin plane (Fig. 3a and b). According to Tamor and Everson (1994), contact twins originate on a {111} surface and share the ⟨111⟩ direction normal to that surface (Fig. 4), whereas penetration twins originate on either

FIG. 5. Re-entrant corner growth of a polycrystalline diamond film (29 sccm CO_2, 31 ccm CH_4; substrate temperature 380 °C: Re-entrant and convex corners at the surface of grain families bounded by {111} facets.

a {111} or a {100} surface, and share a ⟨111⟩ direction which is not normal to that surface (Fig. 6). The angle of 15.8° between the {100} and {111} planes corresponds to the angle between [001] and [115] vectors. Grain boundaries with adjacent {100} and {115} faces have been repeatedly identified and will be treated in greater detail in Section 3.4.1. An atomic scale model of the boundaries which is supposed to be involved in the formation of a {100} ⟨111⟩ penetration twin is shown in Figure 6d (Takami, Suzuki, Kusunoki, Sakaguchi, Nishitani-Gamo and Ando, 1999). The twin boundary of penetration twins thus no longer coincides with the twin plane, as in Figure 4a. As a possible mechanism for the formation of penetration twins, Tamor and Everson invoke the formation of contact microtwins in regions exhibiting microscopic {111} surfaces tilted with respect to the parent face. Contact twins are labeled {111} ⟨111⟩, and penetration twins either {111} ⟨111⟩ or {100} ⟨111⟩, depending on the plane of origin (the indices refer to the parent lattice). Such regions are likely to be associated with growth steps as mentioned before.

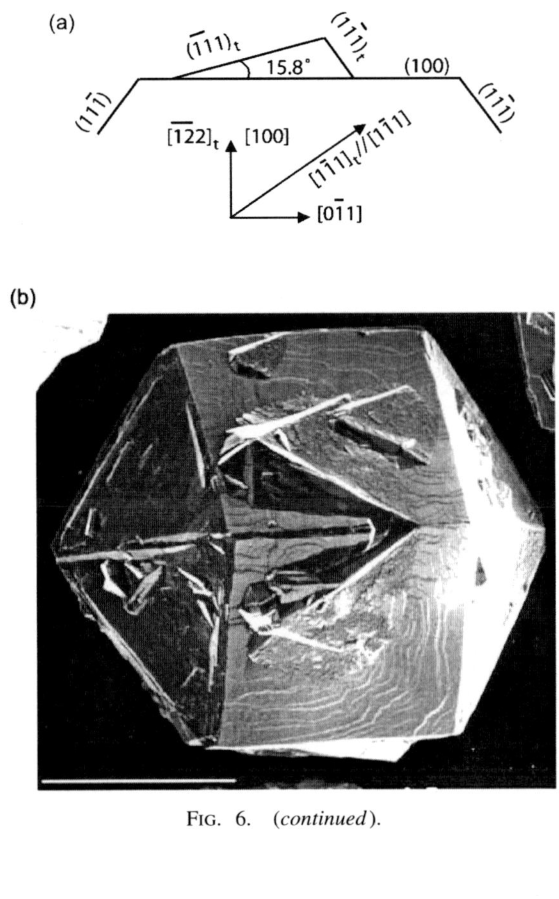

FIG. 6. *(continued)*.

(c)

(d)

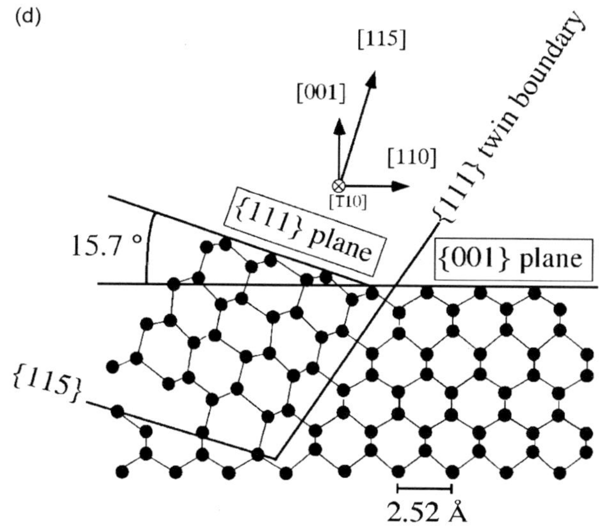

FIG. 6. Penetration twins. (a) Schematic [011] projection of a {100} ⟨111⟩ penetration twin ($\alpha = 1.5$), the twin is bounded by two {111} facets viewed in edge-on projection, and two {100} facets not to be seen; (b) decahedral-prismatic diamond particle exhibiting {100} ⟨111⟩ penetration twins on its cubic facets; the penetration twins have undergone secondary twinning, and one can recognise copious emission of growth terraces emanating from re-entrant corners ($1.5 < \alpha < 2$); (c) octahedral particle with numerous penetration twins protruding from {111} facets ($\alpha \approx 3$); (d) atomic scale model of the grain boundaries involved in the formation of a {100} ⟨111⟩ penetration twin (from T. Takami, K. Suzuki, I. Kusunoki, I. Sakaguchi, Nishitani-Gamo, and T. Ando (1999) *Surf. Sci.* 440, 103–115).

The growth parameter α has also become very important for the characterisation of twin controlled film growth. Assuming that α applies in the same way to parent and twin crystals and applying straightforward geometrical principles, Tamor and Everson (1994) and Wild et al. (1994) have demonstrated that $\{100\}$ $\langle 111\rangle$ penetration twins have a chance to grow for $\alpha < 2$, without being overgrown by the parent phase. If $\alpha = 1.5$, the parent face even is overgrown. By comparison, $\{111\}$ $\langle 111\rangle$ penetration twins can develop only in the range $\alpha > 1.5$, signifying that both kinds of penetration twins coexist for $1.5 < \alpha < 2$ (Fig. 6b). The growth of contact twins does not depend on α, except in approaching $\alpha = 1$ when the octahedral facets disappear. Contact twins are only found on $\{111\}$ facets and they never enter the $\{100\}$ facets. Based on these observations, growth conditions can be selected for which the nuclei of penetration twins will grow or be buried by the parent crystal. By choosing a growth parameter value close to three, diamond films with very sharp $\langle 100\rangle$ textures and covered with $\{100\}$ facets have been deposited (Wild et al., 1994). In fact, the α value must be smaller than three to obtain $\{100\}$ facets at the film surface but greater than two to avoid $\{100\}$ $\langle 111\rangle$ penetration twins to develop. In this α range, the $\{111\}$ growth sectors contain high densities of microtwins because they are unstable against the growth of $\{111\}$ $\langle 111\rangle$ penetration twins (Wang, Bentley, Clausing, Heatherley and Horton (1994). Penetration twins are not equally distributed over the parent facet. In particular, they are not found close to the edges of the facets. Knuyt, Nesládek, Meykens, Quaeyhaegens and Stals (1997) and Knuyt, Nesládek, Quaeyhaegens and Stals (1998) have developed a geometrical concept, called the nucleation surface, which measures the probability to find a twin at a given place on the parent facet. This parameter depends again on the growth parameter α. Its value increases with the distance from the edges between adjacent parent facets.

The models by Wild et al. and by Tamor and Everson refer only to first-order twinning. The same concept has been further developed in order to include first, second, and third order twinning, and the predictions have been confirmed in growth experiments with nitrogen containing homoepitaxial films for which the orientation dependent EPR (electron paramagnetic resonance) signal of P1 defect centres has been measured (Yan and Vohra, 1999). It has been shown that for

(a) $\alpha \geq 2$ only first order twins grow on $\{111\}$; the percentage of twins is near 30%
(b) $\alpha < 1.5$ first and second order twins grow on $\{100\}$
(c) $\alpha = 1.5$ only first order twins grow on $\{100\}$; this condition corresponds to the minimum volume percentage of twins, corresponding to about 20%.
(d) $1.5 < \alpha < 2$ first-order twins grow on $\{100\}$ and $\{111\}$, second-order twins grow on $\{111\}$.

The percentage of second order twins can reach 20%. Third-order twin have not been detected.

Multiple twinning plays an important role in the formation of particles displaying fivefold symmetry, named *multiply twinned particles* (MTP). Icosahedral and decahedral particle shapes with a variety of grooves and cusps at the intersection of facets have been frequently observed in CVD diamond (Fig. 7). The variation of particle shapes correlates with the α value, as shown by Bühler and Prior (2000) who have reproduced the particle contours by using of a 'pseudo'-Wulff polar plot where the facets are obtained by plotting the relative facet growth rates, rather than the free surface energies. The principle of this technique is similar to that by Tamor and Everson but the technique has been extended to include three-dimensional growth and to allow for twin formation in a much more complex growth environment. The morphological grains or islands observed prior to coalescence in polycrystalline film growth frequently represent MTPs (Shechtman, Feldman and Hutchison, 1993a; Shechtman, Feldman, Vaudin and Hutchison, 1993b; Shechtman, Hutchison, Robins, Farabough and Feldman, 1993c). The reader may notice the morphological difference of the particles shown in Figures 6b and 7c, even though their α-values are rather close ($\alpha \lesssim 1.5$ and $\alpha \approx 1.8$, respectively).

2.3. Incorporation of Impurities in Homoepitaxial Films

2.3.1. *Hydrogen*

Except for the deposition of nano-crystalline diamond from argon–methane gas mixtures (Gruen, 1999), hydrogen is the most prominent element in diamond CVD, occupying more than 95% of the gas phase volume. Atomic hydrogen repressents the main driving force for the chemical reactions in the gas phase (Lang and Pang, 1998; Ashfold, May, Petherbridge, Rosser, Smith, Mankelevich and Suetin, 2001), stabilises the sp^3 hybridisation of carbon atoms at the diamond surface and is involved in most surface reactions leading to the incorporation of carbon into the diamond lattice and removal of non-diamond carbon from the surface (Spear and Frenklach, 1994). Diamond films thus are expected to contain substantial amounts of hydrogen which may be bound to structural defects (vacancies, dislocations, grain boundaries), doping elements like boron and nitrogen, and other impurities like oxygen or be located at interstitial sites in the lattice. The concentrations of hydrogen are measured by secondary ion mass spectroscopy (SIMS), nuclear reaction analysis (NRA), Rutherford backscattering (RBS), and elastic recoil detection analysis (ERDA). A defect labeled H1 revealed by electron paramagnetic resonance (EPR) is attributed to a dangling bond on carbon near hydrogen in the vicinity of grain boundaries (Zhou, Watkins, McNamara Rutledge, Messmer and Chawla, 1996). Hydrogen is not a common impurity in HPHT diamond (Kanda and Sekine, 2000) which generally is taken as a substrate for homoepitaxial CVD diamond. The hydrogen content of HPHT

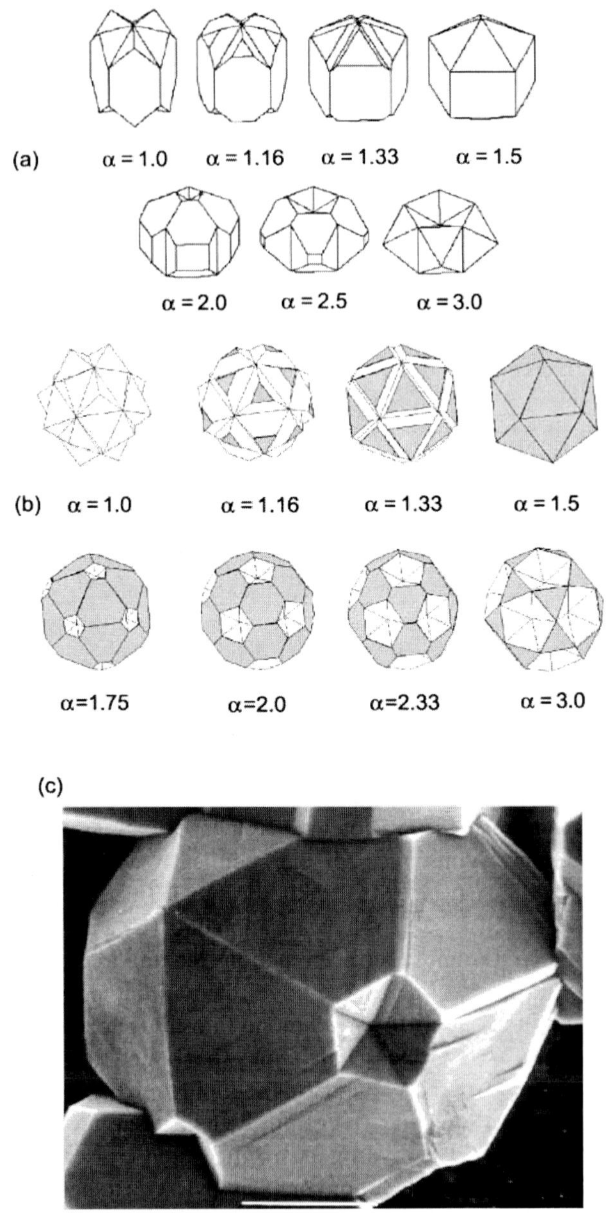

FIG. 7. Morphological evolution of icosahedral and decahedral particles with changing α-parameter. Model for the development of (a) decahedra and (b) icosahedra (after J. Bühler and Y. Prior (2000) *J. Crystal Growth* 209, 779–788); (c) icosahedral particle with $\alpha \leq 2$.

diamond, as determined recently by SIMS, is estimated at less than 10 ppm (Sakaguchi, Nishitani-Gamo, Loh, Haneda and Ando, 1999).

The bulk hydrogen concentration of homoepitaxial films ranges from roughly 0.1–1% (Fuchs, Wild, Schwarz and Koidl, 1995; Samlenski, Haug, Brenn, Wild, Locher and Koidl, 1995; Haug, Gärtner, Portmann, Samlenski, Wild and Brenn, 2001) and depends greatly on film orientation and processing conditions. Based on SIMS measurements with a detection limit estimated at approximately 10 ppm, absence of hydrogen incorporation has been reported for (100) epilayers grown by microwave assisted CVD at substrate temperatures between 600 and 920 °C (Gamo, Loh, Haneda, Hishita and Ando, 1997). Under the same experimental conditions, (111) oriented films incorporate rather high amounts of hydrogen which increase with increasing substrate temperature (0.05% at 600 °C and 0.25% at 920 °C). As the excitation of the gas phase has been kept constant for all these experiments, it has been concluded that the uptake of hydrogen is governed by the same surface chemical reactions, which also occur during (111) and (100) growth. This explanation is corroborated by the observation that the hydrogen content diminishes when more atomic hydrogen becomes available in the gas phase, thus favouring the etching of defective surface sites. A similar effect would be expected if the carbon content in the gas phase was lowered. The influence of film orientation on the uptake of hydrogen has been confirmed by other groups, but the hydrogen concentrations may vary considerably. About 50 ppm hydrogen has been measured by SIMS in (100) epilayers deposited at 820 °C (Chevallier, Lusson, Ballutaud, Theys, Jomard, Deneuville, Bernard, Gheeraert and Bustarret, 2001). In another case, hydrogen concentrations between 0.1 and 0.3% have been analysed by NRA in (100) epilayers, compared to 0.65–0.85% in (111) epilayers (Haug et al., 2001; the NRA measurements have been calibrated against fully hydrogenated polystyrene films).

The surface regions of homoepitaxial films can accumulate impressive amounts of hydrogen. Right at the surface of as-grown (100) epilayers, a concentration of 1.5×10^{15} atoms per cm^2 has been measured by ERDA, indicating that the surface has been reconstructed with a (2×1) pattern and terminated with hydrogen (Bergmaier, Dollinger, Aleksov, Gluhe and Kohn, 2001). This interpretation is supported by the observation that the concentration drops to the bulk value of less than 60 ppm within a distance of less than 0.5 nm below the surface (Bergmaier et al., 2001; Kimura, Nakajima, Yamanaka, Hasegawa and Okushi, 2002). Thus, the existence of a hydrogen-rich subsurface layer, which repeatedly has been alleged in the literature, is not confirmed. This conclusion is indirectly sustained by the absence of deuterium diffusion into IIa single crystals during exposure to a deuterium microwave plasma at 820 K substrate temperature (detection limit for deuterium: about 0.3 ppm; Uzan-Saguy, Cytermann, Fizgeer, Richter, Avigal, Shaanan, Brener, Kalish, Bustarret and Chevallier, 2002). Annealing at 400 °C does not change the hydrogen distribution (Kimura et al., 2002). In the course of ERD measurements, a small increase of the hydrogen concentration in the subsurface region has been detected after plasma rehydrogenation, subsequent to hydrogen desorption (Kimura et al., 2002). This effect does not contradict the aforementioned results because hydrogen can be trapped by lattice defects induced by the bombardment with C^+ ions in ERDA.

Other situations have been reported where the layer with accumulated hydrogen attains a thickness of approximately 100 nm ((100) and (111) layers: Haug et al., 2001; (111) layers only: Sakaguchi et al., 1999). Concerning (100) epilayers, these findings are surprising and the only possible explanation is hydrogen trapping and diffusion along structural lattice defects, or interaction with impurities (see below). In the case of (111) films, the combination of elevated substrate temperature and hydrogen incorporation induces a high density of structural defects at the film surface, and the existence of a relatively thick hydrogen layer may be rationalised in this way. The defects are supposed to be stacking faults and twins, as concluded from scanning electron microscopy (SEM) and RHEED patterns (Sakaguchi et al., 1999). Further more, (111) films exhibit a network of cracks along (111) planes, which do not exist in samples without stacking faults and twins. The same authors report that the hydrogen content of (111) epilayers can be lowered by one to two orders of magnitude if small amounts of oxygen are added to the gas atmosphere during deposition (Sakaguchi, Nishitani-Gamo, Loh, Hishita, Haneda and Ando, 1998). Also the addition of oxygen eliminates silicon impurities and suppresses crack formation. Similar phenomena have been repeatedly observed before (see Sakaguchi et al., 1998, for relevant references).

As far as the theoretical treatment of hydrogen in diamond is concerned, Goss, Jones, Heggie, Ewels, Briddon and Öberg (2002) have carried out first-principle calculations where several interstitial sites are considered as candidate locations for single hydrogen atoms in a defect-free diamond lattice. Most favourable in terms of total energy is the bond-centred location where the hydrogen atom is in the centre of a dilated C–C covalent bond. Positively charged hydrogen ions should prefer a puckered bond-centred location in which the proton precesses about the bond. Its barrier to diffusion is estimated to be at least one order of magnitude lower than that for the other charge states and it has been suggested that the activation energy of 0.35 eV for the diffusion of deuterium in boron-doped layers might be related to the positively charged state (Chevallier, Theys, Lusson, Grattepain, Deneuville and Gheeraert, 1998). In principle, dissolved hydrogen can also be in the form of self-trapped hydrogen complexes but the formation energies are large and positive, rendering their equilibrium concentrations in the bulk insignificant. The first-principle calculations by Goss et al. yield H_2^* to be the most stable dimer with one of the hydrogen atoms being in a bond-centred location and the other one at the anti-bonding site located in prolongation of a C–C bond. Hydrogen dimers are surmised to exist only in the subsurface regions of diamond films exposed to the energetic hydrogen ions of the rf-plasma used for hydrogenation.

The occupation of bond-centred sites by hydrogen should reveal itself by channelling in NRA but there is no such evidence. The absence of any angular dependence of the hydrogen related γ-emission yield has been attributed to high dislocation densities (1×10^{10}–1.3×10^{11} cm^{-2}) providing structural defect sites with random symmetry (Haug et al., 2001). A quite different result has been obtained by Saada, Adler and Kalish (2000) who have identified a six-fold degenerate site (ET) on each C–C bond as the lowest energy configuration. Hydrogen at this site creates a dangling bond that induces a deep donor level at ~0.5 eV above the centre of the energy gap in diamond. Around 600 K, the hydrogen atom finds its way to the nearest ET site. From 1200 K on, the atom becomes able to jump between the six degenerate sites belonging to the same bond. Real diffusion sets in beyond 1700 K. Hydrogen then jumps between ET sites associated with neighbouring bonds. If two hydrogen atoms occupy ET sites located on adjacent bonds, they are unable to leave their respective ET site group. The indicated temperature values are not precise but they provide a hierarchy of the energetic barriers involved in hydrogen migration. Hydrogen atoms on ET sites are in close agreement with the results from EPR measurements by Zhou et al. (1996). Also the hyperfine structure observed in muonium spin resonance can be explained by the model (Patterson, 1988).

In conclusion, there is a general agreement between the theory and experiment concerning the difficulty of hydrogen diffusion in defect-free diamond where the solubility of hydrogen is very small. Significant hydrogen contents only are observed near free surfaces and at defects.

2.3.2. *Boron and Nitrogen*

Boron and nitrogen atoms are similar in size to carbon and can occupy substitutional sites in diamond. Boron-doping is a well established technique to control the conductivity of diamond. It forms an acceptor level with an activation energy of 0.37 eV. Nitrogen that acts as a deep donor level with an activation energy of 1.7 eV influences strongly the growth morphology of CVD diamond (de Theije and van Enckevort, 2000). The levels of boron and nitrogen can be controlled by varying their concentrations in the gas phase. Under identical growth conditions, the incorporation rate for boron in (111) epilayers is more than ten times higher than in (100) epilayers (Samlenski, Haug, Brenn, Wild, Locher and Koidl, 1996). According to lattice spacing measurements carried out on polycrystalline films, diamond seems to form a solid solution with boron, up to a boron content of about 170 ppm (Brunet, Deneuville, Germi, Pernet, Gheeraert and Mambou, 1997; Section 3.1.2.). In homoepitaxial films, metallic conductivity is achieved beyond 1700 ppm boron (Lagrange, Deneuville and Gheeraert, 1998). The onset of metallic conductivity is accompanied by a change in the shape of the zone-centre optical Raman peak and a new signal at lower wave numbers (Pruvost, Bustarret and Deneuville, 2000). The shape change of the zone-centre peak has been associated with a strong coupling of phonons with the continuum of electronic transitions (Pruvost and Deneuville, 2001). Compared to boron, the incorporation of nitrogen is less efficient, but the concentration of nitrogen is still three to four times higher in (111) compared to (100) epilayers (Samlenski et al., 1995a; Samlenski et al., 1996). Ion channeling analysis confirms that 90–100% of the boron atoms and 80–90% of the nitrogen impurities occupy substitutional lattice sites (Samlenski et al., 1996).

Hydrogen in boron-doped diamond causes the boron acceptors to passivate by the formation of (B, H) pairs that are electrically neutral and consist of boron atoms ionised to B^- and hydrogen atoms ionised to H^+ (Chevallier et al., 1998). H^+ is a diffusing species in p-type diamond, as has been demonstrated by reverse bias annealing experiments (Zeisel, Nebel and Stutzmann, 2000). The (B, H) pair begins to dissociate at 550 °C (Chevallier, Ballutaud, Theys, Jomard, Deneuville, Gheeraert and Pruvost, 1999). Correspondingly, in hydrogenation with deuterium below 550 °C, the deuterium concentration matches the boron concentration, i.e., each boron acceptor traps one deuterium atom. At 700 °C, the dissociation is completed and the boron acceptors are no longer passivated (Chevallier et al., 2001). Identical hydrogen concentrations at the surface of boron-doped and undoped (100) epilayers deposited at 800 °C are in agreement with this latter observation (Kimura et al., 2002).

In n-type (111) diamond epilayers doped with phosphorous or nitrogen, the deuterium concentration after deposition from a microwave plasma remains far below the concentration of the dopant and also far below the hydrogen concentration in undoped (111) epilayers. Hydrogen diffusion (deuterium) again does virtually not exist if the films are exposed to a deuterium microwave plasma, even at a film temperature of 1000 °C (Chevallier, Jomard, Teukam, Koizumi, Kanda, Sato, Deneuville and Bernard, 2002). The absence of diffusion has been attributed to the high activation energy for H^- which is in the range 2–2.5 eV according to Goss et al. (2002). Other possible reasons are self-trapping of hydrogen and strong interaction between H and donor atoms.

3. Structural Imperfections at the Meso-Scale and the Atomic Size Scale

The following part mainly deals with linear defects (dislocations, disclinations) and planar defects (stacking faults, twin and grain boundaries). For a detailed treatment of point defects the reader should consult other chapters in this book and refer to earlier reviews by Zhu (1994) and Nazaré and Neves (2001).

3.1. Microstructure Evolution in Polycrystalline Diamond Films

3.1.1. Grain Structure

The nucleation densities at the beginning of diamond deposition commonly are in the range 10^6–10^{10} cm^{-2}, corresponding to a morphological grain size of 0.1–10 μm at the moment of grain coalescence. The coherent domain sizes calculated from peak broadening in X-ray diffraction are much smaller, in the order of a few to some ten nanometers only (see, for example, Ascarelli, Cappelli, Mattei, Pinzari, Fares, Veroli and Martelli, 1996). TEM cross-sections of the interface region indeed confirm that morphological grains, as revealed by SEM, are highly defective and composed of small domains fanning out in three dimensions from well-defined nucleation sites (Williams, Glass, Davis, Kobashi and More, 1990a). The domains are related to each other by multiple twinning. In general, some of them grow larger with increasing film thickness at the expense of others ones which strink. In ⟨100⟩ textured films, most of them grow out and disappear with increasing growth length (Steeds, Gilmore, Bussmann, Butler and Koidl, 1999a). For other film orientations, typically ⟨110⟩ and ⟨112⟩, new ones may be added. The twinned domains then form grain clusters with a common zone axis. They are named *grain families* in the following, because the grains belonging to the same family spring from one another by twinning and have a common origin of nucleation (Shechtman et al., 1993a–c). Examples for grain families are given in Figure 8. In SEM, a grain family appears as one morphological grain (see centre of Fig. 5). In TEM plan views, grains can be almost defect-free or contain large numbers of dislocations and planar faults, i.e., stacking faults and microtwins included in the grain (see, for example, Bozzolo, Barrat, Dieguez and Bauer-Grosse, 1997; Steeds, Gilmore, Wilson and Butler, 1998). Grain families are separated by ordinary large-angle grain boundaries, which tend to be faceted (Ichinose and Nakanose, 1998). At the growth surface, the in-plane diameter of grain families may exceed one hundred micrometers.

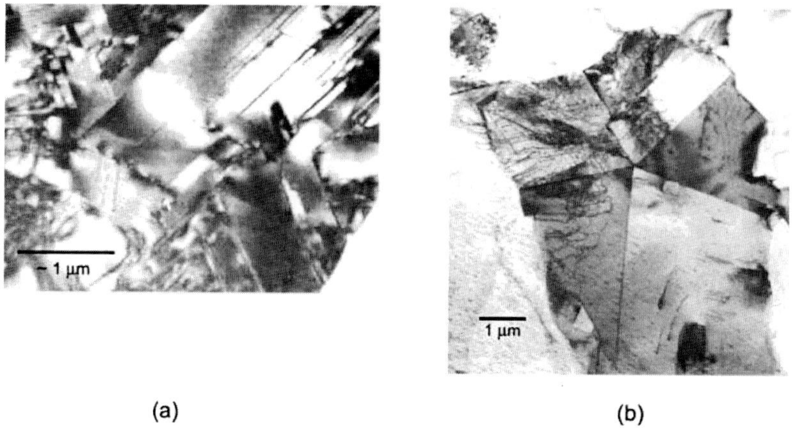

FIG. 8. TEM micrographs showing different morphologies of grain families in polycrystalline diamond. (a) Multiple twinning leading to parallelogram-shaped grains, similar to the structure shown in Figure 5 (after H. Ichinose, and M. Nakanose (1998) *Thin Solid Films* 319, 87–91); (b) Pentagonal centre in a boron-doped diamond film which is at the origin of a grain family consisting of five members; dislocations seem to enter the two bottom grains from the bottom side (after J. W. Steeds, A. E. Mora, J. E. Butler, and K. M. Bussmann (2002) *Philos. Mag. A* 82, 1741–1768, Taylor & Francis Ltd., http://www.tandf.co.uk/journals).

The morphology of grain families originates in the strictly faceted growth mode of diamond. Unlike the bulging boundaries of metal crystals, diamond can grow only perpendicular to the flat {111} and {100} facets. Twinning opens new growth directions and enables the grain family to occupy empty space. If the growth conditions are suitable for twinning to occur, coherent first-order twin boundaries are dominant in polycrystalline films, as repeatedly reported in the literature (Shechtman et al., 1993a–c; Ichinose and Nakanose, 1998; Steeds, Mora, Butler and Bussmann, 2002). A high probability for their formation is expected because of the very low energy of this boundary type.

High nucleation densities lead inevitably to selective grain growth, provided the growth of competing grains is not stopped by the repetitive nucleation of new grains at the crystallisation front, as in nano-crystalline diamond (Gruen, 2000). In competitive grain growth, the two-dimensional grain size in the film plane increases with $\sqrt{d}$ where d is the distance from the interface (von Kaenel, Stiegler, Blank, Chauvet, Hellwig and Plamann, 1996; Paritosh, Srolovitz, Bataille, Li and Butler, 1999). Beginning with an isotropic distribution of nuclei, the growing film adopts a crystallographic texture with increasing distance from the interface. The textures comprise orientations between $\langle 110\rangle$ and $\langle 100\rangle$ according to the growth parameter α chosen for deposition (Wild et al., 1994). The grain families composing a polycrystalline film, however, do not necessarily exhibit the same zone axes. Steeds et al. (2002) have recently reported the coexistence of large $\langle 110\rangle$, $\langle 112\rangle$ and $\langle 111\rangle$ oriented grain families in thick boron-doped films (140–394 μm). The grain boundaries have a tendency to be oriented parallel to the growth direction (Steeds et al., 2002). The grain structure of $\langle 112\rangle$ oriented grain families turns out to be particularly simple because the grain boundaries are coherent {111} twin planes that outcrop as re-entrant corners at the growth surface.

Diamond films in near [001] orientation exhibit a different grain structure. They do not form grain families composed of grains related by twinning. During the early stage of polycrystalline film growth, all grains are highly defective and contain a high density of twins with orientations deviating from the growth direction (Burton, Steeds, Meaden, Shreter and Butler, 1995). After several micrometers thickness growth, however, large {100} growth sectors, which are free of twins and contain only a few dislocations become dominant. In these structures, thin twin plates continue to grow near the boundaries of the columnar grains (Steeds et al., 1999a). Their lateral extension is in the order of one to several micrometers while their thickness is only a few nanometers. The habitus plane is approximately parallel to the growth direction. Nanoscale precipitates seem to decorate the ledges of these thin twins. According to Steeds et al. (1999), grain boundaries and the remainders of {111} growth sectors could act as sinks for impurities (N, H, O) which later on precipitate at the twin sites. The authors surmise that the dip in thermal conductivity observed by Graebner, Reiss, Seibles, Hartnett, Miller and Robinson (1994) in the temperature range 10–100 K might originate in these nanoscale precipitates. Boron-doped samples are free of those defects.

The {111} growth sectors of [001]-oriented films exhibit high densities of planar defects such as stacking faults and microtwins, whilst no twins and, in general, only few dislocations are observed in {100} growth sectors (Wang et al., 1994). Since the beginning of TEM analysis in CVD diamond, it has been recognised that planar defects as well as centres of pentagonal symmetry are fundamental to polycrystalline diamond growth (Matsumoto and Matsui, 1983). The density of these defects increases with the methane content in the gas phase (Zhu, Badzian and Messier, 1989; Bühlmann, Blank, Haubner and Lux, 1999), which means that the twin size decreases, in conformity with the decrease of the domain size observed in X-ray diffraction (Eto, Tamou, Ohsawa and Kikuchi, 1992). Later on, it has been pointed out that this dependence applies only to {111} and not to {100} growth sectors (Samlenski et al., 1995b). Higher defect densities are accompanied by a decrease of the average twin size. The defect density increases also when the substrate temperature is so much lowered that two-dimensional nucleation becomes important (Michler et al., 1997). In such films, microtwins appear as inclusions or as faulted microdomains that are dispersed on the background of a dominant parent crystal defining the outer contours of the crystal (Fig. 9a and b). Fibrous twins of less than 10 nm diameter represent still another morphological growth type which forms in homoepitaxial growth at very low gas pressure on a microcrystalline substrate exhibiting {111} facets in the micrometer range. It can be seen from Figure 9c that the fibres are aligned in parallel and that their orientation changes with the orientation of the underlying substrate facet (Seehofer, 2002). In all these investigations, the distinction between microtwins and faults is not always straightforward.

Steeds et al. (1998) have investigated the morphological evolution of grains terminated by square-facets in ⟨001⟩ oriented diamond films deposited at $2.5 < \alpha < 3$. During the early growth stage, all grains are bounded solely by octahedral facets. If such a grain is viewed in exact [001] growth direction, a pair of defect-free sheets, parallel to (100) and (010), shows up which separates the quadrants occupied by the four {111} growth sectors (Fig. 10a). The sheets intersect along the [001] direction, which coincides with the zone axis of the four {111} growth sectors. The zone of intersection thickens during further growth and eventually develops into a (100) growth sector. The grain has then adopted the shape of a truncated pyramid (Fig. 10b,c).

The octahedral growth sectors contain densely arranged extended defects consisting of faults and twins as well as a particular type of tetrahedral defect with faulted {111} planes or nanotwin platelets, which are inclined with respect to the plane of the {111} growth facet (Wurzinger, Joksch and Pongratz, 1995; Steeds et al., 1998). The tetrahedra point to the grain interior and are open toward the grain exterior. They intersect each other and are nested. The spacing of the faulted planes lies in the range of 10–50 nm. It has been suggested that these defects are induced by the stress field of graphitic inclusions (Steeds et al., 1998). With regard to their orientation, these defects resemble the formerly mentioned penetration twins.

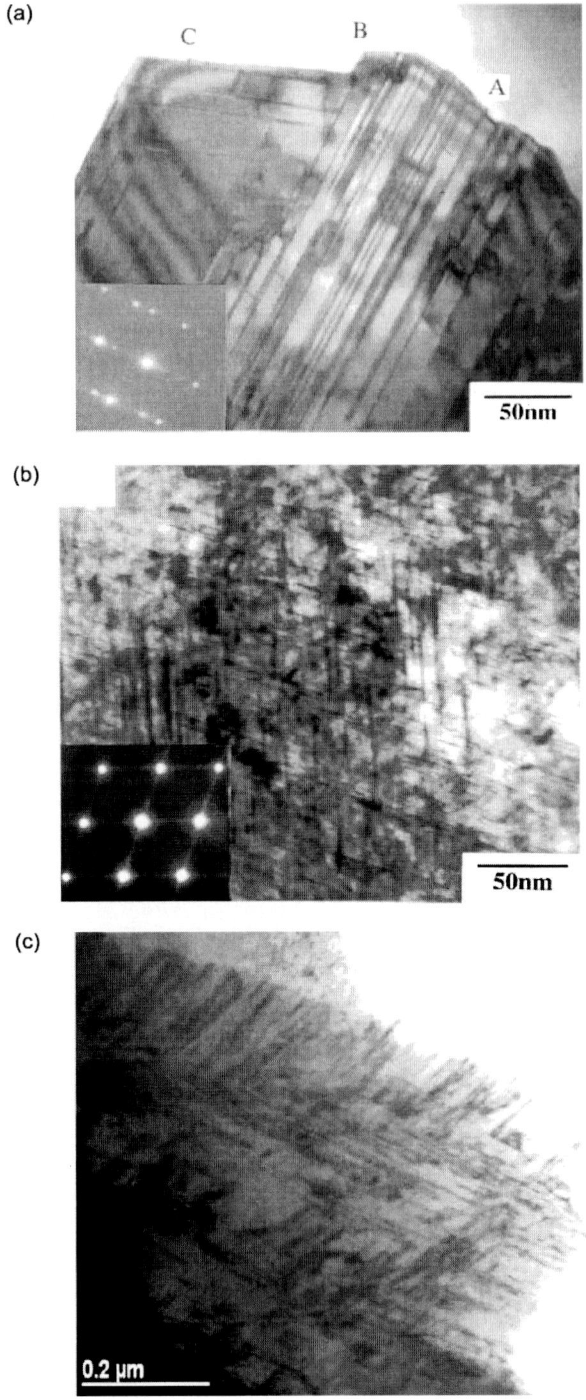
(a)
C
B
A
50nm
(b)
50nm
(c)
0.2 µm

3.1.2. Bond Defects and Impurities in Polycrystalline Films

The impurity content of polycrystalline films follows closely the behaviour of homoepitaxial films. In near-heteroepitaxial [001] oriented films, the incorporation of defects is gradually reduced with increasing film thickness as the fraction of {111} facets diminishes in favour of {100} sectors (von Kaenel et al., 1996; Bozzolo et al., 1997; Michler, von Kaenel, Stiegler and Blank, 1998). The number of dangling bonds measured by EPR increases 3.3 times faster in {111} than in {100} growth sectors (von Kaenel et al., 1996). Although the factor of 3.3 seems to fit perfectly the 3–4 times larger incorporation of nitrogen in (111) compared to (100) homoepitaxial films reported by Samlenski et al. (1996), only 5–10% of the spins revealed by ESR have been attributed to impurity related centres, the overwhelming part being due to carbon bond defects. Graphite-like bond defects have also been identified by EELS in {111} growth sectors where these defects are believed to act like inclusions creating compressive stress fields (Steeds et al., 1998). Stress relaxation is supposed to occur by local atomic displacements that give rise to the formation of the tetrahedral stacking faults mentioned before. This interpretation compares well with earlier ion channelling experiments by Samlenski, Schmälzlin, Brenn, Wild, Müller-Sebert and Koidl (1995b) who found that higher methane contents end up in higher dislocation densities. It is noticed that this latter observation concerns only {111} growth sectors. The quality of the {100} growth sectors is even improved in deposition from higher methane concentrations, as long as the {100} facets are not destabilised by the formation of cauliflower structures.

The uptake of boron has been studied thoroughly because of its relevance for the electrical conductivity of diamond films. Conforming to the different boron incorporation in (100) and (111) homoepitaxial films (Samlenski et al., 1996; Section 2.3), marked fluctuations of the boron content have been found in films with adjacent {100} and {111} growth sectors (Steeds, Gilmore, Charles, Heard, Howarth and Butler, 1999b). There is no evidence for boron segregation to grain boundaries or formation of denuded zones (Wurzinger, Fuchs, Pongratz, Schreck, Hessmer and Stritzker, 1997a; Wurzinger, Pongratz, Hartmann, Haubner and Lux, 1997b; Steeds et al., 1999a,b). Up to a critical boron concentration of about 170 ppm, boron seems to be integrated as a substitutional atom because the spacing of the lattice planes d_{111}, d_{311} and d_{220}, parallel to the substrate widens slightly but continuously (Brunet, Deneuville, Germi, Pernet, Gheeraert and Mambou, 1997). The lattice expansion has been attributed to the larger covalent

---

FIG. 9. Twin formation at low substrate temperature and low pressure. (a) Substrate temperature 340 °C : re-entrant corners at (B) and surface growth steps at (C); a few isolated twins end within the matrix. (b) substrate temperature 310 °C: intense microtwin formation on all four {111} matrix planes (after Michler et al., 1997). (c) twinned nanofibres grown at 1 mbar on microcrystalline diamond substrate at 570 °C (Seehofer, 2002).

radius of boron (Vegard's law). In this composition range, boron doping gives rise to a decrease in the density of point defects as revealed by an increase in phonon lifetime in Raman spectroscopy and a decreasing luminescence background (Colineau, Gheeraert, Deneuville, Mambou, Brunet and Lagrange, 1997). Using TEM, Liao, Zhang, Lee, Tong Lee and Lam (1997) arrived at the conclusion that also the number of planar defects decreases with increasing boron concentration whereas the dislocation density increases. Inspection of TEM micrographs of undoped and doped films in similar orientations, which have been published by

(a)

(b)

FIG. 10. (*continued*)

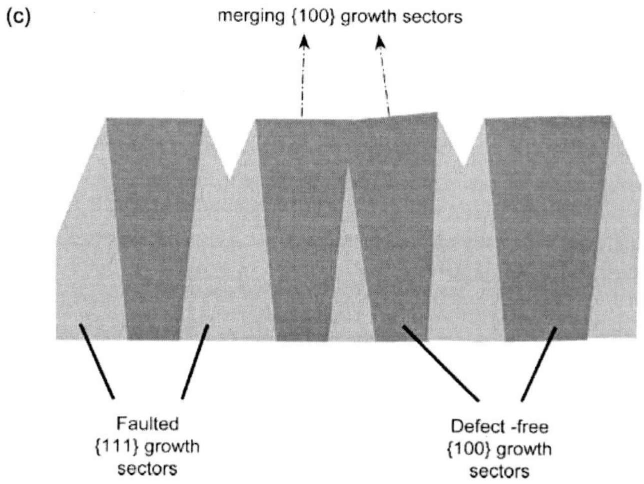

FIG. 10. Morphological evolution of ⟨100⟩ textured films terminated by {100} facets. (a) Early stage of the formation of a (001) facet at the intersection of {100} sheets constituting boundary layers between the {111} growth sectors of octahedral particles. Contrary to the defect-free {100} regions, the octahedral growth sectors are heavily faulted (after J. W. Steeds, A. Gilmore, J. A. Wilson, and J. E. Butler (1998) *Diamond Relat. Mater.* 7, 1437–1450). (b) SEM micrograph of highly oriented, ⟨100⟩ textured grains at 4 μm distance from the interface; the {100} facets have now developed into tiles that begin to touch one another; the fraction of defective {111} growth sectors is reduced. (c) Schematic cross-section of a highly oriented, ⟨100⟩ textured film showing that {100} tiles can merge directly (Section 3.3) or stay separated by grooves forming between them.

various authors, support these observations (*without boron*: Narayan, 1990; Luyten, Van Tendeloo, Amelinckx and Collins, 1992; *boron doped*: Steeds et al., 2002). EELS at the carbon K-edge indicates a decrease of π^*- states in boron-doped films, provided that boron is incorporated *in situ*. A different situation arises if diamond films are doped by ion implantation. The π^*- absorption then is much stronger, apparently due to the creation of bond defects (Waidmann, Bartsch, Endler, Fontaine, Arnold, Knupfer, Leonhardt and Kink, 1999). Beyond the critical boron concentration of about 170 ppm the lattice plane distances increase rapidly (Brunet et al., 1997). Eventually, graphite lamellae show up in the {111} growth sectors of heavily doped films with 2–3% boron, and the lattice parameter increases by 0.3% (Wurzinger et al., 1997a,b). This latter effect has been attributed to the larger bond length of carbon–boron atoms compared to carbon–carbon in pure diamond. It is remarkable that the square {100} growth sectors of boron-rich films are absolutely defect-free but they are heavily strained, owing to the larger lattice constant of the adjacent {111} growth sectors (Wurzinger et al., 1997a,b; Lang and Pang, 1998).

In approaching the limits of diamond growth by lowering the substrate temperature, the propensity for the incorporation of impurities generally is intensified, but there are marked differences concerning the incorporation of hydrogen, oxygen and nitrogen, as measured by ERDA (Stiegler, Bergmaier, Michler, von Kaenel, Dollinger and Blank, 1998; Stiegler, Michler and Blank, 1999). {100} growth sectors disappear completely below 600 °C. If in deposition from nitrogen-rich CH_4/CO_2 gas mixtures the substrate temperature is lowered from 520 to 345 °C, the domain size in XRD decreases from 500 nm to about 10 nm. Concomitantly, the microtwin density in TEM increases tremendously (Michler et al., 1997), as does the incorporation of impurities. While the oxygen content measured by ERDA rises exponentially from 125 to 670 ppm, and hydrogen from 0.05 to 1.8%, the nitrogen content decreases linearly from 140 to less than 80 ppm. In case the gas phase is kept free from nitrogen (< 10 ppm), the hydrogen incorporation is reduced by more than a factor of two. Regarding the distribution of hydrogen, it is generally accepted that hydrogen segregates to the grain boundaries (McNamara, Williams, Gleason and Scruggs, 1994). At such low substrate temperatures, however, substantial amounts of hydrogen finish by being incorporated in the bulk. This follows from the broadening of the diamond Raman peak after subtraction of the contribution from phonon confinement. Line broadening due to internal stress cannot be significant because the measurements have been done on isolated, free-standing grains at the transparent edge of thin foils prepared for TEM. As the solubility of hydrogen in the perfect diamond lattice is very low (Section 2.3) the hydrogen atoms are supposed to be located at twin boundaries, dislocations and amorphous defects commonly encountered in these structures. FTIR spectroscopy in the range 2700–3100 cm^{-1} indicates that the predominant part of hydrogen atoms is bound in sp^3 hybridised CH_x bond configurations (Stiegler et al., 1999).

Grain boundaries thus primarily affect the uptake of hydrogen, while the incorporation of boron and nitrogen turns out to be controlled by the fraction of {111} and {100} facets. This kind of impurity partitioning would be expected because the growing diamond facets are terminated by hydrogen, and the probability for hydrogen being trapped by grain boundaries supposedly is high. Indirect confirmation for this latter assumption has been obtained from bulge tests where the effective Young's modulus can be explained by a serial arrangement of diamond grains interconnected by polymer-type grain boundaries (von Kaenel, 1997).

3.2. Dislocations in Diamond

Dislocations are very common defects in CVD diamond. They are formed during film growth. Dislocations can carry plastic deformation but their mobility is strongly limited in diamond. In semiconductor lattices, dislocations can influence the electrical and optical properties by acting as traps and recombination centres for charge carriers. The latter properties depend very much on the structure of the dislocation core and have been investigated thoroughly for silicon and germanium. In diamond, the situation is different because the sp^3 coordination of carbon atoms may change into sp^2 owing to reconstruction of the dislocation core. The different chemical behaviour of carbon and silicon is believed to originate in the carbon's position in the first row of the periodic table, i.e., its small $1s^2$ core and high electronegativity which both together give rise to the formation of short bonds. These characteristics lead to the ability of carbon to compensate undercoordination by forming π-bonds when the geometry so allows (Heggie, Ewels, Martsinovich, Scarle, Jones, Goss, Hourahine and Briddon, 2002).

The dislocations in diamond generally belong to the $\langle 110\rangle\{111\}$ slip system, meaning that the Burgers vector points in $\langle 110\rangle$ direction and the slip plane is $\{111\}$ as one would expect for fcc structures. Dislocations on $\{001\}$ planes have also been identified. In his classical paper on $(a/2)\langle 110\rangle$ dislocations in diamond, Hornstra (1958) has predicted the geometric structure of dislocation cores by minimising the bond distortions and the number of dangling bonds. In addition to the 0 (screw), 60 and 90° dislocations with dislocation lines along $\langle 110\rangle$, the dissociation of dislocations into partials as well as dislocation reactions giving rise to dislocation lines lying in $\langle 112\rangle$ and $\langle 100\rangle$ directions have been discussed. Dislocations with the Burgers vector $(a/2)\langle 110\rangle$ have indeed been observed by TEM both in natural IIa and CVD diamond (Pirouz et al., 1983; Narayan, 1990; Suzuki et al., 1992; Wurzinger, Joksch and Pongratz, 1993b). The core of the screw and the 60° dislocation turns out to be dissociated. Suzuki et al. as well as Wurzinger et al. estimate the dislocation densities in the investigated regions of CVD films at $5 \times 10^8\ \mathrm{cm}^{-2}$, about two orders of magnitude below the value extracted by Samlenski et al. (1995b) from ion channelling experiments on homoepitaxial films. Since both studies refer to $\{100\}$ growth sectors, there is no straightforward explanation for this discrepancy. The TEM observations cannot validate either the very high dislocation densities of $10^{12}\ \mathrm{cm}^{-2}$ claimed by Graebner et al. (1994) for the interpretation of their phonon scattering experiments. Such high dislocation densities seem to be realistic only on a very local scale in some boron-doped films (Steeds et al., 2002), but their frequent use as a general characteristic of polycrystalline diamond films seems unjustified.

3.2.1. *Atomistic Modelling*

HRTEM has added much to our present knowledge on the atomistic structure of dislocation cores and grain boundaries. Simply by using HRTEM, it is, however, not possible to gain a complete view of the arrangement and the bonding of the atoms in the dislocation core. Much progress in this matter has been made in recent years by atomistic modelling and by the validation of atomistic modelling with HRTEM and nanoprobe EELS. Because of the increasing importance of atomistic modelling in structural analysis, the principles involved in various kinds of models used commonly in the literature are briefly outlined in the following.

3.2.1.1. Theoretical Models. Modelling of the atomic structure of dislocations and grain boundaries faces two major problems (Morris, Fu and Ho, 1996). First, an accurate model for covalent bonding is important. Second, there are many possible grain boundary structures for any given orientation of the grains. Most studies begin, therefore, with a geometrical model that satisfies as many bonds as possible and minimises stretching and bending of bonds. Early geometrical modelling, as carried out by Kohn (1958), Hornstra (1958, 1959) and Papon and Petit (1985), was based on this approach. On the other hand, atomistic models predict forces and energies and allow for efficient local energy minimisation, but the atomic arrangement may relax to a metastable configuration when the initial guess of the structure is not ideal.

First-principle or *ab initio* methods in molecular dynamics simulations are flexible and allow to calculate accurately the forces acting on the atoms as well as the electronic and vibrational properties of the model system (Nakhmanson, 2001). The accurate determination of the atomic forces requires the calculation of the quantum electronic structure by solving the many-body Schrödinger equation for the N electrons of the system. As this is an impossible task for a many-particle system which is required to study the structures and energies of grain boundaries and dislocations, first-principle methods in general use the density-functional theory (DFT) in order to calculate the so-called exchange-correlation potential (Aulbur, 2003). The idea of DFT is to represent the solid with N interacting electrons as a functional depending only on the charge density, meaning that the many-electron problem is mapped into a one-electron problem. DFT gives a good description of the ground-state properties. Within DFT, most first-principle methods make use of the *local density approximation* (LDA) which assumes a homogeneous electron gas for the calculation of the local exchange-correlation energy. While first-principle methods need no fitting to any external parameters, substantial computational effort limits the applicability to relatively small systems.

If larger model systems with a few hundred atoms are to be investigated and the atomic positions to be optimised, as in the calculations of grain boundary structures and energies, methods demanding less computation time are required. The *tight-binding* (TB) Hamiltonian method enables the ground state total energy of the system to be approximated as the sum of the band-structure energy and the repulsive energy (Morris, 1996; Nakhmanson, 2001). The effects of directional covalent bonding are introduced through the underlying electronic structure described by an empirical TB Hamiltonian, which is solved to obtain the electronic eigenvalues. The quantum-mechanical nature of the covalent bonding thus enters the potential in a natural way. The parameters of both the band-structure and the repulsive terms are obtained by fitting to first-principle or experimental data. An important criterion for the usefulness of a TB model is its transferability to other crystalline and disordered structures. In the transferable TB model for carbon by Xu, Wang, Chan and Ho (1992), which is frequently referred to in the literature, both the band-structure and the repulsive energies are represented by terms depending only on the distance between an ad-atom and its individual neighbours. This allows to deal with carbon atoms in a wide range of local environments.

Semi-empirical potentials use a number of adjustable parameters to describe the atomic interactions in a solid. The parameters are fitted to experimental data and to the results of first-principle calculations. While empirical potentials allow much larger systems to be studied owing to their great computational efficiency, their application is restricted to small distortions of the lattice to which they have been fitted. No information on the electronic structure is obtained. The Keating and Stillinger–Weber potentials consider the contributions due to two- and three-body interactions (Keating, 1966; Stillinger and Weber, 1985). By contrast, the binding energy in the more widely used Tersoff potential is written as a sum over repulsive and attractive interactions of a particular atom with its nearest neighbours (Tersoff, 1988a,b). The attractive interactions are weighted by an analytical function for the bond order. The model realistically describes carbon–carbon bond lengths and energies in graphite and diamond (see, for example, Keblinski, Wolf, Cleri, Phillpot and Gleiter, 1998). Intermediate situations where a carbon with three nearest neighbours is bonded to a carbon with four neighbours have been reported to result in non-physical behaviour, and an improved model in which the bond-order function depends on local coordination, bond angles, radical structures, conjugation, and dihedral rotation has been developed by Brenner (1990).

In a few cases, the grain boundary structures and energies calculated by using first-principle, TB and Tersoff potential methods have been compared to each other (Morris, 1996; Shenderova, Brenner and Yang, 1999; Nazarov, Shenderova and Brenner, 2000a). All methods lead to very similar structures, and also the energy ranking of tilt boundaries as a function of misorientation is the same. The grain boundary energies calculated by use of the empirical potential, however, can be significantly higher. Similar conclusions apply to the calculation of dislocation core energies (Blumenau et al., 2002a).

In order to circumvent the computational effort inherent to DFT-LDA and TB methods, multiscale approaches have been developed for calculating dislocation and grain boundary energies from first principles. Since first-principle methods allow to address only a small number of atoms, these techniques are used to calculate the core structures. Appropriate boundary conditions must be chosen to account for long-range elastic interactions. This can occur by imposing the displacement expected from elasticity theory at the cluster boundaries or by imposing periodic boundary conditions (Blase, Lin, Canning, Louie and Chrzan, 2000). In another approach, the grain boundaries are constructed from key structural units accessible to accurate atomic-scale first-principle analysis. If these structural units are treated as disclination dipoles, the elastic energy is controlled by the distance between the dipoles (Shenderova et al., 1998a; Nazarov, Shenderova and Brenner, 2000b).

3.2.1.2. Structure of Dislocation Cores. In order to present the results of modelling, it is convenient to use the analytical expressions for dislocation energies known from linear elasticity theory (Hirth and Lothe, 1968). The energy per unit length of a straight dislocation in a cylinder of radius R is divided into two portions, the energy of the dislocation core, E_c, and the energy of the elastic strain field outside the core. In the isotropic case, this energy is written as

$$E = A \ln(r/r_c) + E_c \tag{1}$$

$$A = b^2 k(\beta)/4\pi \tag{2}$$

$$k(\beta) = G[\cos^2\beta + \sin^2\beta/(1 - \nu)] \tag{3}$$

where b is the Burgers vector, β the angle between the Burgers vector and the direction of the dislocation line, r the radial distance from the dislocation line and r_c the core radius. G is the shear modulus and ν the Poisson ratio. A is a constant. If the energy is plotted as a function of $\ln(r/r_c)$, E becomes a straight line with slope A for $r > r_c$. For $r = r_c$, the dislocation energy equals the core energy that can be obtained by fitting Eq. (1) to the total energy resulting from atomistic modelling. Fitting also provides an independent value for A which may be compared with the 'experimental' value given by Eq. (1). Nandedkar and Narayan (1987, 1990) were the first to calculate the core structures and energies by using Tersoff interatomic potentials. Some of their results are shown in Table I. Nowadays, first-principle calculations replace more and more the use of semi-empirical interatomic potentials which are less well suited to deal with strong bond distortions. The core energies and the $k(\beta)$-values for a large variety of $(a/2)\langle 110\rangle$-type dislocations are given in Table I. They have been calculated by using density-functional-based TB methods (Blumenau et al., 2002b). The satisfying agreement between $k(\beta)$-values obtained from linear elasticity and atomistic modelling, both shown in Table I, confirms that atomistic modelling has achieved a high degree of perfection and may be used to predict the materials' behaviour and support the interpretation of experimental work.

Frequently encountered structural features of dislocation cores are 5-, 7- and 8-membered rings as well as dangling bonds and carbon atoms in threefold coordination (Fig. 11). Different geometric configurations are generated, depending on if one adds or removes a row of atoms along the dislocation line. In this way, the core structures designed by (I) and (V) in Figure 11 are obtained (Blumenau et al., 2002b). Inspection of Table I reveals that dislocations belonging to the shuffle set, i.e., the extra half-plane ends between the widely spaced $\{111\}$ atomic layers in a $\langle 110\rangle$ projection, generally exhibit higher core energies than dislocations of the glide set where the extra half-plane ends between the narrowly spaced $\{111\}$ atomic layers. The geometrical arrangement of the core atoms prior to relaxation is crucial for the results obtained by numerical simulation. In the case of 30 and 90° glide partial dislocations it has been found that the double-period core reconstruction, which involves doubling of the reconstruction period along the dislocation line (DP, Fig. 11d and f), is

TABLE I

DISLOCATION ENERGIES AND CORE RADII CALCULATED BY ATOMISTIC MODELLING (ACCORDING TO BLUMENAU ET AL., 2002a, TABLE III)

Type of dislocation	Core energy (eV Å^{-1})	Core radius (Å)	Atomistic model $k(\beta)^a$ (GPa)	Linear elasticity $k(\beta)^b$ (GPa)
Perfect edge 90° $(a/2)[110](001)\uparrow[1\bar{1}0]$	2.38^c	$\approx 3^c$	659^c	–
Perfect shuffle screw $(a/2)\ [1\bar{1}0]\uparrow[110]$	3.32^d	4.3	582	536
Perfect glide 60° $(a/2)\ [1\bar{1}0](111)$	2.13	4.1	570	566
Partial glide 90° *SP* $(a/2)\ [2\bar{1}\bar{1}](111)$	4.69^c	$\approx 3^c$	664^c	–
	1.32	3	599	576
Partial glide 90° *DP* $(a/2)\ [2\bar{1}\bar{1}](111)$	1.18	3.3	599	576
Partial glide 30° *DP* $(a/2)\ [1\bar{2}1](111)$	0.88	3.1	573	546
Perfect shuffle 60° $(a/2)\ [1\bar{1}0](111)$	3.37^c	$\approx 3^c$	589^c	
	2.60	4.2	591	566
Partial shuffle 30° (V) $(a/2)\ [1\bar{2}1](111)$	2.55	6.0	637	546
Partial shuffle 30° (I) $(a/2)\ [1\bar{2}1](111)$	3.31	5.0	605	546
Partial shuffle 30° (I′) $(a/2)\ [1\bar{2}1](111)$	4.32	5.5	762^e	546
Partial shuffle 90° (V) $(a/2)\ [1\bar{2}1](111)$	1.89	4.4^e	642^e	576
Partial shuffle 90° (I) $(a/2)\ [1\bar{2}1](111)$	3.69^e	4.1^e	602^e	576
Partial shuffle 90° (I′) $(a/2)\ [1\bar{2}1](111)$	2.67^e	4.5^e	640^e	576

The energies of stacking faults subtended by partial dislocations are not included. Column 1: indices following the sign ↑ indicate the direction of the dislocation line. See Figure 11 for the core structures of the corresponding dislocations. (I) and (V) stand for interstitial-like and vacancy-like core structures.

[a] $k(\beta)$ see Eq. (1): fit of the results of atomistic modelling to this equation.

[b] $k(\beta)$ see Eq. (1): Voigt average of elastic constants by Grimsditch and Ramdas (1975).

[c] Data by Nandedkar and Narayan (1990) using Tersoff semi-empirical potentials.

[d] Energy of the unreconstructed core. Recently, it was suggested by Pizzagalli et al. (2002) that sp^2 reconstruction reduces the core energy substantially.

[e] Large error of about 20% due to fitting.

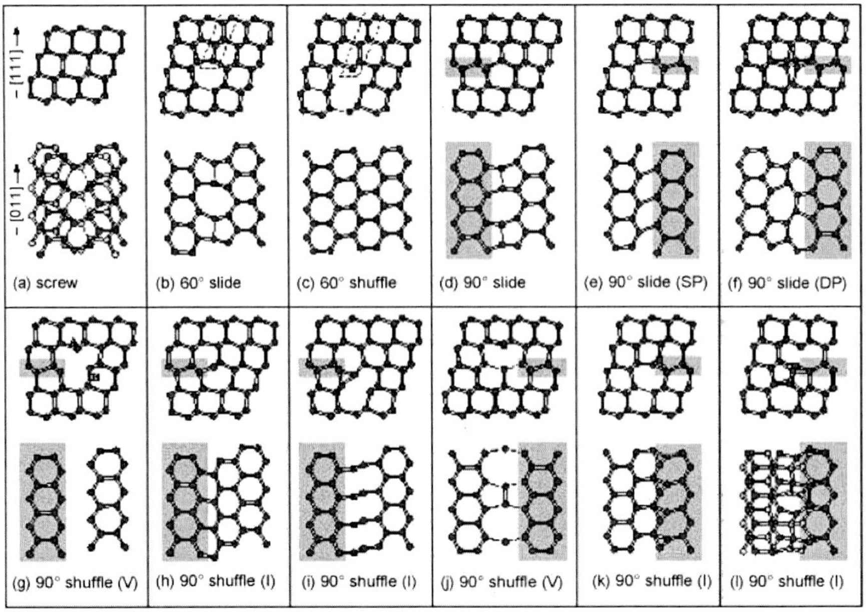

FIG. 11. Fully relaxed core structures of dislocations in diamond: (a) the $[01\bar{1}]$ screw dislocation; (b), (c) the undissociated 60° dislocation; (d)–(f) the glide set of partials; (g)–(l) the shuffle set of partials (according to A. T. Blumenau, M. I. Heggie, C. J. Fall, R. Jones, and T. Frauenheim (2002) *Phys. Rev. B* 65, 205205-1–205205-8). For each dislocation type, the upper figure, which is a ⟨110⟩ projection, shows a view in the direction of the dislocation line whereas the lower figure shows the {111} glide plane. Gray-shaded areas are intrinsic stacking faults. Weak bonds with bond lengths over 1.8 Å are given as dashed lines. In (a), the dislocation core lies between two glide planes which both are shown in projection. In (l), glide plane atoms are depicted in dark gray, the underlying ones in light gray. Additional information is added in parentheses if dislocations have more than one stable core configuration: (SP) and (DP) stand for single and double period, respectively, (V) and (I) for vacancy-type and interstitial-type configurations.

energetically more favourable than single period reconstruction (SP, Fig. 11e) (Blase et al., 2000). Similarly, Pizzagalli, Beauchamp and Rabier (2002) suggest that the sp^2 hybridisation of some core atoms in perfect shuffle screw dislocations is the more stable configuration. Blumenau, Jones, Frauenheim, Willems, Lebedev, Van Tendeloo, Fisher and Martineau (2003) arrive at a similar conclusion for perfect screw dislocations of the glide type.

Dislocations belonging to the ⟨110⟩ {111} system commonly are split (60° *type*: Pirouz et al. (1983); *screw type*: Luyten et al. (1992) and Dorignac et al. (1997); *both* 60° *and screw types*: Suzuki et al. (1992)). For dissociation of dislocations to occur, the Frank criterion stipulates that the strain field energy of the partial dislocations must be smaller than that of the perfect dislocation. Apart from a minor geometrical factor, this criterion depends only on the square of the Burgers vectors and does not take into account the energies of the core and the intrinsic

TABLE II

GEOMETRICALLY AND ENERGETICALLY FAVOURABLE DISSOCIATION REACTIONS (ACCORDING TO BLUMENAU ET AL., 2002a)

Perfect dislocation	Product of dissociation
Screw shuffle	30° glide partial + ISF + 30° glide partial
60° glide	30° glide partial + ISF + 90° glide partial (DP)
60° shuffle	30° glide partial + ISF + 30° shuffle partial (V)

stacking faults subtended by the partials. A complete analysis of a variety of dissociation reactions including the faulted regions recently has been carried out by Blumenau, Heggie, Fall, Jones and Frauenheim (2002b). Energetically favoured dissociation reactions for $(a/2)\langle 110\rangle$ dislocations are given in Table II. The stacking fault widths associated with these reactions amount to 30.4 Å for the dissociated screw and 35.0 Å for the dissociated 60° dislocation, in perfect agreement with the values calculated by classical linear elasticity theory. The core energy of the 60° shuffle is higher by 0.63 eV $Å^{-1}$ compared to the glide type but it is not certain that this dislocation exists. On the other hand, Dorignac et al. (1997) report the observation of screw dislocations split into 30° partials, one belonging to the shuffle, the other one to the glide set. According to Table I, this type of dissociation leads to excess core energies of 1.67 eV compared to the reaction suggested by Blumenau et al. (2002b) (Table II). It is emphasised that such discrepancies do not discredit the theoretical approach, which supposes the presence of a stacking fault to be the result of a dissociation reaction, involving dislocation glide. Since undecorated partial dislocations are immobilised at deposition temperature, owing to the high Peierls potential of diamond, dislocation glide is unlikely to occur in the bulk. It is supposed, therefore, that stacking faults exhibiting a width of several Burgers vector units are created at the growth surface. For such a defect to be formed, one may imagine a partial dislocation buried below the surface that bounds one side of a stacking fault, while the other side of the stacking fault outcrops at the growth surface. In the absence of a second partial, the width of such a stacking fault could become quite large, as it is not subjected to the principle of energy minimisation on which the theoretical calculations are based.

3.2.2. *Electronic Energy States of Dislocations*

Band-A cathodoluminescence observed in undoped CVD diamond in the range 2.8–2.9 eV (λ, 443–427 nm) recently has been attributed to sp^2 bonding states at dislocations and grain boundaries (Takeuchi, Watanabe, Sawada, Yamanaka, Ichinose, Sekiguchi and Okushi, 2001a; Takeuchi, Watanabe,

Yamanaka, Okushi, Sawada, Ichinose, Sekiguchi and Kajimura, 2001b; Takeuchi, Yamanaka, Watanabe and Okusi, 2001c). Also, the natural brown colour of IIa diamond is believed to be due to dislocations in the bulk of the diamond lattice (Wilks and Wilks, 1991). Dislocations are supposed to generate localised energy states in the band gap which allow the absorption of photons with energies in the visible range, a process which results in the excitation of electrons. Low-loss EELS can probe electronic transitions within and above the band gap whilst core-excitation EELS results from electronic transitions between core states and the conduction band (Fall, Goss, Jones, Briddon, Blumenau and Frauenheim, 2001). A major difficulty in the detection of bond defects on dislocations resides in the very small fraction of defective bonds within the probed volume, which amounts to less than 10^{-3} for a 60° shuffle dislocation.

Using first-principle calculations, low-loss and core-excitation EEL spectra have been simulated for various dislocation cores, and compared with experiment (Fall, Blumenau, Jones, Briddon, Frauenheim, Gutiérrez-Sosa, Bangert, Mora, Steeds and Butler, 2001; Fall et al., 2002; Jones, Fall, Gutiérrez-Sosa, Bangert, Heggie, Blumenau, Frauenheim and Briddon, 2002). The outcome of these investigations is that 30 and 90° partials bounding intrinsic stacking faults do not give rise to states in the band gap. Nevertheless, the calculated low-loss EEL spectra differ from the bulk spectra because the onset of absorption moves to lower energies. The absence of gap states is due to core reconstruction, which allows for fourfold coordination of the core atoms and for bond lengths and bond angles similar to those in the bulk. The shuffle screw dislocation also preserves the fourfold coordination of atoms and is devoid of gap states but the band structure is perturbed in the 6–12 eV region. As undercoordinated atoms are retained in the core reconstruction of 60 and 90° shuffle dislocations, it is less surprising that modelling predicts empty states below the conduction band that are localised on the dislocation core. Enhanced EEL absorption is expected for the 60° shuffle dislocation in the 3–12 eV range, and for the 90° partial both in the 0–2 and 6–9 eV range. Support for these simulations is furnished by experimentally measured electron energy states in the 3.5–4.2 eV region of brown IIa diamond which are believed to be correlated with dislocations belonging to the shuffle set (Kolodzie, Murfitt and Bleloch, 2001). For all dislocation types investigated, low loss spectra predict, in comparison with the bulk, additional absorption in the 6–8 eV range, provided the incident electron beam is not parallel to the dislocation line. These predictions have been partially confirmed by experiment (Fall et al., 2002). Since these authors have not detected band gap absorption in CVD diamond, they conclude that band-A luminescence is either due to point defects, such as impurities and jogs, located along the dislocation lines, or to shuffle segments which have escaped the detection by EELS.

3.2.3. Evidence for Dislocation Glide and Climb

TEM investigations of diamond films provide strong evidence for dislocation motion to occur during growth. Both in undoped and boron-doped films, rather high densities of bowed dislocations can be encountered, as illustrated in Figure 8b, where numerous dislocations appear to enter the two bottom grains from the bottom side of the grain family. In boron-free samples deposited by microwave enhanced CVD at 700 °C, many dislocations have been observed to be heavily curved while staying in one and the same (111) glide plane over their whole length (Wurzinger et al., 1993). Dislocations bow-out under stress, otherwise they should remain straight. About one-third of the analysed dislocations do not belong to the ⟨110⟩ {111} slip system. These dislocations have mostly {100} glide planes and are either of the Lomer type (edge dislocation with a ⟨110⟩ line direction) or of the 45° type with line directions near ⟨100⟩ (a dislocation already considered by Hornstra, 1958). Some of the 45° dislocations again are curved, indicating that glide may also occur on {100} planes. Further more, the authors have identified cross slip between the slip systems $[10\bar{1}](100)$ and $[10\bar{1}](1\bar{1}1)$, and analysed the formation of a Lomer lock by the reaction of two glide dislocations:

$$(a/2)[10\bar{1}](111) + (a/2)[\bar{1}\bar{1}0](1\bar{1}1) \rightarrow (a/2)[0\bar{1}\bar{1}](100)$$

All these observations as well as the absence of dislocation climb support the authors' conclusion that dislocation glide must have occurred during deposition. Strong evidence for intense dislocation glide activity has also been delivered by Suzuki et al. (1992) in films deposited at 850 ± 20 °C by dc-plasma jet CVD, but these authors additionally observe dislocations with a characteristic zig-zag shape which they attribute to climb of 60° dislocations. This interpretation has been underscored by numerous features observed in weak beam analysis.

If it is assumed for the moment that in both studies the defect concentrations at the growth front are similar, the main difference between both kinds of films remains the substrate temperature. Interestingly, neutral vacancies required for climb become mobile at 700 °C (Davies, Lawson, Collins, Mainwood and Sharp, 1992). Dislocation climb claimed by Suzuki et al. thus becomes a likely process at 850 °C where vacancies can migrate to the dislocation cores. Transport of matter to and from the dislocations by interstitial diffusion can be ruled out and self-diffusion in diamond is dominated by vacancies (Bernholc, Antonelli, Del Sole, Bar-Yam and Pantelides, 1988). As the vacancy concentration drops within a few micrometers from the growth surface (Allers and Mainwood, 1998), dislocation climb should be restricted to regions near the growth surface, rather than taking place in the bulk. Considerable uncertainty exists in regard to the actual value of vacancy concentration. For the vacancy depth profiles established by Allers and Mainwood, a maximum concentration at the free surface of 1 ppm has been assumed. On the other hand, Dannefaer, Zhu, Bretagnon and Kerr (1996) conclude from positron-lifetime experiments with diamond films grown by the hot-filament technique that a large number of grains in undoped films (35–65%) contain vacancies in excess of 50 ppm, mostly in form of vacancy clusters. In thinner films (<30 μm), single-vacancies have been found. Their concentration increases substantially with decreasing film thickness, an observation, which is in general agreement with the depth profile established by Allers and Mainwood. From the positron-lifetime measurements, one would expect single-vacancy concentrations at the film surface in the order of 50 ppm. The higher the substrate temperature, the faster the vacancies will diffuse and also be trapped, thereby narrowing the concentration profile. Because of the higher deposition temperature, a higher initial vacancy concentration in Suzuki's films also is likely which would enhance the probability for dislocation climb. It is noticed that no vacancies have been detected in boron-doped diamond films (Dannefaer et al., 1996).

As far as dislocation glide is concerned, both Suzuki et al. (1992) and Wurzinger et al. (1993) agree on the dissociated character of the observed glide dislocations. Only in segments where the dislocations have left the glide plane, the partials are constricted. The mobility of the 90° glide partial dislocation which, by analogy with silicon, is considered as the most mobile dislocation (Heggie, Jenkins, Ewels, Jemmer, Jones and Briddon, 2000), has been investigated thoroughly from a theoretical point of view. Sitch, Jones, Öberg and Heggie (1997) have studied the mobility by using an *ab initio* density-functional method applied to a cluster with 110 carbon atoms terminated with hydrogen atoms at the outer sides. The authors find that the dislocation core reconstructs with a C–C bond length comparable to that in defect-free bulk diamond. An activation energy of 3.3 eV has been calculated for dislocation motion by kink formation. Subsequent theoretical studies arrive at similar conclusions (Heggie et al., 2000; Blumenau et al., 2002a). At homologous temperatures (T/T_m) of less than 0.3, corresponding to a substrate temperature of 700 °C at deposition, the contribution of thermal activation to the formation of kinks is not significant and dislocation glide can be ruled out as long as the Peierls potential for dislocation displacement is not substantially lowered. Impurities segregating to the dislocation core can have this effect (George, Escaravage, Champier and Schröter, 1972). Hydrogen, for example, which represents the most abundant impurity in CVD diamond lowers the activation energy considerably. Applying density-functional calculations, Heggie et al. (2000) have shown that H-atoms, once trapped by the dilatational part of the dislocation core, become mobile along the core and assist the formation of kink pairs. The migration of H-catalysed kinks, which eventually becomes the rate-controlling factor for dislocation glide, has a total activation energy of only 1.67 eV, compared to 3.3 eV without hydrogen. But even this value is still too high to account for significant thermally activated dislocation glide. Extremely high stresses would be required for dislocation glide to occur in the bulk. In conclusion, dislocation glide must be restricted to regions near the free growth surface where the reconstruction of the dislocation core is subject to less constraint. Also the migration of hydrogen should be limited to the surface region owing to the extremely small diffusivity of hydrogen in the bulk (Section 2.3.1).

So far, the volume of experimental evidence does not allow one to decide if dislocation glide and climb are general phenomena associated with diamond film deposition. From the foregoing discussion one would expect that both of them can occur only close to the growth surface. Boron doping is not supposed to affect dislocation climb because boron-vacancy pairs are known to dissociate above 500 °C (Chevallier, 2001). Hydrogen trapping by boron should not be a problem at growth temperatures above 700 °C because of the dissociation of (B, H) pairs (Chevallier et al., 1998).

3.3. Low-Angle Boundaries and Disclinations in Near-Heteroepitaxial Films

3.3.1. *Morphology of Near-Heteroepitaxial Films*

The discovery of hetero-epitaxial, bias-enhanced nucleation of diamond on monocrystalline silicon and β-silicon carbide substrates by Stoner, Ma, Wolter and Glass (1992) and Jiang and Klages (1993) has triggered an enormous effort aiming at the development of heteroepitaxial diamond films on silicon. By combining bias-enhanced nucleation with selective grain growth in the α-range beyond 2.5, the spread of grain orientations on silicon substrates has been continuously improved from initially 12 (FWHM) to 3.9° in the azimuthal direction (equivalent to the twist component of the misalignment between the diamond grains and the monocrystalline silicon (001) substrate; Thürer, Schreck and Stritzker, 1998) and to 4.6° with respect to the polar misalignment (corresponding to a tilt of the $\langle 001 \rangle$ axes of diamond grains with respect to the substrate normal; Jiang, Zhang, Yu and Lee, 1998). Much smaller grain misorientations of less than 1° have been obtained on β-silicon carbide (Kawarada, Wild, Herres, Locher and Koidl, 1997). By the use of iridium buffer layers on $SrTiO_3$ (001) substrates, tilt and twist have even been reduced to 0.17 and 0.41°, respectively (Schreck, Schury, Hörmann, Roll and Stritzker, 2002). Perfectly oriented heteroepitaxial films as required for electronic and optical applications have not been achieved yet.

Once the nucleation step is accomplished, selective grain growth results in a rapid elimination of unfavourably oriented grains, a mechanism which is efficient with respect to strongly misaligned grains only. At a later stage of growth, the elimination of grains with small deviations of their $\langle 001 \rangle$ growth axis from the substrate normal would require extreme growth lengths which are incompatible with the CVD growth process. These more geometrical aspects have been discussed in detail by Schreck et al. (2002) in the context of their work on the use of iridium buffer layers for heteroepitaxial films. Even these almost perfect films exhibit mosaic structures composed of slightly misaligned grains. In the central region of the film, the squared {100} facets join each other, whereas in the outer substrate regions they are separated by grooves formed by the octahedral facets of adjacent grains (Fig. 10c). The morphological differences between the substrate centre and the periphery signify a non-uniform α parameter equivalent to non-uniform growth rates of the {100} and {111} facets. The grain boundaries underneath the grooves of highly oriented films are generally flat and parallel to the {110} lattice planes (Wurzinger et al., 1995). Only a few twin lamellae, generally free of microtwins, have been detected. At other places without grooves or surface steps, the grain boundaries are curved, consisting of walls of parallel dislocations that are commonly of the edge-type. The connecting grains must have merged in such regions where dislocations mark the remaining misalignment (Wurzinger et al., 1995). Highly oriented films also exhibit strong non-thermal internal stresses that can cause plastic deformation of the substrate (Michler et al., 1998; Schreck, Roll, Michler, Blank and Stritzker, 2000). In probing the surface of individual {100} facets with confocal microbeam Raman spectroscopy, von Kaenel, Stiegler, Michler and Blank (1997) have measured intense stress fluctuations within individual grains, with stress amplitudes of several GPa.

3.3.2. Incomplete Low-Angle Boundaries and Disclinations

TEM and HRTEM investigations have provided further insight into the merger of adjacent grains which proves to be an efficient mechanism for grain coarsening during film growth. Observations of the cross-sections of heteroepitaxial films with mosaic structures reveal that low-angle boundaries sometimes disappear in order to leave behind a continuous lattice free of defects in the region where one would expect the former boundary to continue (Jiang and Jia, 1996). Structural features of this kind are called disclinations. The missing low-angle boundary is replaced by a lattice distortion, in order to provide a perfectly coherent match of the lattice planes. Disclinations generate considerable residual stress (Romanov and Vladimiro, 1992). Jiang and Jia have localised the merger of two such grains at a 10 nm distance from the interface with the silicon substrate. It appears from the cross-section in Figure 12a, that the tilt low-angle boundary separating the grains DI and DII, which are misaligned by 2°, is discontinued at point B and replaced by a monocrystalline structure. Similar reactions between merging grains have also been observed in TEM plan views taken at much larger distances from the interface (Michler et al., 1998). By correlating TEM and SEM topography, numerous dislocation walls ending within the grains have been disclosed in regions where adjacent grains have merged or penetrated each other (Fig. 12b). The dislocation walls consist of parallel dislocations with line directions parallel to [001]. The dislocations can be of the screw-type with Burgers vectors parallel to [001], but most often they are 45° dislocations with $(a/2)\langle 110\rangle$ Burgers vectors. As the misalignment between neighbouring grains is random, the identification of different types of dislocations is not surprising.

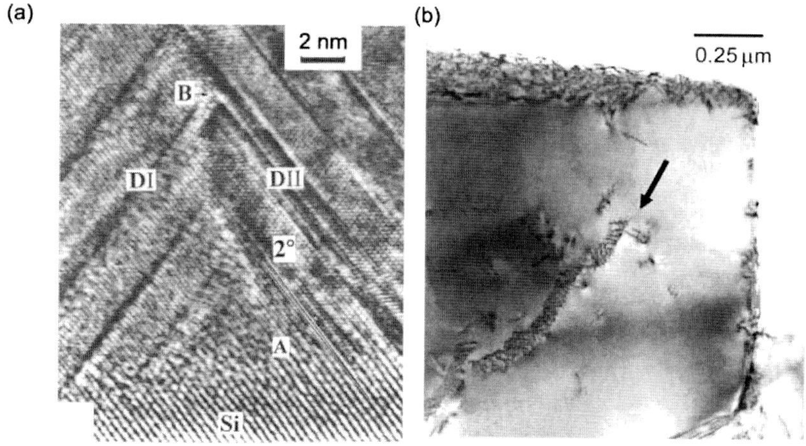

FIG. 12. Disclinations in highly-oriented, mosaic-like ⟨100⟩ diamond films on silicon. (a) HRTEM cross-section near the interface: the tilt boundary between regions DI and DII stops at point B which is located about 10 nm from the interface, and changes into a disclination (according to X. Jiang and C. L. Jia (1996) *Appl. Phys. Lett.* 69, 3902–3904). (b) Dislocation wall in a {100} growth sector: The arrow on the TEM micrograph taken in near [001] projection and under (220) two-beam-condition marks the end of an incomplete tilt subgrain boundary (according to J. Michler (2000) *Thesis Nr. 2115*, Ecole polytechnique Fédérale de Lausanne, Switzerland).

Dislocation walls with tilt character that end within a crystal represent *wedge disclinations* and those with twist character *twist disclinations* (Romanov and Vladimiro, 1992). From a geometrical point of view, one would expect that the domains of small crystallographic misorientation are separated by a low-angle boundary. Under certain circumstances, however, it is for a particle energetically more favourable, to partially suppress the low-angle boundary and accommodate the misorientation by an elastic distortion of the crystal lattice, named disclination. In case of a wedge disclination, the distortion signifies a rotation about an axis located inside the particle, at the end of the incomplete low-angle boundary (Fig. 13). In the mathematical description of wedge disclinations, one considers that the particle first is cut in a radial plane along its length, and that a wedge of material with opening angle ω then is squeezed between the two faces of the cut, or removed from there. In the first case, represented in Figure 13, a negative disclination is generated, and in the second case, a positive one. The latter would correspond to a wall of dislocations with inverse signs compared to the dislocations sketched in Figure 13. Incomplete low-angle boundaries give rise to partial disclinations because the rotation introduced by a wedge of strength ω does not correspond to a lattice symmetry operation. The rotation ω is given by the lattice misorientation across the low-angle boundary, i.e., $\omega\,(\text{rad}) = b/d$ if $\omega << 1$, where b is the Burgers vector and d the dislocation spacing. The misalignment mediated by the dislocation wall shown in Figure 12b amounts to about 0.7°, as calculated from the dislocation spacing.

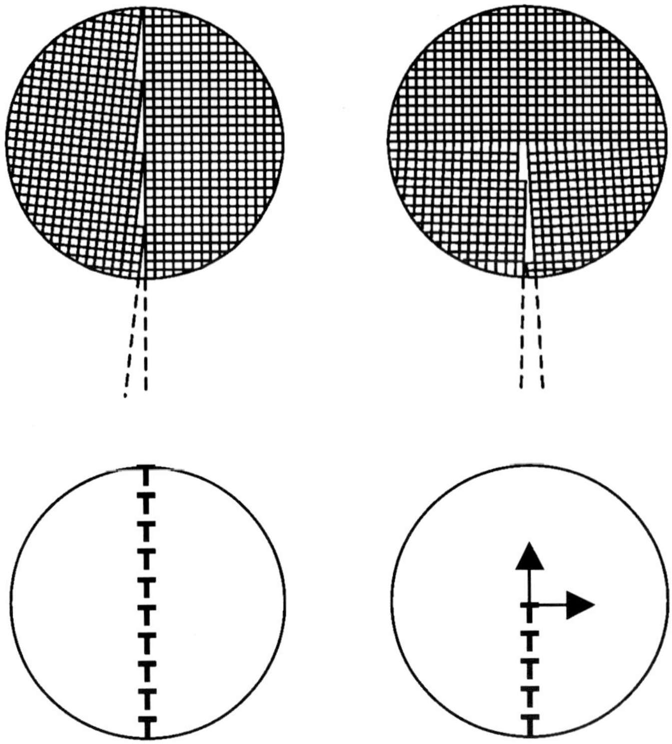

FIG. 13. Schematic view of the substitution of a subgrain boundary separating two misaligned particle domains by a wedge disclination located in the particle centre at the end of the incomplete low-angle boundary.

The elastic energy stored in the strain field of a cylindrical particle containing a partial wedge disclination increases strongly with particle radius r and wedge angle ω. For a particle of length l distorted in plane strain, one obtains for the line energy of a disclination (Romanov and Vladimiro, 1992):

$$E_{\text{discl}} = \frac{G}{16\pi(1 - \nu)}\omega^2 r^2 l \tag{4}$$

G is the shear modulus (574 GPa) and ν the Poisson's ratio (0.07). The energy of the complete low-angle boundary, E_{lagb}, is given as (Hirth and Lothe, 1968):

$$E_{\text{lagb}} = \gamma^{\text{gb}} 2rl \tag{5}$$

where

$$\gamma^{\text{gb}} = \frac{G\vec{b}}{4\pi(1 - \nu)}\omega\left(\ln\frac{\alpha_{\text{core}}}{2\pi\omega} + 1\right) \tag{6}$$

Both energies are shown in Figure 14 as function of the misorientation angle ω for a cylindrical particle of 2 μm diameter. Conforming to the results of

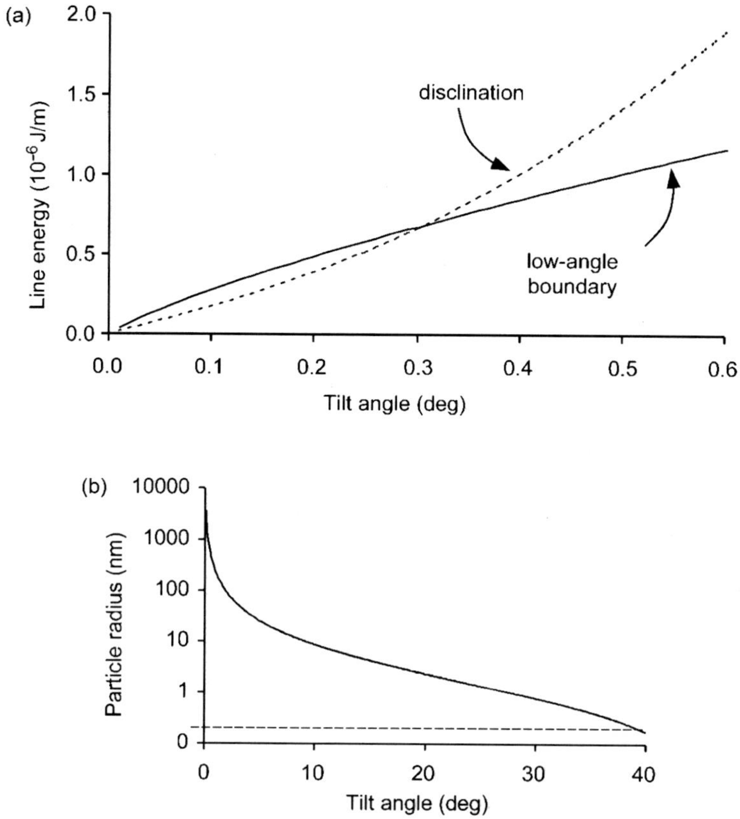

FIG. 14. Energy of a disclination compared to a complete low-angle boundary, both cores consist of $(a/2)\langle 110\rangle$ dislocations: (a) for a cylindrical particle of 2 μm diameter, the disclination is stable only for misalignment smaller than 0.3°; (b) the critical misalignment for the stability of disclinations increases rapidly with smaller particle size.

atomistic modelling (Morris, 1996), a value of 1.8 has been chosen for the core parameter $\alpha_{core} = |\vec{b}|/r_c$ in Eq. (6), where $\vec{b}$ is the Burgers vector and r_c the core radius of the dislocation that terminates the incomplete low-angle boundary and coincides with the centre of the disclination. The disclination energy plotted in Figure 14a includes the energy of the incomplete low-angle boundary, i.e., $E = E_{discl} + 0.5E_{lagb}$. At misorientations inferior to 0.3°, the energy of wedge disclinations becomes smaller than for low-angle boundaries and their formation should be favoured. At larger misorientations, the low-angle boundary requires less energy. This behaviour is related to the stress fields of the two defects, which in case of a low-angle boundary vanishes rapidly with increasing distance from the boundary (within a few boundary dislocation spacings, Hirth and Lothe, 1968), whereas they keep growing with particle size in case of disclinations (for further details see Romanov and Vladimiro, 1992). The upper limit of grain misorientation increases rapidly with decreasing particle size as follows:

$$\frac{r}{|\vec{b}|} = \frac{4\left(\ln \dfrac{e\alpha_{\mathrm{core}}}{2\pi\omega} - \ln \omega\right)}{\omega} \tag{7}$$

Eq. (7) includes the contribution of the incomplete low-angle boundary to the total defect energy. The variation of maximum particle size with grain misalignment is given in Figure 14b. Accordingly, a particle size of about 1 nm corresponds to a grain misalignment of 30°. The horizontal line refers to a dislocation core radius of about 3 Å and should be considered as a lower limit for the application of Eq. (7). Obviously, the disclination shown in Figure 12a has taken place under energetically favourable conditions since the particle size at the moment of disclination nucleation has been approximately 20 nm and the misorientation 2°. Additional constraints, exerted by the substrate, for example, are ignored by the above equations. They would increase the total energy of the disclination. In the present example, it seems reasonable to neglect this influence as it is well known that particles fan out from tiny substrate regions, i.e., only negligeable bending moments should be transferred from the substrate to the particle (Williams et al., 1990a; Tarutani, Shimato, Takai and Shimizu, 1995). Moreover, the strain energy associated with the first continuous layer overgrowing the low-angle boundary, i.e., at the moment of disclination nucleation, must have been very small anyway since the involved volume is extremely small.

3.3.3. *Generation of Disclination Dipoles and Multipoles*

There is no doubt that disclinations identified in thick layers at a distance of more than 10 μm from the interface and joining grains of several micrometers in-plane diameter and several degrees misalignments, fall far outside the range of the energy criterion given by Eq. (7). Their formation either must have been kinetically controlled or very efficient mechanisms of stress relaxation must have been operative. One such mechanism is self-screening of disclination ensembles in dipole, quadrupole and, more generally, multipole configurations that provide vanishing long-range stress fields, similar to dislocation multipoles (Kroupa and Lejcek, 1972; Romanov and Vladimiro, 1992). Multipole configurations are indeed commonly encountered in near-heteroepitaxial diamond films (Fig. 15). They consist of dipoles, i.e., dislocation walls disconnected from grain boundaries, following one another at a distance of sometimes less than one micron. The stresses at the poles are very high and the stress amplitudes alternate between plus and minus values. The Raman shift in Figure 15b and the stress values marked in Figure 15c have been measured on thin films prepared for TEM. Only in-plane stresses are retained during film preparation and even those can partially relax via film bending. Significantly higher stress amplitudes may be expected in the bulk and stress values of several GPa at the disclination core seem realistic.

TEM provides strong evidence that the dislocation walls generating the disclination dipoles are emitted at rather regular intervals from migrating low-angle boundaries. They are similar in shape and their morphology changes only slowly with increasing distance from the nearby low-angle boundary. The spacing of the emitted dislocation walls seems to be correlated with the misalignment of the coalescing grains. While the 'defect bands' in high-quality films with grain alignments of better than 0.5° lie 2–5 μm apart (Schreck et al., 2002), the dislocation walls between grains of several degrees misorientation succeed each other at a distance of typically 0.5–1 μm (Michler, 2000). Although a quantitative model to explain the formation of disclination dipoles in large grains is not available, one may speculate on the controlling mechanisms.

The formation of multipoles with alternating stress amplitudes and vanishing long-range stress fields represents undoubtedly an adequate means to minimise the total elastic strain energy, even in large grains, but the local strain associated with each individual dipole remains very high. Another element largely contributing to strain relaxation at a local scale, is free surface (Romanov and Vladimiro, 1992). There seem to be two different ways how merging grains can make use of free surfaces. If the rotation axis lies parallel to the growth surface, as for the incomplete low-angle boundary shown in Figure 12a, the disclination is nucleated by the first accidental growth step able to bridge the low angle grain boundary. This geometrical configuration seems to represent the most common case of grain merger, occurring when the [001] growth directions of the two grains, which are about to merge, are tilted towards each other. As already mentioned, the nucleation of such a disclination should be kinetically controlled because the distorted volume is extremely small. Once the bridge has been generated, the stress field will build up gradually with further coherent atomic layers being deposited on top of the first one. In the second configuration, the rotation axis is oriented in growth direction and accounts for the azimuthal or twist component of grain misalignment. This defect would show up as an incomplete low-angle boundary composed of dislocations in [001] direction. Disclinations can be formed as long as the merging grains are small, in accordance with Eq. (7), or if the nucleation takes place close to free surfaces oriented parallel to the growth direction. Once such grains have begun to merge, it is difficult to imagine free surfaces, which would assist the nucleation of further disclinations, but such particles, which are distorted in plane strain, can grow in height as columnar grains without increasing the strain energy per unit volume. Both mechanisms cannot, however, explain the formation of disclinations in coarse-grained films with mosaic structure. Still another mechanism must be responsible for continuing grain coarsening during further film growth, as reported by Schreck et al. (2002). The similarity in shape of successive dislocation walls and nearby low-angle boundaries in Figure 15 suggests that

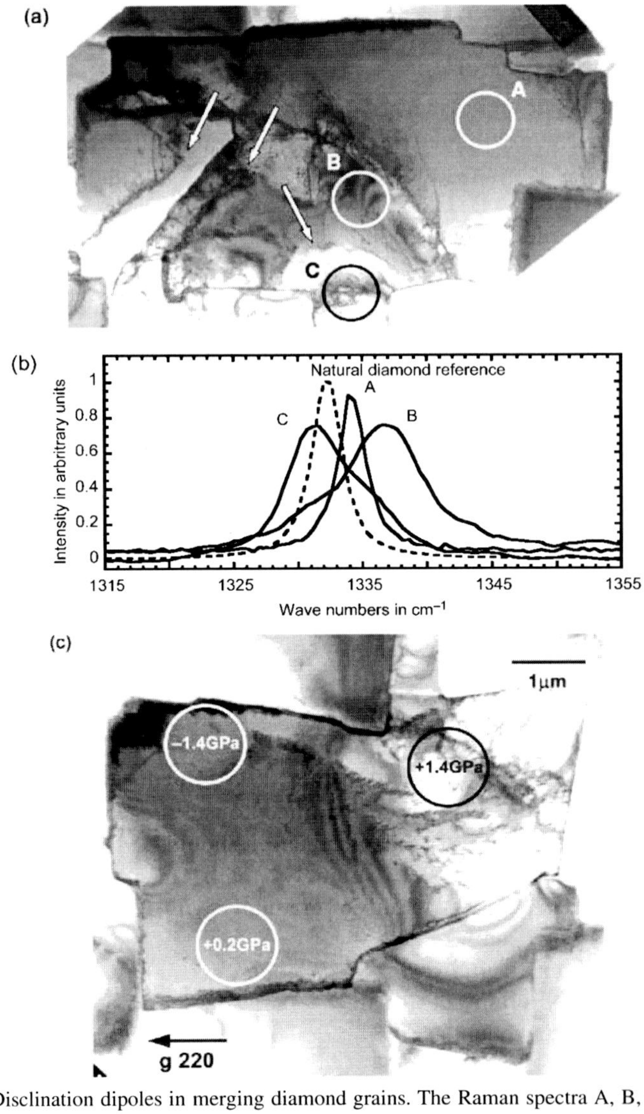

FIG. 15. Disclination dipoles in merging diamond grains. The Raman spectra A, B, C in (b) were taken from the circular regions marked in (a) by corresponding letters. In (b), the line shift is proportional to the local internal stress. (a) Region A is almost free of dislocations; the upper two arrows indicate two incomplete ⟨001⟩ tilt boundaries, one on each side of a complete low-angle boundary. Areas B and C are separated by another incomplete tilt boundary (arrow). In crossing this boundary, the internal stress changes from strongly compressive (B) to strongly tensile (C), see spectra in (b). All these low-angle boundaries tend to be oriented in ⟨100⟩ directions, contrary to the original mosaic boundaries which are oriented in ⟨110⟩, i.e., parallel to the edges of the {100} tiles. (c) TEM micrograph of another grain in near [001] projection and (220) two-beam condition: numerous incomplete low-angle boundaries can be identified which are disclination dipoles. The internal stress amplitude associated with these dipoles fluctuates from plus to minus values. These stresses decrease rapidly with increasing distance from the dipole. The internal stress in defect-free regions is very low.

dislocation walls representing disclination dipoles have been split from the advancing grain boundary in order to save strain and grain boundary energy. It has also been reported that the density of defect bands diminishes steadily (Schreck et al., 2002). Provided the [001] dislocation walls by Michler et al. are identical with the defect bands by Schreck et al. one wonders how dislocation walls oriented in film growth direction should be discontinued if not by the junction with other boundaries. Horizontal grain boundaries, which might stop the continuation of defect bands in growth direction, indeed have been observed occasionally by TEM (Michler, 2000) but reports on horizontal grain boundaries in diamond films are scarce. Horizontal lines luminescing at 630 nm that have been seen in film cross-sections investigated by cathodoluminescence might perhaps be interpreted as edge-on horizontal grain boundaries (Cremades, Dominguez-Adame and Piqueras, 1993).

Successive dislocation walls and low-angle boundaries which initially are parallel to (110) and $(1\bar{1}0)$ planes tend to reorient themselves along cubic planes oriented in growth direction. A comparison between transmission micrographs (TEM) and surface topographs (SEM) of the same grains suggests that the rearrangement along the cubic planes is favoured if grains merge on two sides simultaneously (Fig. 15). Reorientation then would shorten the boundary length and probably also reduce the distortion energy. Schreck et al. (2002) have recently described 'defect bands' disconnected from the grain boundaries of merging grains in almost perfect diamond films. In plan view, these defect bands unveil the same tendency for being oriented along cubic planes as the dislocation walls shown by Michler et al. (1998).

In conclusion, the merger of grains via the formation of disclinations turns out to be an efficacious mechanism of grain coarsening during the growth of $\langle 001 \rangle$ oriented diamond films with mosaic structure. In comparison with other materials exhibiting diamond-type lattices, the frequent formation of disclinations appears to be related not only to the high energy of dislocations and grain boundaries, but also to the low mobility of atoms and vacancies even at deposition temperature, which prevents the reconstruction of defects into low energy configurations once they have nucleated.

3.4. Grain Boundaries

In this section, the structure and properties of grain boundaries generated by twinning and multiple twinning will be addressed first. This kind of boundaries is predominant in polycrystalline diamond films with average grain sizes in the micrometer range. General large-angle boundaries will be addressed later within this chapter. Although many similarities exist with the boundaries in other diamond-lattice materials, several studies clearly indicate that grain boundary structures for silicon and germanium cannot simply be transferred to diamond. Not only the boundary energies of diamond are considerably higher because the covalent bonds are much stiffer, also the reconstruction of boundary atoms and bonds may be different because carbon is prone to change from sp^3 to sp^2 hybridisation.

3.4.1. *Multiple Twinning*

3.4.1.1. Multiply Twinned Particles. The very first reports on diamond grown from the vapour phase already ascertain the importance of multiple twinning for the development of particle morphologies (see Matsumoto and Matsui, 1983, for an acknowledgement of this early work). Particles exhibiting a pentagonal symmetry centre are very frequent. Pentagonal particles have been observed first by Ino (1966, 1969) in vacuum deposition of fcc metals. Their diamond counterparts which also grow directly from the vapour phase have decahedral, decahedral-prismatic or icosahedral shapes (Fig. 7). Perfect pentagonal particles without defects other than first-order twin boundaries are small, typically in the order of a few nanometers (Romanov and Vladimiro, 1992). A decahedral particle composed of five twin units is shown schematically in Figure 16. Each twin units represents a tetrahedron with a dihedral angle of 70°32′ bounded by {111} facets. If five tetrahedra are joined along a common [011] edge, they leave a gap of 7°20′ rather than forming a closed polygon ($5 \times 70°32' = 352°40'$). The gap can be seen sometimes on SEM micrographs (Matsumoto and Matsui, 1983) but the further incorporation of defects in addition to the five star-like twin boundaries allows such particles to compensate for the lacking volume. In very small MTPs of a few nanometer size, the gap can be bridged via rotational strain

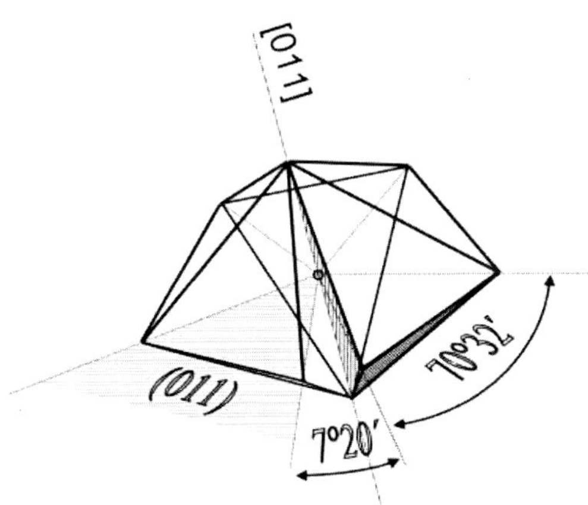

FIG. 16. Schematic of decahedral particle. The [011] axis represents a pentagonal symmetry centre. The basic crystalline unit for such a particle is a tetrahedron bounded by (111) facets exhibiting dihedral angles of 70°32′. Five such units, which are related one to another by successive first-order twinning and join along a common [011] edge, constitute a particle which leaves a gap of 7°20′ between the {111} facets of the first and the last tetrahedral unit.

of the twin units about the [011] axis. Such particles exhibit a planar five-member carbon ring in their centre, which obviously must be distorted (Section 3.4.1.2). The defect associated with this distortion again represents a disclination, more particularly a star disclination, in reference to the centre of pentagonal symmetry with five twin boundaries radiating out from the centre (De Wit, 1972). Compared to monocrystalline particles, the stability of MTPs has been rationalised by a favourable balance between strain energy, twin boundary energy, and surface energy (Ino, 1969). Indeed, very small diamond MTPs are bounded by close-packed {111} facets, and the coherent Σ 3 {111} twin boundaries provide only a minor contribution to the overall energy. Nevertheless, the reason for the formation of pentagonal MTPs is still a matter of speculation. In particular, the origin of the five-member carbon ring in the particle centre remains puzzling. Matsumoto and Matsui (1983) argue that diamond MTPs probably nucleate in the gas phase or from hydrocarbons, as long as the interaction with the substrate is weak. They argue that the precursors of diamond nuclei could be cage compounds composed of five- to seven-member carbon rings which are terminated by hydrogen atoms, as for example, pentadecane and dodecahedrane. Narayan, Srivatsa and Ravi (1989) suggest the nucleation of MTPs at the core of a perfect edge dislocation known to contain a five- and a seven-membered ring (Fig. 11). On the other hand, perfect pentagonal symmetry centres have been reported to be rare (Williams et al., 1990a; with reference to thin films: Dorignac, Delclos and Phillipp, 2001), even if they have often been identified (for example, Narayan, Srivatsa, Peters, Yokota and Ravi, 1988; Narayan et al., 1989). Pentagonal centres tend to dissociate, for example, into a quaternary and a triple junction, in order to alleviate internal stress (Romanov and Vladimiro, 1992). A typical example is the dissociated core region of a fivefold MTP analysed by Williams et al. (1990a) which extends over a region of about 6 nm size. Owing to the limited mobility of atoms and dislocations in diamond, it seems unlikely that such centres first nucleate in a perfect pentagonal mode and then dissociate by atomic rearrangement over such large areas. At least, no such mechanism has been mentioned in the literature. Diamond bonding to the substrate would be a further hurdle for large-scale atomic rearrangement. Thus, the nucleation of pentagonal particles seems more likely to be stabilised according to the scheme of energy minimisation mentioned before (Ino, 1969), without depending on the availability of a precursor with five-fold symmetry.

The hydrostatic stress distribution in a MTP has been calculated by atomistic simulation using the bond-order potential (Sherendova and Brenner, 1999). In this model, the particle is considered as periodic in [011] direction, thus forming a needle-like pentagonal crystal. The stresses are compressive at the inner region and turn to tensile at about one-third of the distance between the centre and the edge of the particles, in agreement with a prediction by mesoscale disclination theory (Romanov and Vladimiro, 1992). For a particle with a radius of 20 Å, a maximum hydrostatic tensile stress at the periphery of about 12 GPa and a hydrostatic compressive stress at the particle centre of about 17 GPa has been calculated. MTPs with radii between 6 and 30 Å, comprising up to 2900 atoms, were found to be stable. At the mesoscale, the elastic energy of the strain field associated with disclinations increases rapidly with particle size and gap angle ($\sim \omega^2 R^2$ for cylindrical elongated particles under plain strain). It is thus not surprising that in pentagonal structures regions without secondary defects are found only near the particle centre. The diameter of the defect-free zone of such a particle analysed by Narayan et al. (1989) is approximately 6 nm. Contrary to Matsumoto and Matsui (1983) who have shown pentagonal MTPs with and without the 7°20′ gap, most other papers focus on particles with closed gaps. Elastic strain, unequally distributed over the five twin units, turns out to be the dominant mechanism of gap closure studied by Narayan et al. (1989). The twin boundaries are coherent and confined to 1–2 atomic planes, and only a few stacking faults are found. Another defect which has been named *tilted twin boundary* permits to remove the crystallographic misorientation across the gap by the superposition of a low-angle boundary and a coherent twin boundary (Williams et al., 1990a). Sometimes, one single boundary may carry the entire misfit (Tamou et al., 1990; Williams et al., 1990a; Williams, Kong and Glass, 1990b). The twin units of MTPs contain numerous defects, mainly intersecting stacking faults and twins parallel to those {111} planes, which bound the respective twin unit toward its neighbours. Although these defects do not provide lattice rotations, they have been reported to reduce the lattice misfit across the gap (Tamou, Ohsawa and Kikuchi, 1990). Inspection of HRTEM micrographs with intersecting twins and stacking faults indeed reveals significant deviations from the parallel orientation of lattice planes, which originate at bond distortions in the faulted regions. The introduction of such defects is known to considerably reduce the strain energy of disclinations (Romanov and Vladimiro, 1992; Oku and Hiraga, 2000). Another common but unexplained feature of diamond MTPs is the absence of defects in bands adjacent to the major boundaries of the twin units (Tamou et al., 1990; Williams et al., 1990a). Although different in orientation, these bands remind morphological and crystallographic similarities with the formation of {100} growth sectors out from defect-free {100} sheets observed at the intersection of {111} growth sectors of single-crystalline bounded particles (Steeds et al., 1998; Section 3.1.1).

3.4.1.2. Multiple Twinning in Polycystalline Grain Growth. In TEM and SEM analysis of polycrystalline films, it has been observed that repeated or multiple twinning leads to the formation of numerous V-shaped boundary configurations (Figs. 5 and 8a). The open side of the Vs is preferentially oriented toward the grain periphery, whereas the tip of a V points back to the grain interior (Shechtman et al., 1993a; Wurzinger et al., 1995). It has been tried to trace back the course of grain growth by following the V-tips. Centres of pentagonal symmetry (quintuplet) surrounded by four $\Sigma = 3$ boundaries and one $\Sigma = 81$ boundary, for example, have been located in this way which have been interpreted as grain nucleation centres. Only one such centre per grain family would be expected. However, several centres with pentagonal symmetry are frequently observed within a single grain (Dorignac et al., 2001), rendering this interpretation questionable. On the other hand, V-shaped boundary configurations are important for the morphological development of grain families. It has been mentioned before (Section 2.2) that V-shaped facets with an opening angle of 109°28′ constitute re-entrant corners favouring the nucleation of new carbon layers. Once a grain has succeeded, via multiple twinning, to be surrounded by Vs open toward the periphery, it can grow in an approximately equiaxed shape which minimises the grain surface and allows the grain to overcome the limitations imposed by the anisotropy of faceted growth. V-shaped grain boundaries thus may be considered as stabilising elements for the formation of grain families.

3.4.1.3. Geometrical Features of ⟨110⟩ Tilt Grain Boundaries. The formation of grain families by multiple twinning is very similar to the growth of MTPs. It is convenient to use the CSL notation to designate the orientations of grains belonging to the same family. Some structural features typical for a grain family with trigonal and pentagonal symmetry centres are sketched in Figure 17. The zone axis of the grains, which again is [011], represents a tilt axis perpendicular to the image plane. The contours of the rhombohedra taken as structural units in Figure 17 denote edge-on {111} planes which can also be considered as free-standing facets. First-order twinning of crystallite p, defined here as the parent phase, leads to a coherent Σ 3 $(11\bar{1})$ twin boundary

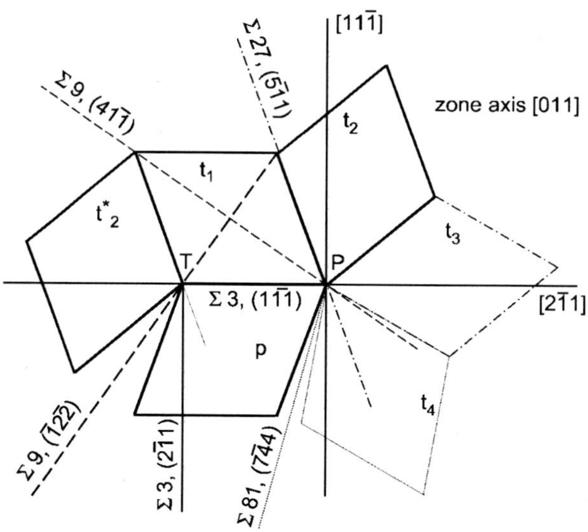

FIG. 17. Schematic representation of a grain family in plan view. The original crystal unit is p, which is bounded by 4-edge-on {111} facets or planes. All other units are obtained by subsequent twinning. See text for the designation of boundaries.

separating p and t_1. The boundary coincides with the twin plane (Section 2.2.2) and represents the median plane with respect to the {111} planes or facets not involved in twin plane formation. A second Σ 3 boundary parallel to $(2\bar{1}1)$ and perpendicular to the aforementioned one corresponds to the median plane which pertains to the complementary angle subtended by the same facets (at point T in Figure 17). Σ 3 {211} boundaries are no longer coherence. If a microtwin is embedded in a defect-free parent phase, the interfaces, which are not twin planes typically are of this kind (see Figure 6 in Takeuchi et al., 2001b). Second-order twinning adds the twin units t_2 and t_2^*, the lattices of which exhibit a Σ 9 relationship with p. Depending on how the free-standing facets of t_2 and t_2^* (heavily drawn) intersect the free-standing facets of p (also heavily drawn), they subtend either an obtuse or an acute angle. Under symmetrical growth conditions, the resulting Σ 9 boundary will then be parallel to $(41\bar{1})$ or $(\bar{1}2\bar{2})$, respectively. Both boundary planes bisect the angles subtended by the free-standing facets and they are again perpendicular to each other. Similarly, the addition of twin t_3 leads to a Σ 27 relationship with p, and the corresponding symmetrical boundary along the median plane is of the Σ 27 $(51\bar{1})$ type. Twin t_4 results in a Σ 81 relationship with a boundary plane parallel to $(7\bar{4}4)$. Table III summarises the geometric elements of these boundaries which are symmetric tilt boundaries because the tilt angle of the lattices on both sides of the boundary is identical. Instead of relating tilt boundary formation to repeated twinning, boundaries frequently are considered as the product of grain boundary reactions. Thus, Σ 9 boundaries become the reaction product of two Σ 3 boundaries at a triple junction. If two Σ 3 {111} boundaries meet under an acute (70°32′) or obtuse (109°28′) angle, they form either a Σ 9 $\{41\bar{1}\}$ or Σ 9 $\{\bar{1}2\bar{2}\}$ boundary, respectively. Similarly, Σ 27 boundaries originate from three Σ 3 boundaries at a quadruple junction or from one Σ 3 and one Σ 9 boundary at a triple junction. Σ 81 can exist at the

TABLE III

SYMMETRIC TILT BOUNDARIES OBTAINED BY MULTIPLE TWINNING

Twin order, *n*	Lattice rotation (angle between equivalent {111} planes)	CSL notation	Symmetric tilt boundary (between parent and *n*th twin)	Angle between the facets	Complem. tilt boundary	Angle between the facets
1	70°32′	$\Sigma = 3$	$(11\bar{1})$	141°03′	$(2\bar{1}1)$	38°57′
2	180° − 2 × 70°32′ = 38°57′	$\Sigma = 9$	$(41\bar{1})$	148°25′	$(\bar{1}2\bar{2})$	31°35′
3	3 × 70°32′ − 180° = 31°35′	$\Sigma = 27$	$(5\bar{1}1)$	77°52′	$(\bar{2}\bar{5}5)$	102°08′
4	360° − 4 × 70°32′ = 77°52′	$\Sigma = 81$	$(7\bar{4}4)$	7°20′	$(87\bar{7})$	172°40′

The indices of the parent crystallite are used for indexation. See text.

intersection of four Σ 3, two Σ 9, or one Σ 27 and one Σ 3 boundary. There is no reason why boundaries of higher orders than $\Sigma = 81$ should not be formed, but they have simply not been observed so far (Shechtman et al., 1993b). An example for the structural complexity of grain families is given in Figure 22, where a pentagonal symmetry centre is embedded in an environment plenty of twins and faults.

It has been repeatedly reported that ⟨011⟩ tilt boundaries prevail in diamond films and that Σ 3 {111} boundaries are predominant within this category (Narayan and Nandedkar, 1991; Luyten et al., 1992; Shechtman et al., 1993b; Zhang, Ichinose, Ito, Ishida and Nakanose, 1996; Steeds et al., 2002). For boron-containing films, the relative frequency of twin boundaries amounts to approximately 65% for Σ 3, 30% for Σ 9 and not more than 5% for Σ 27 boundaries (Mora, Steeds and Butler, 2002). Correspondingly, the large majority of the numerous HRTEM investigations of stacking faults and multiple-order tilt twin boundaries have been performed using diamond films close to ⟨011⟩ orientations (Narayan, 1990; Luyten et al., 1992; Shechtman et al., 1993a–c; Wurzinger et al., 1993, 1995). The predominance of the Σ 3 {111} type among flat boundaries can be explained by the particularly low interfacial energy and the small activation energy required for twin nuclei to flip between stair and boat conformations (Section 2.2.2.2). Their formation thus may occur as a growth accident leading to twinning, but also result from the splitting of higher order boundaries into energetically more favourable configurations (Narayan, 1990; Steeds et al., 2002). Shechtman (1993c) has pointed out that the boundaries of twinned structures can be arranged into two categories, planar boundaries and non-planar ones that rarely lie on well-defined crystallographic planes and whose orientations may vary locally. While $\Sigma = 9$, 27 or 81 boundaries frequently belong to this latter type, also Σ 3 related lattices and even crystals of identical orientation may form non-planar boundaries (Fig. 18). Such deviations from planarity have been associated with the lack of atomic mobility at deposition temperature impeding the rearrangement of boundaries into low energy configurations.

In principle, one would expect flat grain boundaries in diamond because of the high surface energy. Indeed curved boundaries between grains of the same family or between neighbouring families sometimes tend to polygonise into straight boundary segments (Ichinose and Nakanose, 1998; Steeds et al., 2002). Owing to the restricted mobility of carbon atoms in the bulk, one would expect that flattening of boundaries is facilitated at free surfaces where surface reconstruction possibly is supported by reactions with the gas atmosphere.

3.4.1.4. Atomistic Structure of ⟨110⟩ Tilt Boundaries. Our present knowledge concerning atomic grain boundary structures in diamond derives from the combined application of HRTEM, atomistic modelling and EELS. In ⟨011⟩ projections, commonly used for imaging of the atomic columns in HRTEM, the projected distance of the atoms belonging to narrowly spaced {111} layers

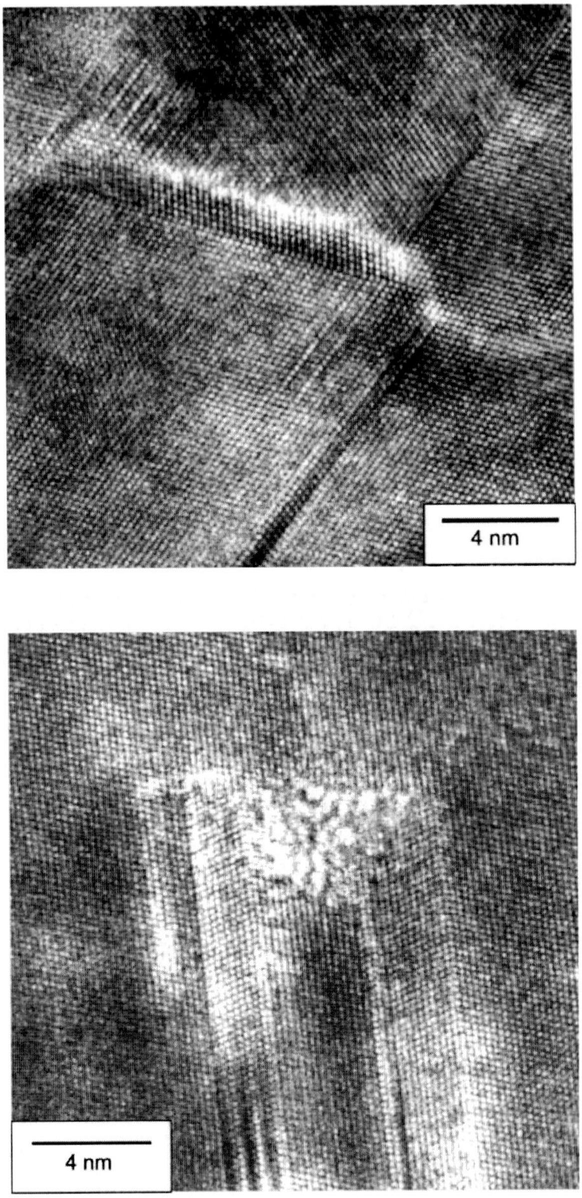

FIG. 18. Tilt boundaries in twinned structures. (a) Contrary to the coherent $\Sigma\,3$ {111} boundaries which are flat, the $\Sigma\,9$ boundary does not follow a well-defined crystallographic plane; (b) a mixture of parallel stacking faults and twin boundaries which all terminate on an amorphous inclusion, even though the central part of the faulted region is perfectly aligned with the surrounding matrix; according to the orientation relationships in this grain, a simple $\Sigma\,3$ {211} boundary could provide the accommodation with the matrix.

is 0.089 nm (corresponding to the projected distance of the encircled atoms in double layer cA of Figure 4a). Since these atom pairs cannot be separated, even in using microscopes with ultra-high resolution, they appear as black or white dots in HRTEM micrographs (Delclos, Dorignac, Philipp, Moulin and Bonnot, 1999; W). Atomistic modelling provides complementary information about coordination numbers, dangling bonds and electronic states which cannot be obtained from the ⟨011⟩ HRTEM projections of tilt boundaries.

The earliest geometric modelling of ⟨011⟩ tilt boundaries was done by Kohn (1958) and Hornstra (1959). The structural motifs were 5-, 6- and 7-member rings of carbon atoms which in the simplest case of a coherent Σ 3 {111} boundary reduced to six-member rings, as in Figure 4. It was pointed out by Hornstra that models without dangling bonds can be obtained for tilt angles smaller than 70°32′ whereas beyond, all models contain dangling bonds. In order to satisfy the sp^3 bonding requirements, Papon and Petit (1985) introduced atomic patterns resulting in double periodicity along the tilt axis. For atomistic grain boundary modelling, both TB and empirical potentials were used. Some of the results for ⟨011⟩ tilt boundaries are compiled in Table IV.

Coherent Σ 3 {111} boundaries are commonly observed in HRTEM (see boundaries t6, t7, and t8 in Figure 22). Several possible configurations have been proposed for Σ 3 {112} boundaries, none of which is coherent (Table IV). In the lowest energy configuration, the boundary atoms are not in a mirror plane position (type pm in Table IV) and the dangling bond associated with the threefold-bonded atoms is eliminated by a 1×2 reconstruction in ⟨011⟩ direction which gives rise to the just mentioned double periodicity along the tilt axis (Kohyama, Ichinose, Ishida and Nakanose, 1996; Morris et al., 1996). Although 1×2 reconstruction is common in germanium and silicon (D'Anterroches and Bourret, 1984), it has not been confirmed in diamond. Type pm boundaries have been observed to bordes twin platelets in non-coherent particles which sometimes are induded in homoepitaxial films (Fig. 6 in Takeuchi et al., 2001b). It cammot be decided however if 1×2 recoustrctional has occurred because of the electron beam orientation parallel to the direction of reconstruction. A Σ 3 {112} boundary with mirror plane symmetry examined by HRTEM is shown in Figure 19 together with two models for the atomic structure. The 1×1 structure (type *p2′mm′*, 1×1 in Table IV), which is higher in energy than the previous one, contains deep electronic states in the band gap. These would be removed in case of a 1×2 reconstruction leading to double periodicity in [011] direction (Kohyama et al., 1996; Morris et al., 1996). Despite large bond distortion due to the 1×2 reconstruction, the energy associated with the three-coordinated sites would be greatly reduced. Experimentally, Σ 3 {112} boundaries have revealed 1s-π^* signals in EELS which have been attributed to a sp^2-like hybridisation of the carbon atoms at three-coordinated sites (Ichinose and Nakanose, 1998; Takeuchi et al., 2001a–c). On the other hand, it has been pointed out that the position of the 1s-π^* edge compared to the 1s-σ^* edge also corresponds to the relative position of unoccupied mid-gap states with respect to the σ^* edge of the conduction band

TABLE IV

CHARACTERISTICS OF DIAMOND ⟨110⟩ TILT BOUNDARIES ACCORDING TO ATOMISTIC MODELLING

	E_{gb} (J m^{-2})	Bond-length distortion[(2)]	Bond-angle distortion[(2)]	Remarks	Band gap states	EELS[a]
Σ 3 {111}	0.12[b]					1s-σ^*
Σ 3 {211}				Without rigid body shift of hemi-crystals		
$p2'mm'(1 \times 1)$	6.02[c], 5.94[d]	−3.5%, −6.3%, −7.7%	−5.4 to 5.6°	With dangling bonds	→ occupied states: 2.9–3.2 eV[e],	1s-σ^*, 1s-π^*
↑		→(3 coordinated site)	(3 coord. site)			
mirror plane symmetry		−3 to 2.3%	−12.7 to 19.5°		→ unoccupied states: 3.3–3.4 eV[e]	
↓						
$p2'mm'(1 \times 2)$						
	3.31[c], 3.04[d]	−5.3 to 8.3%	−16.5 to 24.4°	⟨011⟩ reconstruction of dangling bonds, common in Si, Ge		
Σ 3 {211}				Including rigid body shift of hemi-crystals		1s-σ^*
pm (1 × 1)	5.37[c]			→ with dangling bonds		
pm (1 × 2)	2.43[d]			→ without dangling bonds		
Σ 9 {122}						
Glide plane model (GP)	1.68[c], 1.89[d]	−3.7 to 2.7%	−15.0 to 19.0°	→ without dangling bonds		
Mirror plane model (MP)	3.95[c], 4.02[d]	−7.1 to 8.8%	−18.6 to 26.1°			

Σ 9 {114}	2.73[f]	4 dangling bonds per structural unit composed of a sequence of 6-, 6-, 6-, 7-, 5-member rings	1s-σ^*, 1s-π^*
Σ 9 {115}/{111}			1s-σ^*, 1s-π^*
Σ 27 {511}	2.32[g], 4.0[g], 3.35[h]		

[a]D. Takeuchi, H. Watanabe, S. Yamanaka, K. Okushi, H. Sawada, H. Ichinose, T. Sekiguchi, K. Kajimura; *Phys. Rev. B* 63 (2001) 245328 1–7.
[b]According to Section 2.2.2. Numerical modelling is inappropriate for such small boundary energies, and contributions from second-next nearest neighbours are not taken into account in modelling.
[c]Tight-binding model: M.Kohyama, H. Ichinose, Y. Ishida, M. Nakanose; *Mater. Sci. Forum*, 207–209 (1996) 261–264.
[d]Tight-binding model: J. R. Morris, C. L. Fu, K. M. Ho; *Phys. Rev. B* 54 (1996) 132–138.
[e]In this tight-binding simulation by Kohyama et al. [b], the band gap is 6.3 eV.
[f]Analytical bond-order potential: O. A. Sherendova, D. W. Brenner; *Phys. Rev. B* 60 (1999) 7053–7061.
[g]A. A. Nazarov, O. A. Shenderova, D. W. Brenner; *Mater. Sci. Engng. A* 281 (2000) 148–155: The boundaries were modelled using disclination dipole walls as structural units. The difference in boundary energies resulted from two different approaches to calculate the core energies of the disclination dipole walls. The higher value was obtained using Tersoff's analytical potential, the lower value from a tight-binding simulation.
[h]Using an analytic many-body bond-order potential: O. A. Sherendova, D. W. Brenner; *Phys. Rev. B* 60 (1999) 7053–7061.

in diamond, as those in the (1 × 1) model (Kohyama et al., 1996). At first glance, this interpretation contradicts the former one because a π-bond network has no band gap. Additional information for solving the puzzle has been obtained from the observation of band-A luminescence inside non-epitaxial crystallites in otherwise homoepitaxial films, for which the existence of a band gap is prerequisite. As these particles exhibit non-coherent Σ 3 {112} boundaries, it has been concluded that the defects at the origin of the EELS signals include indeed sp^2-like hybridisation, but the defects must be surrounded by a network of sp^3-hybridised bonds (Takeuchi et al., 2001a–c). Such a model is in satisfying agreement with the grain boundary structure in Figure 19, if it is assumed that the dangling bonds located at the three-coordinated sites reconstruct into π bonding.

A variety of different atomic structures exists also for Σ 9 boundaries. The structural units of the Σ 9 {221} boundaries listed in Table IV are alternating 5- and 7- member carbon rings without dangling bonds whereas {114} and {111}/{115} boundaries consist of sequences of 5-, 6-, and 7-member rings which leave numerous dangling bonds at three-coordinated sites. The presence of 1s-π^* EELS signals indicates that the bonds at these sites have adopted a sp^2-like state, as in case of Σ 3 {112} boundaries. All Σ 9 boundary types noted in Table IV have been identified by HRTEM but it has been repeatedly pointed out that Σ 9 {114} boundaries are much more frequent than the low-energy {221} type (Zhang et al., 1996; Ichinose and Nakanose, 1998). In silicon and germanium, the latter boundary type dominates (Bourret, 1986). Σ 9 {111}/{115} boundaries are asymmetric and their formation seems to be favoured when Σ 3 {111} boundaries meet at an obtuse angle (Section 3.4.1.5).

Detailed structural models of grain boundaries resulting from higher-order twinning in diamond have not been published to the author's knowledge. Their frequency of occurrence decreases with increasing twin-order, as mentioned before. In addition, Σ 27 and Σ 81 boundaries are relatively short (Shechtman et al., 1993b). In order to calculate the energies of symmetric ⟨011⟩ tilt boundaries with misorientations varying between 0 and 70°32′, the boundaries have been modelled as standard small angle boundaries, to which dislocations have been added (Morris et al., 1996), or by using structural units consisting of combinations of differently configured 5- and 7-member carbon rings which allow to tackle boundaries of greater misorientation (Shenderova et al., 1998; Sherendova and Brenner, 1999b; Nazarov, Shenderova and Brenner, 2000a,b). Although the absolute energies calculated by using TB models are significantly smaller than by applying analytic atomic potentials, their relative variation with misorientation is very similar. Analytic many-body bond-order potentials which are based on Tersoff's potential, provide values in between (Sherendova and Brenner, 1999). If the structural units are considered as disclination dipole walls, the boundary energies can even be calculated as a continuous function of grain misorientation (Nazarov et al., 2000a). It turns out from these calculations that the energy of low-angle boundaries rises steeply with growing misorientation to reach a maximum value at a tilt angle of about 20°, corresponding to energies of

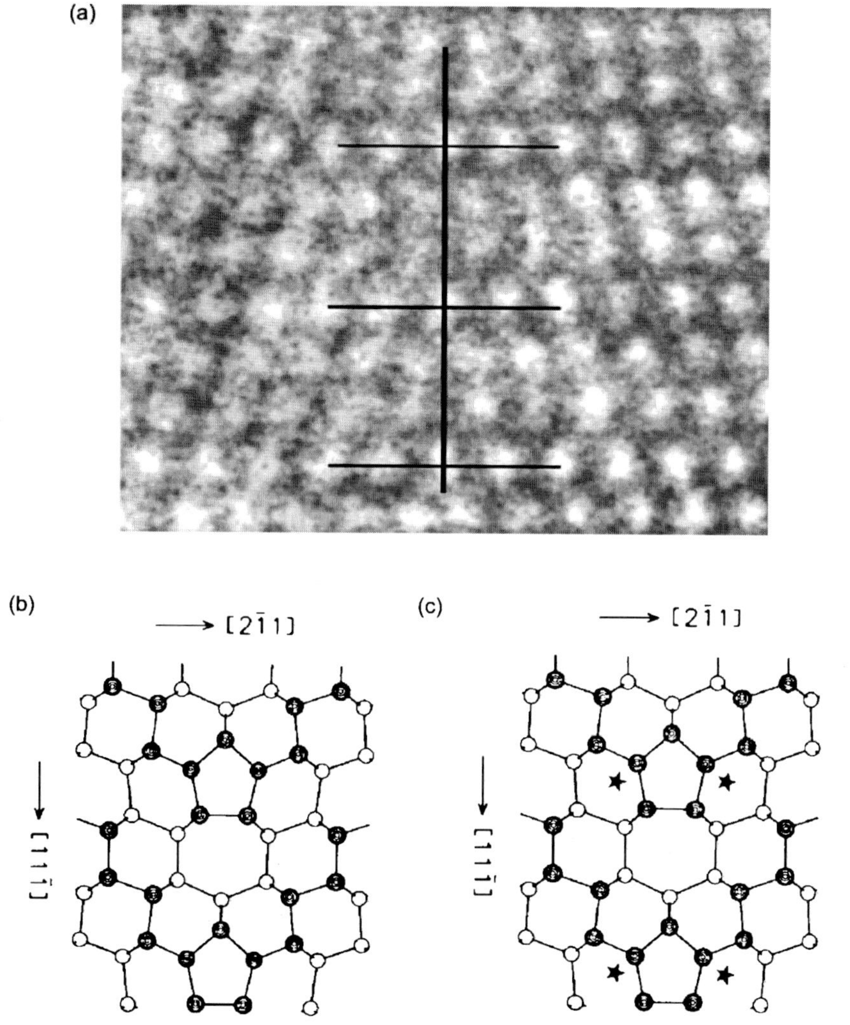

FIG. 19. Σ 3 (112) boundary with carbon atoms in mirror plane position. (a) HRTEM micrograph by H. Ichinose, and M. Nakanose (1998), *Thin Solid Films* 319, 87–91. Atomistic models of relaxed configurations of the Σ 3 (112) boundary: (b) $p2'mm'$ (1 × 2) model and (c) the $p2'mm'$ (1 × 1) model, stars indicate three-coordinated sites (according to M. Kohyama, H. Ichinose, Y. Ishida, and M. Nakanose (1996), *Mater. Sci. Forum* 207–209, 261–264.

about 2.64 J m^{-2} in TB modelling, and 4 J m^{-2} if Tersoff potentials are used (Morris et al., 1996; Nazarov et al., 2000a). Beyond 20°, the boundary energies decrease again to reach deep cusps at the coincidence orientations Σ 19 ($\Theta = 26°32'$; 2.04 J m^{-2}) and Σ 9 ($\Theta = 38°57'$; 1.88 J m^{-2}) (Sherendova and Brenner, 1999b; Nazarov et al., 2000a).

3.4.1.5. Misfit Compensation at Intersections of Stacking Faults, Twin and Grain Boundaries. So far, grain boundaries and stacking faults have been considered as periodic interfaces infinite in extent. Owing to the profusion of interfaces commonly encountered in diamond films, the interface lengths are, however, restricted to the spacing of intersections and junctions of boundaries and stacking faults. Only a few of the numerous possible structural variants have been analysed at the atomistic scale and described in the literature. These studies always refer to reactions between interfaces belonging to the same ⟨110⟩ zone, an exception being stacking fault tetrahedra on octahedral facets of near ⟨001⟩ oriented films discussed in Section 3.1.1. The volumes bounded by such intersecting interfaces constitute the building blocks of the diamond film. Their contiguity along intersections and at junctions is subjected to the constraints exerted by the accommodation of the interfacial atomic structures. Accordingly, the triple junctions of boundaries belonging to grains of the same family, that is to the same zone, and sometimes also the boundaries themselves have been identified as the sites of intense stress formation where various kinds of defects originate which may propagate far into the adjoining grains (Mora et al., 2002; Steeds et al., 2002). Faults on {111} planes which are not part of the dominant ⟨011⟩ zone seem to belong to this category of defects. This view does not ignore the importance of planar faults belonging to the dominant zone for local stress relief. The role of grain boundary junctions becomes particularly important for nano-crystalline films because their spacing is very small. Tilt boundaries are, however, not common in nano-crystalline diamond. The following discussion thus does not directly refer to this kind of films.

When two coherent $\Sigma 3$ {111} boundaries meet under an acute angle (70°32′), they readily form a $\Sigma 9$ {114} boundary with structural units consisting of 5-, 6- and 7- member rings. Such boundaries have been frequently observed by HRTEM (Zhang et al., 1996; Ichinose and Nakanose, 1998). The atoms on both sides of the boundary are in mirror position. Without reconstruction, the boundary contains dangling bonds (Table IV, Fig. 20a). If the Σ 3 {111} boundaries meet under an obtuse angle (109°48′), however, the bisecting plane is no longer a mirror plane and the atoms are in asymmetric position (Zhang et al., 1996; Fig. 20b). The boundary would not lie on the bisecting plane unless a rigid translation in ⟨110⟩ (normal to the image plane in Figure 20b) or ⟨411⟩ (perpendicular to the {221} boundary) of one of twins would occur. This translation would introduce dislocations at the triple junction or lead to distortion of the equilibrium, low-energy Σ 3 {111} boundaries. Zhang et al. (1996), therefore, have proposed two alternative structural arrangements where the two Σ 3 {111} boundaries terminate in a single boat-shaped 6-membered ring which carries a dangling bond (triple point in Figure 20c and d). In one case (Fig. 20c), the resulting boundary continues as an asymmetrical $\Sigma 9$ {111}/{115} boundary with a structural unit composed of a sequence of 7-, 5-, and 6-membered rings containing three dangling bonds. The structure of extended boundaries on this plane has been investigated by Chen and Chen (1993). These asymmetrical

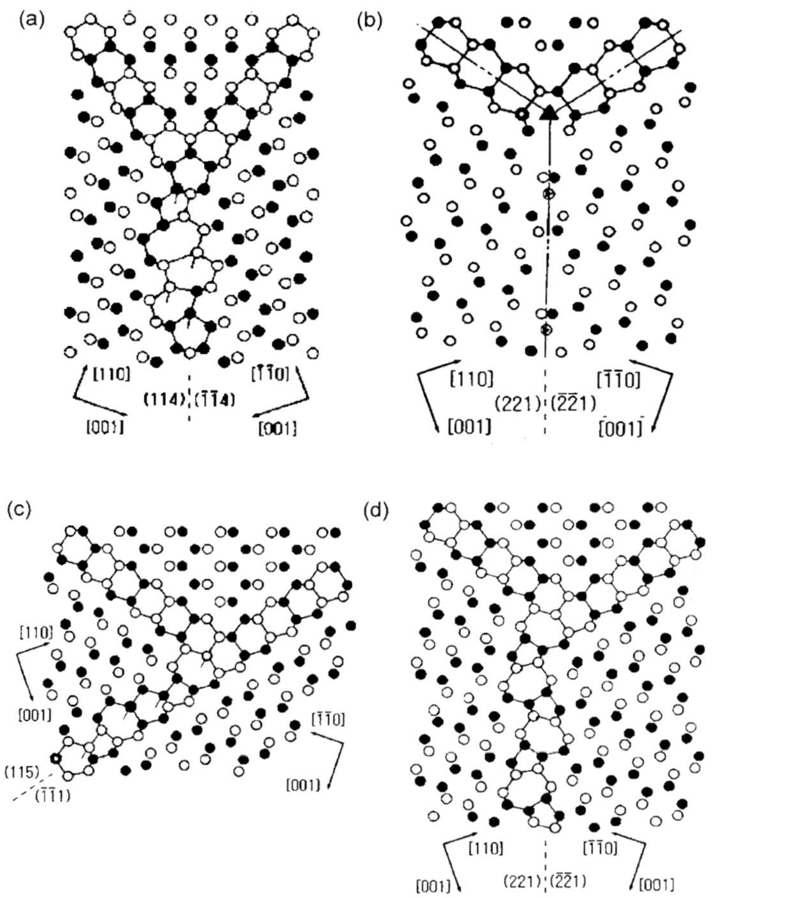

FIG. 20. Geometric models for Σ 9 boundaries generated at the junction of Σ 3 {111} boundaries (after Y. Zhang, H. Ichinose, K. Ito, Y. Ishida, and M. Nakanose (1996) *Mater. Sci. Forum* 204–206, 207–214). (a) A Σ 9 {114} boundary is formed if two Σ 3 {111} boundaries meet under an acute angle (70°32′). (b) The junction of the same boundary type under an obtuse angle (109°48′) leads to a Σ 9 {221} boundary with asymmetric atomic arrangement. (c) Σ 9 {115}/{111} boundary resulting from the junction of Σ 3 {111} boundaries under an obtuse angle. (d) A Σ 9 {221} boundary shifted out of the centre line to avoid the atomic arrangement of (b). At the triple point, the boundary begins like in (c) and then changes its orientation.

boundaries have been observed in undoped and boron-doped diamond films (Zhang et al., 1996; Mora et al., 2002). They are involved in the formation of penetration twins (Fig. 6d; Takami et al., 1999). In case of boron-doping, arrays of extended nano- and microtwins are generated on both sides of this boundary type (Mora et al., 2002). It is not known if the extended twin defects are stress-induced, or if they permit perhaps the reconstruction of dangling bonds.

In the second case of Σ 3 {111} boundary junctions (Fig. 20d), the Σ 9 boundary is parallel to {221} but it is displaced out of the median plane. The first structural unit adjoining the triple junction consists of 7- and 5-membered carbon rings, as for the {111}/{115} type. The boundary then turns parallel to {221} and adopts the minimum-energy configuration, which consists of grouped five- end seven-sided rings in zig-zag configuration. Such Σ 9 {221} boundaries with minimal distortion and without dangling bonds have been observed by TEM. They have also been associated with very high local stress giving rise to a dense stack of nanotwins on either side of the boundary (Mora et al., 2002).

The general problem of two first-order twins meeting at an obtuse angle of 109°28′ has been studied in much detail by King, Chen, Chang and Kai (1997). This simple configuration which is shown schematically in Figure 21, has been chosen within a larger defective region and is frequently met in CVD diamond. Two first-order twin platelets meet under an obtuse angle to form a Σ 9 boundary. The boundary junction at the lower triple point is equivalent to the one discussed just before: For the Σ 9 {221} boundary to adopt the minimum-energy configuration, a displacement, t, is required. As the twins adhere to the matrix via Σ 3 {111} boundaries, the displacement could be created by the addition of a dislocation dipole with a Burgers vector equal to t at one triple junction and $-t$ at the other. An important effect accompanying the removal of the Σ 9 boundary from its actual position is the addition of nano-facets. These correspond to the 5- and 7-membered ring of a beginning Σ 9 {111}/{115} boundary, like in Figure 20d.

Up to this point, only twins of equal thickness have been considered which produce zero lattice displacement on both sides of the twin. When intrinsic or extrinsic stacking faults, which represent twin plates of one or two double-layer thickness, are created in a defect-free crystal (Fig. 4), the matrix on both sides of the twin plates is displaced by $a/6$ $[\bar{1}\bar{1}2]$ and $a/6$ $[11\bar{2}]$, respectively. The displacements are equivalent to Shockley partial dislocations. Only if a third layer is added, the displacement becomes zero. Thus, the relative matrix

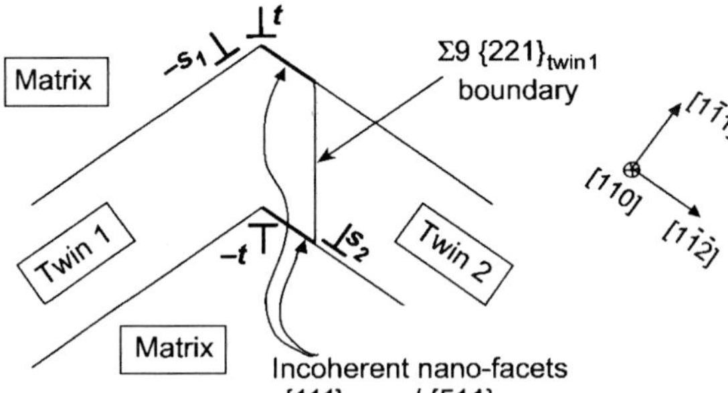

FIG. 21. Accommodation of displacements and translations at the junctions of twins of arbitrary thickness (after A. H. King, F.-R. Chen, L. Chang, and J. J. Kai (1997) *Interface Sci.* 5, 287–303). The crystallographic axes refer to the matrix.

TABLE V

LATTICE DISPLACEMENTS ACROSS TWINS OF VARYING THICKNESS (ACCORDING TO KING ET AL., 1997)

Number of {111} layers	Displacement of the matrix below twin 1 relative to above the twin	Displacement of the matrix below twin 2 relative to above the twin
$3N - 1$	$a/6\ [\bar{1}1\bar{2}]$	$a/6\ [\bar{1}12]$
$3N$	Zero	Zero
$3N + 1$	$a/6\ [1\bar{1}2]$	$a/6\ [1\bar{1}\bar{2}]$

displacement varies periodically with the thickness of the twin plate, as indicated in Table V. where the indices refer to the [110] projection of Figure 4. Consequently, if the thickness of the twin plates in Figure 21 does not represent a multiple of $3N$ double-layers, further dislocations must be added to correct for the matrix displacement. The energy of the system then comprises the energies of the various minimum-energy boundaries, plus that of the dislocation dipole, that of the asymmetric nano-facets and that of the dislocations associated with the matrix displacement, as indicated schematically in Figure 21. Further energy contributions may be expected from the core structure at triple junctions which will be distorted by greater or lesser amounts if the core structures of the joining grain boundaries are not in a low-energy configuration.

The hydrostatic stress distribution for an identical boundary configuration has been calculated by Sherendova and Brenner (1999) using a many-body bond-order potential. The result comes in many respects very close to the schematics given in Figure 21. Both approaches confer a realistic view of the stress problems associated with structural contiguity in twinned diamond. It should be noted that, even for such a simple defect configuration, there are hundreds of different atomic arrangements to accommodate the misfit at the boundary junctions (King et al., 1997). Thus, a high degree of convergence in the explanation of such structures has been reached, defying the great complexity of the problem.

Still more complex extended defect configurations have been analysed by matching HRTEM with three-dimensional computer modelling (Delclos et al., 1999; Delclos et al., 2000; Dorignac et al., 2001). In these calculations, upper limits for the distortion of bond lengths and angles have been fixed which have been taken from *first principles* studies of the atomic arrangement in dislocation cores. Excellent agreement has been reached between the HRTEM images and the computer simulated images (Fig. 22). By using explicit atomic models, it has been demonstrated that the intersection of an intrinsic and extrinsic stacking fault leads to the formation of an edge dislocation called ‘stair-rod’ which is immobile and fixes the stacking faults in their position (Delclos et al., 1999). Additionally, pentagonal twin centres exhibiting 5-member carbon rings at the junction of (a) two $\Sigma 3$ {111} boundaries, one $\Sigma 9$ {221} and an intrinsic stacking fault, (b) three

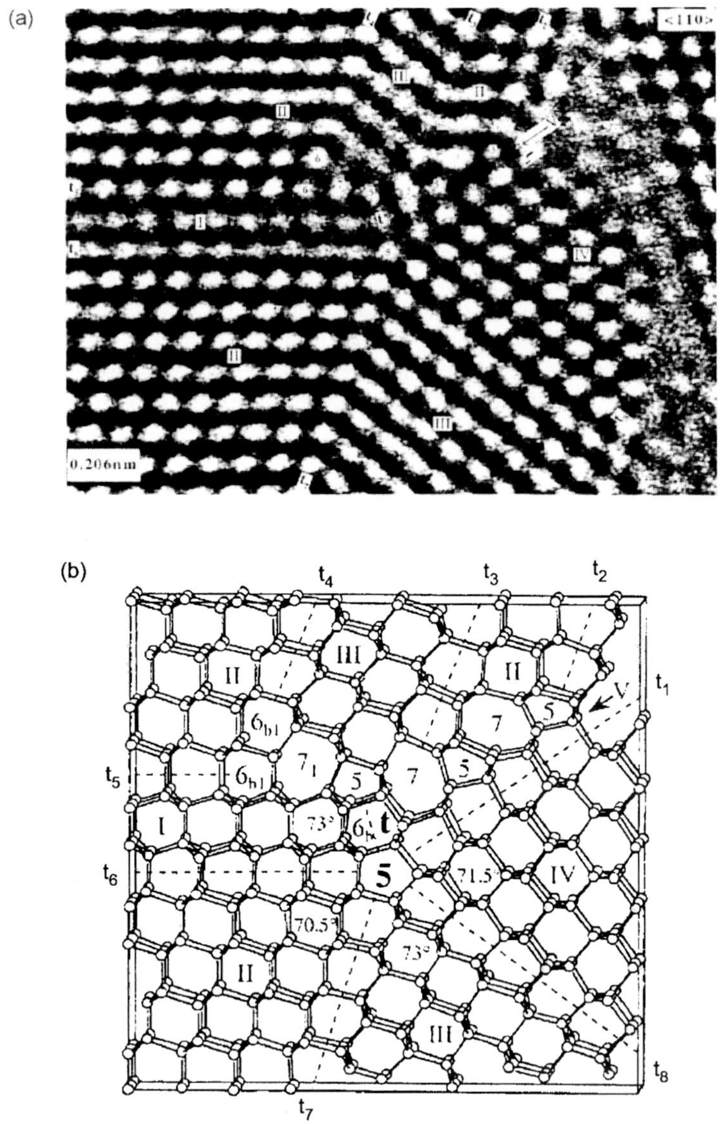

FIG. 22. Nearly perfect fivefold twin centre (5) located in a highly defective atomic environment (a). A three-dimensional atomic model without any dangling bond has been obtained via image matching by computer simulation (b). The pentagonal centre is formed at the intersection of five first-order twin interfaces t_1, t, t_6, t_7, and t_8. The angular mismatch is compensated by elastic distortion of the twins and by the incorporation of defects. Two short segments of higher-order boundaries are revealed : $\Sigma 9$ {122} between the regions III and I, and $\Sigma 27$ {255} between the regions V and II (after D. Dorignac, S. Delclos, and F. Phillipp (2001) *Philos. Mag. B* 81, 1879–1891; Taylor & Francis Ltd., www.tandf.co.uk/journals).

$\Sigma 3$ {111} boundaries, and one two $\Sigma 27$ $\{111\}_{matrix}/\{110\}_{twin}$ boundary, (c) five Σ 3 {111} boundaries have been analysed together with the surrounding structure. Type (a) has delocalised π bonds. The authors emphasise the frequent occurrence of amorphous regions in the proximity of pentagonal centres. Most commonly, pentagonal centres are dissociated into lower-order junctions, thereby reducing the strain field energy of the disclination associated with the pentagonal centre. At the periphery of large grain regions with pentagonal symmetry, radial cracks have been observed (Steeds et al., 2002), which can be explained by the high tensile stress also found at the periphery of MTPs (Sherendova and Brenner, 1999).

Theoretical papers frequently make use of the lowest-energy criterion to justify the choice of certain geometrical models. However, lowest-energy configurations are rarely achieved in CVD diamond where kinetic processes are extremely slow, and diffusion controlled mechanisms of stress relaxation are likely to be confined to regions near the growth front (Section 3.2.3). King et al. (1997) have pointed out that the overall structural incompatibility at junctions and interfaces depends on the translation of the joining components and the position of the interfaces itself. The accommodation of each component of translational incompatibility interacts with all other accommodation components. As a result of the interaction, the interfaces are supposed to adopt structures with different rigid translations, which do not yield the minimum energy configuration known for boundaries of infinite extent. For such an 'incomplete' translation to occur, the energy increase due to having the 'wrong' boundary structure must be smaller than the energy increase due to the strain field of the accommodating dislocation. Depending on the grain size, three different regimes of accommodation are envisaged (King et al., 1997):

Regime 1 In large grained materials, the boundary structures can be predicted by 'infinite-boundary' models, and misfits are accommodated by dislocations at the grain boundary junctions.

Regime 2 In fine-grained materials in which some boundaries may have 'non-infinite' structures or only partially relax as determined by the interactions with other boundaries, energy minimisation derives from the reduction of strain fields that would have been at the origin of junction dislocations in regime 1.

Regime 3 In ultrafine-grained materials, the boundary structure is controlled by the minimisation of atomic misfit at triple junctions.

3.4.2. General Grain Boundaries

By far most of our present knowledge on high-energy boundaries stems from atomistic modelling (Keblinski, Wolf, Phillpot and Gleiter, 1998a; Keblinski et al., 1998b; Cleri, Keblinski, Colombo, Wolf and Phillpot, 1999; Keblinski, Phillpot, Wolf and Gleiter, 1999; Zapol et al., 2001). The cubic {100} planes in the diamond lattice are considered representative of all grain boundary planes other than {111} and {110}. Exactly half of the atomic bonds involve partners on the opposite side of the interface. By contrast, only 25% of the nearest-neighbour bonds cross the interface in {111} and {110} planes. Twist boundaries are chosen to introduce structural disorder. For the simulation of 'infinite' boundaries, periodic three-dimensional cells are used. Their size represents a compromise between the requirements for the description of a general high-angle boundary and the need of remaining computationally tractable. Its size parallel to the boundary plane is given by the CSL notation corresponding to a chosen twist angle. For example, Σ 29 signifies 29 atoms in the planar unit cell. Perpendicular to the boundary plane, two boundaries per repeating cell are used to attain periodicity.

Keblinski et al. using Tersoff interatomic potentials have investigated a Σ 29 {100} twist boundary as representative of an interface between high-energy planes (twist angle $\Phi = 43.6°$), and a Σ 31 {111} boundary (twist angle 42.10°) representative of a interface between low-energy planes. The calculations were done both for diamond and silicon because it was expected that silicon would accommodate structural disorder across the boundaries by a distortion of the tetrahedral bonds while maintaining sp^3-hybridisation, whereas diamond with its greater bond stiffness would change to a sp^2-hybridised state in case of strong bond distortion. The Σ 29 {100} twist boundary was reconsidered by Ceri et al. who used a tight-bonding molecular dynamics simulation in order to get insight into the electrical activity of the boundary. This type of simulation was also applied by Zapol et al. who extended the scope of investigation to other {100} twist boundaries [Σ 5 (2 × 2), $\Phi = 53.1°$; Σ 13, $\Phi = 67.4°$; Σ 29, $\Phi = 43.6°$]. They tackled notably the problem of incorporating hydrogen, nitrogen and silicon impurities in such boundaries. In all of these simulations, the models were allowed to relax at high temperature, corresponding to a high-temperature anneal, prior to being cooled to zero-temperature. This operation led a to significant decrease of grain boundary energy because the number of broken bonds was reduced.

Twist boundaries between {100} planes display a high degree of structural disorder, much stronger than those between {111} planes (Fig. 23). The disorder essentially is confined to the first atomic layer on both sides of the boundary. The values for grain boundary energies are located in the range 7.8–7.96 J m^{-2} which is much higher than for most symmetric tilt boundaries (Zapol et al. 2001). The energy for the {111} boundary is lower by 30% and matches the upper energy range for symmetric tilt boundaries. The two planes facing the boundary contain a high number of threefold coordinated atoms. Depending on the twist angle and the annealing temperature, their fraction ranges from 23 to 35%. Increasing annealing temperature reduces the number of atoms with threefold coordination. Threefold coordinated atoms with planar configuration and bond angles close to 120° are supposed to form π bonds. They constitute 30–75% of the group with threefold coordination, the remaining ones have three bonds in tetrahedral configuration, as typical for carbon atoms with dangling bonds. In {111}

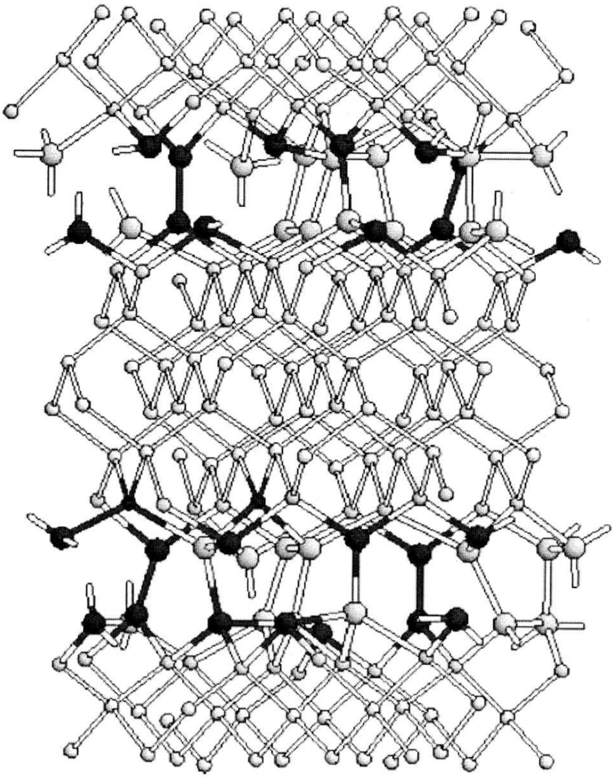

FIG. 23. Simulated Σ 13 {100} twist boundary (twist angle 67.4°) in a periodic cell with 208 atoms after relaxation. Black atoms are three-coordinated and gray atoms are four-coordinated. The larger spheres represent atoms in the first interface layers (from P. Zapol, M. Sternberg, L. A. Curtiss, T. Frauenheim, and D. M. Gruen (2001) *Phys. Rev. B* 65, 045403-1–045403-11).

boundaries, the threefold coordination is much less pronounced. Interesting differences between silicon and diamond appear if one compares their bond angle distributions (Keblinski et al., 1998b): The angular distribution in {111} boundaries of silicon is centred at the tetrahedral bond angle of 109.47°, which is characteristic of sp^3-hybridisation, whereas the distribution in diamond shows a second peak of equal intensity at the sp^2-bond angle of 120°. As far as {100} grain boundaries are concerned, the bond angle distribution approaches that for amorphous structures in both materials, but while the silicon peak is centred at the tetrahedral bond angle, the one for diamond is centred at the angle for planar threefold coordination. In reality, this effect may be less remarkable because the number of sp^2-bonds is likely to be overestimated by a factor of 1.5–2 in this simulation, owing to the use of Tersoff potentials. However, the use of first-principle methods would not change the general trend of high-energy boundaries to adopt amorphous structures within a width corresponding to two- or three-atomic layers. In analysing the bond length distribution by using a TB atomistic model, clearly two groups of bond lengths appear (Zapol et al., 2001). One is centred at about 1.40 Å (variation ±2.1%), which is close to the carbon bond length in benzene the other one is centred at about the diamond bond length of 1.54 Å (variation ±2.5%).

During relaxation of the {111} boundary in diamond, about 50% of the bonds across the boundary are lost, compared to only 35–40% in the {100} boundary (according to simulations with the Tersoff potential). As a consequence, the {100} boundary, which has a higher density of bonds already in the unrelaxed state, exhibits a significantly higher bond density after relaxation, and therefore, a much higher work of adhesion. About three quarters of the threefold coordinated atoms form sp^2-dimers across the interface (Cleri et al., 1999). The π-electron pairing energy in each dimer is weakened due to torsion across the boundary. All studies agree that a volume expansion of the grain boundary region occurs during relaxation which corresponds to a widening of the boundary width by 10–14%. This expansion would explain the sharp increase in compressive residual stress, which has been observed by Raman spectroscopy in thin polycrystalline diamond films at the moment of grain coalescence (Michler, Mermoux, von Kaenel, Haouni, Lucazeau and Blank, 1999). The stress increase exceeds by far any contribution ascribable to the thermal mismatch of film and substrate.

The structural disorder in Σ 29 and Σ 13 {100} boundaries was found to be at the origin of wide subbands within the band gap, the effective gap in the boundary region being reduced to 2.3 eV (Cleri et al., Zapol et al.). The grain boundary features electronic properties which are typical for amorphous carbon, since π and π^* states associated with sp^2-sites are introduced in the $\sigma-\sigma^*$ gap. A wide subband associated with electronic states due to dangling bonds appears deep in the gap and stretches out on both sides of the Fermi level for the grain boundary ($E_F = 3.05$ eV, if E_F for the bulk is zero). The bottom of the conduction band lies at 6.4 eV for these models. According to Cleri et al., the predominant portion of electronic states in the band gap, mostly of π-character, are localised around dimers where they always occur as $\pi-\pi^*$ pairs. Their minimum energy separation of about 2.3 eV is considered as the effective gap. A hopping mechanism for electrical conduction has been formulated where loosely paired π-electrons in distorted sp^2-dimers overcome the $\pi-\pi^*$ band gap (e.g., under an applied field) and hop between dimers through empty π^* states below the mobility edge of the boundary located at 7 eV (Cleri et al., 1999). The densities of states of the Σ 13 and Σ 29 grain boundaries are similar to each other. A schematic band structure for a Σ 13 grain boundary according to Zapol et al. is shown in Figure 24. Simulated EELS spectra present the fingerprints of sp^3 and sp^2 bonding, i.e., 1s-σ^* peaks at 290 eV, and 1s-π^* peaks at 285.5 eV. Further peaks of the simulated spectrum at 282.2 eV have been attributed to dangling bonds and bond distortion at sp^3 sites, and, at 283.7 eV, to conduction-band states. There is no experimental evidence for these details, so far.

Grain boundaries have been shown to strongly influence the photoconductivity of undoped diamond (Gan, Ahn, Zhang, Yoon, Rusli, Kerlit, Huang and

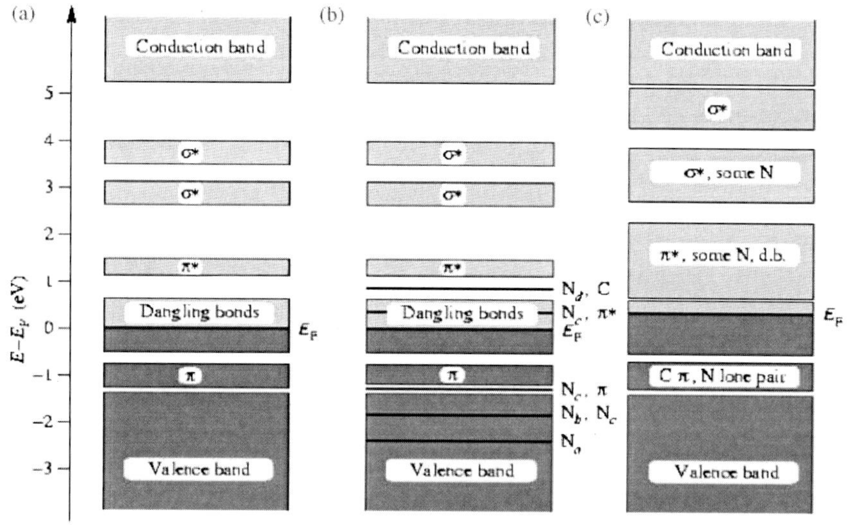

FIG. 24. Schematic band structure for a diamond Σ 13 {100} grain boundary: (a) without nitrogen, (b) with one nitrogen atom per cell, and (c) with sixteen nitrogen atoms per cell (from P. Zapol, M. Sternberg, L. A. Curtiss, T. Frauenheim, and D. M. Gruen (2001) *Phys. Rev. B* **65**, 045403-1–045403-11).

Ligatchev, 2001). In probing different regions of large grains in polycrystalline films by illumination with photons in the subgap energy range between 2.4 and 3.53 eV, the photoconductivity is found to decrease by five orders of magnitude if the grain size increases from 0.5 to 40 μm. It has been concluded that the density of effective recombination centres decreases because the density of mid-gap states decreases, which in undoped diamond is mainly determined by the density of grain boundaries. The existence of mid-gap boundary states has been confirmed by the measurement of carrier lifetimes (Yoneda, Tokuyama, Yamazaki and Ueda, 2000). In irradiating the surface of a polycrystalline film with 4.8 eV photons, the authors have measured a lifetime of several tens of picoseconds, in conformity with former results on photoconductive currents (Yoneda, Ueda, Aikawa, Baba and Shenata, 1998). However, if the photon energy is lowered to 3.3 eV, the carrier lifetimes becomes ultrashort, about 5–7 ps. This effect has been attributed to carrier excitation from mid-gap energy states. In approaching grain boundaries with a focussed laser beam, the lifetime is even further reduced to 0.35–0.5 ps while, simultaneously, the change in reflectivity due to irradiation grows significantly. The authors conclude that the lifetime decrease is related to the changing density of imperfections or mid-gap states inside individual grains of the polycrystalline film, the highest densities being incorporated to grain boundaries. These results are supported by Hall effect measurements on boron-doped diamond films carried out by Fox, Stoner, Malta,

Ellis, Glass and Sivazlian (1994) who found that low-angle boundaries in highly oriented, ⟨100⟩-textured films had a lower density of carrier traps than the large-angle boundaries of polycrystalline films. In single-crystalline films the trap density was reduced by another factor of two.

The incorporation of impurities in grain boundaries has been investigated by Zapol et al. with respect to a Σ 13 {100} boundary. Because of the structural disorder, the sites for impurity atoms are not longer controlled by the translational symmetry of the CSL. The energy of the entire relaxed structure must be minimised to determine if the substitution of a carbon atom by an impurity is stable. In principle, any position can be considered. It has been found that the substitution of carbon at a non-planar threefold site which originally is coupled to a dangling bond, is the most stable nitrogen position. The nitrogen electrons form a lone pair hybridised with the carbon dangling bond. No bond breaking is required, contrary to the bulk case. The substitution energy is -0.6 eV. Also the carbon atoms at planar threefold sites, fourfold sites and the second layer can be replaced, but at the expense of higher substitution energies (up to 2.6 eV). The carbon–nitrogen bond lengths vary from 1.45 to 1.56 Å. Such bond distortions caused by the incorporation of nitrogen are easier to accommodate in the grain boundary than in the bulk. Consequently, the nitrogen solubility in grain boundaries should be higher by several orders of magnitude compared to the bulk. Concerning the electronic structure of individual nitrogen defects, numerous states involving nitrogen have been identified near the top of the valence band (Fig. 24). All sites considered for nitrogen incorporation give also rise to states in a range of about ± 1 eV below and above the Fermi level. They originate from nitrogen lone pairs and are mixed with carbon π states. In case of nitrogen concentrations in the grain boundary of several percent, the nitrogen atoms are found to avoid each other. Altogether, the grain boundary states in the bad gap are significantly more delocalised in the presence of nitrogen impurities. A hopping or similar thermally activated mechanism for electron transport is suggested.

Silicon substitution in a grain boundary always results in the silicon atom being fourfold coordinated, independent of the initial configuration of the carbon atom. In comparison with pure C–C bonds, bond angle distortions are higher, and formation energies range from -0.85 to $+1.45$ eV, which is much lower than that for substitution in a bulk environment (4.45 eV). Since no electronic levels in the diamond band gap have been identified, the electric properties of grain boundaries should not be influenced by the presence of silicon.

Hydrogen atoms saturate the dangling bonds of threefold coordinated carbon atoms in tetrahedral configuration which transform into fourfold coordination under the effect of simulative annealing at 1200 K (Zapol et al., 2001). The number of carbon atoms with modified coordination agrees satisfyingly well with the number of hydrogen atoms added to the grain boundary (four atoms per grain boundary unit cell of 26 atoms, ~ 15%). The average C–H bond length is 1.10 Å, equivalent to what is known from molecular chemistry. Concerning the electronic grain boundary structure, the main effect of hydrogen addition is to reduce the density of states near the Fermi level, as one would expect from the assignment of mid-gap states to the carbon configurations in grain boundaries, explained before. This result is consistent with the well-known fact that hydrogen passivates free surfaces of CVD diamond, and it is consistent with some of the observations on hydrogen diffusion summarised in Section 2.3. Obviously, such a mechanism requires sufficient hydrogen mobility. One may indeed expect lower activation barriers for hydrogen diffusion in grain boundaries than in the bulk: while hydrogen migration in the bulk should occur beyond 1700 K, according to a TB study by Saada et al. (2000), and jumps between equivalent sites on different C–C bonds around 1200 K, the equivalent temperatures in grain boundary regions are reduced to 1200 and 1000 K, respectively.

4. Epitaxy of Diamond Films

4.1. Perfection of Homoepitaxial Films

Crystalline defects, such as dislocations and grain boundaries, deteriorate the electronic, electronic-optical, thermal transport, and mechanical properties of diamond thin films. Mainly due to the abundance of twinning in CVD diamond, it is, however, difficult so far to produce high-quality films with very low defect densities. Even homoepitaxial films grown by plasma assisted CVD on synthetic HPHT diamond substrates of less than 2 mm^2 have been found to exhibit stacking faults and twins in $\{111\}$ growth sectors adjacent to the edges of the substrate, as well as helical defects in the $\{100\}$ growth sector covering the film plane (Tarutani, Takai, Shimizu, Ando, Kamo and Bando, 1996). The helical defects constitute interstitial-type Frank dislocation loops stacked in $\langle 112 \rangle$ directions. These defects, smaller in size, have been also detected at the interface with the substrate where they might be due to Si and Fe impurities, in opposition to the bulk where no such impurities have been found. Similarly, penetration twins have been identified on RHEED and AFM patterns of large (001) homoepitaxial films (5 mm × 5 mm; Takami et al., 1999). Not-withstanding, a multitude of well-defined Kikuchi lines within the RHEED patterns testifies to the otherwise very high degree of crystallinity accomplished in these (100) homoepitaxial films. A particular type of defects in (100) homoepitaxial diamond films are non-epitaxial crystallites, shortly NCs, which are at the origin of band-A emission at 2.8–2.9 eV in cathodoluminecence (Takeuchi et al., 2001a). These particles are formed together with pyramidal hillocks on relatively rough surfaces, and they are surrounded by faulty structures containing $\Sigma 3$ $\{111\}$ boundaries (Takeuchi, Watanabe, Yamanaka, Okushi and Kajimura, 1999). The boundaries separating the NCs from the monocrystalline matrix frequently are of the $\Sigma 3$ $\{112\}$ and $\Sigma 3$ $\{111\}/\{115\}$ type which both show 1s–π^* signals in EELS. They seem to be penetration twins which are stable on $\{100\}$ planes for $\alpha < 1.5$ (Section 2.2.3), i.e., the growth conditions for such films are far from those for $\langle 100 \rangle$ textured polycrystalline films with $2.5 < \alpha < 3$. NCs obstruct hole conduction on H-terminated (100) surfaces (Kasu and Kobayashi, 2003). In high-quality diamond films grown with atomically flat surfaces, growth occurs by a single-domain step-flow mechanism and NCs can be avoided. For that purpose, the methane content must be extremely low (<0.15%), the substrate misorientation angle less than 1.5°, and the substrate temperature be kept within close limits around 750 °C (Takeuchi, Watanabe, Yamanaka, Okushi and Kajimura, 2000; Ri, Yoshida, Yamanaka, Watanabe, Takeuchi and Okushi, 2002; Kasu and Kobayashi, 2003). High quality films of this kind exhibit only band edge emission in cathodoluminescence (Wang, Irie and Ito, 2000) and the Hall mobility in air exceeds $800\ \mathrm{cm^2\,V^{-1}\,s^{-1}}$, in boron-doped films even $1000\ \mathrm{cm^2\,V^{-1}\,s^{-1}}$ (Takeuchi et al., 2001b). These films generally are grown by plasma assisted CVD but they can also been obtained by hot filament CVD (Stammler, Eisenbeiss, Ristein, Neubauer, Gobbels and Ley, 2002). A luminescence signal at 1.68 eV has been assigned to the silicon-vacancy complex in these hot-filament CVD films. Owing to filament evaporation, one would also expect tungsten as an impurity. Much less information is available on (111) homoepitaxial films. Spontaneous fracturing due to residual tensile stress and relatively high surface roughness in the order 10–50 nm has been found in (111) films (Sutcu, Chu, Thompson, Hauge, Margave and D'Evelyn, 1992). When oxygen is added to the gas atmosphere, cracks are suppressed and the surface becomes much flatter. Moreover, the level of hydrogen and silicon impurities is substantially reduced (Sakaguchi et al., 1998). In growing (111) films with and without oxygen, contact twins appear to be easily formed, but no penetration twins have been identified (Rawles, Morris and D'Evelyn, 1996).

The control of growth defects, parts of which have been described in the foregoing sections, has led to an impressive level of crystalline perfection in homoepitaxial films. Until now, the highest degree of crystalline perfection has been achieved for (100) homoepitaxial films deposited in the step-flow growth mode. The ultimate goal of CVD diamond growth are, however, perfect, heteroepitaxial films, which would permit to take advantage of the intrinsic supremacy of diamond thin films, while staying compatible with general semiconductor processing technology. A list of hetero-substrates which have been envisaged for diamond epitaxy is given by Table VI.

4.2. Nucleation on Hetero-Substrates

The nucleation of diamond has kept much of its mystery, in spite of intense research devoted to this subject since the beginning of diamond CVD. It is well known that the nucleation density on silicon and many other materials can be increased by several orders of magnitude simply by abrading the substrate with diamond powder. The diamond grains nucleate preferentially on the seeds left from abrading, as has been demonstrated both by direct microscopic observation

TABLE VI

SUBSTRATE MATERIALS FOR DIAMOND HETEROEPITAXY

Substrate material	Orientation relationship	Lattice constant, registry and misfit (mf)	Angular spread of substrate	Interface	Min. angular spread of diamond twist component	Reference
Diamond (111)		$a_{diam} = 3.567$ Å				–[a]
Graphite	$(111)_{diam} \| (0001)_{graph}$, $\langle 110\rangle_{diam} \| \langle 110\rangle_{graph}$	$2 \times \{0001\}_{graph} \approx 3 \times \{111\}_{diam}$				–[b]
c-BN (100)	$(001)_{diam} \| (001)_{c\text{-}BN}$, $[110]_{diam} \| [110]_{c\text{-}BN}$	$a_{c\text{-}BN} = 3.615$ Å, mf 1.3%		1 dislocation per 82–84 diamond {111} lattice planes		–[c]
Si (100)	$(001)_{diam} \| (001)_{Si}$, $[110]_{diam} \| [110]_{Si}$	$a_{Si} = 5.431$ Å, registry 3:2, mf 1.5%		With and without β-SiC interlayer of several degrees mosaicity[d], interface hillocks 2–3 nm	3.9°	–[e]
Si (111)	$(111)_{diam} \| (111)_{Si}$, $(11-1)_{diam} \| (11-1)_{Si}$[f]		0.03°[g]		0.03° (indivdual particles); film several degrees	–[h]
β-SiC	$(001)_{diam} \| (001)_{Si}$, $[110]_{diam} \| [110]_{Si}$	$a_{SiC} = 4.348$ Å			~1.5°	–[i]
Ir(001)/SrTiO$_3$ (001), Ir[100])/SrTiO$_3$ [100]	$(001)_{diam} \| (001)_{Ir}$, $[100]_{diam} \| [100]_{Ir}$	$a_{Ir} = 3.839$ Å, mf 7%, $a_{SrTiO3} = 3.905$ Å	Ir(002) refl. rocking 0.08°(15)		< 1°(7)	–[j]

(continued)

Table VI (*continued*)

Substrate material	Orientation relationship	Lattice constant, registry and misfit (mf)	Angular spread of substrate	Interface	Min. angular spread of diamond twist component	Reference
Ir/MgO	As before	$a_{Ir} = 3.839$ Å, mf 7%, $a_{MgO} = 4.213$ Å	Ir(004) refl. rocking $<0.2°$[g] roughness[g]: AFM $<$ 0.5 nm	Nucleation along ⟨110⟩ Ir ridges		—[k]
Ni (100)	$(001)_{diam} \parallel (001)_{Ni}$	$a_{Ni} = 3.524$ Å			Unknown	—[l]
Ni (111)	$(111)_{diam} \parallel (111)_{Ni}$				Unknown	—[l]
Pt (111)	$(111)_{diam} \parallel (111)_{Pt}$, $\langle 110\rangle_{diam} \parallel \langle 110\rangle_{Pt}$	$a_{Pt} = 3.923$ Å, mf 10%		With and without graphite interlayer, protrusive Pt interface	$>2°$	—[m]
Pt (111)/Ir (111)/ Pt (111) /sapphire (0001)	As before	$a_{sapphire} = 4.579$ Å, $a_{Pt} = 3.923$ Å	Pt(111) refl. rocking 0.16°[n]		Pt(111) refl. rocking 1.1°	—[o]
Pt (111)/sapphire (0001)	As before				$\sim 2°$	—[p]
Pt (111)/$SrTiO_3$ (111)	As before			Pt (111)/graphite (0001)/diamond (111) or without interlayer	Unknown	—[q]
Sapphire (0001)	$(111)_{diam} \parallel (0001)_{sapph}$	mf 6.4%			Unknown	—[n]

[a]C. K. Suzuki, A. H. Shinohara, P. H. Godoy, N. Watanabe, and M. Kamo, *Diamond Relat. Mater.* 7 (1998) 289.

[b]Z. Li, L. Wang, T. Suzuki, A. Argoitia, P. Pirouz, and J. C. Angus, *J. Appl. Phys.* 73 (1992) 711–715; W. R. Lambrecht, C. H. Lee, B. Segall, J. C. Angus, Z. Li, and M. Sunkara, *Nature* 364 (1993) 607–610.

[c]S. Koizumi, T. Inuzuka, T. Imuzuka, and K. Suzuki, *Appl. Phys. Lett.* 57 (1990) 563; M. Yoshikawa, H. Ishida, A. Ishitani, S. Koizumi, and T. Inuzuka, *Appl. Phys. Lett.* 58 (1991) 1387; T. Tomikawa, and S. Shikata, *Jpn. J. Appl. Phys.* 32 (1993) 3938–3942.

[d]K.-H. Thürer, M. Schreck, and B. Stritzker, *Phys. Rev. B* 57 (1997) 15454.

[e]K.-H. Thurer, M. Schreck, R. Klarmann, and B. Stritzker, *J. Appl. Phys.* 81 (1997) 3096.

[f]M. Schreck, and B. Stritzker, *Phys. Status Solidi (a)* 154 (1996) 197–217.

[g]Prior to deposition.

[h]H.-J. Ryu, J.-L. Lee, and J. H. Je, *Chem. Vap. Deposit.* 7 (2001) 22–24; S. N. Polyakov, A. T. Rakhimov, N. V. Suetin, M. A. Timofeev, and A. A. Pilevskii, *JETP Lett.* 65 (1997) 434–438.

[i]H. Kawarade, in "Low-Pressure Synthetic Diamond" (B. Dischler, and C. Wild (Eds.)), Springer, Berlin, 1998.

[j]Initial film mosaicity: M. Schreck, H. Roll, and B. Stritzker, *Appl. Phys. Lett.* 74 (1999) 650; substrate: F. Hormann, H. Roll, M. Schreck, and B. Stritzker, *Diamond Relat. Mater.* 9 (2000) 256–261.

[k]T. Fujisaki, M. Tachiki, N. Taniyama, M. Kudo, and H. Kawarada, *Diamond Relat. Mater.* 11 (2002) 478–481; T. Fujisaki, M. Tachiki, N. Taniyama, M. Kudo, and H. Kawarada, *Diamond Relat. Mater.* 12 (2003) 246–250.

[l]P.C. Yang, W. Zhu, and J.T. Glass, J. Mater. Res. 8 (1993) 1773–1776; W. Zhu, P.C. Yang, and J.T. Glass, *Appl. Phys. Lett.* 63 (1993) 1640–1642.

[m]Y. Shintani, *J. Mater. Res.* 11 (1996) 2955–2956; T. Tachibana, Y. Yokota, K. Miyata, K. Kobashi, and Y. Shintani, *Diamond Relat. Mater.* 5 (1996) 197; T. Tachibana, Y. Yokota, K. Kobashi, and Y. Shinatani, *J. Appl. Phys.* 82 (1997) 4327; T. Tachibana, Y. Yokota, K. Miyata, T. Onishi, k. Kobashi, M. Tarutani, Y. Takai, R. Shimizu, and Y. Shintani, *Phys. Rev. B* 56 (1997) 15967–15981; M. Tarutani, G. Zhou, Y. Takai, R. Shimizu, T. Tachibana, K. Kobashi, and Y. Shintani, *Diamond Relat. Mater.* 6 (1997) 272–276; K. A. Cherian, S. D. Wolter, and R. Roy, *Diamond Relat. Mater.* 6 (1997) 1747–1752.

[n]M. M. M. Kuypers, R. D. Pancost, and J. S. S. Sinninghe Damsté, *Nature* 399 (1999) 340–342.

[o]T. Tachibana, Y. Yokota, K. Hayashi, and K. Kobashi, *Diamond Relat. Mater.* 10 (2001) 1633–1636.

[p]T. Tachibana, Y. Yokota, K. Kobashi, and M. Yoshimoto, *J. Crystal. Growth* 205 (1999) 163.

[q]H. Zhou, M. Tarutani, Y. Takai, and R. Shimizu, *Jpn. J. Appl. Phys.* 36 (1997) 2298–2302.

and by analysis of the Avrami-type growth behaviour (Iijima, Aikawa and Baba, 1990; Maeda, Ikari, Okubo, Kusakabe and Morooka, 1993; Stiegler, von Kaenel, Cans and Blank, 1996; Arnault, Demuynck, Speisser and Le Normand, 1999). As pointed out by Stoner et al. (1992), diamond-seeding cannot be the only mechanism for grain nucleation because high nucleation densities have also been obtained by (i) abrading the substrate with cubic boron nitride and silicon carbide, (ii) pre-depositing carbonaceous layers or oil residue, (iii) biasing the substrate negatively while immersed in a methane–hydrogen plasma, as originally proposed by Yugo, Kanai, Kimura and Muto (1991). Heterogeneous nucleation indeed has been identified as a separate process, in addition to seeding (Stiegler et al., 1996; Arnault et al., 1999). Concerning the substrate materials suitable for diamond nucleation, the formation of a thin carbide layer has been postulated to be prerequisite for diamond nucleation (Haubner, Lindlbauer and Lux, 1996). Refractory metals, which all are carbide forming elements, thus are preferential substrate materials while other materials with a high diffusivity for carbon are not. This criterion constitutes a useful concept for the selection of substrate materials but it is incomplete in as far as a silicon carbide substrate without pre-treatment does not yield higher nucleation densities than untreated silicon. Protrusions with sharp edges and points which are created by scratching or etching prior to CVD processing are favourite sites for grain nucleation (see, for example, Stoner et al., 1992). As regards bias-enhanced nucleation (BEN), Yugo, Kimura and Kanai (1993) have suggested that bombarding the substrate with carbon ions should facilitate the formation of carbon clusters and transform their sp^2- into sp^3- structure, provided the ion energy is appropriate and atomic hydrogen helps etching the sp^2 components. Clusters rich in sp^3-hybridised bonds then should become the precursors for diamond nucleation. This type of model would allow an explanation of the nucleation experiments carried-out by Yugo, Semoto, Hoshina, Kimura and Nakai, 1995. The physical mechanisms of BEN leading to the enhancement of nucleation have been investigated by Wang, Wang, Liao and Xiao (2001) who have analysed the effects of the ion distribution and ion energy distribution in the cathode sheath of the glow discharge induced by biasing. According to this study, (i) the formation energy of critical nuclei decreases owing to the production of numerous pits created by ion bombardment; (ii) the substrate temperature increases, thus enhancing ion diffusion at the substrate surface; and (iii) the bond strength of diamond nuclei on the substrate is strengthened. Earlier, Chen and Lin (1996) have argued that the intense electron emission of diamond should allow to increase the ion flux to the substrate and thus intensify nucleation.

4.3. The Heteroepitaxial Interface with Silicon

In 1992–1993, it was found that BEN of diamond on silicon and β-SiC substrates not only gave rise to very high nucleation densities ($\sim 10^{10}\,cm^{-2}$), a large fraction of the nuclei had also adopted an epitaxial growth relationship with the substrate (Table VI; references for Si: Jiang, Klages, Zachai, Hartweg and Füsser, 1993a; Jiang et al., 1993b; Wolter, Stoner, Glass, Ellis, Buhaenko, Jenkins and Southworth, 1993; for β-SiC: Stoner and Glass, 1992; Kohl, Wild, Herres, Koidl, Stoner and Glass, 1993). The crystal lattices of film and substrate were parallel but the individual grains were slightly misoriented with respect to the substrate and not exactly parallel to each other. After grain coalescence, the diamond films displayed a mosaic structure with an orientation distribution of about 9° FWHM. While the tilt component of grain misalignment could be reduced to about 2° by competitive grain growth leading to overgrowth of strongly tilted grains (Section 2.1), the twist component was hardly affected by this process (Thürer et al., 1998). Since, the nucleation on silicon has been investigated thoroughly because silicon would constitute an almost ideal substrate material for heteroepitaxial diamond films owing to its crystalline perfection and because of its importance in semiconductor industry. Research has focussed on: (i) the effect of the lattice mismatch of the participating phases; (ii) the role of β-SiC as interlayer in diamond nucleation; (iii) the role of protrusions as nucleation sites; (iv) the effect of ion bombardment associated with BEN, and, eventually; (v) atomistic modelling of nucleation.

In spite of the structural similarity between the lattices of Si, β-SiC and diamond, epitaxial growth is surprising because of the enormous lattice mismatch ($a_{\mathrm{Si}} = 0.543$ nm, $a_{\beta\text{-SiC}} = 0.436$ nm; $a_{\mathrm{diamond}} = 0.357$ nm). The excess energy associated with the heteroepitaxial interface is composed of the strain energy in the film and the misfit energy, which is composed of bond distortions, dangling bonds and bond reconstruction, much alike the energy of dislocations and grain boundaries, but with a rather abrupt phase change across the boundary. Based solely on strain energy, it is impossible to predict any preference for epitaxial growth of diamond on β-SiC which would agree with experiment (Zhu, Wang, Stoner, Ma, Kong, Braun and Glass, 1993). Thus, the existence of a β-SiC layer intercalated between diamond and silicon cannot be rationalised by its mediating role in reducing the strain energy compared to the direct transition between silicon and diamond.

There is plenty of evidence that β-SiC can be deposited in epitaxial orientation with silicon during biasing and that diamond forms on top of it (see, for example, Stoner et al., 1992; Chen, Chang, Lin and Chen, 1995; Wurzinger et al., 1997b). The same sequence of phase formation has been proposed for the nucleation of polycrystalline diamond on substrates abraded with diamond powder (Arnault et al., 1999). Back to BEN, it has been shown that high quality films can be grown with an angular spread of only 0.6° which is close to the mosaicity of the β-SiC substrate alone (Kawarada et al., 1997). By a proper choice of BEN parameters, it is possible to keep the orientation distribution of the diamond grains and the substrate in the same range, as evidenced by RHEED. By contrast, under optimum conditions for diamond nucleation on silicon, the surface of epitaxial β-SiC layers is rough, and its angular spread of 5–6° strongly limits the quality of a diamond film growing epitaxially on top of it (Thürer et al., 1998). Meanwhile, the existence of a β-SiC interlayer is no longer considered as a prerequisite for diamond nucleation on silicon. Growth of a continuous β-SiC layer seems to depend on the choice of processing parameters. Independent nucleation of diamond and β-SiC particles has also been observed on silicon surfaces prepared by heat treatment under UHV conditions (Arnault, Pecoraro, Werckmann, Le Normand, Motta and Polini, 2001).

A further aspect important for nucleation concerns the interface topology. Atomically flat interfaces provide nucleation rates in the order $10^4\ cm^{-2}$ which are insufficient for film growth. Recently, it has been demonstrated that multi-vicinal surfaces existing near edges, protrusions and tips are required for noticeable nucleation to occur (Arnault et al., 2001), and observations pointing into the same direction have been reported before (Wurzinger et al., 1997a,b). In TEM on cross-sections of silicon substrates decorated with dispersed diamond particles deposited by CVD, the locus of nucleation has always been identified as a tip protruding from the adjacent silicon–diamond interface (Tarutani et al., 1995; Jiang, Jia, Szameitat and Rickers, 2001). It is, however, difficult to decide if these tips already existed prior to nucleation. As the uncovered substrate between the laterally growing diamond particles is continuously etched by the hydrogen-rich gas atmosphere, the interface moves gradually back into the substrate, and the nucleation site which is protected by the diamond particle, finishes by standing out from the interface. Nevertheless, the fore-mentioned investigations by Arnault et al. (2001) on the effects of multi-vicinal surfaces confirm the importance of surface topology. These problems had been already tackled by Yagyu, Mori, Hatta, Ito and Hiraki (1993) who were able to prove by TEM and HRTEM that single diamond particles deposited on ultrasonically abraded substrates can grow directly, i.e., without β-SiC interlayer, on protuberances etched into the silicon surface. No indications have been found in the literature pointing to preferential nucleation of β-SiC at protrusions.

The processing parameters for BEN strongly influence the angular spread of the diamond mosaic structure. Schreck, Thurer and Stritzker (1997) have determined the twist components of grain misalignment by X-ray diffraction. Under the experimental conditions of −200 V bias voltage, 1% methane and 790 °C substrate temperature, the twist component exhibits a minimum angular spread after 10 min biasing. The time window for optimum alignment varies strongly with bias voltage and gas phase composition. At shorter than optimum biasing durations, the nucleation density is insufficient while beyond, the epitaxial relationship with the substrate is increasingly lost. It has been reported that the ion bombardment during biasing causes misalignment of the nuclei by roughening of the substrate surface, and generates defects and secondary nucleation at the surface of the growing crystallites (Jiang et al., 2001). Even in homoepitaxial diamond growth, that is without any lattice mismatch between film and substrate, biasing leads to misalignment (Jiang, Zhang, Paul and Klages, 1996). Under the same biasing conditions as before (−200 V), a biasing time 20 s is sufficient to entirely destroy the epitaxial growth relationship between a diamond film and a diamond substrate (Thürer et al., 1998).

Several papers on BEN now prove that diamond can grow directly and epitaxially on silicon protrusions (Lee, Peng, Zhou, Wang, Lee, Bello and Lifshitz, 2000; Jiang and Jia, 2000). In these studies, deposition was stopped prior to particle coalescence. Figure 25 shows two diamond particles of, respectively, 2 and 6 nm size growing on steps at the surface of the silicon substrate composed of {111} and {100} facets. Under BEN conditions, the critical size for a diamond nucleus apparently is less than 2 nm. Another particle shown in the same paper and considered to be randomly oriented, seems to have nucleated in a twinned position with respect to the substrate (Fig. 2 in Lee et al., 2000). None of these small particles deposited by a hot filament technique under biasing conditions

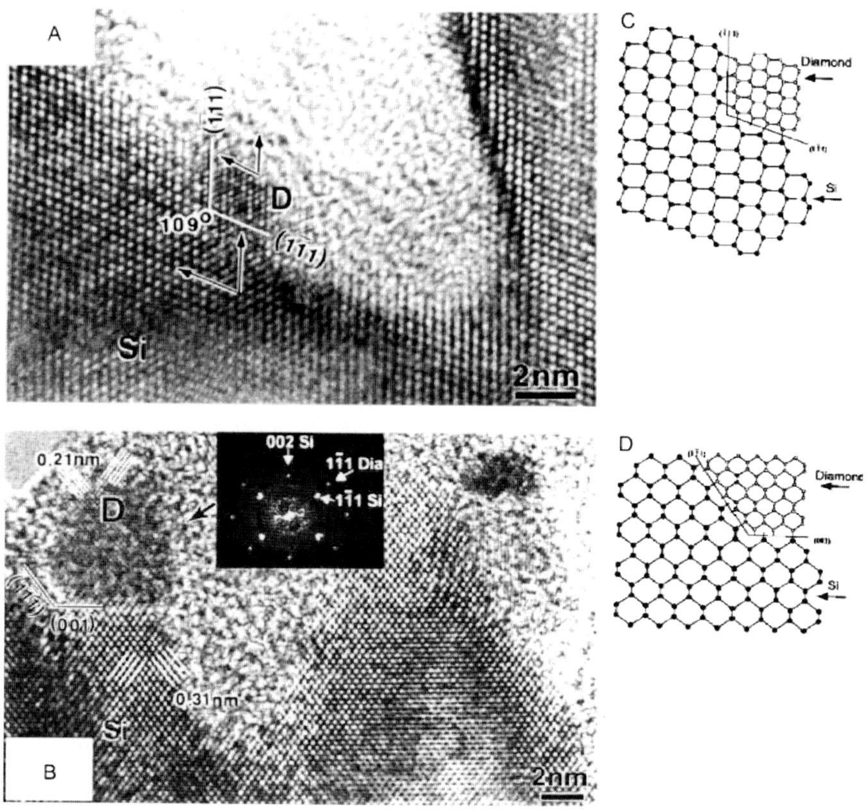

FIG. 25. HRTEM micrographs of direct epitaxial growth of diamond nuclei on silicon. (A) Epitaxial nucleation of diamond on silicon step bounded by ($\bar{1}11$)and ($1\bar{1}1$) lattice planes; the diameter of the crystallite is ~2 nm. (B) Epitaxial nucleation of diamond on Si (001) surface; diameter ~6 nm. (C), (D) Schematic of the interfaces between Si and diamond in (A) and (B). According to S. T. Lee, H. Y. Peng, X. T. Zhou, N. Wang, C. S. Lee, I. Bello, and Y. Lifshitz, (2000) *Science* 287, 104–106.

contains a defect (estimated ion dose of 3×10^{19} ions per cm^2), but the authors have noticed deviations by about 1–2° between corresponding {111} planes, even in this early stage of growth. By contrast, the larger particles investigated by Jiang et al. (2000) which have been deposited by microwave assisted CVD are twinned. Almost perfect heteroepitaxy has been reported for diamond islands deposited by plasma assisted CVD on ultrasonically abraded Si (111) substrates (Ryu, Lee and Je, 2001). The mosaic distributions of both the tilt and twist components of the diamond islands, both being measured by synchrotron X-ray scattering, amount to 0.03° FWHM. The particles are strained by 0.14% perpendicular to the substrate plane (increase of the lattice constant). According to facet orientation on SEM photographs, some of the particles seem to be strongly misoriented, however. Once the particles begin to coalesce, the lattice strain is released, and the angular spread increases drastically (14° FWHM after 10 h deposition). From the comparison of these results one may conjecture that many of the defects associated with particle misorientation are related to ion bombardment in microwave assisted CVD under biasing conditions.

Molecular dynamics simulations have been carried out in order to better understand the misalignment of diamond clusters nucleating on a Si (100) 2×1 surface (Jiang et al., 1998; Zhang, Wang, Estevea and Bertran, 1998; Jiang and Jia, 2000). This step-by-step optimisation procedure is shown schematically in Figure 26. Diamond nucleation begins by the adsorption of a CH_2 species bridging two silicon atoms in the substrate surface, while two hydrocarbons, or a C_2 species, are forming a seven-membered ring with the neighbouring Si-atoms in the surface (Fig. 26a). This configuration brings the carbon atoms close enough together to form a six-membered and a five-membered carbon ring which by periodic lateral continuation establishes the 3 : 2 lattice registry required by the lattice constants (Fig. 26b). Already the next layer on top of the first one consists of six-membered rings exclusively. The fivefold/sevenfold bonding pattern obviously conforms to the core structures of dislocations and grain boundaries. According to the authors, the Si–C and C–C bond lengths are rather close to the bulk bond lengths of β-SiC and diamond, once the structure of the geometric model has been given the opportunity to relax in simulation. Linking the carbon atoms is, however, associated with bond angle distortion (deviation from the tetrahedral angle of 109°28′). The distortion can induce tilting of the diamond nucleus, as shown by these simulations, especially if the adsorption of carbon atoms on a silicon protrusion proceeds in an asymmetric manner, and if the silicon substrate is etched simultaneously by H^+ ions (Fig. 6c). In the model, the diamond clusters of the model propagate automatically in $\langle 110 \rangle$ and $\langle 1\bar{1}0 \rangle$ directions, and misalignment of the new grain can occur by tilting about the two axes. The simulated tilt angles are smaller than in reality. One would expect the tilt angle to depend on substrate roughness. Additionally, a remaining misfit of 1.5% in addition to the perfect 3 : 2 lattice registry is not accounted for by the model. This strain is likely to contribute to the grain misalignment as it is susceptible to relax via the generation of defects. In another paper, the authors

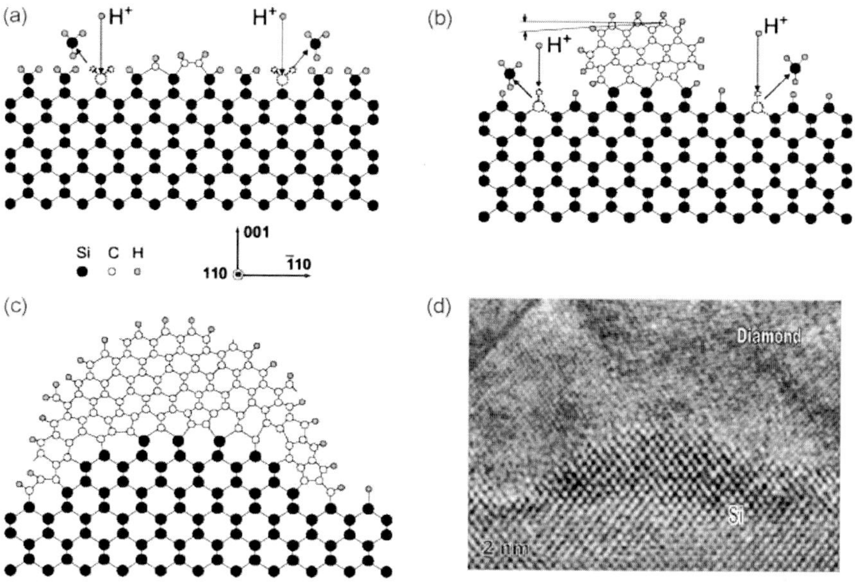

FIG. 26. Schematics of the nucleation of diamond on silicon. (a) Beginning deposition of carbon or hydrocarbon species at the silicon (001) surface, and H^+ ion etching. (b) Further etching and formation of a tilted nucleus. (c) Continuing hillock formation due to etching of the silicon surface and further growth of the nucleus. (d) HRTEM cross-section of a hillock at the substrate-film interface where diamond nucleation is believed to have occurred (from X. Jiang and C. L. Jia (2000) *Phys. Rev. Lett.* 84, 3658–3660).

deal with the twist component of nucleus misalignment (Zhang, Zhang, Sun, Jiang and Lee, 1999). They have observed that the grains are not randomly distributed but exhibit a maximum at twist angles of 20 and 30°, besides the main peak at 0°. Simulation indeed has permitted to identify these orientations as metastable configurations. In conclusion, the results from modelling add qualitatively new information about the origin of grain misalignment. Based on the experience with atomistic modelling of grain boundary structures outlined in the preceding sections, one may, however, wonder about the appropriateness of semi-empirical interatomic potentials to tackle interphase boundaries involving different elements, in addition to carbon sp^3/sp^2 reconstruction. Further on, the assumption of a sharp interface between silicon and diamond might be physically unrealistic, as subplantation of carbon ions is likely to induce distortion of the interface. Atomic displacement by subplantation cascades might be at the origin of the blurry interface structures seen on most HRTEM micrographs of nucleation sites on silicon. Subplantation models focussing on an environment consisting of carbon atoms solely have been brought forward to explain BEN of diamond (Sattel, Gerber and Erhardt, 1996).

4.4. Promising Substrate Materials for Heteroepitaxy

Cubic boron nitride (c-BN) has a diamond-type type lattice with a lattice mismatch of only 1.3% (Table VI) and its physical properties are similar to those of diamond. Thus, c-BN would be an ideal candidate for diamond heteroepitaxy, if the problem of synthesising large c-BN wafers were solved. Only small c-BN crystals have been used so far as hetero-substrates for diamond CVD. Diamond has been deposited on {100}, {111}, and {221} facets of c-BN (Koizumi, Inuzuka, Imuzuka and Suzuki, 1990; Argoitia, Angus, Ma, Wang, Pirouz and Lambrecht, 1994; Chai, Gu, Li, Mai, Li, Yuan and Pang, 1994) and its lattice orientation turns out to be always parallel to the c-BN lattice. Despite a high surface energy of c-BN, diamond deposition proceeds by nucleation of three-dimensional islands followed by coalescence. Most information is available for growth on c-BN {111} facets, which must be terminated by boron atoms in order to enable epitaxial growth of diamond (Argoitia et al., 1994; Ito, Abe, Inuzuka and Koizumi, 1996). During the initial growth stage, carbon atoms are adsorbed to the boron-terminated surface where they are arranged in fourfold coordination (Koizumi, 1998). Diamond grows on top of the carbon layer. Defects at the c-BN surface are preferential sites for diamond nucleation. The interface is characterised by numerous defects both on the c-BN and diamond side. Despite interface dislocations which are supposed to relieve strain caused by the lattice mismatch, residual tensile strain of 0.29% has been revealed in a thin film of 200 nm thickness.

High-quality heteroepitaxial (100) diamond films have also been achieved by deposition on Ir(100)/$SrTiO_3$(100) (Schreck, Roll and Stritzker, 1999) and Ir(100)/MgO(100) substrates (Ohtsuka, Suzuki, Sawabe and Inuzuka, 1996). Iridium acts as a diffusion barrier for carbon without forming carbides. Even if the lattice mismatch with diamond amounts to 7.6%, iridium has become a serious candidate for diamond heteroepitaxy. Less than 1° misalignment has been attained for both the tilt and the twist components of very thin films (thickness 600 nm; Schreck, Hörmann, Roll, Lindner and Stritzker, 2001). If these films are grown to greater thickness, the angular spread is further reduced. Slightly misoriented grains then merge to form larger grains via the disclination mechanism discussed in Section 3.3. This technique seems promising for the achievement of single-crystalline films. In the best case, an angular spread of only 0.38° for the crucial twist component has been obtained (Schreck et al., 2001). Flat substrates with low mosaicity are prerequisite to obtain such films. The tilt components of both the Ir/MgO and Ir/$SrTiO_3$ surfaces do not exceed 0.2° (Hörmann, Roll, Schreck and Stritzker, 2000; Fujisaki, Tachiki, Taniyama, Kudo and Kawarada, 2002). The average AFM roughness prior to BEN is about 0.5 nm but height variations of a few nanometers may be observed within a lateral range of 0.5 μm. Numerous ridges with a maximum height of 2 nm run in $\langle 110 \rangle$ directions at 10–40 nm intervals and diamond appears to nucleate on these ridges

(Fujisaki, Tachiki, Taniyama, Kudo and Kawarada, 2003). Although the substrate surface is more corrugated after BEN, the sharpness of the Kikuchi lines in electron back-scatter diffraction (EBSB) indicates that only little surface damage occurs by BEN induced ion bombardment (Schreck, Bauer, Gsell, Hörmann, Bielefeldt and Stritzker, 2003). The ridges at the substrate surface, which seem to be intensified by BEN, become the nucleation sites (Sawabe, Fukuda, Suzuki, Ikuhara and Suzuki, 2000). Growth on ridges would allow the diamond nuclei to minimise their strain energy (Schreck et al., 2000) but this effect cannot explain the distribution of nucleation centres: Minor topographical variations at the substrate surface seem to strongly influence the local efficiency of BEN and lead to well-defined domains where the ridges suitable for nucleation are clustered. So far, this phenomenon is not understood. At the interface between the substrate and the diamond film, the transition between iridium and diamond occurs abruptly, and structural integrity appears to be better preserved than in case of silicon substrates, even if numerous defects can be identified (Sawabe et al., 2000; Hörmann, Peng, Bauer, Li, Schreck, Lifshitz, Lee and Stritzker, 2002).

5. Concluding Remarks

After two decades of intensive research, the state of perfection accomplished in the field of CVD diamond is impressive. All along this period, the control of growth imperfections has held a key position in the search for superior film properties. Growth mechanisms, which favour or can avoid the formation of imperfections have been investigated, and fundamental defect structures influencing materials properties have been identified. Depending on the final application, imperfections may be tolerated or not, or they may even be desirable.

The highest level of crystalline perfection has been achieved by applying the step-flow growth mode which deposits (100) homoepitaxial diamond films with atomically flat surfaces. These materials have a high potential for use in electronic devices. According to their performance grading, they are classed in device-grade one (Okushi, Watanabe, Ri, Yamanaka and Takeuchi, 2002). The alternative route of mono-crystalline films by heteroepitaxy has not yet reached the same level of perfection but the degree of mosaicity in nucleation has been continuously reduced to actually less than 1° by using iridium substrates. Although grain boundaries are annihilated during film growth via disclination formation, relatively thick films of some tens of micrometers are required for a further significant reduction of the angular spread. Several hurdles thus remain to be leapt. Better understanding and control of BEN appears to be one of them. Another one refers to the suppression of octahedral facets in the early stage of diamond growth, which are prone to twinning and disappear only after growth lengths of several micrometers. Even without {111} facets, surface roughening due to ion bombardment seems to hinder the formation of atomically flat films, which are required for step-flow growth over wide areas. The advantage of heteroepitaxial film growth obviously would be a much higher degree of compatibility with existing semiconductor processing technology, and this factor alone represents a major driving force for further development.

By a combination of microscopic and spectroscopic techniques with atomistic modelling, a thorough understanding of the formation and characteristics of many kinds of growth defects in CVD diamond has been reached. The results of these investigations demonstrate the marked differences of diamond compared to other materials with diamond-type lattices. Not only carbon readily changes its state of bond hybridisation, also kinetic processes in diamond are extremely slow, so that defects generally cannot adopt minimum energy configurations. They are the result of growth accidents occurring during film deposition. Since reconstruction of the defect core structures is kinetically hindered, boundaries with dangling bonds are common, contrary to what occurs in silicon where the occurrence of dangling bonds in general can be ruled out. According to King et al. (1997), the observed defect configurations in diamond represent kinetically accessible structures rather than true energy minima.

Defects of the type discussed in this section induce local residual strain and stress fields. Many of them directly affect the electronic band structure, and therefore, the electronic and optical properties. Disclinations have been identified as the sites of intense local stress generation, which may even cause cracking. Although disclinations tend to be arranged in multipole configurations, thereby annihilating long-range stress fields, the local stress concentrated at each pole remains high. Other sources of constraint susceptible to induced residual strain and stress are grain boundaries and grain boundary junctions with their contiguity problems, or volume changes accompanying the formation of grain boundaries at the moment of grain coalescence. The stress components associated with these defects exceed by far any thermal stress contribution. All stress components are superimposed in a diamond film. Long range residual stresses causes film distortion while long and short-range stresses weaken the mechanical properties and obstruct adhesion. Extended defect structures such as disclinations, which are at the origin of intense residual stress formation, or intrinsically weak grain boundaries, for example, of the {111} type, are probably responsible for the comparatively low rupture strength of highly oriented and polycrystalline diamond films. Commonly observed rupture strengths of the order of 2–4 GPa are almost two orders of magnitude below the theoretical strength which is of the order of one-tenth of Young's modulus. From a purely mechanical point of view, nano-crystalline diamond films are therefore, expected to be superior to microcrystalline films because the much smaller size of extended defects and the lower residual stresses associated with these defects will probably improve fracture toughness significantly.

References

L. Allers and A. Mainwood, Surface vacancies in CVD diamond, *Diamond Relat. Mater.* 7, 261–265 (1998).

S. L. Altmann, K. W. Lodge, and A. Lapiccirella, The stacking-fault energy in diamond, *Philos. Mag. A* 47, 827–834 (1983).

J. C. Angus, F. A. Buck, M. Sunkara, T. F. Groth, C. C. Hayman, and R. Gat, Diamond growth at low pressures, *MRS Res. Bull.* October, 38–47 (1989).

J. C. Angus, M. Sunkara, S. R. Sahaida, and J. Glass, Twinning and faceting in early stages of diamond growth by chemical vapor deposition, *J. Mater. Res.* 7, 3001–3009 (1992).

A. Argoitia, J. C. Angus, J. S. Ma, L. Wang, P. Pirouz, and W. R. L. Lambrecht, Heteroepitaxy of diamond on c-BN-growth mechanisms and defect characterisation, *J. Mater. Res.* 9, 1849–1865 (1994).

J. C. Arnault, L. Demuynck, C. Speisser, and F. Le Normand, Mechanisms of CVD diamond nucleation and growth on mechanically scratched Si (001) surfaces, *Eur. Phys. J. B* 11, 327–343 (1999).

J. C. Arnault, S. Pecoraro, J. Werckmann, F. Le Normand, N. Motta, and R. Polini, Earlystages of the HFCVD process on multi-vicinal silicon surfaces studied by electron microscopy probes (SEM, TEM), *Diamond Relat. Mater.* 10, 1612–1616 (2001).

P. Ascarelli, E. Cappelli, G. Mattei, F. Pinzari, V. Fares, C. Veroli, and S. Martelli, Relation among growth rate, microstructure and the physical properties of diamond films, *Diamond Relat. Mater.* 5, 308–311 (1996).

M. N. Ashfold, P. W. May, J. R. Petherbridge, K. N. Rosser, J. A. Smith, Y. A. Mankelevich, and N. V. Suetin, Unravelling aspects of the gas phase chemistry involved in diamond chemical vapour deposition, *Phys. Chem. Chem. Phys.* 3, 3471–3485 (2001).

Aulbur (2003). Density Functional Theory. Reference: http://www.physics.ohio-state.edu/~aulbur/dft.html.

C. C. Bataille, D. J. Srolovitz, and J. E. Butler, Morphologies of diamond films from atomic-scale simulations of chemical vapor deposition, *Diamond Relat. Mater.* 6, 1198–1206 (2001).

A. Bergmaier, G. Dollinger, A. Aleksov, P. Gluhe, and E. Kohn, Deuterium depth profiles at CVD diamond surfaces, *Surf. Sci.* 481, L433–L436 (2001).

J. Bernholc, A. Antonelli, T. M. Del Sole, Y. Bar-Yam, and S. T. Pantelides, Mechanism of self-diffusion in diamond, *Phys. Rev. Lett.* 61, 2689–2692 (1998).

X. Blase, K. Lin, A. Canning, S. G. Louie, and D. C. Chrzan, Structure and energy of the 90° partial dislocation in diamond: a combined *ab initio* and elasticity theory analysis, *Phys. Rev. Lett.* 80, 5780–5783 (2000).

A. T. Blumenau, M. I. Heggie, C. J. Fall, R. Jones, and T. Frauenheim, Dislocations in diamond: core structures and energies, *Phys. Rev. B* 65, 205205-1–8 (2002a).

A. T. Blumenau, C. J. Fall, R. Jones, M. I. Heggie, P. R. Briddon, T. Frauenheim, and S. Öberg, Straight and kinked 90° partial dislocations in diamond and 3C-SiC, *J. Phys.: Condens. Matter* 14, 12741–12747 (2002b).

A. T. Blumenau, R. Jones, T. Frauenheim, B. Willems, O. I. Lebedev, G. Van Tendeloo, D. Fisher, and P. M. Martineau, 2003. Dislocations in diamond: dissociation into partials and their glide motion, *Phys. Rev.* B68, 014115-1–9 (2003).

A. Bourret, Atomic structure of (011) and (001) pure tilt grain-boundaries in germanium and silicon, *J. Phys.* 46, 27–38 (1986).

N. Bozzolo, S. Barrat, I. Dieguez, and E. Bauer-Grosse, Crystalline quality of highly oriented diamond films grown on ⟨100⟩ silicon studied by conventional TEM, *Diamond Relat. Mater.* 6, 41–47 (1997).

D. W. Brenner, Empirical potential for hydrocarbon for use in simulating the chemical vapor deposition of diamond, *Phys. Rev.* 9458–9471 (1990).

F. Brunet, A. Deneuville, P. Germi, M. Pernet, E. Gheeraert, and J. Mambou, Effect of boron incorporation on the structure of polycrystalline diamond films, *Diamond Relat. Mater.* 6, 774–777 (1997).

J. Bühler and Y. Prior, Study of morphological behavior of single diamond crystals, *J. Crystal Growth* 209, 779–788 (2000).

S. Bühlmann, E. Blank, R. Haubner, and B. Lux, Characterization of ballas diamond depositions, *Diamond Relat. Mater.* 8, 194–201 (1999).

F. P. Bundy and J. S. Kasper, Hexagonal diamond—a new form of carbon, *J. Chem. Phys.* 46, 3437–3446 (1967).

N. C. Burton, J. W. Steeds, G. M. Meaden, Y. G. Shreter, and J. E. Butler, Strain and microstructure variation in grains of CVD diamond films, *Diamond Relat. Mater.* 4, 1222–1234 (1995).

W. P. Chai, Y. S. Gu, M. Li, Z. H. Mai, Q. Z. Li, L. Yuan, and S. J. Pang, Epitaxial growth of diamond films on the (221) and (100) surfaces of c-BN with microstructures full of (100) facets, *Appl. Phys. Lett.* 64, 1941–1943 (1994).

C. J. Chen, L. Chang, T. S. Lin, and F. R. Chen, Microstructural evolution of diamond/Si (100) interfaces with pretreatments in chemical vapor deposition, *J. Mater. Sci.* 10, 3041–3049 (1995).

C. J. Chen, L. Chang, T. S. Lin, and F. R. Chen, Direct observation of heteroepitaxial diamond on a silicon (110) substrate by microwave plasma chemical vapor deposition, *J. Mater. Res.* 11, 1002–1010 (1996).

W. J. Chen and F. R. Chen, The atomic structure of $\sigma = 1$ and $\sigma = 3$ $NiSi_2$/Si interfaces, *Philos. Mag. A* 68, 605–630 (1993).

Q. J. Chen and Z. D. Lin, Electron-emission-enhanced diamond nucleation on Si by hot filament chemical vapor deposition, *Appl. Phys. Lett.* 68, 2450–2452 (1996).

J. Chevallier, Passivation of acceptors in diamond: B–H interaction in "Properties, Growth and Applications of Diamond" Emis Data Reviews Series (M. H. Nazaré and A. J. Neves, Eds.), pp. 236–242. INSPEC, The Institution of Electrical Engineers, London, UK, 2001, ISBN 0 85296 785 3.

J. Chevallier, D. Ballutaud, B. Theys, F. Jomard, A. Deneuville, E. Gheeraert, and F. Pruvost, Hydrogen in monocrystalline CVD boron doped diamond, *Phys. Status Solidi (a)* 174, 73–81 (1999).

J. Chevallier, F. Jomard, Z. Teukam, S. Koizumi, H. Kanda, Y. Sato, A. Deneuville, and M. Bernard, Hydrogen in n-type diamond, *Diamond Relat. Mater.* 11, 1566–1571 (2002).

J. Chevallier, A. Lusson, D. Ballutaud, B. Theys, F. Jomard, A. Deneuville, M. Bernard, E. Gheeraert, and E. Bustarret, Hydrogen-acceptor interactions in diamond, *Diamond Relat. Mater.* 10, 399–404 (2001).

J. Chevallier, B. Theys, A. Lusson, C. Grattepain, A. Deneuville, and E. Gheeraert, Hydrogen–boron interactions in *p*-type diamond, *Phys. Rev. B* 58, 7966–7970 (1998).

H. R. Christen and F. Vögtle, "Organische Chemie: Von den Grundlagen zur Forschung", 2 Auflage Band I, p. 1992. Verlag Salle und Sauerländer, Frankfurt/Main, 1992.

C. J. Chu, R. H. Hauge, J. L. Margrave, and M. P. D'Evelyn, Growth kinetics of (100), (110), and (111) homoepitaxial diamond films, *Appl. Phys. Lett.* 61, 1393–1395 (1992).

R. E. Clausing, L. Heatherly, L. L. Horton, E. D. Specht, G. M. Begun, and Z. L. Wang, Textures and morphologies of chemical vapor deposited (CVD) diamond, *Diamond Relat. Mater.* 1, 411–415 (1992).

F. Cleri, P. Keblinski, L. Colombo, D. Wolf, and S. R. Phillpot, On the electrical activity of sp^2-bonded grain boundaries in nanocrystalline boundaries, *Europhys. Lett.* 46, 671–677 (1999).

E. Colineau, E. Gheeraert, A. Deneuville, J. Mambou, F. Brunet, and J. P. Lagrange, Minimization of the defect concentrations from boron incorporation in polycrystalline diamond films, *Diamond Relat. Mater.* 6, 778–782 (1997).

A. Cremades, F. Dominguez-Adame, and J. Piqueras, Study of defects in chemical vapor deposited diamond films by cross-sectional cathodoluminescence, *J. Appl. Phys.* 74, 5726–5728 (1993).

S. Dannefaer, W. Zhu, T. Bretagnon, and D. Kerr, Vacancies in polycrystalline diamond films, *Phys. Rev. B* 53, 1979–1984 (1996).

C. D'Anterroches and A. Bourret, Atomic structure of [011] and [001] near-coincident tilt boundaries in germanium and silicon, *Philos. Mag. A* 49, 783–807 (1984).

G. Davies, S. C. Lawson, A. T. Collins, A. Mainwood, and S. J. Sharp, Vacancy-related centers in diamond, *Phys. Rev. B* 46, 13157–13170 (1992).

S. Delclos, D. Dorignac, F. Philipp, S. Moulin, and A. M. Bonnot, UHREM investigation of stacking fault interactions in the CVD diamond structure, *Diamond Relat. Mater.* 8, 682–687 (1999).

F. K. de Theije and W. J. P. van Enckevort, Influence of nitrogen on the surface morphology of CVD grown diamond in "Properties, Growth and Applications of Diamond" Emis Datareviews Series no. 26 (M. H. Nazaré and A. J. Neves, Eds.), pp. 149–154. INSPEC, The Institution of Electrical Engineers, London, 2000, ISBN 0 85296 785 3.

M. P. D'Evelyn, J. D. Graham, and L. R. Martin, [100] versus [111] diamond growth from methyl radicals and/or acetylene, *J. Crystal Growth* 231, 506–519 (2001).

R. De Wit, Partial disclinations, *J. Phys. C: Solid State Phys.* 5, 529–534 (1972).

S. Delclos, D. Dorignac, F. Phillipp, F. Silva, and A. Gicquel, Ultra-high resolution microscopy investigation of growth defects in CVD diamond: twin interactions and fivefold twin centers, *Diamond Relat. Mater.* 9, 346–350 (2000).

D. Dorignac, S. Delclos, and F. Phillipp, Atomic structure of a complex defect configuration in synthetic diamond: a fivefold twin centre connected to two high-order grain boundaries, *Philos. Mag. B* 81, 1879–1891 (2001).

D. Dorignac, V. Serin, S. Delclos, F. Philipp, D. Rats, and I. Vandenbulcke, HREM and EXELFS investigation of local structure in thin CVD films, *Diamond Relat. Mater.* 6, 758–762 (1997).

S. Ergun and L. E. Alexander, Crystalline forms of carbon—a possible hexagonal polymorph of diamond, *Nature* 195, 765–767 (1962).

H. Eto, Y. Tamou, Y. Ohsawa, and N. Kikuchi, TEM observations of diamond films prepared by microwave plasma CVD, *Diamond Relat. Mater.* 1, 373–379 (1992).

C. J. Fall, A. T. Blumenau, R. Jones, P. R. Briddon, T. Frauenheim, A. Gutiérrez-Sosa, U. Bangert, A. E. Mora, J. W. Steeds, and J. E. Butler, Dislocations in diamond: electron energy-loss spectroscopy, *Phys. Rev. B* 65, 205208-1–7 (2002).

C. J. Fall, J. P. G. Goss, R. Jones, P. R. Briddon, A. T. Blumenau, and T. Frauenheim, Modelling electron energy-loss spectra of dislocations in silicon and diamond, *Physica B* 308–310, 577–580 (2001).

B. A. Fox, B. R. Stoner, D. M. Malta, P. J. Ellis, R. C. Glass, and F. R. Sivazlian, Epitaxial nucleation, growth and characterization of highly oriented (100)-textured diamond films on silicon, *Diamond Relat. Mater.* 3, 382–387 (1994).

M. Frenklach, R. Kematick, D. Huang, W. Howard, K. E. Spear, A. W. Phelps, and R. Koba, Homogeneous nucleation of diamond powder in the gas-phase, *J. Appl. Phys.* 66, 395–399 (1989).

C. Frondel and U. B. Marvin, Londsdaleite, a hexagonal polymorph of diamond, *Nature* 214, 587–689 (1967).

F. Fuchs, C. Wild, K. Schwarz, and P. Koidl, Hydrogen-related IR absorption in chemical-vapor deposited diamond, *Diamond Relat. Mater.* 4, 652–656 (1995).

T. Fujisaki, M. Tachiki, N. Taniyama, M. Kudo, and H. Kawarada, Fabrication of heteroepitaxial diamond thin films on Ir(001)/MgO(001) substrates using antenna-edge-type microwave plasma-assisted chemical vapor deposition, *Diamond Relat. Mater.* 11, 478–481 (2002).

T. Fujisaki, M. Tachiki, N. Taniyama, M. Kudo, and H. Kawarada, Initial growth of heteroepitaxial diamond on Ir(001)/MgO(001) substrates using antenna-edge-type microwave plasma-assisted chemical vapor deposition, *Diamond Relat. Mater.* 12, 246–250 (2003).

M.-N. Gamo, K. P. Loh, H. Haneda, S. Hishita, and T. Ando, Silicon incorporation into CVD diamond: a role of oxygen, *Appl. Phys. Lett.* 71, 629–631 (1997).

B. Gan, J. Ahn, Q. Zhang, S. F. Yoon, K. Rusli, C. Kerlit, Q. F. Huang, and V. A. Ligatchev, Influence of grain boundaries on the photoconductivity of undoped CVD diamond, *Radiat. Effect Defects Solids* 156, 203–207 (2001).

A. George, C. Escaravage, G. Champier, and W. Schröter, Velocities of screw and 60°-dislocations in silicon, *Phys. Status Solidi (b)* 53, 483–496 (1972).

J. P. Goss, R. Jones, M. I. Heggie, C. P. Ewels, P. R. Briddon, and S. Öberg, Theory of diamond, *Phys. Rev. B* 65, 115207-1–13 (2002).

J. E. Graebner, M. E. Reiss, L. Seibles, T. M. Hartnett, R. P. Miller, and C. J. Robinson, Phonon scattering in chemical-vapor-deposited diamond, *Phys. Rev. B* 50, 3702–3713 (1994).

M. H. Grimsditch and A. K. Ramdas, Brioullin-scattering in diamond, *Phys. Rev. B* 11, 3139–3148 (1975).

D. M. Gruen, Nanocrystalline diamond films, *Annu. Rev. Mater. Sci.* 29, 211–259 (1999).

D. M. Gruen, Growth of nanocrystalline diamond films in "Properties, Growth and Applications of Diamond" Emis Datareviews Series No. 26 (M. H. Nazaré and A. J. Neves, Eds.), pp. 303–306. INSPEC, The Institution of Electrical Engineers, London, 2000, ISBN 0 85296 785 3.

R. Haubner, A. Lindlbauer, and B. Lux, Diamond nucleation and growth on refractory metals using microwave plasma deposition, *Int. J. Refract. Met.* 14, 119–125 (1996).

C. Haug, H. Gärtner, J. Portmann, R. Samlenski, C. Wild, and R. Brenn, Ion channeling studies of hydrogen in homoepitaxially grrown CVD diamond, *Diamond Relat. Mater.* 10, 411–415 (2001).

M. I. Heggie, C. P. Ewels, N. Martsinovich, S. Scarle, R. Jones, J. P. Goss, B. Hourahine, and P. R. Briddon, Glide dislocations in diamond: first-principles calculations of similarities with and differences from silicon and the effects of hydrogen, *J. Phys.: Condens. Matter* 14, 12689–12696 (2002).

M. I. Heggie, S. Jenkins, C. P. Ewels, P. Jemmer, R. Jones, and P. R. Briddon, Theory of dislocations in diamond and silicon and their interaction with hydrogen, *J. Phys.: Condens. Matter* 12, 10263–10270 (2000).

J. P. Hirth and J. Lothe, "Theory of Dislocations". McGraw Hill, New York, 1968.

F. Hörmann, H. Y. Peng, Th. Bauer, Q. Li, M. Schreck, Y. Lifshitz, S. T. Lee, and B. Stritzker, Flat epitaxial diamond/Ir(001) interface visualised by high resolution transmission electron microscopy, *Surf. Sci.* 513, 525–529 (2002).

F. Hörmann, H. Roll, M. Schreck, and B. Stritzker, Epitaxial Ir layers on $SrTiO_3$ as substrates for diamond nucleation: deposition of the films and modification in the CVD environment, *Diamond Relat. Mater.* 9, 256–261 (2000).

J. Hornstra, Dislocations in the diamond lattice, *J. Phys. Chem. Solids* 5, 129–141 (1958).

J. Hornstra, Models of grain boundaries in the diamond lattice, *Physica* 25, 409–422 (1959).

W. Howard, D. Huang, J. Yuan, M. Frenklach, K. E. Spear, R. Koba, and A. W. Phelps, Synthesis of diamond powder in acetylene oxygen plasma, *J. Appl. Phys.* 68, 1247–1251 (1990).

H. Ichinose and M. Nakanose, Atomic and electronic structure of diamond grain boundaries analysed by HRTEM and EELS, *Thin Solid Films* 319, 87–91 (1998).

S. Iijima, Y. Aikawa, and K. Baba, Early formation of chemical vapour-deposition diamond films, *Appl. Phys. Lett.* 57, 2646–2648 (1990).

S. Ino, Epitaxial growth of metals on rocksalt faces cleaved in vacuum. II: orientation and structure of gold particles formed in ultrahigh vacuum, *J. Phys. Soc. Jpn* 21, 346–362 (1966).

S. Ino, Stability of multiply-twinned particles, *J. Phys. Soc. Jpn* 22, 941–954 (1969).

T. Ito, T. Abe, T. Inuzuka, and S. Koizumi, Observation of the carbon layer formed on a c-BN surface at the very early stage of diamond growth, *Diamond Relat. Mater.* 5, 53–55 (1996).

G. Janssen, W. J. P. von Enckefort, W. Vollenberg, and L. J. Giling, Anisotropy in monocrystalline CVD diamond growth, II. Defect structure of hot filament grown films deposited on planar substrates, *J. Crystal Growth* 148, 355–364 (1995).

X. Jiang and C. L. Jia, The coalescence of [001] diamond grains heteroepitaxially grown on (001) silicon, *Appl. Phys. Lett.* 69, 3902–3904 (1996).

X. Jiang and C. L. Jia, Direct local epitaxy of diamond on Si (100) and surface-roughening-induced crystal misorientation, *Phys. Rev. Lett.* 84, 3658–3661 (2000).

X. Jiang, C. L. Jia, M. Szameitat, and C. Rickers, Epitaxy of diamond on Si (100) and surface-roughening-induced crystal misorientation, *Phys. Rev. B* 64, 245413-1–5 (2001).

X. Jiang and C.-P. Klages, Heteroepitaxial diamond growth on (100) silicon, *Diamond Relat. Mater.* 2, 1112–1113 (1993).

X. Jiang, C.-P. Klages, R. Zachai, M. Hartweg, and H.-J. Füsser, Epitaxial diamond thin films on (001) silicon substrates, *Appl. Phys. Lett.* 62, 3438–3440 (1993).

X. Jiang, W. J. Zhang, M. Paul, and C.-P. Klages, Diamond film orientation by ion bombardment during deposition, *Appl. Phys. Lett.* 68, 1927–1929 (1996).

X. Jiang, R. Q. Zhang, G. Yu, and S. T. Lee, Local strain in interface: origin of grain tilting in diamond (001)/silicon (001) heteroepitaxy, *Phys. Rev. B* 58, 15351–15354 (1998).

R. Jones, C. J. Fall, A. Gutiérrez-Sosa, U. Bangert, M. I. Heggie, A. T. Blumenau, T. Frauenheim, and P. R. Briddon, Calculated and experimental low-loss electron energy loss psectra of dislocations in diamond and GaN, *J. Phys.: Condens. Matter* 14, 12793–12800 (2002).

H. Kanda and T. Sekine, High temperature high pressure synthesis of single crystal diamond in "Properties, Growth and Applications of Diamond" Emis Datareviews Series No. 26 (M. H. Nazaré and A. J. Neves, Eds.), pp. 247–255. INSPEC, The Institution of Electrical Engineers, London, 2000, ISBN 0 85296 785 3.

M. Kasu and N. Kobayashi, High mobility and high crystalline-quality chemical-vapor-deposition grown homoepitaxial diamond, *Diamond Relat. Mater.* 12, 413–417 (2003).

H. Kawarada, C. Wild, N. Herres, R. Locher, and P. Koidl, Heteroepitaxial growth of highly oriented diamond on cubic silicon carbide, *J. Appl. Phys.* 81, 3490–3493 (1997).

P. N. Keating, Effect of invariance requirements on elastic strain energy of crystals with application to diamond structure, *Phys. Rev.* 145, 637–645 (1966); P. N. Keating, Theory of 3rd-order elastic constants of diamond-like crystals, *Phys. Rev.* 149, 674–678 (1966).

P. Keblinski, S. R. Phillpot, D. Wolf, and H. Gleiter, *NanoStruct. Mater.* 12, 339–344 (1999).

P. Keblinski, D. Wolf, F. Cleri, S. R. Phillpot, and H. Gleiter, On the Nature of Grain Boundaries in Nanocrystalline Diamond, *MRS Bull.* 23, 36–41 (1998).

P. Keblinski, D. Wolf, S. R. Phillpot, and H. Gleiter, Role of bonding nd coordination in the atomic structure and energy of diamond and silicon grain boundaries, *J. Mater. Res.* 13, 2077–2099 (1998).

K. Kimura, K. Nakajima, S. Yamanaka, M. Hasegawa, and H. Okushi, Hydrogen analysis of CVD homoepitaxial diamond films by high-resolution elastic recoil detection, *Nucl. Instrum. Meth. Phys. Res. Section B—Beam Interact. Mater. Atoms* 190, 689–692 (2002).

A. H. King, F. R. Chen, L. Chang, and J. J. Kai, Toward understanding polycrystalline aggregate structure: analysis of a twin intersection and the intersection between interfaces in diamond, *Interf. Sci.* 5, 287–303 (1997).

D. S. Knight and W. B. White, Characterization of diamond films by Raman spectroscopy, *J. Mater. Res.* 4, 385–393 (1989).

G. Knuyt, M. Nesládek, K. Meykens, C. Quaeyhaegens, and L. M. Stals, On the {111}⟨111⟩ penetration twin density in CVD diamond films, *Diamond Relat. Mater.* 6, 1697–1706 (1997).

G. Knuyt, M. Nesládek, C. Quaeyhaegens, and L. M. Stals, On the behaviour of the penetration twin abundance on {100} diamond facets, *Diamond Relat. Mater.* 7, 1095–1099 (1998).

R. Kohl, C. Wild, N. Herres, P. Koidl, B. R. Stoner, and J. T. Glass, Oriented nucleation and growth of diamond films on beta-Sic and Si, *Appl. Phys. Lett.* 63, 1792–1794 (1993).

J. A. Kohn, Twinning in diamond-type structures—a proposed boundary-structure model, *Am. Miner.* 43, 263–284 (1958).

M. Kohyama, H. Ichinose, Y. Ishida, and M. Nakanose, Tight-binding calculation of grain boundaries in diamond, *Mater. Sci. Forum* 206–209, 261–264 (1996).

S. Koizumi, Heteroepitaxy of diamond on cubic boron nitride, *Isr. J. Chem.* 38, 33–40 (1998).

S. Koizumi, T. Inuzuka, T. Imuzuka, and K. Suzuki, Epitaxial growth of diamond thin-films on cubic boron notride (111) surfaces by DC plasma chemical vapour deoposition, *Appl. Phys. Lett.* 57, 563–565 (1990).

A. T. Kolodzie, M. Murfitt, and A. L. Bleloch, Detecting local energy states at dislocations in diamond, *Inst. Phys. Conf. Ser.* 168, 247–250 (2001) (Section 6).

F. Kroupa and L. Lejcek, Elastic interaction between wedge disclinations, *Phys. Status Solidi (b)* 51, K121–K124 (1972).

Y. Kuang, A. Badzian, T. T. Tsong, N. Lee, T. Badzian, and C. Chen, Scanning tunneling microscopy study of antiphase boundaries on the (001) surface of homoepitaxial diamond films, *Thin Solid Films* 272, 49–51 (1996).

J. P. Lagrange, A. Deneuville, and E. Gheeraert, Activation energy in low compensated homoepitaxial boron-doped diamond films, *Diamond Relat. Mater.* 7, 1390–1393 (1998).

A. R. Lang and G. Pang, On the dilatation of diamond by nitrogen impurity aggregated by A defects, *Philos. Trans. R. Soc. Series A* 156, 1397–1419 (1998).

S. T. Lee, H. Y. Peng, X. T. Zhou, N. Wang, C. S. Lee, I. Bello, and Y. Lifshitz, A Nucleation Site and Mechanism Leading to Epitaxial Growth of Diamond Films, *Science* 287, 104–106 (2000).

X. Z. Liao, R. J. Zhang, C. S. Lee, S. Tong Lee, and Y. W. Lam, The influence of boron doping on the structure and characteristics of diamond thin films, *Diamond Relat. Mater.* 6, 521–525 (1997).

W. Luyten, G. Van Tendeloo, S. Amelinckx, and J. L. Collins, Electron microscopy study of defects in synthetic diamond layers, *Philos. Mag. A* 66, 899–915 (1992).

H. Maeda, S. Ikari, T. Okubo, K. Kusakabe, and S. Morooka, Effect of substrate pre-treatment on diamond deposition in a microwave plasma, *J. Mater. Sci.* 28, 129–137 (1993).

H. Maeda, K. Ohtsubo, M. Irie, N. Ohya, K. Kusakabe, and S. Morooka, Determination of diamond [100] and [111] growth rate and formation of highly oriented diamond films by microwave plasma-assisted chemical vapor deposition, *J. Mater. Res.* 10, 3115–3123 (1995).

S. Matsumoto and Y. Matsui, Electron microscopic observation of diamond particles grown from the vapour phase, *J. Mater. Sci.* 18, 1785–1793 (1983).

K. M. McNamara, B. E. Williams, k. K. Gleason, and B. E. Scruggs, Identification of defects and impurities in chemical-vapor-deposited diamond through infrared-spectroscopy, *J. Appl. Phys.* 76, 2466–2472 (1994).

J. Michler, 2000. "Diamond and diamond-like carbon films: microstructure and assessment of mechanical properties." Ecole Polytechnique Fédérale de Lausanne, Switzerland, PhD Thesis No. 2115.

J. Michler, M. Mermoux, Y. von Kaenel, A. Haouni, G. Lucazeau, and E. Blank, Residual stress in diamond films: origins and modelling, *Thin Solid Films* 357, 189–201 (1999).

J. Michler, J. Stiegler, Y. von Kaenel, P. Moeckli, W. Dorsch, D. Stenkamp, and E. Blank, Microstructure evolution and non-diamond carbon incorporation in CVD diamond thin films grown at low substrate temperatures, *J. Crystal Growth* 172, 404–415 (1997).

J. Michler, Y. von Kaenel, J. Stiegler, and E. Blank, Complementary application of electron microscopy and micro-Raman spectroscopy for microstructure, stress, and bonding defect investigation of heteroepitaxial chemical vapor deposited diamond films, *J. Appl. Phys.* 83, 187–197 (1998).

A. E. Mora, J. W. Steeds, and J. E. Butler, Grain boundaries in boron-doped CVD diamond films, *Diamond Relat. Mater.* 11, 697–702 (2002).

J. Morris, 1996. Tight-binding model and calculational details. Reference: http://www.public.iastate.deu/edu/~jrmorris/grain/node2.html.

J. R. Morris, C. L. Fu, and K. M. Ho, Tight-binding study of tilt grain boundaries in diamond, *Phys. Rev. B* 54, 132 (1996).

S. Nakhmanson, 2001. Modeling methodology for amorphous silicon. Reference: http://plato.phys.ohiou.edu/~drabold.serge/node9.html.

A. S. Nandedkar and J. Narayan, Atomic structure of dislocations and dipoles in silicon, *Philos. Mag. A* 56, 625–639 (1987).

A. S. Nandedkar and J. Narayan, Atomic structure of dislocations in silicon, germanium and diamond, *Philos. Mag. A* 61, 873–891 (1990).

J. Narayan, Dislocations, twins, and grain boundaries in CVD diamond thin films: Atomic structure and properties, *J. Mater. Res.* 5, 2414–2422 (1990).

J. Narayan and A. S. Nandedkar, Atomic structure and energy of grain boundaries in silicon, germanium and diamond, *Philos. Mag. B* 63, 1181–1192 (1991).

J. Narayan, A. R. Srivatsa, M. Peters, S. Yokota, and K. V. Ravi, On epitaxial growth of diamond films on (100) silica substrates, *Appl. Phys. Lett.* 53, 1823–1825 (1988).

J. Narayan, A. R. Srivatsa, and K. V. Ravi, Mechanism of formation of ⟨110⟩ oriented microscrystallites in diamond films, *Appl. Phys. Lett.* 54, 1659–1661 (1989).

M. H. Nazaré and A. J. Neves, Eds., in "Properties, Growth and Applications of Diamond" Emis Datareviews Series No. 26 INSPEC, The Institution of Electrical Engineers, London, 2001, ISBN 0-85296-785-3.

A. A. Nazarov, O. A. Shenderova, and D. W. Brenner, On the disclination-structural unit model of grain boundaries, *Mater. Sci. Engng A* 281, 148–155 (2000a).

A. A. Nazarov, O. A. Shenderova, and D. W. Brenner, Elastic models of symmetrical ⟨001⟩ and ⟨011⟩ tilt grain boundaries, *Phys. Rev. B* 61, 928–936 (2000b).

K. Ohtsuka, K. Suzuki, A. Sawabe, and T. Inuzuka, Epitaxial growth of diamond on iridium, *Jpn. J. Appl. Phys.* 236, L1072–L1074 (1996).

T. Oku and K. Hiraga, Atomic structures and stability of hexagonal BN diamoond and Au multiply-twinned nanoparticles with five-fold symmetry, *Diamond Relat. Mater.* 10, 1398–1403 (2000).

H. Okushi, H. Watanabe, S. Ri, S. Yamanaka, and D. Takeuchi, Device-grade homoepitaxial diamond film growth, *J. Crystal Growth* 237–239, 1269–1276 (2002).

A. M. Papon and M. Petit, A survey of the geometrical reconstruction of [011] defects in semiconductor-grain-boundaries and dislocations, *Scripta Metall. Mater.* 19, 391–396 (1985).

D. J. Paritosh, C. C. Srolovitz, C. C. Bataille, X. Li, and J. E. Butler, Simulation of faceted film growth in two-dimensions: microstructure, morphology and texture, *Acta Mater.* 47, 2269–2281 (1999).

B. D. Patterson, Muonium States in Semiconductors, *Rev. Mod. Phys.* 60, 69–159 (1988).

A. U. Persson, A first-principle calculation of the intrinsic stacking fault energy in diamond, *Philos. Mag. A* 47, 835–839 (1983).

P. Pirouz, D. J. H. Cockayne, N. Sumida, P. Hirsch, and A. R. Lang, Dissociation of dislocations in diamond, *Proc. R. Soc. London A* 386, 241–249 (1983).

L. Pizzagalli, P. Beauchamp, and J. Rabier, Stability and core structure of undissociated screw dislocations in group IV materials investigated by means of atomistic calculations, *J. Phys.: Condens. Matter* 14, 12681–12688 (2002).

F. Pruvost, E. Bustarret, and A. Deneuville, Characteristics of homoepitaxial heavily boron-doped diamond films from their Raman spectra, *Diamond Relat. Mater.* 9, 295–299 (2000).

F. Pruvost and A. Deneuville, Analysis of the Fano in Diamond, *Diamond Relat. Mater.* 10, 531–535 (2001).

R. E. Rawles, W. G. Morris, and M. P. D'Evelyn, Effect of growth-rate ratio on surface morphology of homoepitaxial diamond (100) and (111), *Appl. Phys. Lett.* 69, 4032–4034 (1996).

S. G. Ri, H. Yoshida, S. Yamanaka, H. Watanabe, D. Takeuchi, and H. Okushi, Misorientation angle dependence of surface morphology in homoepitaxial diamond film growth at al CH_4/H_2 ratio, *J. Crystal Growth* 235, 300–306 (2002).

S. D. Rigby, K. A. Cherian, J. D. Lister, V. Rudolph, and E. T. White, Quantifying CVD diamond growth and morphology: a useful technique for studying growth kinetics, *Diamond Relat. Mater.* 9, 328–332 (2000).

A. E. Romanov, V. I. Vladimiro, Disclinations in solids in "Dislocations in Solids" (F. N. R. Nabarro, Ed.), Vol. 9, pp. 191–407, 1992.

H.-J. Ryu, J.-L. Lee, and J. H. Je, Near Perfect Heteroepitaxy of Diamond Islands on Si (111), *Chem. Vapor Deposit.* 7, 22–24 (2001).

D. Saada, J. Adler, and R. Kalish, Lowest-energy site for hydrogen, *Phys. Rev. B* 61, 10711–10715 (2000).

I. Sakaguchi, M. Nishitani-Gamo, K. P. Loh, H. Haneda, and T. Ando, Hydrogen incorporation control in high quality homoepitaxial diamond (111) growth, *Diamond Relat. Mater.* 8, 1291–1293 (1999).

I. Sakaguchi, M. Nishitani-Gamo, K. P. Loh, S. Hishita, H. Haneda, and T. Ando, Suppression of surface cracks on (111) homoepitaxial diamond through impurity limitation by oxygen addition, *Appl. Phys. Lett.* 73, 2675–2677 (1998).

R. Samlenski, C. Haug, R. Brenn, C. Wild, R. Locher, and P. Koidl, Incorporation of nitrogen in chemical vapor deposition diamond, *Appl. Phys. Lett.* 67, 2798–2800 (1995a).

R. Samlenski, C. Haug, R. Brenn, C. Wild, R. Locher, and P. Koidl, Characterisation and lattice location of nitrogen and boron in homoepitaxial CVD diamond, *Diamond Relat. Mater.* 5, 947–951 (1996).

R. Samlenski, J. Schmälzlin, R. Brenn, C. Wild, W. Müller-Sebert, and P. Koidl, Characterization of homoepitaxial diamond films by nuclear methods, *Diamond Relat. Mater.* 4, 503–507 (1995b).

S. Sattel, J. Gerber, and H. Erhardt, Ion induced nucleation of diamond, *Phys. Status Solidi (a)* 154, 141–153 (1996).

A. Sawabe, H. Fukuda, T. Suzuki, Y. Ikuhara, and T. Suzuki, Interface between CVD diamond and iridium films, *Surf. Sci.* 467, L845–L849 (2000).

M. Schreck, Th. Bauer, S. Gsell, F. Hörmann, H. Bielefeldt, and B. Stritzker, Domain formation in diamond nucleation on iridium, *Diamond Relat. Mater.* 12, 262–267 (2003).

M. Schreck, F. Hörmann, H. Roll, J. K. N. Lindner, and B. Stritzker, Diamond nucleation on iridium buffer layers and subsequent textured growth: a route for the realization of single-crystal diamond films, *Appl. Phys. Lett.* 78, 192–194 (2001).

M. Schreck, H. Roll, J. Michler, E. Blank, and B. Stritzker, Stress distribution in thin heteroepitaxial diamond films on Ir/SrTiO3 studied by X-ray diffraction, Raman spectroscopy, and finite element analysis, *J. Appl Phys.* 88, 2456–2466 (2000).

M. Schreck, H. Roll, and B. Stritzker, Diamond/Ir/$SrTiO_3$: a material combination for improved heteroepitaxial diamond films, *Appl. Phys. Lett.* 74, 650–652 (1999).

M. Schreck, A. Schury, F. Hörmann, H. Roll, and B. Stritzker, Mosaicity reduction during growth of heteroepitaxial diamond films on iridium buffer layers: experimental results and numerical simulations, *Phys. Rev. B* 91, 676–685 (2002).

M. Schreck, K.-H. Thurer, and B. Stritzker, Limitations of the process window for the bias enhanced nucleation of heteroepitaxial diamond films on silicon in the time domain, *J. Appl. Phys.* 81, 3092–3095 (1997).

H. Seehofer, 2002. Perturbation of the low-pressure diamond growth process by carbon supersaturation and radiofrequency plasma, PhD Thesis EPFL Nr. 2594.

D. Shechtman, Twin-determined growth of diamond films, *Mater. Sci. Engng A* 184, 113–118 (1994).

D. Shechtman, A. Feldman, and J. Hutchison, High-order twin boundaries in CVD diamond films, *Mater. Lett.* 17, 211–216 (1993a).

D. Shechtman, A. Feldman, M. D. Vaudin, and J. L. Hutchison, Moiré fringe images of twin boundaries in chemical vapor deposited diamond, *Appl. Phys. Lett.* 62, 487–489 (1993b).

D. Shechtman, J. L. Hutchison, L. H. Robins, E. N. Farabough, and A. Feldman, Growth defects in diamond films, *J. Mater. Res.* 8, 473–479 (1993c).

O. A. Sherendova and D. W. Brenner, Atomistic simulation of structures and mechanical properties of ⟨011⟩ tilt grain boundaries and their triple jonctions in diamond, *Phys. Rev. B* 60, 7053–7061 (1999).

O. A. Shenderova, D. W. Brenner, A. A. Nazarov, A. E. Romanov, and L. H. Yang, Multiscale modeling approach for calculating grain-boundary energies from first principles, *Phys. Rev. B.* 57, R3181–R3184 (1998).

O. A. Shenderova, D. W. Brenner, and L. H. Yang, Atomistic simulation of structures and mechanical properties of polycrystalline diamond: symmetrical ⟨001⟩ tilt grain boundaries, *Phys. Rev. B* 60, 7043–7052 (1999).

S. R. P. Silva, G. A. J. Amaratunga, E. K. H. Salje, and K. M. Knowles, Evidence of hexagonal diamond in plasma-deposited carbon films, *J. Mater. Sci.* 29, 4962–4966 (1994).

P. K. Sitch, R. Jones, S. Öberg, and M. I. Heggie, An *ab initio* study of the 90° partial dislocation core in diamond, *J. Phys. III Fr.* 7, 1381–1387 (1997).

K. E. Spear and M. Frenklach, Mechanisms for CVD Diamond Growth in "Synthetic Diamond: Emerging CVD Science and Technology" (K. E. Spear and J. P. Dismukes, Eds.), pp. 243–304. Wiley, New York, 1994, ISBN 0-471-53589-3.

K. E. Spear, A. W. Phelps, and W. B. White, Diamond polytypes and their vibrational spectra, *J. Mater. Res.* 5, 2277–2285 (1990).

M. Stammler, H. Eisenbeiss, J. Ristein, J. Neubauer, M. Gobbels, and L. Ley, Growth of high-quality homoepitaxial diamond films, *Diamond Relat. Mater.* 11, 504–508 (2002).

J. W. Steeds, A. Gilmore, K. M. Bussmann, J. E. Butler, and P. Koidl, On the nature of grain boundary defects in high quality CVD diamond films and their influence on physical properties, *Diamond Relat. Mater.* 8, 996–1005 (1999).

J. W. Steeds, A. Gilmore, S. Charles, P. Heard, B. Howarth, and J. E. Butler, Use of novel methods for the investigation of the boron distribution in CVD diamond, *Acta Mater.* 47, 4024–4030 (1999).

J. W. Steeds, A. Gilmore, J. A. Wilson, and J. E. Butler, On the nature of extended defects in CVD diamond and the origin of compressive stresses, *Diamond Relat. Mater.* 7, 1437–1450 (1998).

J. W. Steeds, A. E. Mora, J. E. Butler, and K. M. Bussmann, Transmission electron microscopy investigation of boron-doped polycrystalline chemically vapour deposited diamond, *Philos. Mag. A* 82, 1741–1768 (2002).

J. Stiegler, A. Bergmaier, J. Michler, Y. von Kaenel, G. Dollinger, and E. Blank, Impurity and defect incorporation in diamond films deposited at low substrate temperature, *Diamond Relat. Mater.* 7, 193–199 (1998).

J. Stiegler, J. Michler, and E. Blank, An investigation of structural defects in diamond films grown at low substrate temperature, *Diamond Relat. Mater.* 8, 651–656 (1999).

J. Stiegler, Y. von Kaenel, M. Cans, and E. Blank, Space filling by nucleation and growth in chemical vapour deposition of diamond, *J. Mater. Sci.* 11, 716–726 (1996).

F. H. Stillinger and T. A. Weber, Computer simulation of local order in condensed physes of silicon, *Phys. Rev. B* 31, 5262–5271 (1985).

B. Stoner and J. T. Glass, Textured diamond growth on (100) beta-SiC via microwave plasma chemical vapor-deposition, *Appl. Phys. Lett.* 60, 698–700 (1992).

B. R. Stoner, G. H. M. Ma, S. D. Wolter, and J. T. Glass, Characterization of bias-enhanced nucleation of diamond on silicon by *in vacuo* surface analysis and transmission electron microscopy, *Phys. Rev. B* 45, 11067–11084 (1992).

H. Sun, S. Yu, Z. Jiang, D. Li, L. Liu, Z. Jin, and G. Zou, Twin-step-related growth phenomena of thick diamond film, *Diamond Relat. Mater.* 5, 1308–1312 (1996).

H. Sun, S. Yu, Z. Jiang, D. Li, L. Liu, Z. Jin, and G. Zou, Dissociation of {111} faces into {110} faces in thick diamond films, *J. Crystal Growth* 177, 162–166 (1997).

L. F. Sutcu, C. J. Chu, M. S. Thompson, R. H. Hauge, J. L. Margave, and M. P. D'Evelyn, Atomic force microscopy of (100), (110), and (111) homoepitaxial films, *J. Appl. Phys.* 71, 5930–5940 (1992).

K. Suzuki, M. Ichihara, S. Takeuchi, N. Ohtake, M. Yoshikawa, K. Hirabayashi, and N. Kurihara, Electron microscopy studies of dislocations in diamond synthesized by a CVD method, *Philos. Mag. A* 65, 657–664 (1992).

T. Takami, K. Suzuki, I. Kusunoki, I. Sakaguchi, M. Nishitani-Gamo, and T. Ando, Diamond thin film grown homoepitaxially on diamond (001) substrate by microwave plasma CVD method studied by reflection high-energy electron diffraction and atomic force microscopy, *Surf. Sci.* 440, 103–115 (1999).

D. Takeuchi, H. Watanabe, S. Yamanaka, H. Okushi, and K. Kajimura, Defects in device grade homoepitaxial diamond thin films grown with ultra-low CH_4/H_2 conditions by microwave-plasma chemical vapor deposition, *Phys. Status Solidi (a)* 174, 101–115 (1999).

D. Takeuchi, H. Watanabe, S. Yamanaka, H. Okushi, and K. Kajimura, Homoepitaxial diamond films grown by step-flow mode in various misorientation angles of diamond substrates, *Diamond Relat. Mater.* 9, 231–235 (2000).

D. Takeuchi, H. Watanabe, H. Sawada, S. Yamanaka, H. Ichinose, T. Sekiguchi, and H. Okushi, Origin of band-A emission in homoepitaxial diamond films, *Diamond Relat. Mater.* 10, 526–530 (2001a).

D. Takeuchi, H. Watanabe, S. Yamanaka, H. Okushi, H. Sawada, H. Ichinose, T. Sekiguchi, and K. Kajimura, Origin of band-A emission in diamond thin films, *Phys. Rev. B* 63, 245328-1–245328-7 (2001b).

D. Takeuchi, S. Yamanaka, H. Watanabe, and H. Okusi, Device grade B-doped homoepitaxial diamond thin films, *Phys. Status Solidi (a)* 186, 269–280 (2001c).

M. A. Tamor and M. P. Everson, On the role of penetration twins in the morphological development of vapor-grown films, *J. Mater. Res.* 9, 1839–1848 (1994).

Y. Tamou, Y. Ohsawa, and N. Kikuchi, Study of diamond deposition from the gas phase using transmission electron microscopy in "Science and Technology of New Diamond" (S. Saito, O. Fukunaga, and M. Yoshikawa, Eds.), pp. 25–28. KTK Scientific Publishers/Terra Scientific Publishing Company, 1990.

M. Tarutani, Y. Shimato, Y. Takai, and R. Shimizu, Cross-sectional transmission electron microscopy study of diamond particles grown on a mirror-polished Si substrate, *Appl. Phys. Lett.* 67, 632–634 (1995).

M. Tarutani, Y. Takai, R. Shimizu, T. Ando, M. Kamo, and Y. Bando, Transmission electron microscopy study of interface and internal defect structures of homoepitaxial diamond, *Appl. Phys. Lett.* 68, 2070–2072 (1996).

J. Tersoff, New empirical approachg for thr structure and energy of covalent systems, *Phys. Rev. B* 37, 6991–7000 (1988).

J. Tersoff, Empirical interatomic potential for carbon, with application to amorphous-carbon, *Phys. Rev. Lett.* 61, 2879–2882 (1988).

K. H. Thürer, M. Schreck, and B. Stritzker, Limiting processes for diamond epitaxial alignement on silicon, *Phys. Rev. B* 57, 15454–15464 (1998).

W. Tiller, "The Science of Crystallization: Microscopic Interfacial Phenomena". Cambridge University Press, Cambridge, UK, 1991.

C. Uzan-Saguy, C. Cytermann, B. Fizgeer, V. Richter, Y. Avigal, M. Shaanan, R. Brener, R. Kalish, E. Bustarret, and J. Chevallier, Diffusion of hydrogen from a microwave plasma into diamond and ist interaction with dopants and defects, *Diamond Relat. Mater.* 11, 316–322 (2002).

A. van der Drift, Evolutionary selection, a principle governing growth orientation in vapour-deposited layers, *Philips Res. Rep.* 22, 267–288 (1967).

W. J. P. von Enckefort, G. Janssen, J. J. Schermer, and L. J. Giling, Step-related growth phenomena on exact and misoriented {001} surfaces of CVD-grown single-crystal diamonds, *Diamond Relat. Mater.* 4, 250–255 (1995).

W. J. P. von Enckefort, G. Janssen, W. Vollenberg, and L. J. Giling, Anisotropy in monocrystalline CVD diamond growth, III. Surface morphology of hot filament grown films deposited on planar substrates, *J. Crystal Growth* 148, 365–382 (1995).

Y. von Kaenel, 1997. Défauts et propriétés des couches minces de diamant texturées élaborées par dépôt chimique en phase vapeur, Ecole Polytechnique Fédérale de Lausanne, Thesis Nr. 1624.

Y. von Kaenel, J. Stiegler, E. Blank, O. Chauvet, Ch. Hellwig, and K. Plamann, Microstructure Evolution and Defect Incorporation in Highly Oriented and Textures CVD Diamond films, *Phys. Status. Solidi (a)* 154, 219–238 (1996).

Y. von Kaenel, J. Stiegler, J. Michler, and E. Blank, Stress distribution in heteroepitaxial chemical vapor deposited diamond films, *J. Appl. Phys.* 81, 1726–1736 (1997).

S. Waidmann, K. Bartsch, I. Endler, F. Fontaine, B. Arnold, M. Knupfer, A. Leonhardt, and J. Kink, Electron energy-loss spectroscopy in transmission od undoped and doped diamond films, *Carbon* 37, 823–827 (1999).

Z. L. Wang, J. Bentley, R. E. Clausing, L. Heatherley, and L. L. Horton, Direct correlation of microtwin distribution with growth face morphology of diamond films by a novel TEM technique, *J. Mater. Res.* 9, 1552–1565 (1994).

C. L. Wang, M. Irie, and T. Ito, Growth and characterization of hillock-free high quality homoepitaxial diamond films, *Diamond Relat. Mater.* 9, 1650–1654 (2000).

B. B. Wang, W. L. Wang, K. J. Liao, and J. L. Xiao, Experimental and theoretical studies of diamond nucleation on silicon by biased hot filament chemical vapor deposition, *Phys. Rev. B* 63, 085412-1–085412-12 (2001).

C. Wild, N. Herres, and P. Koidl, Texture formation in polycrystalline diamond films, *J. Appl. Phys.* 68, 973–978 (1990).

C. Wild, P. Koidl, W. Müller-Sebert, H. Walcher, R. Kohl, N. Herres, R. Locher, R. Samlenski, and R. Brenn, Chemical vapour deposition and characterisation of smooth {100}-faceted diamond films, *Diamond Relat. Mater.* 2, 158–168 (1993).

C. Wild, R. Kohl, N. Herres, W. Müller-Sebert, and P. Koidl, Oriented CVD diamond growth twin formation, structure and morphology, *Diamond Relat. Mater.* 3, 373–381 (1994).

J. Wilks, E. Wilks, "Properties and Applications of Diamond". Butterworth-Heinmann Ltd, Oxford, 1991, p. 82.

B. E. Williams, J. T. Glass, R. F. Davis, K. Kobashi, and K. L. More, Electron microscopy of diamond films and diamond substrate/interfaces in "Science and Technology of New Diamond" (S. Saito, O. Fukunaga, and M. Yoshikawa, Eds.), pp. 319–326. KTK Scientific Publishers/Terra Scientific Publishing Company, (1990a).

B. E. Williams, H. S. Kong, and J. T. Glass, Electron microscopy of vapor phase deposited diamond, *J. Mater. Res.* 5, 801–810 (1990b).

S. D. Wolter, B. R. Stoner, J. T. Glass, P. J. Ellis, D. S. Buhaenko, C. E. Jenkins, and P. Southworth, Testured growth of diamond on silicon via in-situ carburization and bias-enhanced nucleation, *Appl. Phys. Lett.* 62, 1215–1217 (1993).

P. Wurzinger, N. Fuchs, P. Pongratz, M. Schreck, R. Hessmer, and B. Stritzker, TEM investigations on the heteroepitaxial nucleation of CVD diamond on (001) silicon substrates, *Diamond Relat. Mater.* 6, 752–757 (1997a).

P. Wurzinger, M. Joksch, and P. Pongratz, Defects in polycrystalline CVD diamond layers analysed with the TEM, *Inst. Phys. Conf. Ser. No. 135, Section 3*, 157–164 (1993).

P. Wurzinger, M. Joksch, and P. Pongratz, Defects and microstructure of CVD diamond as observed with the TEM, *Int. Phys. Conf. Ser. No. 146*, 523–528 (1995).

P. Wurzinger, P. Pongratz, P. Hartmann, R. Haubner, and B. Lux, Investigation of the boron incorporation in polycrystalline CVD diamond films by TEM, EELS and Raman spectroscopy, *Diamond Relat. Mater.* 6, 763–768 (1997b).

C. H. Xu, C. Z. Wang, C. T. Chan, and K. M. Ho, A transferable tight-binding potential for carbon, *J. Phys.: Condens. Matter.* 4, 6047–6054 (1992).

H. Yagyu, Y. Mori, A. Hatta, T. Ito, and A. Hiraki, Observations of the initial stage of chemical-vapor-deposited diamond growth using transmission electron microscopy, *Jpn. J. Appl. Phys.* 32, L1775–L1777 (1993).

C. S. Yan and Y. K. Vohra, Multiple twinning and nitrogen defect center in chemical vapor deposited homoepitaxial diamond, *Diamond Relat. Mater.* 8, 2022–2031 (1999).

H. Yoneda, K. Tokuyama, R. Yamazaki, and K.-I. Ueda, Effect of grain boundaries on carrier lifetime in chemical-vapor-deposited diamond film, *Appl. Phys. Lett.* 77, 1425–1427 (2000).

H. Yoneda, K. Ueda, Y. Aikawa, K. Baba, and N. Shenata, The grain size dependence of the mobility and lifetime in chemical vapor deposited diamond photoconductive switches, *J. Appl. Phys.* 83, 1730–1733 (1998).

S. Yugo, T. Kanai, T. Kimura, and T. Muto, Generation of diamond nuclei by electric-field in plasma chemical vapour-deposition, *Appl. Phys. Lett.* 58, 1036–1038 (1991).

S. Yugo, T. Kimura, and T. Kanai, Nucleation mechanisms of diamond in plasma chemical-vapor-deposition, *Diamond Relat. Mater.* 2, 328–332 (1993).

S. Yugo, K. Semoto, K. Hoshina, T. Kimura, and H. Nakai, A modelling of diamond nucleation, *Diamond Relat. Mater.* 4, 903–907 (1995).

P. Zapol, M. Sternberg, L. A. Curtiss, T. Frauenheim, and D.M. Gruen, Tight-binding molecular-dynamics simulation of impurities in ultrananocrystalline diamond grain boundaries, *Phys. Rev.* B 65, 045403-1–11 (2001).

R. Zeisel, C. E. Nebel, and M. Stutzmann, Capacitance–voltage profiling of deuterium passivation and diffusion in diamond Schottky diodes, *Diamond Relat. Mater.* 9, 413–416 (2000).

Y. Zhang, H. Ichinose, K. Ito, Y. Ishida, and M. Nakanose, Grain boundary structure and growth sequence of diamond thin film, *Mater. Sci. Forum Vols.* 204–206, 207–214 (1996).

R. Q. Zhang, W. L. Wang, J. Estevea, and E. Bertran, Molecular mechanics simulation of the diamond nucleation and growth on silicon (001) and (111) surfaces, *Thin Solid Films* 317, 6–9 (1998).

R. Q. Zhang, W. J. Zhang, C. Sun, X. Jiang, and S.-T. Lee, Theoretical study on misorineted diamond nucleations on Si(001) surface, *Diamond Relat. Mater.* 8, 1418–1422 (1999).

X. Zhou, G. D. Watkins, K. M. McNamara Rutledge, R. P. Messmer, and S. Chawla, Hydrogen-related defects in polycrystalline CVD diamond, *Phys. Rev. B* 54, 7881–7890 (1996).

W. Zhu, Defects in Diamond in "Diamond: Electronic Properties and Applications" (L. S. Pan and D. R. Kania, Eds.), pp. 175–239. Kluwer Academic Publishers, 1994, Chapter 5, ISBN 0-7923-9524-7.

W. Zhu, A. R. Badzian, and R. Messier, Structural imperfections in CVD diamond films, *J. Mater. Res.* 4, 659–663 (1989).

W. Zhu, X. H. Wang, B. R. Stoner, G. H. M. Ma, H. S. Kong, M. W. H. Braun, and J. T. Glass, Diamond and β-SiC heteroepitaxial interfaces: a theoretical and experimental study, *Phys. Rev. B* 47, 6529–6542 (1993).

CHAPTER 3

DOPING DIAMOND BY ION-IMPLANTATION

R. Kalish

DEPARTMENT OF PHYSICS, SOLID STATE INSTITUTE, TECHNION-ISRAEL INSTITUTE OF TECHNOLOGY, HAIFA 32000, ISRAEL

1. Introduction

Ion-implantation is a technique in which ions are forced into the near surface region of a solid target. It, therefore, enables one to introduce almost any impurity into any desirable material thus modifying its properties, regardless of the phase diagram, solubility and thermodynamic equilibrium considerations. Ion-implantation is widely used for the controlled doping of commonly used semiconductors, rendering them either p- or n-type. It is an essential part in semiconductor processing, and without it no state-of-the-art micro-electronic devices could be manufactured. In the case of doping diamond, for which impurity in-diffusion is most unlikely, the practical modes for the introduction of dopants are either during growth or by ion-implantation. The variety of dopant atoms that can be introduced into the crystal during CVD growth is, so far, rather limited; hence ion-implantation offers a convenient way of forcing any desired dopant into diamond. However, ion-implantation into diamond is particularly complicated due to the metastability of diamond with respect to graphite, as will be explained in detail below.

Ion-implantation is a high-energy process in which the ion energies employed typically exceed the binding energies of atoms in a solid by several orders of magnitude; hence ion-implantation is always accompanied by severe damage to the target. This damage is of particular interest in the case of a diamond target, since its sp^3 bonding configuration is metastable. Broken bonds may, upon annealing, reform as diamond sp^3 bonds or may transform into the more stable sp^2 graphitic bonding configuration.

The physical processes which slow down an ion moving in a solid can be roughly divided into two dominant mechanisms: electronic energy loss, involving the interaction between the incident ion and electrons of the host material, and nuclear energy loss, involving the interaction between the incident ions and the atoms (considered as shielded point charges) of the host material. It is the latter process which inflicts most damage to the implanted material. Once the energy loss of the moving ion in the target material is known, both the ion range and the defect distributions can be calculated. For a review of the ion–solid

ISBN: 0-12-752185-2
ISSN: 0080-8784

interactions which describe the physics of the energy loss of ions moving in a solid, see for example Eckstein (1991).

Practical information about the implant distribution in the target and the related damage inflicted to it is commonly obtained from computer simulations. The most widely used code is the TRIM program (Transport of Ions in Matter) developed by Biersack and Haagermark (1980) in which the best known semi-empirical values for the stopping powers as well as the statistical nature of the slowing down process are incorporated. The TRIM (or the later SRIM version) program is a Monte Carlo program which follows the collisions that individual ions undergo while in motion in the target material. The input parameters to the program are the required experimental conditions (ion type and energy, target material, composition and density) and an intrinsic parameter E_d, which is the energy required to displace a target atom far enough from its lattice site so that it will not fall back into the vacancy that it has left behind.

The procedure followed by the program is to shoot the required ion into the target; the ion will then proceed its motion in the target while losing energy continuously by electronic processes until it collides with a target atom with a randomly selected impact parameter. The TRIM program calculates the kinematics of the collision, and then follows the trajectory of the recoiling target atom which itself undergoes collisions that may put new target atoms in motion, thus creating a branch of recoiling target atoms. Once all recoiling target atoms have attained energies below a certain cutoff energy the program will return and follow the motion of the original implant. The projectile may now proceed with its new energy and direction of motion until another statistically selected collision occurs with its branch of recoils. The procedure continues until the energy of the implant itself is below E_d, and hence the implant cannot dislodge any more target atoms and it comes to rest. The program now stores the collision history of the primary ion and of all secondary events and proceeds to follow a new ion shot into the target.

Even though the TRIM simulation gives excellent insight into the implantation process and the damage that accompanies ion-implantation, it ignores several features which are of importance in the real implantation process. The program, for example, assumes a statistically homogeneous distribution of the host atoms, according to the average density of the target materials; hence, effects that are due to the periodicity of atoms in a single crystal (channeling) are neglected by the program. TRIM does not allow for any recombinative annihilation of defects (i.e., no vacancy–interstitial recombination is included in the program), so that TRIM always overestimates the absolute magnitude of the damage. The program also does not allow for the rearrangement of atoms in the implantation-affected volume to form new phases, such as the conversion of diamond to graphite. Nevertheless, TRIM and similar computer simulation programs give excellent estimates of both the implant and the damage profiles, and hence offer very valuable information on the state of a target after ion-implantation.

Typical damage cascades created by a single light (C) and a single heavy (Xe) ion, implanted at 100 and 320 keV, respectively, into diamond are shown in Figure 1a and b. Figure 2a and b shows the statistical distributions of the implanted ions and of the vacancies created by the above C and Xe implantations into diamond (using $E_d = 52$ eV). Note the large difference in the average number of vacancies created by each carbon or xenon ion is 92 and 627 vacancies per ion, respectively. Hence, the total damage and the density of the damage cascades of heavy ion implants are much larger than for light ion implants.

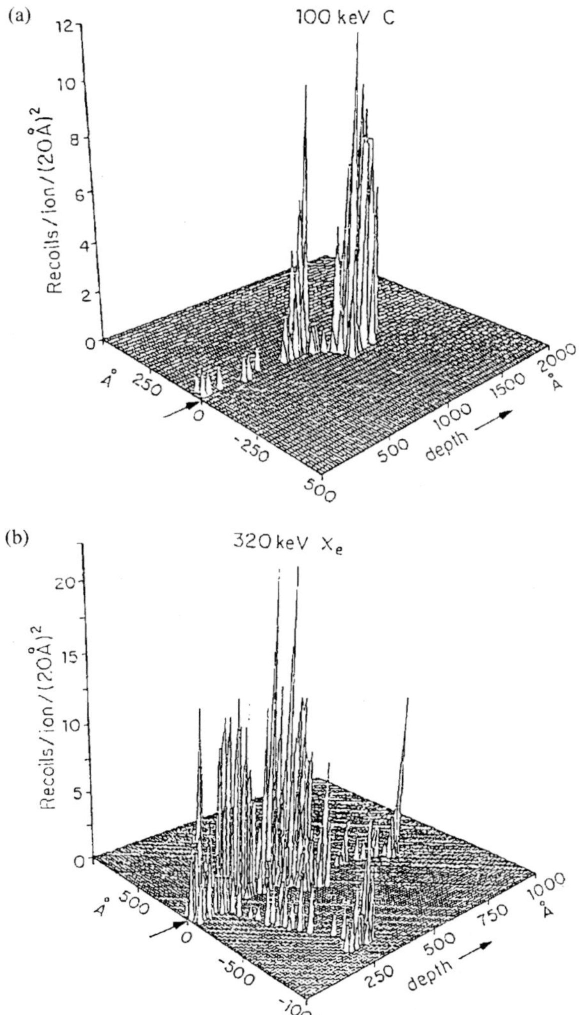

FIG. 1. Collision cascade following the penetration into diamond of a single 100 keV C ion (a) and a single 320 keV Xe ion (b) (from Prawer and Kalish (1995)).

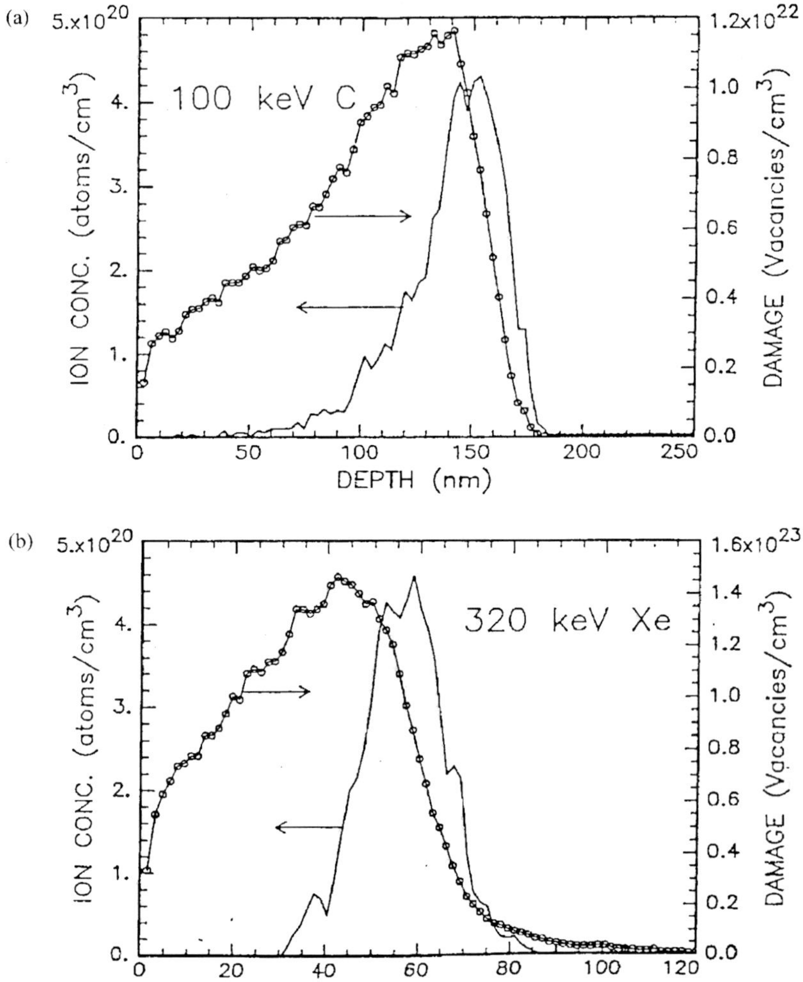

FIG. 2. TRIM calculated depth profiles of implanted species and vacancy distributions for a disc of 1×10^{15} ions per cm^2 of 100 keV C (a) and 320 keV Xe implantations (b) (from Prawer and Kalish (1995)).

In most cases the damage that accompanies ion-implantation is undesirable and needs to be annealed out. This is particularly important when the purpose of the implantation is doping of semiconductors, where ion-implantation offers many advantages over other doping techniques. The removal of the damage requires the annihilation of the implantation related defects. This is achieved by imparting sufficient energy to the defects to allow the recombination of vacancies and interstitials, for the case of elementary point defects, and to break up complex defect agglomerates, such as dislocation loops. The first of these two requires the diffusion of the defects so that they can meet and annihilate, or disappear at the surface, whereas the second requires the break up of the defect complex and then the disappearance of the liberated defects. Annealing is usually achieved by heating the sample to sufficient temperatures to allow these processes to occur. It can be carried out in furnaces, in systems which allow rapid heating of the sample, or in situ, during the implantation. In the latter case the formation of some defects is inhibited from the very beginning. In the particular case of diamond, annealing the damage is complicated due to the high temperatures required to achieve complete defect removal, as will be mentioned below.

Another requirement of the annealing for effective doping of semiconductors is to drive the implants into sites in the crystal (usually the substitutional site) where they are expected to act as donors or acceptors. This, again, is achieved by thermal treatment of the specimens.

As will be shown below, diamond poses particular complications but it also offers particular possibilities regarding annealing of post-implantation defects. Both have to do with the tendency of damaged diamond to graphitize.

It is important to realize that for the case of diamond the defects that accompany the implantation process may give rise to electrical and optical phenomena. These must be understood and avoided by proper implantation or annealing schemes, if real chemical doping due to the presence of the implanted impurities is sought for.

Below we review the electrical phenomena associated with the presence of native defects in diamond, i.e., those which accompany the implantation process. We then review the transformations that damaged diamond undergoes as a result of thermal annealing. We briefly survey the dopant elements to be considered to yield p- and n-type conduction in diamond as concluded from various doped diamond growth experiments.

With these at hand we turn to discuss the accepted ways of achieving chemical doping of diamond by ion-implantation.

Several reviews were written over the years on this important subject, and the reader is referred to those for further, though not always up-to-date information (Prins, 1992a; Dresselhaus and Kalish, 1992, 1997; Davies, 1994a; Kalish and Uzan-Saguy, 2001; Kalish and Prawer, 2003).

2. Ion-Implantation Related Damage in Diamond

2.1. General Considerations

The slowing down of energetic ions in diamond is accompanied by massive bond breakage and by the creation of interstitials and vacancies. The density of these grows with increasing ion dose until the whole volume affected by the ions (usually extending from the surface to about the end of the ion range) is amorphized.

The transformation of diamond to graphite, which occurs naturally, but at an extremely slow rate, can be much enhanced when a sufficient number of sp^3 bonds are broken. This change in bonding of the carbon atoms is noticeable by drastic changes in physical and chemical properties of the material. It is noticeable by a large decrease in density, a tremendous increase in electrical conductivity, a mechanical weakening of the material, the appearance of new optical absorption and Raman lines and by changes in chemical properties (graphite being etchable, while diamond is chemically extremely inert). These changes, all of which are enhanced upon heating, have indeed been observed in implanted diamond and have been used to learn about damage mechanisms in diamond. Hence the basic physics of passage of ions through matter in general can be beautifully studied through the particular case of diamond.

The implantation-induced damage in diamond and its annealing is influenced by the different diffusivities of the two elementary damage products, namely, vacancies and interstitials. Carbon interstitials seem to become mobile in diamond at about 50 K (Koike, Mitchelle and Parkers, 1991) or around room temperature while vacancies probably start to diffuse only at about 700 K (Davies, 1994b). This may cause an imbalance between interstitials and vacancies in the vicinity of the implant end-of-range (depending on the implantation temperature) which has important implications regarding the possibility of annealing-implanted diamond, as will be discussed below.

The physics that leads to the amorphization of ion-damaged diamond and its eventual graphitization is still under dispute. The fact that the electrical conductivity of diamond rises dramatically when exposed to ion-implantation at or above a particular dose has been interpreted by Kalish, Bernstein, Shapiro and Talmi (1980) and Prawer and Kalish (1995) within a percolation picture.

By performing in situ measurements of the resistivity as function of temperature $R(T)$ of intrinsic diamonds damaged by different ions, different implantation doses at different implantation temperature (T_i), information about the graphitization process and about conduction mechanism in the damaged diamond could be obtained (Prawer and Kalish, 1995).

Based on the assumption that at a particular implantation dose, D_c, a percolative conductive pathway is formed between graphitized (or partially graphitized) electrically conductive regions around each ion track (Kalish et al., 1980) the following conclusions can be drawn about the creation of damage by single ion implant in diamond and their dependence on target temperature:

(1) For low implantation temperatures, clusters of point defects between which hopping conduction occurs are formed around each ion track.
(2) At higher implantation temperatures, these clusters (assumed for simplicity to be spheres) shrink to a smaller average radius due to dynamic annealing

until, at high enough implantation temperatures, they never form to begin with. At low doses nearest-neighbor hopping occurs between these spheres, but at a critical dose percolation takes place to form a conductive pathway with graphite-like conductivity.

(3) The temperature dependence of the shrinkage of the radii of the conducting spheres, as deduced from percolation theory, is given in Figure 3. It is well described by the Morehead and Crowder (1970) approach that has been developed to describe the ion-beam induced amorphization of Si. This model assumes that annealing takes place in a sheath surrounding the damage spheres, and that the thickness of the sheath increases with increasing implantation temperature. The fit of this model to the data for damaged diamond predicts the complete disappearance of the conducting spheres due to instantaneous annealing for implantation temperatures exceeding 815 K (Figure 3 and point (2) above). Under these conditions, graphitization of the diamond will never occur regardless of the ion dose. The model also predicts an activation energy for defect (probably interstitial) diffusion during ion irradiation of about 0.2 eV.

(4) The functional dependences of the electrical resistance (R) of the implanted diamond on implantation dose D are remarkably similar in general shape for different implanted ions (Xe and C), apart from the doses required to produce a given conductivity. These scale according to the density of energy deposited by nuclear collisions, thus demonstrating that it is the density of collisionally induced defects which governs the electrical conductivity in ion-damaged diamond.

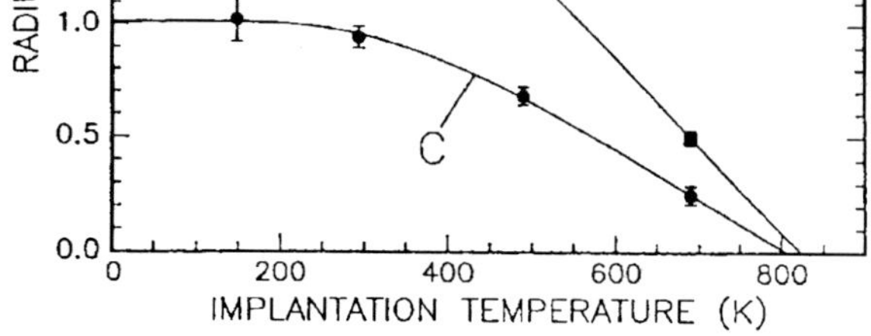

FIG. 3. The radii of the damage spheres as a function of implantation temperature for 100 keV C and keV Xe irradiations. The solid lines are fits to the data which allow the extraction of an activation energy of about 0.2 eV for defect diffusion during ion-implantation (from Prawer and Kalish (1995)).

Further experimental evidence for the process of damage formation with increasing ion dose is found in recent work of Orwa, Nugent, Jamieson and Prawer (2000). In that work, a deeply buried damage layer was created in diamond by MeV He ion-implantation. Spatially resolved confocal Raman spectra were taken along the ion pathway, from the surface down to and beyond the end-of range. Hence information about the nature of damage in diamond as a function of the density of the damage cascades could be obtained. The picture that emerges from that work is that radiation damage creates tiny amorphous regions along each ion track which are isolated and far apart for low-dose implantations. Interspersed between these are point defects in an otherwise damage free diamond lattice. With increasing ion dose, the amorphized regions grow in number until, at sufficiently high damage levels, the amorphized regions overlap to yield a continuous amorphized damage region. Hence the Raman results fully support the percolation picture proposed to explain the electrical behavior of the ion-damaged diamond. In addition to the general picture offered by the Raman data, the annealing behavior of different lines in the spectra could be identified with the annealing of particular kinds of defects in the diamond lattice (Orwa et al., 2000): the Raman line at 1490 cm^{-1} has been associated with vacancies whereas the line at 1630 cm^{-1} has been proposed to be due to an interstitial related defect, presumably the $\langle 100 \rangle$ split interstitial dumbbell defect.

Molecular dynamic calculations were recently performed by Saada, Adler and Kalish (1999) to test the above model for the formation of defects around an ion track in diamond. In these computations, momentum was imparted to a particular C atom in the diamond lattice and the evolution of the damage around the track of the recoiling ion was simulated. The transformation that the thus disrupted bonds undergo upon annealing was also studied. Indeed, the $\langle 100 \rangle$ dumbbell defect was found in the damaged region and graphitic sheets are observed to form upon annealing, confirming the above picture.

In several recent publications Prins (Prins, 2001a and references therein) has objected to the above graphitization model based on the Crowder and Morehead model. Instead, Prins has proposed that the amorphization and eventual graphitization of ion-damaged diamond is a gradually occurring phenomenon. Prins distinguishes between the probability of an interstitial 'meeting-up' with a vacancy and the probability of 'recombining' with it. Based on random walk statistical considerations, Prins comes up with ideas about the optimal annealing conditions for achieving best recovery of the diamond crystal. For the argument behind this theory and the assumptions underlying this approach the reader is referred to Prins (2001a).

Regardless of the exact physical picture that describes the evolution of damaged diamond to graphite, it is clear, and undisputed, that there exists a certain critical density of defects N_c (which is caused by a certain critical implantation dose D_c which depends on ion species, energy and implantation temperature) for which, when exceeded, the damaged diamond will collapse upon annealing to graphite (Uzan-Saguy, Cytermann, Brener, Shaanan and

Kalish, 1995a). For implantation doses below this critical damage density the damaged diamond can be annealed back to its original diamond structure whereas for damage densities in excess of the critical value graphitization of the damaged layer occurs upon annealing. This has been verified by several experiments which hinge on the fundamental differences between diamond and graphite as observed by optical absorption, Raman spectroscopy, electrical conductivity, chemical inertness and atomic density. Clearly, for doping purposes by implantation of acceptor or donor ions, the low implantation dose regime, for which the diamond can be annealed, is required.

2.2. Electrical Properties of Ion-damaged Diamond

Those aspects by which the ion-induced damage influences the electrical properties of diamond are of great relevance to implantation doping experiments as their presence may cause electrical effects which may be mistaken for real chemical doping. These electrical effects can be manifested in the following measurable properties: (a) creation of defect-related energy levels within the band gap which may either give rise to donor or acceptor states, or may compensate dopant levels; (b) creation of conduction pathways (by hopping, percolation or by the formation of new energy bands within the diamond forbidden band-gap), in particular via sp^2 bonded states or clusters; (c) creation of scattering centers which limit the carrier mobilities; (d) creation of dopant-defect complexes which passivate the dopants.

Regarding the type of the conductivity induced by the implantation damage Fang, Hewett, Fernandes and Lau (1989) and Uzan-Saguy, Richter, Prawer, Lifshitz, Grossman and Kalish (1995b) found that low temperature low dose Xe and Kr implantations yields p-type conductivity with poor mobilities ($\sim 10\ cm^2\ V^{-1}\ s^{-1}$) whereas implantation at higher temperature or post-implantation annealing of the damage (Uzan-Saguy et al., 1995b) as well as electron irradiation (Clark, Kemmey and Mitchell, 1961) were found to yield n-type conductivity. Since dopant ion-implantation always requires annealing, one can expect the residual damage following incomplete annealing to act as amphoteric centers in the diamond. These can give rise to n-type conduction, or compensate acceptor states (Kalish, Uzan-Saguy, Philosoph, Richter and Prawer, 1997b). Indeed this damage related n-type conduction was observed (Prins, 1982; Fang et al., 1989) and even used (Prins, 1988).

Below, we briefly review the electrical properties of defects in ion-damaged diamond and their annealing.

2.2.1. *Damaged Above the Graphitization Limit and Its Annealing*

Diamond damaged above the graphitization limit has been shown (Reznik, Richter and Kalish, 1997) to exhibit variable range hopping (VRH) conduction which is characterized by a $T^{-1/4}$ dependence of the resistivity on measuring temperature. The slope of the line of log R vs. $T^{-1/4}$ is related to the product of the density of hopping states near the Fermi level ($N(E_f)$) and the radius of the localized hopping states (a).

This fact was used in the work of Reznik et al. (1997) to learn about the kinetics of the conversion of broken diamond bonds to graphitic bonds in ion-implantation-damaged diamond.

In that work a type IIa diamond has been implantation-damaged at low temperature to a defect density above the critical density for graphitization.

The sample was contacted in the van der Pauw configuration and was mounted onto a thermal stage which could be cooled and heated under the flow of nitrogen under computer control at different rates for different times. Hence this setup could be used both to anneal the sample as well as to measure, in situ, the temperature dependence of its resistivity. The results of an experiment in which the sample was heated up to a preset annealing temperature (T_a) during 2.5–4.5 h, while measuring its electrical resistivity every 10 °C, are shown in Figure 4, plotted as $\log_{10} R$ vs. $T^{-1/4}$. Evidently, this presentation of the data follows an excellent linear dependence, proving that the conductivity is well described by the VRH formalism. The change in slope of the lines in Figure 4 is indicative of a change in the coefficient T_a of the VRH formalism, i.e., with advanced annealing temperature $N(E_f)a$ increases until at $T_a \sim 600$ °C complete graphitization of the damaged layer is reached. This is evident from both the absolute (low) value of the resistivity and from its very weak dependence on measurement temperature that is typical for metallic conductivity. Measurements of the time evolution of the conductivity has proven this to be

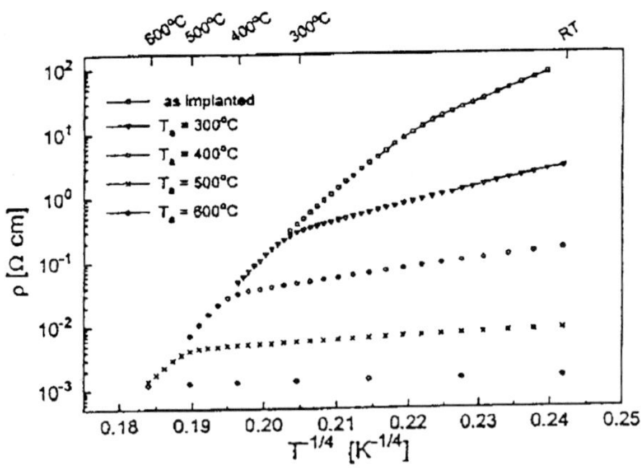

FIG. 4. Temperature dependence of the resistivity of diamond, damaged above the critical dose, as function of $T^{-1/4}$ for different annealing temperatures (T_a). The excellent linear dependence proves the conduction to be well described by VRH. The changes in the slopes of the lines is interpreted as being due to an increase in the density of hopping sites, i.e., an increase in sp^2 bonded sites (from Reznik et al. (1997)).

exponentially increasing with time, indicating that a 'barrier penetration' process (similar to that which governs alpha decay in nuclear physics) may be used to describe the transformation of broken diamond bonds to graphitic bonds. The characteristic 'life time' for this transformation was found to decrease with increasing annealing temperature. It was thus concluded in Reznik et al. (1997) that the mechanism of the transformation of broken sp^3 bonds to sp^2 bonds is governed by the penetration of the system through the potential barriers which separates these two bonding configurations.

2.2.2. *Damaged Below the Graphitization Limit and Its Annealing*

The transformation that diamond, damaged below the graphitization limit undergoes as a result of gradual annealing from damaged to undamaged diamond was studied both structurally (by confocal Raman measurements) and electrically and the results of these two kinds of experiments were correlated (Reznik, Richter and Kalish, 1998; Reznik, Kalish, Nugent and Prawer, 1999). In these studies a type IIa diamond was damaged by carbon-ion-implantation to the total concentration of vacancies of 8×10^{21} cm^{-3} (i.e., below the critical damage level (N_c) of 10^{22} vacancies per cm^3 (Uzan-Saguy et al., 1995a)) and was subjected to annealing at ever-increasing temperatures (T_{ann} = RT−1200 °C). The temperature dependence of the electrical resistivity of this diamond was measured in situ during and following annealing. The measured temperature dependences of the electrical resistivity R determined in the experiment are plotted in Figure 5. It is obvious that (a) the data, when plotted as $\log R$ vs. $1/T$, follow a linear dependence indicating a *thermally activated* behavior of the conductivity and

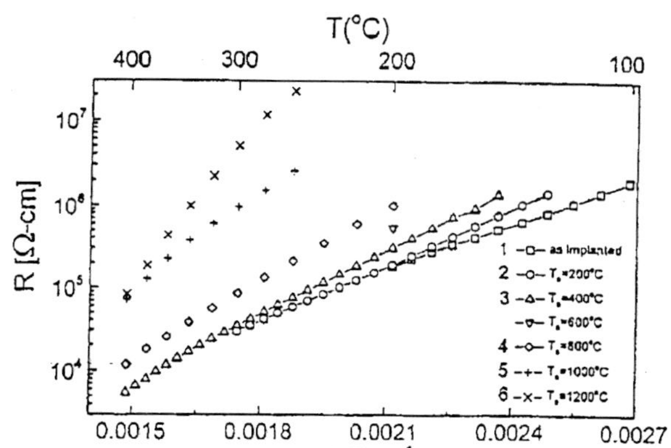

FIG. 5. Temperature dependence of the resistivity of diamond, damaged below the critical dose, as function of T^{-1} for different annealing temperatures (T_a). The excellent linear dependence proves the conduction to be well described by an activated process. The changes in the slopes of the lines represent changes in the activation energies (from Reznik et al. (1998)).

(b) the *activation energy of the conductivity increases* with increasing T_{ann}, i.e., with *decrease in defect concentration.* Activation energies ranging from 0.35 V up to 1.15 eV for annealing temperature 200–1200 °C were deduced as shown in Figure 6. No Hall effect could be measured, even for the more conductive samples, at any temperature (up to 500 °C), presumably because of very low carrier mobility. Recently a model to account for the strange behavior of 'activation energy' on defect density has been proposed (Baskin, Reznik, Saada and Adler, 2003). It is based on the fact that the vacancy in diamond, when neutral, has a deep localized state occupied by an electron, with higher lying states which can trap an additional electron. The states thus formed give rise to an energetically higher lying band in which electrical conduction may take place. The energy gap between the Fermi level and the mobility edge of this band is related to the observed activation energy. Since both shift of the Fermi level and the width and shape of this band depend on the density of defects, the activation energy for conduction depends on defect concentration, i.e., on the degree of defect annealing. Good agreement between the measured activation energies and the predictions of this model was obtained. Raman measurements were performed on a sample similar to that used for the electrical measurements, and the changes in the Raman spectra were monitored (Reznik et al., 1999) following annealing at ever-increasing temperatures. The results reveal the creation and disappearance of defect states in the lightly damaged diamond. The results of the experiment on near surface defect evolution as a result of annealing show that at 1000 °C all Raman active defects have vanished; nevertheless full electrical activation of implanted dopants cannot be achieved following annealing at this temperature and substantially higher temperatures (~ 1450 °C) are required (as will be discussed below). Interestingly, the defects observed are similar to those found in Orwa et al. (2000) in the Raman spectra of deeply buried defects following high-energy ion-implantation of diamond described above.

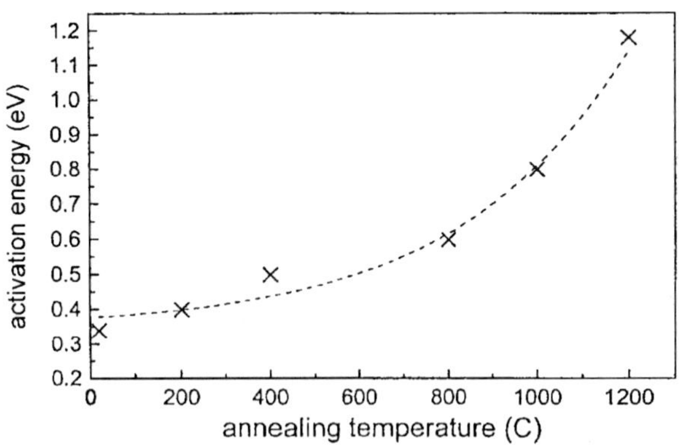

FIG. 6. Activation energies of the conductivity for diamond damaged below the critical dose (as extracted from the slopes of the lines of Figure 5) plotted as function of annealing temperature. The dotted line is drawn to guide the eye (from Baskin et al. (2003)).

It is hoped that experiments of the kind described above, in which structural properties are correlated with electrical or optical features of defect containing diamond will enable to identify the particular defects that give rise to specific electrical and optical properties of damaged diamond. The damage related conductivity is always characterized by extremely low mobilities; it can thus not be used for most device applications. As mentioned above, this conduction may exhibit some 'n-type' features, hence it may be mistaken for real chemical doping, especially for cases in which new potential dopants, whose expected activation energy is unknown, are searched for. In order to prove that indeed electrical effects due to the presence of dopant impurities are obtained following ion-implantation-doping experiments, it is imperative to perform control experiments in which non-dopants (i.e., C or noble gas) ions are implanted into similar diamond substrates under identical damaging and annealing conditions to those applied in the doping experiments. Only if the former do show electrical effects that are absent in the latter, 'damage alone' control experiments, can conclusive doping be proven. Not all implantation-doping experiments reported in the literature do include such controls.

3. Implantation/Annealing Schemes

3.1. General Considerations

It is obvious from the above that special care must be given to the removal of implantation-related defects, in order to assure that the electrical effects measured in diamond implanted with potential dopant ions are directly related to chemical doping of the diamond, and are not an artifact caused by the presence of defects. The electrical effects caused by the defects thus need to be understood, and control experiments have to be performed. For those cases in which the expected effect due to the doping is known, either from previous experiments on semiconducting diamond doped by equilibrium, non-implantation methods (such as achieved in nature, during artificial HPHT or CVD diamond growth) or from theory, the search for affective doping by ion-implantation is facilitated, since the electrical effects expected for real chemical doping are known. Indeed, this is the case for p-type boron doped diamond. However, for the case of n-type diamond, for which only recently indications of what dopants do act as donors, with clear signatures regarding the expected activation energies are known, the situation is less clear. It is in this regime the CVD methods are limited by the availability of suitable sources for the desired dopants, where, in principle, ion-implantation doping could be most useful thanks to the flexibility of this method which permits to shoot any desired ion into the matrix. Hence the importance of the ongoing efforts to achieve diamond doping (in particular n-type) by ion-implantation.

The evaluation methods employed for the assessment of the success and the quality of the doping are, naturally, based on detailed electrical measurements. Among these, the most reliable are Hall effect measurements covering a wide range of temperatures, as these yield information on the conduction type, the activation energy, the carrier type (holes or electrons), the carrier mobility and the compensation ratio. Measurements of the resistivity as function of temperature alone do not provide a full picture of the conduction properties. In all electrical measurements attention should be given to the quality of the contacts, hence the use of the van der Pauw geometry in the electrical measurements is preferable. The determination of the carrier type by 'hot probe' (Seebeck effect) measurements is not always completely reliable. This is particularly important when the properties of a very thin (thickness of only several hundreds of nanometers) implantation modified layer on top of a very thick (thickness of several hundreds of micrometers) highly thermally conducting substrate, such as diamond is, are to be determined. Furthermore, it should be borne in mind that 'hot-probe' measurements of highly resistive diamond are always biased towards indicating 'n-type' features due to the higher mobility of electrons to that of holes in diamond. Hence indications of 'n-type' by the use of this method does not necessarily imply that real n-type doping has been achieved.

Luminescence measurements yield information on the various kinds of excitons that may exist in the material under study. When cathodoluminescence (CL) measurements are performed it is possible to restrict the volume from which the optical information is obtained to just a thin layer, which can be tuned to mainly include the implantation affected near-surface region. Such measurements are contactless; they may yield information on the various states within the gap that exist in the material, and identify those that can be correlated with the presence of the implanted species. Ion beam probing experiments, in particular light ion channeling experiments can yield information on the presence of point defects in the near-surface region and on the lattice site occupied by the implants. However, the sensitivity of these methods is by far insufficient to yield information on the electrical properties of the implantation-damaged and annealed material. Thus, these methods can serve only as a crude guideline for the required annealing to be used. Some other nuclear methods such as those relying on the implants being radioactive (emission channeling and perturbed angular correlation experiments) can yield useful information on the immediate environment of the implant in the lattice. All these have been applied, for some selected suitable cases, to the study of the structure of implanted and annealed diamond.

Other kinds of measurements probe the whole volume of the sample. Hence these are not particularly suitable to assess the results of ion-implantation, since a method sensitive only to the near surface volume, within the ion range (which is usually restricted to less than a micrometer from the surface), is needed. In order to extract information regarding the implantation from such measurements, it is necessary to perform identical measurements on non-implanted specimens of similar bulk properties and subtract the results obtained for the implanted from those of the reference sample, a procedure that usually introduces large errors. This applies, among others, to optical measurements performed in transmission and to resonance measurements such as EPR or NMR. Nevertheless, such measurements were performed also on implanted and annealed diamonds and valuable information was obtained.

Most recently, direct observation of the defects remaining in diamond following deep ion-implantation and subsequent annealing has been obtained from high resolution TEM experiments (Lai, Prawer and Bursil, 2001). It was found in that work that the outermost layer of the implanted diamond, closer to the surface than the implant range, has nearly completely recovered upon annealing. On the other hand, the end-of-range region, where the defect density has exceeded the critical dose for graphitization, both i-carbon (an amorphous form of carbon observed in C ion beam irradiated materials) and highly curved graphitic layers were seen in the HRTEM images. Interestingly, such layers were found (Saada et al., 1999) to be formed as a result of annealing in simulated annealing computations of a highly damaged regions embedded in an intact diamond crystal.

Finally, it should be noted that there must not necessarily exist one unique specific recipe for optimal annealing of diamond implanted with different ions, to different doses. Rather, it can be that each dopant may require its own annealing procedure, in particular when interstitial or substitutional, and donor or acceptor dopants are concerned.

3.2. Room Temperature Implantation Followed by Furnace Annealing

The simplest and most straightforward way of doping a semiconductor by ion-implantation is to shoot the ions into the target at room temperature to the desired depth (i.e., at the required energy) and dose and then to heat the sample in a furnace, under vacuum or inert gas atmosphere. This procedure, however, does not result in optimal defect removal when applied to diamond, as mentioned above. Hence other annealing schemes have been developed and tried out.

It is important to note that following any high-temperature thermal treatment the diamond surface is likely to be covered with a thin graphitic layer that will give rise to electrical conduction. It is thus imperative to thoroughly clean the surface by suitable graphite etching procedures to remove this layer prior to electrical measurements.

3.3. Cold-implantation Followed by Rapid Annealing (CIRA)

Perhaps the best implantation/annealing scheme that yields highly reproducible very good results when applied to p-type doping of diamond by B ion-implantations is the so-called CIRA (Cold-Implantation followed by Rapid

Annealing) method devised by Prins (1988a). The idea behind this method is to enable as many interstitial–vacancy recombinations to occur and to allow the implant to find a vacancy so that it will occupy a substitutional site in an undamaged diamond environment. Implantation at low temperature (usually liquid nitrogen temperature) will 'freeze-in' all point defects since the temperatures required for their diffusion are much higher. Subsequent rapid heating of the sample, (directly from the low implantation temperature to temperatures in excess of 500 °C) will initiate diffusion of all defects in the sample. Hence the likelihood of their meeting each other and recombining is high, thus reducing the density of defects. It has, however, been shown (Fontaine, Uzan-Saguy and Kalish, 1996) that rapid annealing at temperatures in excess of the diffusion temperatures of the vacancies and interstitials does not suffice to lead to good electrical properties of the doped layer; another, very high-temperature annealing stage (1450 °C) is required to yield high-doping efficiency and high-carrier mobility, at least for the case of B p-type doping.

3.4. Low Damage Drive-In Implantation (LODDI)

Another method proposed by Prins, though to the best of our knowledge, never implemented by others, is the Low Damage Drive-In implantation Doping (LODDI) method (Prins, 1998b). Here the idea is to create a vacancy rich region by a deep ion-implantation (for example, by H or He implantation) and adjacent to it, though not overlapping, implant the desired dopant. It is thus hoped that the interstitials will find vacancies created by the deep implantation to recombine within the damaged region.

3.5. Implantations into Pre-heated Diamond (Hot Implantations)

As was shown above (Section 2.1), implantation into diamond held at elevated temperatures does not leave many point defects due to the instantaneous annealing of the damage created by every impinging ion. Hence advantage can be taken of 'hot implantations' to circumvent the need for post-implantation annealing. This doping method, however, does leave extended defects in the implanted layer which are very difficult to remove by post-implantation annealing; their presence degrades the electrical properties of the implanted layer, as reflected in particular the inferior carrier mobility and in the relatively poor doping efficiency (Zeidler, Hewett and Wilson, 1993). The presence of these defects also seem to inhibit the diffusion of impurities (even hydrogen) in diamond that has been subjected to ion-implantation regardless of post-implantation annealing applied (Fizgeer, Uzan-Saguy, Cytermann, Richter, Avigal, Shaanan, Brener and Kalish, 2001; Uzan-Saguy, Reznik, Cytermann, Brener, Kalish, Bustarret, Bernard, Deneuville, Gheeraert and Chevallier, 2001a).

3.6. High Dose Implantations Followed by Chemical Etching

As was shown above (Section 2.1), implantation of diamond to doses which exceed the graphitization limit leads, following a high-temperature anneal, to the formation of a graphitic layer in those regions in which the total defect density exceeds the critical defect density of $N_c = 10^{22}\,cm^{-3}$. Following chemical removal of this graphitic layer, a near surface region of intact, well-annealed diamond, which contains the 'tail' of the implant distribution (i.e., those ions which reside in a region in which the defect concentration was below N_c) remains. These implants are expected to be electrically active. Indeed Braunstein and Kalish (1983) and Venkatesan, Melta, Das and Belu (1993) could unambiguously show that p-type activity due to B doping has been achieved by employing this method for heavily ion-implanted diamond. Unfortunately by the very nature of this method it offers no control on dopant concentration as it relies on a self-controlling process determined primarily by D_c. Also, the carriers are located in a very shallow near surface region (several tens of nanometers thick) at high concentrations (typically 6×10^{21} B cm^{-3}). Hence the carrier mobility in this layer is very low.

The technique of high dose implantation/annealing/etching for the doping of diamond does, however, find application in the fabrication of excellent ohmic contacts to p-type diamond.

3.7. High Dose Rate (High Current) Implantations

A combination of the above two methods was recently devised (Kalish, Uzan-Saguy, Ran, Ferber, Guettler and Zachai, 1999) in which heating of the diamond is achieved during implantation by the high power delivered to the target by high current ion beams. By implanting a diamond with sub-MeV ions at a current of several hundred micro-amps over an area of about 1 cm^2, power densities of about 100 W cm^{-2} can be dumped into the sample. This power density can heat the sample to very high temperatures by the implantation process itself. It has been shown (Kalish et al., 1999) that, depending on heat-sinking of the target diamond, either implantations that lead to near surface graphitization (Section 3.6) or hot implantations which inhibit graphite formation altogether (Section 3.5) could be achieved. This method does yield p-type layers when used with B ion-beams; however, due to the high ion fluxes, it is very difficult to control the exact ion dose which has been implanted, which is usually very high even when exposing the diamond to the ion beam for only a few seconds. No n-type doping by implanting Li and P ions using this method could be observed (as is the case also for most other donor implantation doping attempts. Section 4.2.2).

3.8. High Energy (Deep) Implantations

The phase diagram of diamond indicates that diamond synthesis requires high temperatures and high pressures. Hence it is conceivable that annealing defects in diamond under ‘internal’ pressure (caused by an intact diamond layer surrounding the defective region) may favor diamond regrowth over graphite formation. Such conditions can be achieved by performing high energy (‘deep’) ion-implantations. In this case a diamond top layer, above the implant containing and highly defective layer, will remain mainly undamaged, since in this region most ions undergo predominantly electronic stopping which inflicts only little damage. Indeed it has been shown (Prawer, Jamieson and Kalish, 1992) that graphitization can be avoided for deep ion-implantation induced damage, even if the critical dose for graphitization is exceeded.

In a series of experiments in which native defects were created in diamond by high energy light ion-implantations followed by rapid annealing achieved by pulsed laser irradiation, Prawer and co-workers (Prawer et al., 1992; Jamieson, Prawer, Dooley and Kalish, 1993) have studied the annealing behavior of damage deeply buried in the diamond. In this case the damage was created by MeV ion-implantations; hence the defects were subjected to internal pressures exerted on the damaged layer by the intact diamond cap. The nature of the defects and their annealing was studied by performing confocal Raman measurements on a ‘sideway’ implanted diamond. This has enabled to obtain Raman spectra along various parts of the damage profile extending from the surface down and beyond the end of range of the implants. The annealing behavior of such samples was studied (Orwa et al., 2000). Most recently the deep implantation method was used to enable the preparation of HRTEM (Lai et al., 2001) samples thus allowing the direct observation of the defects remaining in implantation damaged diamond after annealing. These could be correlated with other, more macroscopic structural evaluation methods such as EELS and Raman.

Hence annealing deeply implanted diamond may yield better results than comparable ‘shallow’ implantations. This indeed was verified for some cases for which this implantation mode has yielded improved doping properties, as will be described below.

4. Doping Diamond by Ion-Implantation

As mentioned above, proof for real chemical doping requires parallel processing of dopant and inert ion implanted and annealed samples. Also, in order to obtain optimal results, high-quality undoped single crystal diamond should be used. It is most useful to have a guideline as to what properties may be expected for diamond containing different dopants, and what contacts to use for the those. Hence we first summarize the know-how that exists to date on natural diamonds and on diamonds doped during CVD growth. We then review the more recent and more promising results obtained so far on doping diamond by ion-implantation.

4.1. Available Data on Doping Diamond by Thermodynamic Equilibrium Methods

4.1.1. p-Type Doping

p-Type, boron containing, diamonds exist in nature (type IIb natural diamond), and their optical and electrical properties have been extensively studied and documented (Collins, 1994). These have shown that the activation energy of the B acceptor is 0.37 eV. Very high hole mobilities, in the range of 1000–2000, were reported. More recently B doped p-type diamonds were realized both by HPHT and CVD techniques. The best p-type artificial diamonds are obtained nowadays by very careful and slow CVD homo-epitaxial growth on ⟨100⟩ natural diamond substrates. These have yielded a record mobility of 1800 $cm^2\ V^{-1}\ s^{-1}$ and a carrier concentration of $2.3 \times 10^{14}\ cm^{-3}$ at room temperature (Takeuchi, Yamanaka, Watanabe and Okushi, 2001).

The optimal properties of p-type diamond due to the boron acceptor, are thus well known, and they serve as a guideline for the p-type properties of B ion-implantation doped diamond to be aimed for. Also, the best ways of obtaining good ohmic contacts to p-type diamond have been extensively investigated and these can be directly implemented also to contacting implantation doped diamond layers. It should be noted that the realization of very low resistivity contacts to p-type diamond is based on ion-implantation (Venkatesan et al., 1993) (Section 3.6). Nevertheless, satisfactory ohmic contacts to B doped diamonds are routinely obtained by just evaporating Ti/Au contacts.

Because of the excellent properties of B containing p-type diamond, no further extensive studies on doping attempts of other potential group III acceptors in diamond have been undertaken, to the best of our knowledge. Very old results on p-type doping attempt of diamond with Al ion-implantations, were not promising enough to justify further study, and have not been further pursued.

4.1.2. n-Type Doping

Unfortunately, the situation of the existence of a well-known and much studied dopants in diamond, like the B acceptor impurity, does not exist for the case of n-type dopants.

The obvious candidates to act as donors in group IV semiconductors are group V impurities, i.e., N, P, As and Sb, when occupying substitutional sites in the crystal. Among those only the lighter elements, whose atomic radii are not too large to fit substitutionally into the tight diamond crystal, are likely to enter the crystal and occupy the undisturbed substitutional site where they are expected to act as donors.

Nitrogen is an abundant substitutional impurity in diamond. It is found in natural diamonds (type Ib) and it can be synthesized in the laboratory by HPHT and by CVD techniques. Unfortunately N forms a deep donor in diamond ($E_a =$ 1.7 eV) which does not donate a sufficient number of electrons to conduction at room temperature. N has also been predicted theoretically to have a deep donor level in diamond (Kahijira, Antonelli, Bernholc and Car, 1991).

Phophorus, when residing on a substitutional site, has been predicted to be a donor in diamond with a rather shallow level; however, it was also predicted (Kahijira et al., 1991) to have a very low solubility in diamond. Indeed P has been proven to be a donor in diamond, however, with a rather deep level of about 0.5–0.6 eV (Koizumi, Kamo, Sato, Ozaky and Inuzuka, 1997a; Kamo, Sato, Mita, Sawabe, Reznik, Uzan-Saguy, Ran and Kalish, 1997b). Unfortunately, the doping of diamond with P by CVD methods has turned out to be rather difficult, and only very few groups in the world have succeeded, so far, to grow P doped homo-epitaxial diamond layers with reasonable electrical properties (the RT mobilities are typically of the order of 100 $cm^2 V^{-1} s^{-1}$ or less). Although the electrical properties of the n-type P doped diamond are not as good as expected, many studies on various properties of P in diamond have been published, and most recently a first pn homo-epitaxially grown junction, based on B and P impurities has been reported (Koizumi, Watanabe, Hasegawa and Kanda, 2001/2).

First principal calculations have predicted that the donor level associated with P can be lowered when another impurity (H, B or N) is located next to the P atom (Katayama-Yoshida, Nishimatsu, Motroshiromizu and Orita, 1999). However, no successful results on such co-doping experiments have been published so far.

Sulfur and Oxygen. Some reports on the electrical effects of incorporation of group VI elements (O and S) in diamond have recently been published. In particular excitement has been struck by the report that S has a fairly shallow donor level in diamond (0.37 eV) with a good electron mobility (Sakaguchi, Gamo, Kikuchi, Yasu, Haneda, Suzuki and Ando, 1999). This report has unfortunately been shown to be doubtful (Kalish, Reznik, Uzan-Saguy and Cytermann, 2000) and the encouraging results of Sakaguchi et al. (1999) await to be confirmed. Till now, despite many continuing reports on the properties of S doped diamond, no conclusive proof of real n-type properties of S doped diamond have been published. Nevertheless, many studies on the CVD growth of S containing diamond and on their properties have been published recently and much important work is currently being done on this topic.

Ab initio calculations on the state of S in diamond (Saada, Adler and Kalish, 2000) have shown that the likelihood of S residing on a substitutional site in the diamond crystal at the ionization state required for it to act as a donor is very low.

Some reports on the effects of inclusion of oxygen in the CVD growth atmosphere on the quality of diamond and on growth rates were published;

however, no clear electrical n-type features which can be attributed to the presence of O in the diamond were reported.

Lithium and Sodium. An interesting prediction that group I elements (Li and Na), when occupying interstitial sites in diamond, may donate their outermost electron for electron conduction (Kahijira et al., 1991) have triggered experiments attempting to introduce Li and Na into diamond during CVD growth (Sternschulte, Schreck, Stritzker, Bergmaier and Dollinger, 1999a), by various diffusion modes (Popovici, Wilson, Sung, Prelas and Khasawinah, 1995; Uzan-Saguy et al., 2001b) and by ion-implantation (see below). None of these have yielded, so far, convincing n-type characteristics.

4.2. Doping Diamond by Ion-Implantation

As stressed above, ion-implantation offers a non-thermodynamic-equilibrium mode of forcing any desired impurity into diamond. However, ion-implantation is usually accompanied by severe damage to the lattice which has to be removed by proper annealing and the implants have to be driven onto the required lattice sites to electrically act as desired. Control experiments with inert implants are required to assure that the measured effects are indeed due to chemical doping, and not to residual defects.

4.2.1. p-Type Doping by Ion-Implantation

The most promising acceptor impurity in diamond is boron; hence much work has been devoted to the study of B ion-implantation into diamond and the annealing schemes required to render the implanted layer p-type.

As these studies have been going on for over a decade, and good results were obtained, this topic has been extensively reviewed (Dresselhaus and Kalish, 1992; Prins, 1992a; Davies, 1994a; Kalish, 1997; Kalish and Uzan-Saguy, 2001; Kalish and Prawer, 2003), hence we only give here the major points.

Already in the early 70s, first attempts to dope IIa diamond by implanting boron and other ions were reported by Vavilov, Gukasyan, Guseva, Karatygina and Konorova (1974a). These pioneering studies, however, suffered from the lack of control experiments; hence effects due to the presence of incompletely annealed implantation related defects could not be properly eliminated. Almost 10 years later, in 1983 Braunstein and Kalish (1983) reported on the fabrication of highly conducting thin p-type diamond layers by high dose B implantation followed by high-temperature annealing (1450 °C) and graphite removal. This method (Section 3.6) has later been further perfected, and it is now commonly used as a method to obtain a p^{++} ohmic contact to p-type diamonds.

Prins (1988a) has considered the importance of the presence of vacancies to permit the B implant to find a substitutional site. As a result of this study Prins devised the above-mentioned CIRA technique (Section 3.3), which is till now the most commonly used method for implantation doping of diamond with B. In this method, the diamond is implanted at low temperature, then rapidly heated to a high temperature ($\leq$ 1300 K), to induce interstitial–vacancy recombination, followed by an ex situ high-temperature anneal (1700 K) to further reduce the radiation damage and increase the electrical activation of the dopants.

In the last few years, many attempts to optimize the implantation/annealing procedure for doping diamond p-type with boron have appeared in the literature. These are mainly based on the CIRA method, or on the high-dose graphite removal method. The various implantation and annealing parameters which influence the quality (mainly mobility) of the p-type diamond obtained by B ion-implantation were investigated (Paul, de la Houssaye, Penchina, Zeidler and Wilson, 1992; Hewett, Nguyen, Zeidler and Wilson, 1993). The need for in situ rapid annealing of the implanted diamond to medium temperatures (up to ~1300 K) was not conclusively proved to be essential to improve the doping. Most recently Prins has suggested that a low-temperature RTA (~500 °C) to follow the cold implantation may be sometimes advantageous to higher instantaneous annealing (Prins, 2001b). Post-implantation, post-RTA ex situ annealing to various temperatures was also studied (Prins, 1991, 1992b). Work of the Technion group (Fontaine et al., 1996) indicates that the highest possible temperatures (1700 K under high vacuum) following the CIRA procedure is essential to further increase the carrier concentration and their mobility and to reduce the compensation ratio. Typical results of Hall effect measurements (as function of temperature) obtained for B implanted diamonds according to the CIRA procedure followed by a further high-temperature anneal at 1450 °C are: $p = 6 \times 10^{13}\ \text{cm}^{-3}$, $\mu = 400\ \text{cm}^2\ \text{V}^{-1}\ \text{s}^{-1}$ and $N_d/N_a = 0.005$ where N_a and N_d are the acceptor and donor concentrations, respectively (Fig. 7) (Fontaine et al., 1996). Other values of higher mobilities obtained in very specially selected type IIa substrates are quoted in the literature (Prins, 2001a), but are not supported by any detailed data.

Prawer, Nugent and Jamieson (1997) have shown that by creating a deeply buried B rich layer in diamond by mega-volt (MeV) ion-implantation improved mobilities can be obtained (Section 3.8). By implanting 2 MeV B ions into a type IIa diamond held at LN temperature to a dose of 10^5 B cm^{-2} (just below the graphitization limit) a 100 nm thick ‘delta doped’ B containing layer, buried about 1.5 μm under the diamond surface was obtained. Contact to this layer was realized by laser-induced graphitization at the desired points. The sample has undergone two annealing steps: the first, at a temperature of 950 °C, has removed most defects but has not sufficiently activated the B, hence the sample turned very resistive; a second anneal at 1450 °C has drastically reduced the resistivity of the sample due to activation of the implants and the correct activation energy could be deduced from $R(T)$ measurements. A novel method for making ohmic contacts to the buried layer, based on local graphitization by laser pulses was devised. The electrical properties of these B doped, buried layers, reported by Uzan-Saguy, Kalish, Walker, Jamieson and Prawer (1998), are characterized by high

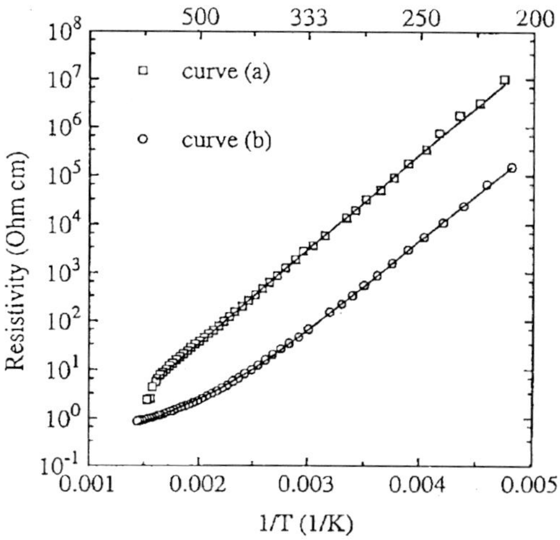

FIG. 7. Measured resistivity vs. inverse temperature after step I (in situ rapid annealing, 1050 °C, 10 min: curve (a), open squares) and after step II (subsequent annealing, 1450 °C, 10 min: curve (b), open circles). The solid lines are fits to the full expression for the carrier concentrations of a partially compensated semiconductor (from Fontaine, Deneuville, Gheeraert and Gonon (1994)).

mobilities (600 $cm^2 V^{-1} s^{-1}$) and a low compensation ratio (5%). The reason for these very good properties of the so-implanted sample is possibly the superior annealing due to the internal pressure exerted on the damaged layer by the overlying, only lightly damaged, diamond layer.

The optical signature of B in implanted and annealed diamond has been investigated (Prins, 1994; Sternschulte, Tonke and Sauer, 1999b) by CL measurements at 77 K. New luminescence lines, one related to the exciton bound to B in substitutional sites at 5.22 eV (observable only after annealing at 1000 °C) and one at 4.5 eV, were found. Prins (1994) showed that the 4.5 eV band is quenched when introducing defect related donor centers into the implantation-doped layer. Moreover, Prins (1998c) also showed that the two bands related to B are observable in the CL spectrum of diamond samples, even though their electrical measurements indicate that $N_a < N_d$. This is in contradiction with previous conclusions that claim that these lines appear only if $N_a > N_d$.

The electrical properties of the B implantation doped diamond were correlated with Raman, EPR and CL results by the Grenoble group (Fontaine et al., 1994). These studies have shown that the implantation doped p-type diamond very much resembles in its structural properties those of B doped CVD diamond. However, some compensation caused by implantation related residual unannealed defects was noticed.

It is important to mention that *no* major differences were found in the properties of implantation doped single crystal (type IIa) and polycrystalline (CVD) diamonds (Kalish, Uzan-Saguy, Samoiloff, Locher and Koidl, 1994) apart from differences in mobility which can be related to the presence of grain boundaries in the polycrystalline material. This result indicates that the optimal implantation/annealing methods developed for the 'pure case' of single crystal diamonds are applicable also to CVD diamond layers. This conclusion is not a priori obvious since one could have expected that some graphitization along grain boundaries due to the high-temperature annealing stage could occur, thus modifying the electrical properties of the implanted layer. Luckily, this does not seem to be the case, at least for annealing temperatures below 1000 °C.

4.2.2. n-Type Doping by Ion-Implantation

It is obvious from the above that one of the most important topics in research related to diamond electronics is to obtain n-type diamond, by any doping method, in a reliable and reproducible way. Hence, no wonder that much effort has been devoted to the search for suitable dopants and suitable annealing procedures to obtain impurity related donor conductivity by ion-implantation.

All of the 'potential' donor elements described above, which have been experimentally tried by various growth methods, or have been theoretically predicted to act as donors, have also been subjected to ion-implantation attempts. These have, unfortunately, not yielded, till now, completely reliable and widely reproducible results.

An important key issue in the search for implantation doping methods of diamond with donor impurities is the question whether the information gained over the years in ion-implantation p-type doping of diamond with B is applicable also to n-type dopants. In general, annealing depends on thermal treatments which are meant to set the defects (interstitials and vacancies) in motion. However the diffusivities of vacancies and interstitials depend on their charge state, which, in turn, depend on the position of the Fermi level, hence on the kind of doping. It is therefore not obvious that annealing modes that have proven to work well for acceptor impurities will also be applicable to donor impurities.

An important work in proving that perhaps implanted donors do behave similarly to acceptors as far as annealing implantation related defects is concerned was done by Kalish, Uzan-Saguy, Philosoph, Richter, Lagrange, Gheeraert, Deneuville and Collins (1997a): in the absence of any well-known donor in diamond other than nitrogen, the donor level of which is too deep to affect the electrical properties of N 'doped' diamond up to high temperatures, indirect methods had to be applied to check whether diamond implanted with N and subjected to annealing that works well for B resembles natural N containing diamonds.

Type IIa diamonds were implanted (Kalish et al., 1997a) with N ions and were subjected to different annealing schedules. The resulting N containing diamonds were checked by resistivity $\rho(T)$, EPR and CL measurements. It was found that following CIRA implantation and a high-temperature anneal, almost 50% of the implanted N atoms are located in substitutional sites. The EPR signatures due to defects and N which are well known from studies of type Ib diamond were found to exist also in the N implantation-doped diamond. The conclusions drawn from these results are that the CIRA procedure is quite suitable for N, hinting that perhaps CIRA can lead to good annealing and electrical activation also for potential, shallower, donors implanted into diamond. However, it is conceivable that atomic size and solubility considerations may play an important role in the possibility to incorporate impurities by ion-implantation into sites where they may be electrically active in diamond. Hence these may affect the optimal annealing applicable for elements much different from B.

4.2.2.1. Implantation of Group I Elements. Lithium (and Na). Even before the prediction of Kajihara that interstitial Li should exhibit a shallow donor level in diamond (Kahijira et al., 1991) Li implantation doping attempts were made by the Russian group (Vavilov, Gukasyan, Guseva, Karatygina and Konorova, 1976) and excellent results were reported ($E_a = 0.1$ eV and mobility about 1000 $cm^2 V^{-1} s^{-1}$). However, with no proper control experiments having been done to eliminate radiation damage effects, these results could, unfortunately, not be verified in later experiments. Prawer, Uzan-Saguy, Braunstein and Kalish (1993) have implanted, at low temperatures, Li and C (as a control non-dopant ion), into type IIa diamonds to doses below and above the threshold for graphitization. For the low dose implanted and annealed samples, no significant electrical signature could be detected. The high dose implanted sample, annealed and acid etched (Section 3.6) did show electrical conductivity. However, no Hall effect signal could be found and the temperature dependence of the resistivity could be described equally well by an activated process (with $E_a = 0.22$ eV) as by VRH conduction. The fact that no Hall effect measurements were possible supports the latter conduction mechanism.

Job, Werner, Denisenko, Zaitsev and Fahrner (1996) have performed high dose Li implantations into diamonds heated to 850 °C (Section 3.5) and have subsequently annealed the sample at 900 °C. As expected, no radiation-induced graphitization was observed. n-Type conductivity could be verified by hot probe measurements only. An activation energy of 0.23 eV was deduced from $R(T)$ measurements, however, over only a very narrow temperature range (Fig. 8). This experiment, like many other experiments, does not include a 'damage alone' control experiment, hence the possibility that the observed effects are defect related cannot be ruled out.

Buckley-Golder, Bullough, Hayns, Willis, Piller, Blamires, Gard and Stephen (1991) have indirectly deduced n-type indications for hot Li implanted (Section 3.6) p-type diamond by observing rectifying pn features. However, these must not necessarily be associated with a definite donor level related to the implanted Li ions and may well be defect related effects. As mentioned above, implantation related defects in diamond may give rise to 'donor-like' conduction; hence implantation of any damaging ion into p-type diamond can exhibit rectifying

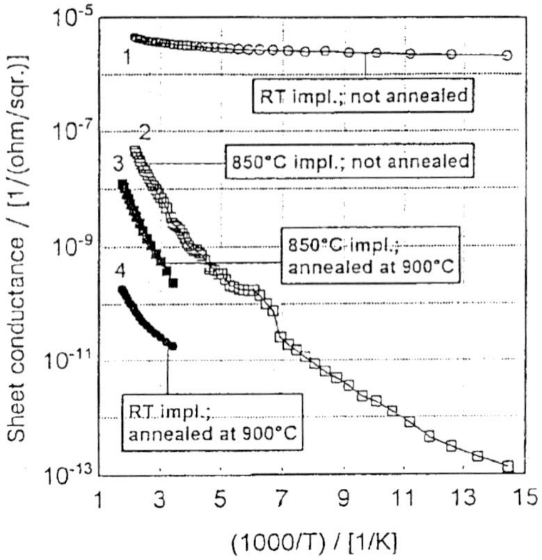

FIG. 8. Temperature dependence of the conductance of synthetic diamond crystals implanted with lithium ions. *Curve 1* (○). Sample 1 40 keV, 4×10^{16} cm^{-2} Li implantation at RT. The surface amorphous layer being removed by electro erosion. *Curve 2* (□). Sample 2 50 keV, 2×10^{16} cm^{-2} Li implantation at 850 °C. *Curve 3* (■). Sample 2 after 900 °C annealing and acid cleaning. *Curve 4* (●). Sample 1 after 900 °C annealing and acid cleaning (from Job et al. (1996)).

features (Prins, 1982). The conduction mechanism for the hot-implanted Li ions may well be due to VRH and not to a specific donor level (Prawer et al., 1993).

High current density implantations (Kalish et al., 1999) into CVD diamonds of Li (C as a control) ions (Section 3.7) to high doses were performed in a fashion similar to that successfully applied to the high dose p-type B implantation-doping. No clear indication of doping due to the presence of Li could be identified.

A most careful and comprehensive study of implantation doping of diamond with Li ions, with all the required checks (using C ion-implantations as controls), was done by Ran (1998) using most of the implantation/annealing schemes which were successfully applied to p-type doping by B ion-implantation, or which were reported by others to have resulted in n-type features of the implanted layer. We briefly report here on these experiments and on their results, some of which should be repeated and verified.

In the first experiment, a type Ib (N containing diamond) was used. The idea was that the use of a donor containing diamond might reduce compensation, hence perhaps enhance the appearance of donor activity due to the implanted Li ions. The CIRA procedure was employed and multiple energy implantations were performed to yield a rather wide box profile of the implants. These were followed by post-CIRA annealing at 1600 K. Concentrations ranging from 3.7×10^{18} to 1×10^{20} Li cm^{-3} were employed in consecutive implantation experiments. Comparable C control implantations were performed on the same sample (Fig. 9b).

(a)

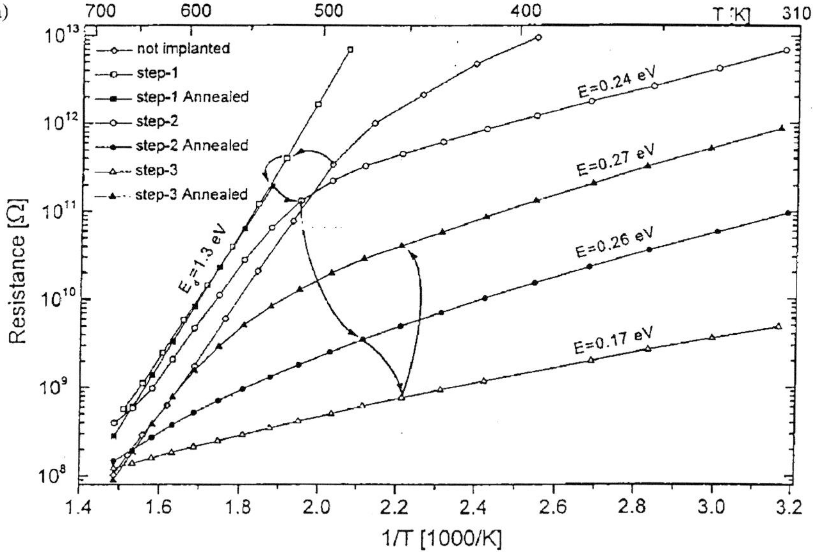

(b)

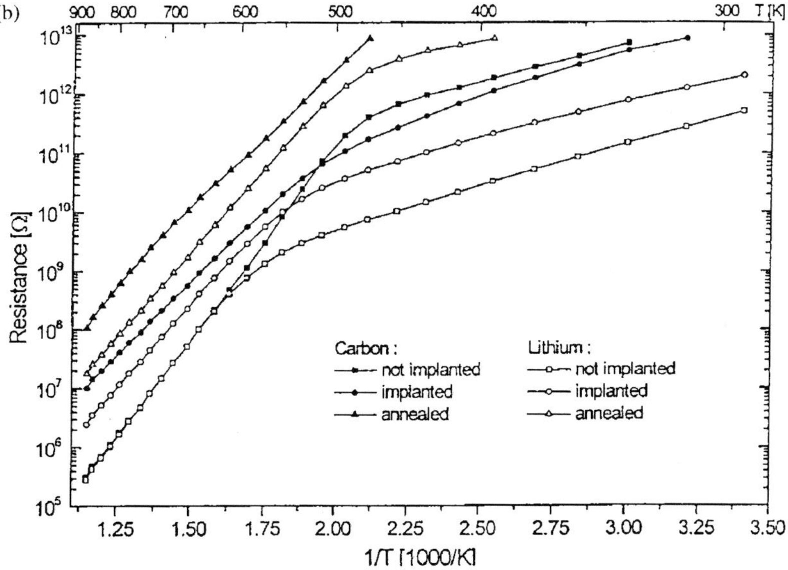

FIG. 9. (a) Evolution of the resistance of a type Ib diamond implanted with Li ions at ever-increasing doses (10 × and 30 ×). Some possible indication of Li related conductivity may be noticed (from Ran (1998)). (b) Temperature dependence of the resistance of a type Ib diamond implanted on one half with Li and on the other half with C ions (as a control, under identical damaging conditions). A consistently lower resistance is evident for the Li implanted part of the sample (from Ran (1998)).

Figure 9a shows the $R(T)$ results of these experiments. It seems as if after the second and third (highest dose) implantation, some conduction, which can be described by an activation energy of 0.26–0.27 eV remained. However, the fact that the resistance following the third (highest dose) implantation (and hence the third annealing) increased as compared to the second implantation, and the fact that the C implanted and annealed part was not very significantly different from the Li part (the Li implanted part of the sample being only $\sim 10\times$ more conductive than its control partner) has shed some doubt on the interpretation of the effects to be due to the presence of the implanted Li.

In the second set of experiments, similar experiments were performed with Li and C implantations into a type IIa diamond held at LN temperature. Following annealing at ever-increasing temperatures, a small difference between the C and Li implanted sections of the sample could again be noticed (following annealing at 630 K, the Li part was about $5\times$ more conductive). Unfortunately the resistivities of both parts were very high, and did not permit Hall effect measurements.

In a third set of experiments Ran (1998) has implanted Li into a type IIa diamond held at 1070 K. As expected, no graphitization has occurred; however, no changes in electrical properties of the samples could be noticed, even after a further high-temperature anneal (to 1550 K). This is in contrast to the results of Job et al. (1996) (Fig. 8).

In order to verify that the implanted Li was indeed contained in the sample and did not out-diffuse during or after the hot implantation, SIMS profiling was carried out. Interestingly, all implanted Li was retained inside the sample and no change whatsoever in its profile could be noticed. This is in accordance with more recent experiments of the Technion group searching for the diffusion of implanted Li in diamond, which have also not shown any diffusion (Cytermann, Brener and Kalish, 1994).

The lattice location of Li implanted into diamond at low doses was reported by Restle, Bharut-Ram, Quintel, Ronning, Hoffsass, Jahn, Wahl and ISOLDE-Collaboration (1995) using the emission channeling technique and by Braunstein and Kalish (1981) using nuclear reaction channeling analysis. Both measurements show that a fraction of about 40% of the implanted Li occupies interstitial lattice sites, where Li was predicted to act as a donor, the remaining Li seems, however, to occupy either substitutional or ‘random’ sites where they may act as compensating acceptors. No clear electrical n-type signature could be identified in any of the implanted samples.

Data for ion-implantation doped diamond with Na ions is scarce, and, once more, no significant indication for the existence of a donor level associated with Na in the crystal could be found (Prawer et al., 1993). Hunn, Parikh, Swanson and Zuhr (1993) have implanted Na (and Ne as a control) into type IIa diamonds at different temperatures ranging from 77 to 820 K. The samples were subjected to ion-channeling and resistance measurements. Whereas some electrical effects were noticeable for low-temperature annealing, these have vanished upon higher

T anneals. Hence the authors conclude that defect related conductivity was most likely responsible for the observed conduction.

4.2.2.2. Group V Implantation Doping. Nitrogen. The results on the implantation and annealing of N into diamond were reported above; they were not intended for doping purposes.

Arsenic and Antimony. Very early reports on electrical activities due to implanted heavy group V elements into diamond (As and Sb) (Vavilov, Krasnopevtsev, Miljutin, Gorodetsky and Zakharov, 1974b) all seem to have suffered from non-annealed defects which were misinterpreted as being due to the chemical doping of these potential (very large!) donor atoms. These are unlikely to resume an unperturbed substitutional site in the diamond lattice (Kalish, Deicher, Recknagel and Wichert, 1979).

Phosphorus. Phosphorus is the most 'natural' n-type dopant in group IV semiconductors, and many attempts to incorporate P into the diamond crystal and to electrically activate it were done over the last decades even preceding the theoretical predictions of Kahijira et al. (1991) and the success in doping CVD diamond with P of Koizumi et al. (1997a,b) mentioned above.

Ion channeling combined with PIXE experiments on P implanted diamond have proven the implanted P to predominantly occupy substitutional sites in the diamond lattice where they are expected to act as donors (Venkatesan et al., 1993). Much later, EPR experiments on P implanted type IIa diamond following the CIRA method have confirmed (Casanova, Gheeraert, Deneuville, Uzan-Saguy and Kalish, 2001) that a large fraction of the implanted P does occupy a substitutional site, and the characteristic EPR signals were identified. Interestingly, as also seen in other works (see below) these signals have diminished upon high-temperature annealing (Casanova et al., 2001). However, no electrical effects which can be attributed to P acting as a donor in diamond could be detected in these studies.

Prins (1995) has proposed to obtain n-type doping by using low dose implantations followed by low-temperature annealing (LDLT), hoping that the reduced number of defects due to the very low doses employed will facilitate the annealing and will allow P to find an undisturbed substitutional site. In that work B, C and P ions were implanted at very low dose increments (from 10^9 to 10^{11} cm^{-2}) in consecutive steps. Following each implantation dose the samples were annealed at 500 and 1200 °C. The so-implanted diamond was reported to be n-type, as determined by hot probe measurements, with an activation energy of the order of 0.1 eV. No Hall measurements were reported, and the resistance vs. *T* measurements (Fig. 10) suffered from some hysteresis, perhaps indicating non-ohmicity and curing of the contacts. It is also difficult to account for the measured *R*(*T*) data of Prins (1995) (Fig. 10) which show no saturation at high temperatures, as would be expected to occur for a case of such low doping levels and low activation energy (even when assuming the unlikely case of complete electrical activation of all the implanted ions and of no compensation).

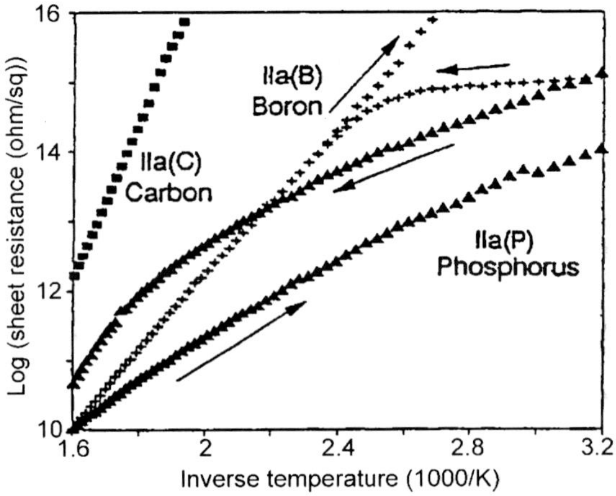

FIG. 10. A comparison of the sheet resistances for CIRA-doped type IIa diamonds using low dose C, B and P ions, followed by rapid thermal annealing to 500, then to 1200 and then to 1600 °C (from Prins (1995)).

Ran (1998) has attempted, in a very systematic way, to obtain n-type doping of type IIa diamonds by implanting P ions following various implantation/annealing schemes: (i) by using the CIRA procedure followed by high-temperature annealing, (ii) by using implantation into a hot diamond followed by various annealing steps, (iii) by repeating the LDLT procedure utilized by Prins. In all these experiments, controls were included by subjecting one face of the diamond to P implantation and the second face to identical Ar (a non-dopant ion producing the same damage profile as P) implantation. The two implanted parts were therefore obviously subjected to exactly the same annealing conditions. The resistivities of the two parts of each sample were measured by using the van der Pauw configuration, so that possible non-ohmic contacts did not influence the results. In all the above described procedures, no significant difference could be found between the resistivity of the P and Ar implanted samples, although, at some stage of the annealing sequence a slightly lower resistivity in the P implanted part, as compared to the Ar control part could be noticed (Fig. 11).

Interestingly, work by the Technion group (Uzan-Saguy, Reznik and Kalish, unpublished) on P doped homo-epitaxial layers (grown by the NIRIM group) have shown that too high-temperature treatments of the as-grown or H implanted (through the layer) samples has deteriorated the electrical properties of the n-type layer.

The deep, MeV ion-implantation method that did show promise for improving the mobility of B implanted into diamond (Prawer et al., 1997; Uzan-Saguy et al., 1998), was tried by Allen, Prawer, Jamieson and Kalish (1993) also for the case of P. The implantation was carried out such that a buried P containing layer was formed about 1 μm below the type IIa diamond surface. Laser graphitized

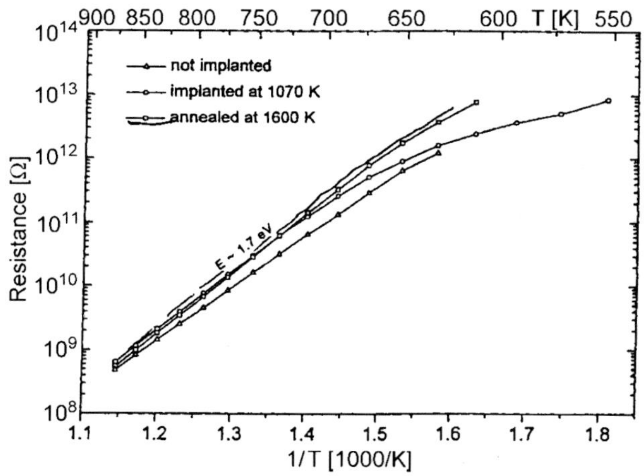

FIG. 11. Temperature dependence of the resistance of a type IIa diamond implanted with P ions at 1070 K and further annealed at 1600 K. At the high annealing temperatures the little conductivity that might have existed seems to be lost (from Ran (1998)).

columns reaching all the way from the surface to the P containing layer contacts were realized at four corners of the implanted region. The sample has been subjected to sequential heating at ever-increasing temperatures. Resistivity and Hall measurements performed in the van der Pauw geometry did not reveal any signs of conductivity that can be attributed to donor states due to the electrical activation of P.

Another attempt to achieve n-type conduction by P ion-implantation into a very specially selected type IIa diamond was recently reported by Prins (2001a). In that work a near surface P containing layer was created by fairly shallow P ion-implantations according to the CIRA method with the rapid heating part reaching only 500 °C. This was followed by a much deeper hydrogen implantation so as to load the P layer with point defects. A further annealing was carried out with the hope that the defects adjacent to the P ions will assist its electrical activation. Indeed a reduction of about one order of magnitude was found in $R(T)$ measurements for the H co-implantation case. This was attributed in Prins (2001a) to a larger number of P being electrically activated in the latter case. However, no proof for n-type features of the sample is provided. It should be noted that some increase in the conductivity is expected to appear in the sample and also includes the defective (hence perhaps conductive) layer due to the deep H 'damaging' implantation.

It is conceivable that the introduction of the large P atom into the tight diamond crystal will induce strain which is likely to attract defects to form complexes with the P which are electrically inactive. The fact that the implantation process, in its very nature, involves the formation of defects in the immediate vicinity of the stopping point of each implanted ion, which should happen for even a single implanted ion, may enhance the formation of such complexes; thus reduce the n-type doping efficiency of P implanted diamond at any dose, as low as it may be. Whether low-dose implantations or further loading the P containing region with more defects due to a deep co-implantation can overcome this inherent problem, as proposed by the recent experiments of Prins (2001a) is unclear, and more convincing data have to be provided on the success of the various methods in activating large ions (P) implanted into the tight diamond crystal.

4.2.2.3. Group VI Implantation Doping.

Oxygen and Sulfur. Group VI elements, when in the +1 ionization state, may act as donors, provided that the required ionization state is sufficiently occupied. The obvious candidates for achieving such doping are O and S, the former being a rather small ion may have some advantages over the latter, larger ion.

CVD diamond growth in the presence of oxygen was attempted and the effect that the addition of O to the gas mixture has on the growth rate and layer quality was studied. No electrical effects which can be attributed to O acting as a donor in the CVD diamond layer were, however, reported. Prins has tried to apply his low-dose low-temperature CIRA method to achieve n-type doping of type IIa diamond by O ion-implantation (Prins, 2000). C and C + B implantations to doses which induce comparable defect densities to the O implanted case were used as controls. Significant differences between the controls and the O doped sample were found in the resistivities, with the O implanted sample being most conductive (Fig. 12). Hot point measurements have indicated that the O implanted sample is n-type while the B implanted one was p-type. An activation energy of 0.32 eV was deduced from the $R(T)$ data, though from a rather narrow temperature range. Hall effect has yielded $[n] = 5 \times 10^{11}$ cm^{-3} and a mobility of ~40 cm^2 V^{-1} s^{-1}. This is a remarkable, encouraging, result which should be repeated and confirmed.

The reports that S when introduced into a CVD grown diamond layer is a n-type dopant in diamond (Sakaguchi et al., 1999), even though put in doubt (Kalish et al., 2000; Saada et al., 2000), has triggered attempts to achieve donor activity in diamond by S ion-implantation.

Hasegawa, Takeuchi, Yamanaka, Ogura, Watanabe, Kobayashi, Okushi and Kajimura (1999) have implanted S into an undoped and into a B doped homo-epitaxial ⟨100⟩ diamond held at 400 °C, at various energies to yield S layers of different thicknesses (up to 270 nm) and high S concentrations (around 10^{20} cm^{-3}). No further annealing was applied. Implantation-graphitized contacts covered with Ti/Au were used. The $R(T)$ data exhibit a wide variety of activation energies for the differently implanted samples, ranging from 0.19 to 0.33 eV. Hall effect measurements were non-conclusive regarding the conduction type. To verify that indeed n-type was obtained due to chemical doping by the S implantation, a pn junction was realized by implanting S into a B doped p-type homo-epitaxial layer. However, as pointed out above, rectification due to an

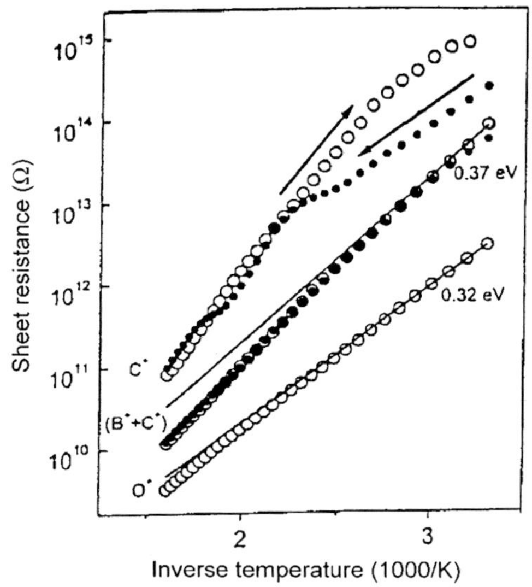

FIG. 12. A comparison of the electrical resistance of three diamonds implanted according to the CIRA process rapidly heated to 500 °C for O, C and B + C implantations. Filled points were measured during the heating, open data points were measured during subsequent cooling (from Prins (2000)).

implanted layer in a p-type diamond, does not necessarily prove that chemical donor activity due to the implanted impurities is responsible for the observed junction behavior, as mentioned above.

Prins (2001a) has applied the same procedure he has used for the O doping of type IIa diamond also to S ion-implantations, with C implantations as a control. The $R(T)$ results for the S implanted sample resemble very much those of the O implanted samples, only the S sample was somewhat more resistive than the O implanted one. Both group VI implants showed an activation energy of ~0.28 eV at low temperature and 0.32–0.33 eV at higher temperatures. No proof for n-type features are given in Prins (2001a).

4.2.2.4. Summary of n-Type Attempts. As is obvious from the above, n-type doping of diamond by ion-implantation is, so far, inconclusive. Many contradictory reports appear in the recent literature, and the situation needs to be clarified by further, well-controlled and reproducible experiments. The complications that may arise from electrical effects due to residual, not completely annealed out, defects must be carefully considered. At the moment it seems as if none of the so far reported 'successful' n-type doping methods is general and proven enough, to enable its wide application as a tool for achieving n-type diamond. Either very selected diamond substrates, such as are not commonly available, were used, or very small (perhaps insignificant) electrical effects, much too small for electronic applications, were found.

One feature, common to several publications, does seem, however, to emerge: the advantages of very high post-implantation annealing, proven to be important for achieving good p-type activities in B implanted diamond, may not hold for n-type doping when P, and perhaps other large impurity atoms are used. The reason for this may well be the large atomic (ionic) size of required impurity atom which induces strain in the tight diamond crystal, thus serving as an attractive potential-well for point defects, which are inherently in the vicinity of the implanted ion, thus diminishing the doping effects.

5. Conclusions

Ion-implantation has proven to be a most important technique for doping semiconductors towards the realization of electronic devices. Without ion-implantation the advancement in the miniaturization of devices, mainly in Si, would be impossible. For the case of diamond, even though its use as a material for electron devices is promising due to the unique physical and chemical properties, its application in practical devices is still rather scarce.

Among the difficulties encountered are problems related to doping diamond under well-defined spatial restrictions and to well controllable doping levels, for which ion-implantation doping is particularly suitable.

Unfortunately, due to the very special properties of diamond (tight crystal, metastability of the sp^3 vs. the sp^2 bonding, high annealing temperatures, wide band gap) doping diamond by ion-implantation and correctly assessing the results is particularly complicated.

The need to find a way of obtaining n-type conductivity in diamond by identifying a suitable dopant with a shallow enough donor level, and devising methods of incorporating it into the diamond crystal are still unsolved. Ion-implantation may offer unconventional ways of solving the above problem. In order to achieve this goal, theoretical predictions and guidance by computational methods are required.

The problems and advantages that ion-implantation in diamond offers have been reviewed above. Difficulties and pitfalls one has to deal with and be aware of in implantation experiments into diamond have been stressed. It is important to critically review some of the published results in light of the particular complexity of diamond and its defects. Nevertheless, the large body of work devoted to the studies of defects in diamond and ways of avoiding their formation or removing them by proper annealing, as reviewed above, has contributed much, not only to the particular efforts in obtaining an implantation related method for doping diamond, but also to the fundamental understanding of diamond and its metastability. Perhaps, some avenues paved by the implantation studies of diamond reviewed here, may be applicable also to other wide-band-gap semiconductors, and these should be tried in the future.

Some aspects of ion-implantation of diamond do find application even today, regardless of the many problems still open. These include the realization of high quality ohmic contacts by high-dose ion-implantation and the use of the electrical properties of defects in diamond, which, even if not 'pure', are well understood and do show promising features. The understanding of defects related to ion-impact on diamond, which is fundamental to the implantation doping of diamond, and which has been somewhat reviewed here, is directly related to various ion-beam deposition techniques, such as biased nucleation of CVD diamond growth and ion-beam deposition of various carbon based films.

It is very difficult to give a comprehensive unbiased review of a field as big as that of ion-implantation into diamond, and which has been the subject of many studies for several decades.

The author wishes to apologize to those whose contribution to the field has not been mentioned here due to lack of space.

And, as a final word, it is very important to continue the ion-implantation work into diamond as it may yield practical results; it deepens our physical insight into diamond, and related materials and it may find analogy in implantation-doping studies and the understanding of defect formation and annealing in other wide band-gap semiconductors.

Acknowledgements

Partial support for this work by the Israeli Ministry of Science and by the European Community (contract HPRN-CT-1999-00139; DoDDS network) is acknowledged. Discussions with Dr C. Uzan-Saguy are greatly appreciated.

References

M. G. Allen, S. Prawer, D. N. Jamieson, and R. Kalish, *Appl. Phys. Lett.* 3, 2062 (1993).

E. Baskin, A. Reznik, D. Saada, J. Adler, and R. Kalish, *Phys. Rev. B* 64, 224110 (2001).

J. P. Biersack and L. G. Haagermark, *Nucl. Instr. Meth. Phys. Res.* 174, 257 (1980).

G. Braunstein and R. Kalish, *Appl. Phys. Lett. (USA)* 38, 416–418 (1981).

G. Braunstein and R. Kalish, *J. Appl. Phys.* 54, 2106 (1983).

I. M. Buckley-Golder, R. Bullough, M. R. Hayns, J. R. Willis, R. C. Piller, N. G. Blamires, G. Gard, and J. Stephen, *Diamond Relat. Mater.* 1, 43 (1991).

N. Casanova, E. Gheeraert, A. Deneuville, C. Uzan-Saguy, and R. Kalish, *Diamond Relat. Mater.* 10, 580 (2001).

C. D. Clark, P. J. Kemmey, and E. W. J. Mitchell, *Discuss. Faraday Soc.* 31, 96 (1961).

A. T. Collins, in "Properties and growth of Diamond" EMIS Data reviews Series no. 9 (G. Davies, Ed.), pp. 261–288, INSPEC. The Institute of Electrical Engineers, London, UK, 1994, Chapter 9.

C. Cytermann, R. Brener, and R. Kalish, *Diamond Relat. Mater.* 3, 677 (1994).

G. Davies (Ed.), Ion implantation into diamond, Invited Chapter in "Properties of Diamond,", pp. 191–214, INSPEC (The Institute of Electrical Engineering), 1994a.

G. Davies (Ed.), Diffusion in diamond, Invited Chapter in "Properties of Diamond," pp. 215–219, INSPEC (The Institute of Electrical Engineering), 1994b.

M. S. Dresselhaus and R. Kalish, "Ion Implantation into Diamond, Graphite and Related Materials". Springer, Berlin, 1992.

W. Eckstein, "Computer Simulation of Ion–Solid Interactions". Springer Series in Material Sciences, Vol. 10. Springer, Heidelberg, 1991.
F. Fang, C. A. Hewett, M. G. Fernandes, and S. S. Lau, *IEEE Trans. Electr. Devices* 36, 1783 (1989).
B. Fizgeer, C. Uzan-Saguy, C. Cytermann, V. Richter, I. Avigal, M. Shaanan, R. Brener, and R. Kalish, *Phys. Status Solidi (a)* 186, 281 (2001).
F. Fontaine, A. Deneuville, E. Gheeraert, and P. Gonon, *Diamond Relat. Mater.* 3, 623 (1994).
F. Fontaine, C. Uzan-Saguy, and R. Kalish, *Appl. Phys. Lett.* 68, 2264 (1996).
M. Hasegawa, D. Takeuchi, S. Yamanaka, M. Ogura, H. Watanabe, N. Kobayashi, H. Okushi, and K. Kajimura, *Jpn. J. Appl. Phys.* 38, L1519 (1999).
C. A. Hewett, R. Nguyen, J. R. Zeidler, and R. G. Wilson, *Diamond Relat. Mater.* 3, 17 (1993).
J. D. Hunn, N. R. Parikh, M. L. Swanson, and R. A. Zuhr, *Diamond Relat. Mater.* 2, 847 (1993).
D. N. Jamieson, S. Prawer, S. P. Dooley, and R. Kalish, *Nucl. Instrum. Meth. B* 77, 457 (1993).
R. Job, M. Werner, A. Denisenko, A. Zaitsev, and W. R. Fahrner, *Diamond Relat. Mater. (Netherlands)* 5, 757–760 (1996).
J. F. Prins, *Appl. Phys. Lett.* 197, 950 (1982).
J. F. Prins, *Phys. Rev. B* 38, 5576 (1988).
S. A. Kahijira, A. Antonelli, J. Bernholc, and R. Car, *Phys. Rev. Lett.* 66, 2010 (1991).
R. Kalish, Ion-implantation into diamond: damage, annealing and doping, "Proceedings of the International School of Physics Enrico Fermi Course CXXXV" (A. Pauletti and A. Tocciarone Eds.), p. 373, IOA press, Amsterdam, 1997.
R. Kalish, T. Bernstein, B. Shapiro, and A. Talmi, *Rad. Effects* 52, 153 (1980).
R. Kalish, M. Deicher, E. Recknagel, and T. h. Wichert, *J. Appl. Phys.* 50, 6870 (1979).
R. Kalish and S. Prawer, Ion implantation of diamond and diamond films: fundamentals and applications in "Industrial Handbook for Diamond and Diamond Films" (M. A. Prelas, G. Popovici, and L. K. Bigelow, Eds.). Marcel Dekker, New York, 2003, in press.
R. Kalish, A. Reznik, C. Uzan-Saguy, and C. Cytermann, *Appl. Phys. Lett.* 76, 757 (2000).
R. Kalish and C. Uzan-Saguy, Doping of diamond using ion-implantation Invited Chapter in "Properties Growth and Applications of Diamond," (M. H. Nazare and A. J. Neves, Eds.), Emis Data review series No. 26, INSPEC (The Institute of Electrical Engineering), 2001.
R. Kalish, C. Uzan-Saguy, B. Philosoph, V. Richter, J. P. Lagrange, E. Gheeraert, A. Deneuville, and A. T. Collins, *Diamond Relat. Mater.* 6, 516 (1997).
R. Kalish, C. Uzan-Saguy, B. Philosoph, V. Richter, and S. Prawer, *Appl. Phys. Lett.* 70, 999 (1997).
R. Kalish, C. Uzan-Saguy, B. Ran, H. Ferber, H. Guettler, and R. Zachai, *Diamond Relat. Mater.* 8, 877 (1999).
R. Kalish, C. Uzan-Saguy, A. Samoiloff, R. Locher, and P. Koidl, *Appl. Phys. Lett.* 64, 2532 (1994).
H. Katayama-Yoshida, T. Nishimatsu, Y. Motroshiromizu, and N. Orita, "The Sixth NIRIM International Symposium on Advanced Materials (ISAM'99)" p. 35, 1999.
J. Koike, T. E. Mitchelle, and D. M. Parkers, *Appl. Phys. Lett.* 59, 2515 (1991).
S. Koizumi, M. Kamo, Y. Sato, S. Mita, A. Sawabe, A. Reznik, C. Uzan-Saguy, B. Ran, and R. Kalish, *Diamond Relat. Mater.* 7, 540 (1997).
S. Koizumi, M. Kamo, Y. Sato, H. Ozaky, and T. Inuzuka, *Appl. Phys. Lett.* 71, 1065 (1997).
S. Koizumi, K. Watanabe, M. Hasegawa, and H. Kanda, *Science* 292, 1899 (2001).
P. F. Lai, S. Prawer, and L. A. Bursil, *Diamond Relat. Mater.* 10, 82 (2001).
F. F. Morehead and B. L. Crowder, *Rad. Effects.* 6, 25 (1970).
J. O. Orwa, K. W. Nugent, D. N. Jamieson, and S. Prawer, *Phys. Rev. B* 62, 5461 (2000).
R. Paul, de la Houssaye, C. M. Penchina, J. R. Zeidler, and R. G. Wilson, *J. Appl. Phys.* 71, 3220 (1992).
G. Popovici, R. G. Wilson, T. Sung, M. A. Prelas, and S. Khasawinah, *J. Appl. Phys.* 77, 5103 (1995).
S. Prawer, D. N. Jamieson, and R. Kalish, *Phys. Rev. Lett.* 69, 2991 (1992).
S. Prawer and R. Kalish, *Phys. Rev. B* 51, 15711 (1995).
S. Prawer, K. W. Nugent, and D. N. Jamieson, *Diamond Relat. Mater.* 7, 106 (1997).
S. Prawer, C. Uzan-Saguy, G. Braunstein, and R. Kalish, *Appl. Phys. Lett.* 63, 2502 (1993).

J. F. Prins, *Nucl. Instrum. Meth. B* 59/60, 1387 (1991).
J. F. Prins, *Mater. Sci. Rep.* 7, 271 (1992).
J. F. Prins, *Nucl. Instrum. Meth. B* 64, 210 (1992).
J. F. Prins, *Adv. New Diam. Sci. Technol.*, 443 (1994).
J. F. Prins, *Diamond Relat. Mater.* 4, 580 (1995).
J. F. Prins, *Phys. Rev. B* 38, 5576 (1988).
J. F. Prins, *Diamond Relat. Mater.* 7, 545 (1998).
J. F. Prins, *Appl. Phys. Lett.* 73, 2308 (1998).
J. F. Prins, *Phys. Rev. B* 61, 7191 (2000).
J. F. Prins, INCDST7 meeting, *Diamond Relat. Mater.* (2001) (in press).
J. F. Prins, *Diamond Relat. Mater.* 10, 463 (2001).
B. Ran, Technion MSc Thesis, Unpublished, 1998.
ISOLDE-Collaboration, M. Restle, K. Bharut-Ram, H. Quintel, C. Ronning, H. Hoffsass, S. G. Jahn, and U. Wahl, *Appl. Phys. Lett. (USA)* 66, 2733–2735 (1995).
A. Reznik, R. Kalish, K. W. Nugent, and S. Prawer, *Nucl. Instr. Meth. B* 148, 626 (1999).
A. Reznik, V. Richter, and R. Kalish, *Phys. Rev. B* 56, 7930 (1997).
A. Reznik, V. Richter, and R. Kalish, *Diamond Relat. Mater.* 7, 317 (1998).
D. Saada, J. Adler, and R. Kalish, *Phys. Rev. B* 59, 6650 (1999).
D. Saada, J. Adler, and R. Kalish, *Appl. Phys. Lett.* 77, 878 (2000).
I. Sakaguchi, M. N. Gamo, Y. Kikuchi, E. Yasu, H. Haneda, T. Suzuki, and T. Ando, *Phys. Rev. B* 60, R2139 (1999).
H. Sternschulte, M. Schreck, B. Stritzker, A. Bergmaier, and G. Dollinger, *Phys. Status Solidi (a)* 174, 65 (1999).
H. Sternschulte, K. Tonke, and R. Sauer, *Phys. Status Solidi* 172, 37 (1999).
D. Takeuchi, S. Yamanaka, H. Watanabe, and H. Okushi, *Phys. Status Solidi (a)* 186, 269 (2001).
C. Uzan-Saguy, C. Cytermann, R. Brener, M. Shaanan, and R. Kalish, *Appl. Phys. Lett.* 67, 1194 (1995).
C. Uzan-Saguy, R. Kalish, R. Walker, D. N. Jamieson, and S. Prawer, *Diamond Relat. Mater.* 7, 1429 (1998).
C. Uzan-Saguy, A. Reznik, R. Kalish, Unpublished.
C. Uzan-Saguy, V. Richter, S. Prawer, Y. Lifshitz, E. Grossman, and R. Kalish, *Diamond Relat. Mater.* 4, 569 (1995).
C. Uzan-Saguy, A. Reznik, C. Cytermann, R. Brener, R. Kalish, E. Bustarret, M. Bernard, A. Deneuville, E. Gheeraert, and J. Chevallier, *Diamond Relat. Mater.* 10, 453 (2001).
V. S. Vavilov, M. A. Gukasyan, M. I. Guseva, T. A. Karatygina, and E. A. Konorova, *Sov. Phys. Semicond.* 8, 471 (1974).
V. S. Vavilov, M. A. Gukasyan, M. I. Guseva, T. A. Karatygina, and E. A. Konorova, *Sov. Phys. Semicond.* 9, 962 (1976).
V. S. Vavilov, V. V. Krasnopevtsev, Y. V. Miljutin, A. E. Gorodetsky, and A. P. Zakharov, *Rad. Effects* 22, 141 (1974).
V. Venkatesan, D. M. Melta, K. Das, and A. M. Belu, *J. Appl. Phys.* 74, 1179 (1993).
R. Zeidler, C. A. Hewett, and R. G. Wilson, *Phys. Rev. B* 47, 2065 (1993).

This page is intentionally left blank

CHAPTER 4

BORON DOPING OF DIAMOND FILMS FROM THE GAS PHASE

A. Deneuville

UNIVERSITY JOSEPH FOURIER - GRENOBLE I AND LABORATOIRE D'ETUDES DES PROPRIÉTÉS ELECTRONIQUES DES SOLIDES, CENTRE NATIONAL DE LA RECHERCHE SCIENTIFIQUE, BP 166, 38 042, GRENOBLE CEDEX 9, FRANCE

1. Introduction and Historical

Diamond has many extreme properties which makes it attractive for lot of applications. The properties include the highest thermal conductivity at room temperature (five times that of copper), a high mobility for the two types of carriers and a high electric field breakdown. It is the best semiconductor for high-temperature electronics and from the classical figure of merit of Johnson and Keyes, for power and high frequency devices (Davis, Sitar, Williams, Kong, Kim, Palmour, Edmond, Ryu, Glass and Carter, 1988).

However, the practical achievement of diamond devices suffers from an insufficient control of its technology. Besides the limited size of the substrates, the n-type doping was demonstrated without ambiguity only recently from the gas phase [koi 97] and in a few laboratories, and there is still continuous progress for p-type doping (Okushi, 2001). Besides the classical p-type doping obtained by boron incorporation, superficial p-type doping after interaction with hydrogen plasma is also used. We will restrict this review to p-type doping by boron incorporation.

Natural and synthetic crystals as well as polycrystalline, highly oriented and homoepitaxial films, can be obtained as boron-doped samples. Boron doping was first studied in natural IIb diamond, then in synthetic crystals and finally in CVD films. A brief description of this historical evolution will be given. Boron can be incorporated in the films from the gas phase or by implantation (Fontaine, Uzan-Saguy and Kalish, 1996). Boron incorporation from the gas phase gives at that time, better electrical characteristics (Uzan-Saguy, Reznik, Cytermann, Brener, Kalish, Bustarret, Bernard, Deneuville, Gheeraert and Chevallier, 2001). The chapter 3 of this book is devoted to boron doping by implantation.

The boron incorporation induces striking effects on the physicochemical and the electronic properties of diamond that will be described here and compared with models. We will look for relationships between these properties in polycrystalline, highly oriented and homoepitaxial films. The results will be

ISBN: 0-12-752185-2
ISSN: 0080-8784

compared to those on IIb natural crystals. Even about boron-doped homoepitaxial films, there are still differences between the experimental results that might originate from different deposition techniques and/or deposition conditions. We will show to what extent the reported results have been obtained independently by different groups. For clarity, we will recall briefly the main deposition techniques and the usual preparation conditions for polycrystalline, highly oriented and homoepitaxial films. Finally, the role of boron doping in some applications will be described.

Several impurities exist in IIb natural crystals, and at the very beginning some groups ascribed their p-type doping to aluminum (Crowther, Dean and Sherman, 1967). Boron on isolated substitutional sites was unambiguously recognized after checking its concentration with the electrical properties of the IIb crystals (Collins and Williams, 1971). As described, for instance, in Field (1979), the study of the boron-doped diamond started in the 1950s on natural IIb, then on synthetic IIb crystals. However, it remained limited as the boron concentration was low and cannot be controlled in IIb natural crystals, and was very difficult to control in synthetic IIb crystals. Nevertheless, a basic knowledge was already obtained about the physicochemical and electronic properties of boron in diamond, and a lot of complex defects were identified from optical and electro-optical measurements. Although somewhat deep (with a fundamental level of the bound hole 0.378 eV above the valence band edge), isolated substitutional boron appeared convenient for p-type doping (Collins and Lightowlers, 1979).

The deposition of diamond thin films on diamond started in the 1960s, mainly in Soviet Union (Derjaguin, Fedosev, Lukyanovich, Spitsyn, Ryabov and Lavrentyev, 1968) also in United States (Angus, Will and Stanko, 1968), and then on foreign substrates (Spitsyn, Bouilov and Deryagin, 1981). In both the countries, homoepitaxial undoped and p-type boron-doped (Poferl, Gardner and Angus, 1973; Vavilov, 1975; Spitsyn et al., 1981; Alexenko and Spitsyn, 1992) diamond films were obtained. The Spitsyn group introduced atomic hydrogen during the deposition and obtained homoepitaxial diamond films of good quality (Spitsyn et al., 1981).

During the 1970s, a program on diamond started at NIRIM in Japan. The role of atomic hydrogen for diamond deposition was systematically studied. This laboratory introduced CVD techniques (hot filament, microwave plasma, DC plasma, etc.) with a very high concentration of atomic hydrogen during the deposition of diamond, which resulted in undoped and boron-doped diamond films with low (polycrystalline films) or negligible (homoepitaxial films) residual concentration of foreign parasitic phases (Matsumoto, Sato, Kamo and Sekata, 1982a; Matsumoto, Sato, Tsutsumi and Sekata, 1982b; Kamo, Sato, Matsumoto and Setaka, 1983). Boron doping of the diamond films from the gas phase was fully demonstrated by several groups (Fujimori, Imai and Doi, 1986; Okano, Nakuri, Akiba, Kurosu, Iida and Hirose, 1988; Mort, Kuhman, Machonkin, Morgan, Jansen, Okumura, LeGrice and Nemanich, 1989; Okano, Nakuri, Akiba, Kurosu, Iida, Hirose and Nakamura, 1989) at the end of the 1980s.

Due the high potential for applications of diamond from its intrinsic properties, large efforts on the deposition of p-doped polycrystalline and monocrystalline diamond thin films started in United States in the late 1980s and the beginning of the 1990s and then in Europe in the beginning of the 1990s. Demonstrators using thin films were built in most of the interesting area—high temperature electronics (pn junctions (Borst, Strobel and Weis, 1995), Schottky diodes (Geis, Rathman, Ehrlich, Murphy and Lindey, 1987), MESFET (Shiomi, Nihsibayashi and Fujimori, 1989), MISFET (Dreyfus, Tessmer, Holmes, Kao, Malta, Plano and Stoner, 1994), basic logic circuits (Fox, Hartsell, Malta, Wynand, Kao, Plano, Tessmer, Henard, Holmes, Tessmer and Dreyfus, 1995), etc.), sensors (UV solar blind (Marchywka, Hochedez, Geis, Socher, Moses and Goldberg, 1991), high energy particles (Kania, Landstrass, Plano, Pan and Han, 1993), pressure (Taher, Aslam, Tamor, Potter and Elder, 1994), temperature (Aslam, Yang and Masood, 1994), etc.), cold cathode (Wang, Garcia, Ingram, Lake and Kordesch, 1991), electrodes (Xu, Granger, Chen, Strojek, Lister and Swain, 1997) for various electrochemical reactions. Now, research is still very active in Japan and Europe.

2. Influence of the Deposition Technique and of the Deposition Conditions on the Quality of the Undoped Films

The diamond films contain parasitic phases, crystallographic defects, impurities and complexes. Their concentrations in the undoped films vary according to the deposition technique and the deposition conditions. The effect of boron doping on their concentration will be checked in Section 3. We first recall the influence of the deposition technique and of the deposition conditions on the quality of the undoped films.

2.1. Influence of the Deposition Technique

2.1.1. The Problems for Diamond Deposition

The diamond films are deposited by CVD from carbon containing gases, mainly from carbon hydrides, and most often from methane. Although graphite is the thermodynamically stable phase of carbon in usual temperature and pressure conditions, the formation energy of diamond is only 30 meV per atom higher than that of graphite (Angus and Hayman, 1988). Small differences in the formation energies of the cubic and hexagonal phases are quite common for wide band gap semiconductors (Yeh, Lu, Froyen and Zunger, 1992), and quite generally co-deposition of the two phases occurs on foreign substrates (Yeh et al., 1992; Daudin, Feuillet, Hübner, Samson, Widman, Philippe, Bru-Chevallier, Guillot, Bustarret, Bentoumi and Deneuville, 1998; Gherthsen, Neubauer, Dieker, Lantier, Rizzi and Luth, 1999). When the differences between the formation energies are very small (for instance, 10 meV per atoms for GaN), the nature (GaAs, SiC, etc.) and the orientation (100) of the substrates are sufficient to favor the deposition of a cubic phase with a higher formation energy (Daudin et al., 1998; Gherthsen et al., 1999). In the case of carbon the difference is too high (30 meV) and the graphite has to be removed to obtain diamond. In the earlier works (Angus et al., 1968; Derjaguin et al., 1968; Spitsyn et al., 1981) purely thermal chemical vapor deposition (CVD) was used. For instance, the groups of Soviet Union prepared the carbon hydrides by in situ etching around 2000 °C of graphite by hydrogen, while diamond was deposited in a cooler zone (900–1000 °C) (Spitsyn et al., 1981). Alternate sequences of deposition and etching by atomic hydrogen were used to remove graphite (Angus et al., 1968; Spitsyn et al., 1981). A striking improvement came from the Spitsyn group (Spitsyn et al., 1981) that introduces atomic hydrogen and then from the NIRIM laboratory (Matsumoto et al., 1982a,b; Kamo et al., 1983) where very high atomic hydrogen concentrations—which preferentially etches the co-deposited graphite or amorphous phases and more generally promotes the sp^3 bonding by saturating the carbon dangling bonds with hydrogen—was used during the deposition process of the carbon films. Almost all the up-to-date deposition techniques (Hot filament, microwave, DC plasma, etc.) use decomposition of mixtures of hydrogen and some percent or less of a gas containing carbon, to generate such a high concentration of atomic hydrogen. Obviously the concentration of atomic hydrogen in contact with the film surface during the growth process depends on the deposition techniques, and with the same technique on the deposition conditions. Therefore, various 'qualities' of diamond films (from foreign residual phases, structural defects and impurities incorporation) are obtained by the different groups not only for polycrystalline and highly oriented films on foreign substrates, but also for homoepitaxial films.

2.1.2. *Impurities Introduced by the Different Deposition Techniques*

Lower concentrations of impurities and structural defects are obtained with the Hot Filament (HFCVD) (Matsumoto et al., 1982a) and the Microwave Plasma (MPCVD) (Kamo et al., 1983) deposition techniques. However, up to 10^{20} cm^{-3} of refractory metals from the filaments are incorporated in the films prepared by the hot filament deposition technique (Jansen, Machonkin and Kuhman, 1990; Cifre, Lopez, Morenza and Esteve, 1992; Hinneberg, Heck and Schmidt, 1992; Gheeraert, Deneuville, Brunel and Oberlin, 1992; Griesser, Stingeder, Grasserbauer, Baumann, Link, Wurzinger, Lux, Haubner and Lux, 1994; Mehta Menon, Edwards, Feigerle, Shaw, Coffey, Heatherly, Clausing, Robinson and Glasgow, 1999). The microwave plasma deposition technique is preferred to study the electronic properties from its higher stability and reproducibility, and its lower impurities contents in the films. Microwave plasma assisted CVD apparatus for the diamond deposition include the simpler quartz tube crossing a microwave guide originally developed by the NIRIM group (Kamo et al., 1983) to more or less sophisticated apparatus where a larger volume of plasma is produced within a glass chamber or a metallic chamber.

There are only few measurements about residual impurities in MPCVD undoped diamond films: (i) [H] is found around $3 \times 10^{20}\ cm^{-3}$ in (100) homoepitaxial films (Samlenski, Schmälzlin, Brenn, Wild, Muller-Sebert and Koidl, 1995; Borst and Weis, 1996)—decreasing down to $2 \times 10^{19}\ cm^{-3}$ after 10 min at 950 °C annealing under vacuum (Borst and Weis, 1996)—and around $10^{21}\ cm^{-3}$ in polycrystalline films (Kimura, Nakatani, Yamada and Suzuki, 1999) somewhat dependent on the preparation conditions (Dollinger, Bergmaier, Frey, Roesler and Verhoeven, 1995) (ii) O around $10^{20}\ cm^{-3}$ in polycrystalline films (Dollinger et al., 1995) (iii) N $< 10^{18}\ cm^{-3}$ from the residual vacuum in (100) homoepitaxial films (Borst and Weis, 1996) (iv) Si (Collins, Kamo and Sato, 1990) about $3 \times 10^{17}\ cm^{-3}$ in (100) homoepitaxial films (Borst and Weis, 1996) ascribed to the plasma etching of the quartz deposition chamber or the quartz window of the metallic chamber.

By analogy with Si microelectronics, contamination with metallic impurities is expected to occur with metallic walls or metallic substrate holder.

2.2. Defects and Parasitic Phases in Polycrystalline, Highly Oriented and Epitaxial Films

2.2.1. *Homoepitaxial Films*

Homoepitaxial films are obtained on the (100), (110) and (111) surfaces of natural or synthetic crystals. However, systematic studies are now performed on the (100) surface which gives the best quality for homoepitaxial films (Shiomi, Tanabe, Nishibayashi and Fujimori, 1990; Visser, Bauhis, Janssen, Vollenberg, van Enckevort and Gilling, 1992) in agreement with the general trend when cubic and hexagonal phases exist (Yeh et al., 1992; Daudin et al., 1998; Gherthsen et al., 1999). Homoepitaxial films are obtained when the substrate temperature is in the range 750–950 °C and up to about 10% methane (Kiyota, Matshushima, Sato, Okushi, Ando, Tanaka, Kamo and Sato, 1997) can be used allowing deposition rates in the $\mu m\ h^{-1}$ range. The 'quality' of the undoped homoepitaxial films is similar to that of the diamond crystal from the full width at half maximum (FWHM) of the $1332\ cm^{-1}$ Raman line of diamond (Field, 1992; Sato and Kamo, 1992) (Section 5) and even better from their concentration of structural defects for the best homoepitaxial films (Okushi, 2001).

2.2.2. *Polycrystalline Films*

The nucleation rate of diamond on foreign substrates is very low. It is enhanced by scratching the surface of these substrates by diamond powder.

The growth rate depends widely not only on the deposition temperature (750–950 °C), but also on the direction of the growth sector. As the thickness of the film increases, there is progressively dominance of the crystals with the fastest growth rate (van der Drift process), and columnar growth appears (Yarbrough and Messier, 1990; Wild, Khol, Herres, Muller-Sebert and Koidl, 1994). The selected direction depends critically on the deposition conditions (Wild et al., 1994). The polycrystalline films are usually deposited on (100) silicon. They have grain boundaries containing foreign phases. Grains with size of the micron range are expected to exhibit poorer crystalline quality than the homoepitaxial films. Low methane concentrations ($\leq 0.5\%$) have to be used in order to minimize the concentration of the foreign phases and to improve the quality of the film (Field, 1992; Sato and Kamo, 1992). The nature and concentration of the foreign phases as well as a first approximation of the quality of the diamond part of the films are checked by Raman experiments (Section 5). They are more sensitive to the foreign phases when the excitation of the film is done by a light of low energy (Wagner, Wild and Koidl, 1991). The diamond quality seems better for isolated crystallites than for polycrystalline films. The quality of the films improves as the size of the crystallites increases. It is poorer on the usual void-free and continuous 1–10 μm thick films with crystallites size ≤ 1 μm where the FWHM of the diamond Raman line is in the range 9–12 cm^{-1} (Robin, Farabaugh and Feldman, 1990; Field, 1992; Sato and Kamo, 1992). The 0.5–1 mm-thick free standing films with crystallite size of the order of 100 μm have FWHMs of the order of 3 cm^{-1} (Coe and Sussmann, 2000), nearby that of homoepitaxial thin films.

2.2.3. *Towards Heteroepitaxy: Highly Oriented Diamond Films on Si*

The size of the synthetic diamond substrates is one of the limitation for the use of homoepitaxial diamond films. Therefore, numerous works have looked for heteroepitaxy of diamond, on metals (Zhu, Yang and Glass, 1993) and semiconductors (BN (Koizumi, Murakami, Inuzuka and Suzuki, 1990), SiC (Stoner and Glass, 1992) and Si (Jiang, Klages, Zachai, Hartweg and Füsser, 1993; Stoner, Sahaida, Bade, Southworth and Ellis, 1993)). Despite the very high mismatch (52%) of the silicon with diamond, when particular in situ treatment of the (100) surface and particular growth conditions are used, highly oriented films with local heteroepitaxy ($[001]_{diamond}//[001]_{silicon}$ and $[110]_{diamond}//[110]_{silicon}$ (Jiang et al., 1993; Stoner, Kao, Malta and Glass, 1993)) are obtained inside columns which have a top surface of several microns. The misorientation between the columns can be $<2°$ (Fox, Stoner, Malta, Ellis, Glass and Sivazlian, 1994).

The deposition of these highly oriented diamond films on Si (100) uses three steps, and a small nitrogen content (N/C up to 25 ppm) in the gas phase increases significantly the growth rate and stabilizes the formation of smooth (Kiyota et al., 1997) faceted surfaces. The quality of the films is better than that of (randomly oriented) polycrystalline films, with FWHM of the '1332 cm^{-1}' Raman line around 7 cm^{-1} for film in the micron range thicknesses (Locher, Wild, Harris, Behr and Koidl, 1994; Muller-Sebert, Wörner, Fuchs, Wild and Koidl, 1996).

2.3. Improvement of the Quality of the Deposited Monocrystalline Diamond

Although good diamond films are obtained with the usual deposition conditions (Field, 1992; Sato and Kamo, 1992), several groups further attempt to improve their quality.

After the first results of several groups (Chen, Huang, Hosomi and Yoshida, 1989; Liou, Weimer, Knight and Messier, 1990; Muranaka, Yamashita, Sato and Miyadera, 1990), Backmann tries to generalize (Bachmann, Leers and Lydtin, 1991) the use of mixture containing not only hydrogen and carbon, but also oxygen. On one hand, oxygen etches even more drastically than hydrogen the parasitic phases, but on the other hand, it does not promote the sp^3 bonding. The addition of oxygen in the gas mixture seems to result in a better structural quality of isolated crystallites (Gonon, Gheeraert, Deneuville and Abello, 1995b). It decreases the Si incorporation in the film (Sakaguchi, Nishitani-Gamo, Loh, Haneda, Hishita and Ando, 1997), but it also decreases by a factor of 10–100 the efficiency of boron incorporation in the diamond films (Mort et al., 1989; Ruan, Kobashi and Choyke, 1992; Sakaguchi, Nishitani-Gamo, Loh, Haneda and Ando 1998), even when oxygen is contained in the carbon vector (Erz, Dotter, Jung, and Ehrardt, 1995).

Recent works show that low misorientation angle ($<1°$) of the substrate surface with the (100) plane and very low methane contents (down to 0.015%) in hydrogen gives homoepitaxial films of excellent quality. These conditions minimize the number of growth defects (unepitaxial crystallites and pyramidal hillock) (Takeuchi, Watanabe, Yamanaka, Okushi and Kajimura, 2000), which results in free exciton cathodoluminescence without the usual A band (Watanabe, Hayashi, Takeuchi, Yamanaka, Okushi, Kajimura and Sekiguchi, 1998) and near ideal Schottky diodes with $n = 1.01$ over seven orders of magnitude at room temperature (Yamanaka, Watanabe, Masai, Kawata, Hayashi, Takeushi, Okushi and Kajimura, 1998a; Takeuchi et al., 2000). However, the deposition rate is drastically reduced (19 nm h^{-1} with 0.025% of methane) and the incorporation of B (which, on the other hand, is reported to promote unepitaxial crystallites and hillocks) becomes very weak (Okushi, 2001). Methane concentrations $\geq 0.3\%$ are used for boron doping of films deposited on a buffer layer of undoped diamond deposited from gas mixtures with very low methane contents (Okushi, 2001).

3. Boron in Diamond Films

The lack (at that time) of successful p-doping with atoms other than boron can tentatively be ascribed to the small covalent radius of carbon (0.077 nm). While the small covalent radius of boron (0.088 nm) allows its substitution for carbon, space would not be sufficient for the substitution of another atom of the third column Al (0.118 nm), Ga (0.126 nm) and In (0.144 nm). Boron can be introduced in the gas phase from various precursors. From its covalent radius (0.088 nm $>$ 0.077 nm), it is expected to expand the diamond lattice. Actually, the boron concentration has striking effect on the structural properties of the films, including ultimately the stability of the diamond phase.

Some characteristics of the films seem independent of the deposition technique used and on the boron precursor, while others depend on both of them. The depositions of polycrystalline and homoepitaxial films are usually done with 0.5% and up to some percent of methane, respectively.

3.1. From the Gas Phase to the Solid Phase

3.1.1. The Precursors

Not only diborane (Fujimori et al., 1986; Mort et al., 1989; Gheeraert, Gonon, Deneuville, Abello and Lucazeau, 1993; Ager, Walukiewicz, McCluskey, Plano and Landstrass, 1995; Kiyota et al., 1997; Sonoda, Won, Yagi, Hatta, Ito and Hiraki, 1997; Inushima, Matsushjta, Ohya, Shiomi, 2000; Kim, Choi, Wakimoto, Usami, Tomokage and Ando, 2000), but also boron oxide dissolved in acetone (Okano et al., 1988) or in ethanol (Show, Matsukawa, Ito, Iwase and Izumi, 2000) or in methanol (Sugino, Karasutani, Mano, Kataoka, Shirafuji and Kobashi, 1994) or equivalently trimethylborate (Chen, Chen, Hong and Wang, 1994; Locher, Wagner, Fuchs, Maier, Gonon and Koidl, 1995), trimethylboron (Cifre, Puigdollers, Polo and Esteve, 1994; Spicka, Griesser, Hutter, Grasserbauer, Bohr, Haubner and Lux, 1996; Maeda, Ohtsubo, Kameta, Saito, Kusakabe, Morooka and Asano, 1998; Yamanaka et al., 1998a), triethylboron (Haubner, Bohr and Lux, 1999) were used to introduce boron in the gas phase. Boric acid (Freitas, 1994), boron trichloride (Dorsch, Holzner, Werner, Obermeir, Harper, Johnston, Chalker and Buckley-Golder, 1993; Jackman, Baral, Kingley and Foord, 1996; Werner, Johnston, Chalker, Romani and Buckley-Golder, 1996b), borane trimethylamine (Jackman et al., 1996), etching of boron (Gildenblat, Grot, Hatfield, Wronski, Badzian, Badzian and Messier, 1990) or boron oxide (Borst and Weis, 1995) powder, boron etched from the BN substrate holder (Janssen, Enckervort, Vollenberg and Gilling, 1992) or of a boron rod more or less deeply introduced in the plasma (Borst, Münzinger and Weis, 1994) were also used.

Diborane and at a lesser extent trimethylboron are the most usual boron precursors.

3.1.2. Boron Concentration in the Solid Phase

Quite generally, the boron concentration in the solid phase depends on the B/C ratio in the gas phase. The incorporation rate of boron from the gas phase to the solid film is more than 10 times higher for the growth in the (111) than in the (100) direction (Spitsyn et al., 1981; Gheeraert et al., 1993; Jin and Moustakas, 1994; Locher et al., 1995; Wurzinger, Pongratz, Hartmann, Haubner and Lux, 1997; Ushizawa, Watanabe, Ando, Nishitani-Gamo, Sato and Kanda, 1998) (as in synthetic crystals (Burns, Cvetkovic, Dodge, Evans, Rooney, Spear and Welbourn, 1990)). Therefore only mean values of [B] can be given in polycrystalline films. A large boron concentration up to $5-8 \times 10^{21}\ cm^{-3}$ (i.e., 5 at.%) can be incorporated by the hot filament (polycrystalline films) (Spicka et al., 1996) and MPCVD (polycrystalline (Cifre et al., 1994; Werner, Job, Zaitsev, Fahrner, Sieffet, Johnston and Chalker, 1996a; Werner et al., 1996b) and up to $8 \times 10^{20}\ cm^{-3}$ in homoepitaxial films (Gheeraert, Deneuville and Mambou, 1998)) by MPCVD techniques. There is no boron segregation at the grain boundaries of polycrystalline films (Zhang, Lee and Lam, 1996; Wurzinger et al., 1997). As already stated (Section 2.3), the introduction of oxygen in the gas mixture and the use of very low methane content reduces drastically the boron incorporation. Boron contamination of the deposition chamber walls might result in boron background concentrations in the HFCVD (Spicka et al., 1996) and MPCVD (Nishimura, Das and Glass, 1991) films in the $10^{17}\ cm^{-3}$ range.

The various deposition techniques and the various precursors can reach similar maximum values for the boron incorporation in the diamond films.

3.1.3. Boron Concentration and Phase Stability

The diamond phase exists at least up to about 8×10^{21} B-cm^{-3} (Werner et al., 1996a,b; Lagrange, Deneuville and Gheeraert, 1998). In polycrystalline films grown by the hot filament and the MPCVD techniques and doped from diborane and trimethylboron, $[B] \geq 10^{21}\ cm^{-3}$ results in films containing an additional graphite phase (Kobashi, Nishimura, Miyata, Kawate, Glass and Williams, 1988) well identified from its $1350-1580\ cm^{-1}$ Raman signals (Bernard and Deneuville (unpublished); Nishimura et al., 1991; Miyata, Kumagai, Nishimura and Kobashi, 1993; Cifre et al., 1994) (Fig. 1, Bernard and Deneuville (unpublished))). More precisely, there are honeycomb like structures with graphite lamellae containing no boron (Wurzinger et al., 1997) around diamond containing a high-boron concentration. For polycrystalline films deposited by the hot-filament technique and doped from triethylboron, graphite appears above the metallic conductivity threshold on the (111) face, below this threshold and with a more important concentration on the (100) face (Wurzinger et al., 1997). For MPCVD (100) homoepitaxial films boron doped from diborane, graphite appears only above 10^{21} B-cm^{-3} (Bernard and Deneuville (unpublished)). Very heavy

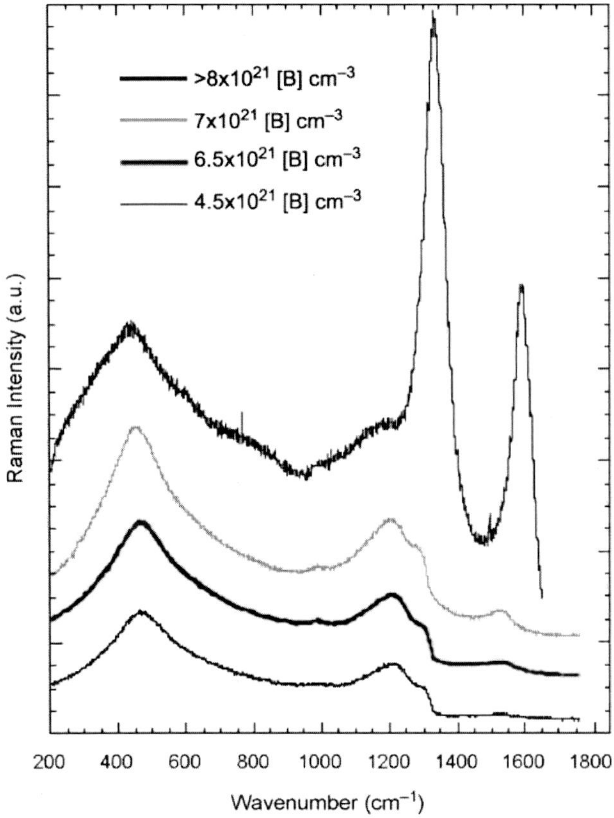

FIG. 1. Raman spectra of heavily boron-doped polycrystalline diamond films.

B/C in the gas phase induces lower B incorporation and ill-defined ballas structures in hot filament (Haubner et al., 1999) and MPCVD films (Cifre et al., 1994).

In polycrystalline and homoepitaxial films deposited by HFCVD or MPCVD with all precursors, an additional graphite phase appears at heavy boron concentration, well above the threshold boron concentration corresponding to the metallic conduction on the boron impurity band. The critical boron concentration is around 10^{21} *cm*$^{-3}$ *when B is incorporated from diborane and trimethyboron. It is lower and depends on the growth sector when boron is incorporated from triethylboron in films deposited by HFCVD.*

3.1.4. Incorporation Efficiency from the Gas Phase to the Solid Phase

The nature of the boron-containing species of the gas phase in contact with the growth surface is not so clear. Thermodynamics (Haubner et al., 1999) predict

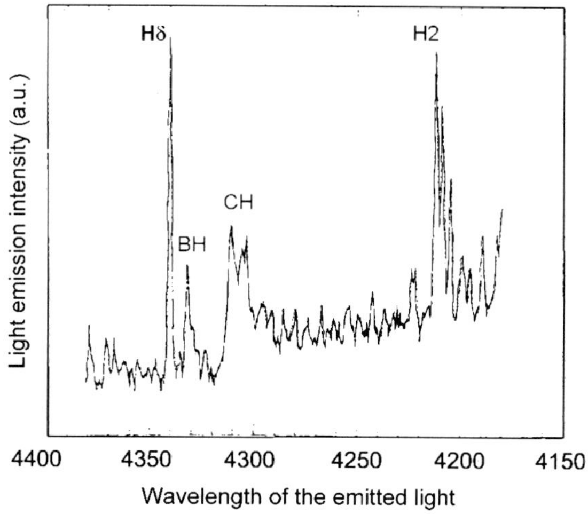

FIG. 2. UV emission spectrum of BH radicals in the gas phase.

BH_3, BH_2 and BH radicals in the gas phase whose concentrations modify those of the CH_x radicals and of the atomic H. At very high boron concentrations, the appearance of ballas is ascribed to the formation of hydrogen–carbon–boron compounds (Haubner et al., 1999). From its UV emission, the gas phase contains BH radicals (Fig. 2) and the ratio of CH/H, C_2/H and H/H_2 decreases as the B/C concentration in the gas phase increases (Gonon et al., 1995a).

Boron incorporation from the gas phase to the solid phase has a high efficiency. In polycrystalline films the incorporation efficiency (defined as $(B/C)_{solid}/(B/C)_{gas}$) is reported constant, around 7 with HFCVD and using triethylboron (Spicka et al., 1996), and around 400 with MPCVD and using trimethylboron (Cifre et al., 1994), but increasing with B/C in the gas phase from 0.06 (Fujimori, Nakata and Imai, 1990) to 2.5 (Nishimura et al., 1991) with MPCVD and using B_2H_6. In homoepitaxial films deposited by MPCVD and using diborane, incorporation around one is obtained in the whole B/C range, in agreement with theoretical models (Jin and Moustakas, 1994).

The incorporation efficiency of boron from the gas to the solid phase depends on the deposition technique, the precursors and the polycrystalline or homoepitaxial nature of the films.

3.1.5. *Deposition Rate and Boron Concentration*

The deposition rate of the boron-doped films are usually in the range 0.1–1 $\mu m\ h^{-1}$. It decreases as the B/C ratio increases (Miyata et al., 1993; Cifre et al., 1994; Gonon et al., 1995a) for MPCVD films, but increases up to 5×10^{21} B-cm^{-3} and then slightly decreases for deposition by the hot-filament technique

(Wurzinger et al., 1997). In MPCVD films, the decrease is ascribed to the lower concentration of CH_x radical in the gas phase for a growth controlled by surface reactions (with may be electronic effects as a faster decrease is reported around boron concentration in the solid phase corresponding to the appearance of metallic conductivity) (Gonon et al., 1995a).

As the boron concentration increases in the films, their deposition rates decreases and increases, respectively, when the MPCVD and HFCVD deposition techniques are used.

3.2. Structural Characteristics of the Films

3.2.1. *Morphology of the Polycrystalline Films*

Boron compounds in the gas phase induce secondary nucleation (Show et al., 2000) in polycrystalline films. Therefore the size of the crystallite decreases with increasing boron content (Wang, Ma, Zhu, Glass, Bergman, Turner and Nemanich, 1992; Gheeraert et al., 1993; Miyata et al., 1993; Cifre et al., 1994; Liao, Zhang, Lee, Tong Lee and Lam, 1997). As mentioned previously, the van der drift selection process induces a columnar growth resulting rapidly in texture of the films. The effect of boron on the texture varies according to the authors. For 10 μm thick undoped films deposited by MPCVD at 730 °C on Si (100), $\{111\}$ poles figures give $\langle 111 \rangle$ fiber texture. This texture is kept up to the metallic transition on the boron impurity band, but completely lost very rapidly for higher boron contents (Chateigner, Brunet, Deneuville, Germi, Pernet, Gheeraert and Gonon, 1995). This is ascribed to the loss of differences in the growth rates according to the crystallographic direction when there is metallic conductivity on the boron impurity band whose physical origin (electronic effect, increase of the defect density) remains unclear. On the contrary with increasing boron content, surface morphology from random, to triangular (111), pyramidal (111), before the appearance of graphite has also been reported (Miyata et al., 1993; Wurzinger et al., 1997) for MPCVD and HFCVD films.

The size of the crystallite of the polycrystalline films decreases when their boron content increases. Their morphology varies with their boron content, in a way dependent on the detail of the deposition process.

3.2.2. *Lattice Parameter of the Polycrystalline and Homoepitaxial Films*

As already mentioned, the covalent radius of boron (0.088 nm) is larger than that of carbon (0.077 nm) which suggests local expansion of the lattice. The lattice parameter of diamond has been studied against boron concentration in CVD powder, in polycrystalline and in homoepitaxial films.

3.2.2.1. Experimental Results. The ASTM card gives a lattice parameter of 0.35667 ± 0.00002 nm for IIa natural crystals. For powders obtained by CVD, starting from a lattice parameter of 0.35659 nm for undoped diamond, nearly isotropic expansion is obtained from about 1.76×10^{19} B-cm^{-3} up to 1.76×10^{21} cm^{-3} where the lattice parameter reaches 0.35767 nm (Voronov and Rakhmanina, 1993). Lattice expansion with boron incorporation is reported for polycrystalline films deposited by the MPCVD (Cifre et al., 1994; Brunet, Deneuville, Germi, Pernet and Gheeraert, 1997) and the HFCVD technique (Wurzinger et al., 1997). When the films are deposited on Si (100), differences were found according to the crystallographic directions of diamond (Brunet et al., 1997). Values similar to those for CVD powder were obtained in the (220) direction, while starting from 0.35669 nm the lattice parameter expands up to 0.35762 and 0.3575 nm around 8×10^{21} B-cm^{-3}, respectively, in the (111) and (311) directions. Actually, the lattice parameter increases slowly from about 10^{17} to about 10^{20} B-cm^{-3}, then more rapidly for higher B values (Brunet et al., 1997; Cifre et al., 1994). For homoepitaxial films, different behaviors were reported according to the (110) (Spitsyn et al., 1981) or (100) (Brunet, Germi, Pernet, Deneuville, Gheeraert, Laugier, Burdin and Rolland, 1998) direction of the growth surface on, respectively, natural IIa and synthetic Ib diamond crystals. On the (110) surface of IIa crystals, starting from a lattice parameter of 0.35664 ± 0.00001, the lattice parameter decreases first to 0.35659 ± 0.00001 nm up to 1.76×10^{20} B-cm^{-3}, then increases at 1.76×10^{21} B-cm^{-3} up to the parameter value of the IIa substrate 0.35668 ± 0.00001 nm. For deposition on the (100) surface of Ib crystals, deformation of the unit cell was reported. After correction to obtain the parameter of the equivalent isotropic material, starting from a value of 0.35670 ± 0.00001 nm, the cell parameter increases nearly linearly, first slowly up to 0.35677 nm up to 2.7×10^{20} B-cm^{-3}, then more rapidly to 0.35728 nm at 8×10^{20} B-cm^{-3} (Brunet et al., 1998), Fig. 3.

3.2.2.2. Comparison of the Experimental Results with Models. The simpler model of Vegard (1921) for a monocrystalline compound containing two atoms predicts a lattice parameter varying linearly proportionally to their concentrations in the mixture. Later, Bardeen and Shockley (1950) introduces variation of the lattice parameters with high carrier concentrations through the deformation potentials. Usually (Pietsch and Unger, 1983), there is a contraction of the lattice proportional to the hole concentration. Finally, some defects introduce non-isotropic deformation of the lattice and usually expansion of the lattice proportional to the number of those defects (Cargill, Segmuller, Kuech and Theis, 1992).

All these models have been invoked to explain the experimental results on boron-doped diamond film. The lattice parameter of CVD powder lower than that of natural crystal was ascribed to the presence of impurities in natural crystals (Voronov and Rakhmanina, 1993). For instance, incorporation of nitrogen, which is expected to decrease the diamond lattice from its lower covalent radius

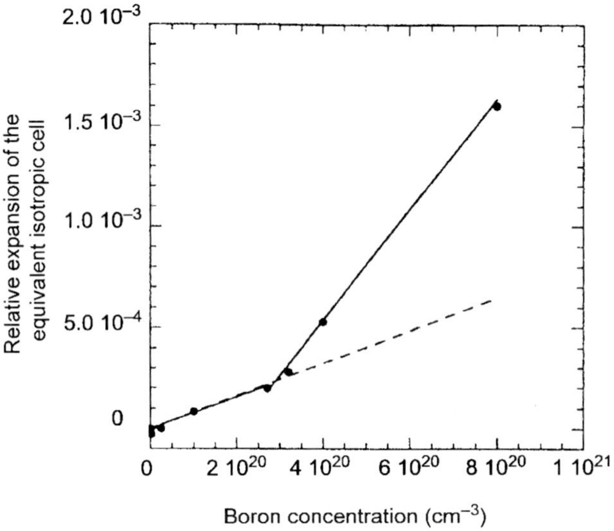

FIG. 3. Relative expansion of the equivalent isotropic cell vs. homoepitaxial diamond films vs. their boron content.

(0.07 nm) than carbon (0.077 nm) according to the Vegard law, but which actually increases it (Voronov, Rakhmanina and Khlybov, 1995) because of the anisotropy of its site (Smith, Sorokin, Gelles and Lasher, 1959) might support the third model.

For the homoepitaxial films deposited on the (110) diamond surface, the decrease of the lattice parameter up to 1.76×10^{20} B-cm^{-3} and its expansion for higher B incorporation were ascribed to incorporation of boron in substitutional and interstitial site, respectively (Spitsyn et al., 1981). However, as we shall see in Section 4.1, the incorporation of boron in substitutional sites appears quite general, even for concentrations higher than those corresponding to the metallic conductivity on the boron impurity band ($3 \times 10^{20} > 1.76 \times 10^{20}$ cm^{-3}).

For homoepitaxial films deposited on the (100) surface of diamond (Brunet et al., 1998), the slow lattice expansion up to B $\cong 3 \times 10^{20}$ B-cm^{-3} (threshold for the metallic conductivity on the boron impurity band) matches with the Vegard's law using the covalent radii of carbon and boron. The lattice expansion at higher (heavy) B was ascribed to the interaction between the holes on the metallic boron impurity band and the lattice as for boron-doped CVD powder (Voronov and Rakhmanina, 1993) and polycrystalline films (Brunet et al., 1997). However, the deformation potential of the valence band is usually negative for holes in semiconductors (Pietsch and Unger, 1983), which would result in a contraction rather than an expansion of the lattice. In the case of diamond, calculations give positive (53.8–106 eV) (Cardona and Christensen, 1986) as well as negative (−2.6 eV) (Cardona and Christensen, 1987) deformation potential, while a very high positive value (99.1 eV) was obtained indirectly from an experiment on undoped diamond (Reggiani, Waechter and Zukotynski, 1983). Deformation potentials of 16 and 19.5 eV are deduced from the fit of the lattice expansion with the deformation potential model, respectively, for homoepitaxial (Brunet et al., 1998) and polycrystalline (Brunet et al., 1997). The theoretical models do not take into account the occurrence of a boron impurity band outside the valence band (Section 4.2.1), and this might tentatively explain the differences obtained between the models and the experimental results.

The shape of the diffraction spots can give information on the local strain (Gaussian profile) and the size of the coherent domain for X-ray diffraction (Cauchy profile). For homoepitaxial films deposited on the (100) surface of diamond, such analysis indicates shapes originating from the size of the coherent domains (Brunet et al., 1998) which agrees with the nature and the concentration of the crystallographic defects found in these films (Section 3.2.3.2). On the contrary, the strain dominates in the boron-doped polycrystalline films (Brunet et al., 1997).

Up to about 3×10^{20} *B-cm*$^{-3}$ *(threshold for the metallic conduction on the boron impurity band) almost all works indicate an expansion of the diamond lattice with boron incorporation. At higher boron concentration all works indicate a faster expansion of the lattice. It is ascribed to holes–lattice interactions through a positive deformation potential.*

3.2.3. *Quality and Structural Defects of Polycrystalline Films*

The boron incorporation modifies quite generally the quality of the polycrystalline films, and more precisely, the concentration of the parasitic phases and of the extended and point defects.

3.2.3.1. Quality. In polycrystalline films deposited by MPCVD with 0.5% of CH_4 and doped from diborane, the decrease of the FWHM of the 1332 cm^{-1} Raman line as the boron concentration increases was ascribed to a decrease of the global concentration of defects (Nishimura et al., 1991; Wang et al., 1992; Gheeraert et al., 1993; Deneuville and Gheeraert, 1996) (Section 5.2). In the same time, the slope of the background of the Raman signal from luminescence decreases (Wang et al., 1992; Deneuville and Gheeraert, 1996), and the signal around 1550 cm^{-1} from parasitic phases decreases (Nishimura et al., 1991; Wang et al., 1992; Deneuville and Gheeraert, 1996), which are coherent with such a picture. This picture was shown to hold up to the B concentration around that inducing the metallic conductivity (Nishimura et al., 1991; Deneuville and Gheeraert, 1996). With deposition by the same MPCVD technique, different results can be obtained with other boron precursors. Increasing the boron concentration also decreases the concentration of parasitic phases when B is supplied by trimethylboron (Cifre et al., 1994), but increases it when B is supplied from trimethylborate (Chen et al., 1994).

While the B concentration increases in MPCVD films doped from diborane, other experiments confirm independently the decrease of particular point defects.

First, the 1.681 eV luminescence signal due to Si associated with a vacancy (Collins et al., 1990) decreases with increasing B incorporation (Wang et al., 1992).

Then, polycrystalline diamond films exhibit ESR signals (Watanabe and Sugata, 1988) with Landé factors around $g = 2.0026$ and 2.0029 and width of, respectively, 2.5 and 8 G (Jia, Shinar, Lang and Pruski, 1993; Holder, Rowan and Krebs, 1994). Their physical origin is not so clear. They have been ascribed to carbon dangling bonds in different environments (Jia et al., 1993; Holder et al., 1994). More specifically, characteristic wings on the ESR signal of undoped films were ascribed to interaction between the C dangling bond and hydrogen (Jia et al., 1993). Such wings do not appear in MPCVD films boron doped from diborane (Colineau, Deneuville, Mambou and Gheeraert, 1996) and HFCVD films doped from boron trioxide (Show et al., 2000). Quite generally, the same ESR signals are found, with a much lower concentration in IIa crystals, and with increasing concentration with the dose after any ion implantation in IIa crystals (Brosious, Corbett and Bourgoin, 1974; Morhange, Beserman and Bourgoin, 1975) or diamond polycrystalline films (Fontaine, Gheeraert, Deneuville, Khong and Collins, 1994). When the polycrystalline films are prepared by HFCVD and doped from boron trioxide (Show et al., 2000) or by MPCVD and doped from diborane (Colineau et al., 1996), the density of such defects decreases as the boron concentration increases up to a concentration in the $10^{21}\ \mathrm{cm}^{-3}$ range. The $g = 2.0023$ and 2.0029 signals originate from two different defects as their spin concentrations in MPCVD films decrease from the almost $10^{18}\ \mathrm{cm}^{-3}$ down to $10^{16}\ \mathrm{cm}^{-3}$ with different variation laws (Fig. 4 and Colineau et al., 1996). The appearance of the $g = 2.000$ ESR signal for very heavy B content (Colineau, Gheeraert, Deneuville, Mambou, Brunet and Lagrange, 1997) might to be connected with that of graphite.

3.2.3.2. Structural Defects from TEM. Information about structural defects can be deduced from TEM. The main defects of diamond crystals and films are twinning, stacking faults in the (111) direction and dislocations.

Boron-doped HFCVD polycrystalline films have a lower concentration of structural defects than the undoped films (Yacobi, Badzian and Badzian, 1991). Strain and square defects resulting from the differences in the boron incorporation rate according to the growth sector are reported in HFCVD films (Wurzinger et al., 1997). With the usual gas mixture containing 0.5% of methane, the twin planes were reported to be mainly on the growing (111) surface and decrease (Wang et al., 1992; Liao et al., 1997) in MPCVD films with increasing B. Their reduction is ascribed to an increasing mobility of C atoms on the surface or to a lower probability of a 60° rotation of C atoms on the (111) surface

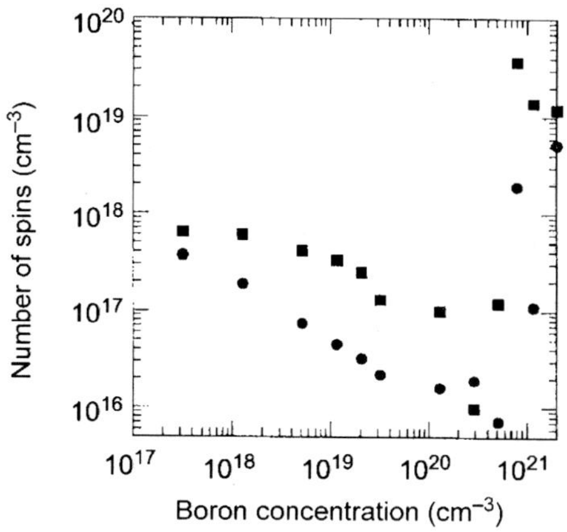

FIG. 4. Number of spins in the EPR signals at $g = 2.0026$ and 2.0030 of polycrystalline diamond films vs. their boron content.

(Wang et al., 1992). The effect on the concentration of other defects of increasing boron incorporation is less clear. It was reported to decrease (Wang et al., 1992) and to increase (Liao et al., 1997) the number of dislocations.

In the best homoepitaxial films deposited with a concentration $\geq 0.3\%$ of methane on buffer undoped films deposited with $\leq 0.03\%$ of methane, incorporation of boron is reported to increase the concentration of unepitaxial and pyramidal crystallites (Okushi, 2001).

Boron incorporation improves the quality of the polycrystalline films by reducing the concentration of the planar and of some point defects independently of the deposition techniques and of the precursor excepting trimethylborate. It increases the concentration of unepitaxial and pyramidal crystallite in the best homoepitaxial films.

3.2.3.3. The Bernholc Model. The decrease of the concentration of extended defects with increasing boron doping was predicted by the Bernholc model (Bernholc, Antonelli, Del Sole, Bar-Yam and Pantalides, 1988). According to this model, in the wide band gap semiconductors (i) the formation energies of the positively or negatively charged vacancies decrease and their diffusion coefficients increase as the doping level increases; (ii) relaxation of the extended defects occur through vacancies displacement. The concentration and the mobility of charged vacancies would be high during the boron-doped diamond deposition and higher when the boron concentration increases. This would result in a concentration of extended defects which decreases as the doping level increases.

The results on polycrystalline boron-doped films can be understood through the Bernholc model.

4. Infrared Absorption from Boron-doped Films

4.1. The Experimental Results

The occurrence of boron in diamond induces absorption in the infrared and visible region. Its concentration remains in the 10^{16} cm^{-3} range in the natural IIb crystals and cannot be sufficiently controlled in the synthetic boron-doped crystals. Similar results are found for the infrared absorption of homoepitaxial thin films with [B] $< 10^{18}$ cm^{-3} and IIb natural crystals. They are more easily obtained in crystals that are several orders of magnitudes thicker than the thin films.

At room temperature infrared absorption of the natural IIb crystals (Austin, 1956; Smith and Taylor, 1962) with [B] $\leq 10^{17}$ cm^{-3} exhibit absorption peaks around 0.305 eV (2459 cm^{-1}), 0.347 eV (2806 cm^{-1}), 0.363 eV (2935 cm^{-1}), 0.462 eV (3725 cm^{-1}, phonon replica of the 0.305 eV), 0.508 eV (4096 cm^{-1}, phonon replica of the 0.347 eV) superimposed on a step with an edge around 0.378 eV (3050 cm^{-1}). Uniaxial stress induces shifts and modifies these bands (Anastassakis, 1969). Fine structures appear in the peaks (except that around 0.305 eV) at liquid nitrogen temperature (Charette, 1961; Smith and Taylor, 1962), and more clearly at liquid helium temperature (Charette, 1961; Kim, Vogelgesang, Ramdas, Rodriguez, Grimsditch and Anthony, 1997).

While the boron concentration in IIb crystal remains weak, it can be varied continuously from 10^{16} to 10^{21} cm^{-3} in boron-doped films. At room temperature and up to the metallic conductivity on the boron impurity band (about 3×10^{20} B-cm^{-3}, see Section 6.3.1), the same peaks and step are obtained from homoepitaxial films (Janssen et al., 1992; Ogasawara, Inushima, Shiraishi, Ohya, Karasawa and Shiomi, 1997; Gheeraert et al., 1998) (Fig. 5a, B/C $\leq$ 1000 ppm (Ogasawara et al., 1997)). The energy positions of the peaks and of the step remain nearly constant (Ogasawara et al., 1997; Gheeraert et al., 1998), while the FWHM at room temperature of the 0.305 and 0.347 eV peaks increases from 7 to about 35 meV as the boron concentration increases from 10^{18} to 3×10^{20} cm^{-3} (Gheeraert et al., 1998). Around liquid nitrogen temperature, fine structures similar to those of IIb crystals are found in homoepitaxial films with [B] in the 10^{17} cm^{-3} range (Ribot, Gheeraert and Deneuville (unpublished)). At room temperature, for heavy boron concentrations corresponding to metallic conductivity on the boron impurity band, a wide absorption band appears around the 0.378 eV step with structures around 0.347 and its phonon replica around 0.508 eV (Ager et al., 1995; Ogasawara et al., 1997) (Fig. 5b, B/C $\geq$ 5000 ppm (Ogasawara et al., 1997)).

The step and less and less clearly defined peaks are obtained for polycrystalline films when they are highly textured (Erz et al., 1995), textured (Locher et al., 1995) and randomly oriented (Mort, Machonkin and Okumura, 1991).

Quite generally the IIb crystals (Clark, Kemmey, Mitchell and Henvis, 1960) and the homoepitaxial (Ager et al., 1995; Gheeraert et al., 1998) or

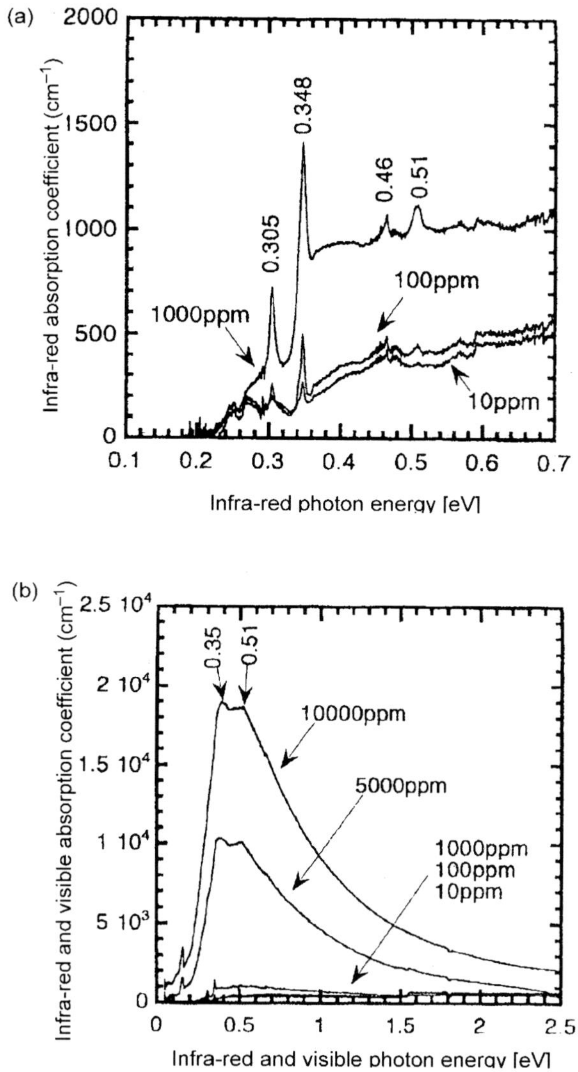

FIG. 5. Absorption coefficient of boron-doped homoepitaxial diamond films vs. photons energy (a) Highly boron-doped films. (b) Heavily boron-doped films inducing metallic conductivity ($B/C \geq 5000$ ppm in the gas phase).

polycrystalline (Mort et al., 1991) thin films at room temperature also exhibit another almost rectangular band between 0.14 eV (1150 cm^{-1}) and 0.16 eV (1290 cm^{-1}). It widens (Ager et al., 1995) while another structure appears around 0.1 eV (800 cm^{-1}) as the boron concentration increases, and when there is metallic conductivity on the boron impurity band (Ager et al., 1995).

4.2. The Origins of the Infrared Absorption

Although first order phonon absorption is forbidden by the symmetry of diamond, the 0.16 eV appears as a boron-induced one phonon absorption as we will see later, while the other signals arise from electronic transitions involving the bound hole of boron.

4.2.1. Electronic Transitions

From the low infrared dielectric constant of diamond (5.7) and a mean holes effective mass relatively high ($0.7m_0$ (Rauch, 1961; Collins and Lightowlers, 1979)) at the top of the valence band, (although it is not expected to hold) a simple hydrogen-like model (Gheeraert et al., 1999) predicts a high value (0.31 eV) for the ionization energy of any p-doping atom in the diamond, and in particular for boron which is easily incorporated in the diamond lattice.

The electrical measurements on natural IIb crystals give an ionization energy of boron of 0.368 eV (Collins and Williams, 1971) close to the values measured by the photothermal ionization spectroscopy (0.382 eV) or predicted by the hydrogen-like model (Gheeraert et al., 1999). For low boron concentrations, because of this high ionization energy of boron in diamond, only about 10^{-3} of the boron atoms are ionized at room temperature. Therefore, the step around 0.378 eV (3050 cm^{-1}) is ascribed to the transition between the fundamental level of the bound hole and the valence band edge (ionization energy threshold). The peaks around 0.305 eV (2459 cm^{-1}), 0.347 eV (2806 cm^{-1}) and 0.363 eV (2935 cm^{-1}) might be ascribed to transitions between the fundamental and the first (0.305 eV), the second (0.347 eV) and the third excited states (Austin, 1956; Smith and Taylor, 1962). This implies a substitutional site for boron in diamond, which is kept for metallic conductivity on the boron impurity band, from Figure 5b. The peaks at higher energies (0.462 and 0.508 eV) are ascribed to phonons (shift of about +0.160 eV) replica of the previous fundamental to excited level transitions (Smith and Taylor, 1962). These replicas are specific to diamond and indicate a large interaction between the electronic transitions and the diamond lattice. The widening of the 0.305 and 0.347 eV peaks as the boron concentration increases, indicates an increasing coupling between the hole levels (Gheeraert et al., 1998). As [B] increases, it induces the formation of the boron impurity band and then the metallic conductivity on this band. The absorption band of Figure 5b obtained for metallic conductivity (Ager et al., 1995; Ogasawara et al., 1997) is ascribed (Ager et al., 1995) to transitions between the boron impurity band centered around 0.378 eV above valence band and a mixture of this valence band with the excited states of the bound holes. The transition of the hole from the fundamental level to the second excited state around 0.347 eV and its phonon replica around 0.508 eV still appear. These last results imply a boron impurity band about 0.38 eV above the valence band and a substitutional site for boron in the diamond lattice even for metallic conductivity on the boron impurity band.

The spin–orbit interaction splits the top of the valence band ((Rauch, 1961) by 2 meV (Kim et al., 1997)) and then the energies corresponding to transitions between the ground and the bound hole levels (Anastassakis, 1969). The fine structures in the absorption bands of IIb crystals at low temperature and the variations of the IR spectrum at room temperature under uniaxial stress are ascribed to transitions involving these split levels (Charette, 1961; Smith and Taylor, 1962; Anastassakis, 1969; Kim et al., 1997). The shapes and widths of the measured infrared signals at room temperature and low boron contents are ascribed to the overlapping between transitions on the same boron atom involving these sub-levels (Smith and Taylor, 1962; Anastassakis, 1969).

While the step around 0.378 eV is still there, the difficulty to distinguish the 0.305 and 0.347 eV peaks in thin polycrystalline films (Mort et al., 1991) can be ascribed to the different boron concentrations in the film according to the growth sector. However, these peaks appear for thick polycrystalline films (10–60 μm) (Erz et al., 1995).

The structures found between 0.3 *and* 0.63 *eV in the infrared absorption and the models used for their interpretation imply a substitutional site for boron in the diamond lattice for all [B] including those corresponding to the metallic conduction on the boron impurity band. This band is centered around* 0.38 *eV above the valence band edge of diamond.*

4.2.2. *Phonon Absorption*

While the selection rules forbid (one) phonon absorption in intrinsic diamond up to 0.165 eV (1332 cm^{-1}), a square-like absorption band appears for boron-doped diamond below 0.16 eV. Its physical origin is still not clear. It was first associated to the bound hole (Hardy, 1962). This band might be more generally ascribed to local removal of the selection rules by impurities (Smith and Taylor, 1962; Ager et al., 1995; Gheeraert et al., 1998). Its structures which appear around 0.1, 0.14 and 0.16 eV (800, 1150 and 1290 cm^{-1})—at the same energies than those in the total one phonon density of state of diamond (Tubino, Piseri and Zerbi, 1972)—supports such an interpretation.

The 0.16 *eV absorption band is ascribed to boron-induced one phonon absorption.*

4.3. Determination of the Boron Concentration from the Infrared Absorption Bands

It is important to know the boron concentration in thin films. The infrared absorption coefficients α are proportional to the concentration of unionized (electronic transitions) boron or to the total concentration of boron (impurity-induced phonon absorption band). The usual secondary ion mass spectroscopy is destructive, and there were several attempts to determine the boron concentrations in homoepitaxial films from their infrared absorption coefficients. The one phonon absorption coefficient around 0.16 eV (α value) (Davies, 1977), and (assuming negligible compensation, see Section 6.2) the infrared absorption coefficients for the transitions of the bound hole from the fundamental to the second excited state around 0.347 eV (area under the peak) (Collins and Williams, 1971) or from the fundamental level to the valence band around 0.378 eV (α value) (Burns et al., 1990), were used for such a purpose. With $[C] = 1.76 \times 10^{23}\ \text{cm}^{-3}$ in diamond, they give for infrared measurements at 300 K, before the formation of the boron impurity band

[B] ppm $= 5.53 \times 10^{-4}\ \text{area}_{(\text{at } 0.348\ \text{eV})}\ \text{cm}^{-2}$ (Collins and Williams, 1971).
[B] $\text{cm}^{-3} = 8.08 \times 10^{15}\ \text{cm}^{-3} \times \alpha_{(\text{at } 0.378\ \text{eV})}\ \text{cm}^{-1}$ (Burns et al., 1990).
[B] $\text{cm}^{-3} = 2.1 \times 10^{17}\ \text{cm}^{-3} \times \alpha_{(\text{at } 0.16\ \text{eV})}\ \text{cm}^{-1}$ (Davies, 1977).

Different concentrations of boron are sometimes obtained from these expressions. This can originate from differences in the concentration of compensating donors N_d, as the first and second expressions actually give the concentration N_a of unionized acceptor, therefore $N_A = N_a(= [B]) - N_d$, while an impurity-induced one phonon absorption would give the boron concentration in the third expression. Recent measurements on homoepitaxial films (Gheeraert et al., 1998) confirm the calibration by the one phonon mode around 0.16 eV.

The infrared absorption bands induced by boron in diamond can be used to determine its concentration in homoepitaxial films.

5. Raman Diffusion and B Content

5.1. Non-diamond Phases

The concentrations of the defects and of the parasitic phases has to be minimized in polycrystalline, highly oriented and homoepitaxial diamond films. In monocrystalline diamond, the selection rules allow only a narrow line around $1332\ \text{cm}^{-1}$ from the zone center optical phonon whose FWHM increases with the concentration of point defects. In polycrystalline films, some other defects and the parasitic phases are easily detected from their contributions that appear at other wavenumbers (Knight and White, 1989; Robin et al., 1990). If the signals around $1580\ \text{cm}^{-1}$ and the pair around $1350-1580\ \text{cm}^{-1}$ are well known to originate from graphite and nanocrystallites of graphite, respectively (Knight and White, 1989), the physical origin of most of the other signals remains unclear (Ramsteiner and Wagner, 1987). The more usual 1550 and $1500\ \text{cm}^{-1}$ signals are ascribed, respectively, to amorphous carbon (Bou and Vandenbulcke, 1989) or so called ‘graphitic phases’ (Bachmann et al., 1991). The relative intensity of the Raman signals from these parasitic phases increases as the excitation energy of the Raman signal decreases (Wagner et al., 1991), which is ascribed to resonance with carbon phases that have lower forbidden gaps than diamond (Ramsteiner and Wagner, 1987; Wagner et al., 1991).

5.2. The Raman Signal at High B Concentrations

The concentration of the parasitic phases decrease as the boron content increases both in the polycrystalline films as already mentioned (Wang et al., 1992; Gheeraert et al., 1993; Miyata et al., 1993; Deneuville and Gheeraert, 1996), and in the growth sector of isolated diamond crystallites (Ushizawa et al., 1998), except for the polycrystalline films deposited by MPCVD and doped from trimethylborate (Chen et al., 1994). Moreover, the FWHM of the 1332 cm^{-1} is often used as a rough probe of the total amount of defects in polycrystalline and homoepitaxial diamond films. This Raman line has a Lorentzian shape, i.e., whose FWHM depends on the lifetime of the $k = 0$ optical phonon of diamond. This lifetime depends on the perfection of the crystal lattice, i.e., is limited by the global concentration of defects in the material. The best natural crystals have a FWHM of 1.8–2.6 cm^{-1}. While the FWHM of the best homoepitaxial films is similar to that of the best diamond crystals (Field, 1992; Sato and Kamo, 1992), FWHM of the (some μm thick) polycrystalline films is significantly larger 12–15 cm^{-1} from their higher stress and their higher concentration of defects (Robin et al., 1990; Field, 1992; Sato and Kamo, 1992) (the large crystallites of the 0.5–1 mm thick free standing films have a FWHM around 3 cm^{-1} (Coe and Sussmann, 2000) similar to that of monocrystalline diamond).

This FWHM decreases as the boron content increases in polycrystalline films deposited by MPCVD and doped from diborane (Nishimura et al., 1991; Wang et al., 1992; Gheeraert et al., 1993) up to the metallic conductivity on the boron impurity band (Deneuville and Gheeraert, 1996). As the lifetime of the phonon depends on a lot of defects, extended defects, point defects and strain, the Raman spectra gives only a convenient quick overview of the effect of boron incorporation. Measurements of the concentration of some specific defects (Section 3.2.3.1), or of the concentration of structural defects (Section 3.2.3.2) confirm the Raman results. On the contrary, the Raman FWHM widens and the crystallographic quality of the polycrystalline films deteriorates (Chen et al., 1994) when the boron is introduced from trimethylborate in HFCVD films. The model of Bernholc ascribes the improvements to a better recovery from the extended defects during the film growth (Bernholc et al., 1988) (Section 3.2.3.3).

The Raman spectra show quite easily that both the concentration of the parasitic phases and of the defects decrease as the boron incorporation in polycrystalline films increases up to the boron concentration corresponding to the metallic conductivity on the boron impurity band.

5.3. The Raman Signal at Heavy Boron Concentrations

When the boron concentration induces a metallic conductivity on the boron impurity band (above about 3×10^{20} cm^{-3}), an increasing deformation appears on the signal of the zone center optical phonon peak around 1332 cm^{-1}, and simultaneously a continuum with two new bands rises around 1220 and 500 cm^{-1}.

5.3.1. *The Raman Signal from the Zone Center Optical Phonon*

5.3.1.1. Experimental Results. Looking first to the signal of the zone center phonon, it decreases, widens and shifts to lower wavenumbers as [B] increases in polycrystalline MPCVD films (Gheeraert et al., 1993; Cifre et al., 1994; Ager et al., 1995; Gonon, Gheeraert, Deneuville, Fontaine, Abello and Lucazeau, 1995c), in polycrystalline HFCVD films (Zhang et al., 1996; Liao et al., 1997; Wurzinger et al., 1997), in textured polycrystalline MPCVD films (Locher et al., 1995) and in MPCVD homoepitaxial films (Pruvost, Bustarret and Deneuville, 2000; Pruvost and Deneuville, 2001) as well as from the (100) and (111) faces of isolated crystallites (Ushizawa et al., 1998). According to the nature of the films and the energies of excitation, various positions of the anti-resonance-like minimum in regard to the maximum were reported. On polycrystalline films, on the high and low-energy side when they were excited, respectively, at 1.95 eV (633 nm) (Zhang et al., 1996) and 2.54 eV (488 nm) (Gheeraert et al., 1993; Ager et al., 1995); first on the high-energy side at low energy (excitation energy ≤ 1.65 eV), then on the low-energy side (excitation energy ≥ 2.18 eV) of $\langle 110 \rangle / \langle 111 \rangle$ textured of polycrystalline films (Locher et al., 1995); on the high-energy side of homoepitaxial films (Pruvost et al., 2000; Pruvost and Deneuville, 2001); on the high-energy side for both the (100) and (111) faces of isolated crystallites (Ushizawa et al., 1998).

5.3.1.2. Comparison with Models. All works ascribed this deformation to the Fano effect (Fano, 1961), i.e., to a quantum interference between a discrete transition (here the phonon) and a continuum of electronic transitions around the same energy. Such effects appear, for instance, in heavily doped Si and Ge (Abstreiter, Cardona and Pinczuk, 1984), in intercalated graphite (Ohana, Yacoby and Schmeltzer, 1986) and in C60 (Daneli, Denisov, Ruani, Zamboui, Taliani, Zakhidov, Ugawa, Imaeda, Yakushi, Inokuchi, Kikuchi, Ikemoto, Suzuki and Achiba, 1992). In Si, the model fits with a continuum of interband electronic transitions between the lower (filled) and the higher (empty) splitted valence band (Abstreiter et al., 1984). In diamond, the 0.002 eV splitting of the top of the valence band (Kim et al., 1997) is too small for a similar continuum of interband electronic excitations at 0.165 eV. On the other hand from (i) the similarity of the Raman spectra at room and low temperature (Ager et al., 1995) (ii) the infrared absorption of these heavily boron-doped films (Section 4.2.1), the boron impurity band does not merge in the valence band. This continuum is therefore ascribed in diamond to transitions involving the impurity band, hole excited levels and the valence band (Ager et al., 1995; Pruvost et al., 2000).

After subtraction of the contribution of the 1220 cm^{-1} band, the Fano parameters (Abstreiter et al., 1984) can be deduced from the fit of the Raman spectra by the following expressions.

$$I = \frac{(q+\varepsilon)^2}{1+\varepsilon^2} \qquad \text{with} \qquad \varepsilon = \frac{\varpi - \Omega}{\Gamma} \quad \text{and } q = \frac{V\dfrac{T_p}{T_e} + V^2 R(E)}{\pi V^2 \rho(E)} \tag{1}$$

and

$$h\Gamma = \pi V^2 \rho(E) + h\gamma \qquad h\Omega = h\omega_0 - V^2 R(E) \tag{2}$$

In these expressions, ω_0 and γ are, respectively, the real and imaginary part of the self energy of the phonon in the undoped material, Ω and Γ are the new real and imaginary parts of this self energy after coupling between the discrete phonon transition and a continuum of electronic transitions around the energy of the phonon. V is the matrix element of this coupling, $\rho(E)$ is the density of the continuum of states, and $\pi^{-1}R(E)$ its Hilbert transform.

Ω and Γ vary with the square V^2 of strength V of the coupling (Eq. (2)) that depends on the doping level of the semiconductor and, in principle, do not depend on the Raman excitation energy. On the contrary q depends on the magnitude and sign of V and of the Raman scattering amplitudes of the decoupled phonon T_p and of the electronic continuum T_e (Eq. (1)). T_p, T_e and the T_p/T_e ratio are expected to have different variation laws with the energy of excitation of the Raman diffusion.

The distortion of the '1332 cm^{-1}' signal depends mainly on the q value that is positive or negative, respectively, when the anti-resonance-like minimum energy is lower or higher than the energy of the maximum. As recalled previously, this position has been reported at higher energy for (100) homoepitaxial films (Pruvost and Deneuville, 2001) and the (100) and (111) faces of isolated crystals (Ushizawa et al., 1998), but at higher or lower energy according the excitation for polycrystalline films (Gheeraert et al., 1993; Ager et al., 1995; Locher et al., 1995; Zhang et al., 1996; Liao et al., 1997). As the concentration of boron is about 1 order of magnitude higher in (111) than in (100) sectors of growth, the polycrystalline films mix the contributions of (111) and (100) sectors with different boron content. This might explain the differences in the position of the anti-resonance minimum between the polycrystallines films and the other samples. We will keep here the high energy positions ($q < 0$) obtained on the monocrystalline surfaces (Ushizawa et al., 1998; Pruvost et al., 2000).

With excitations at 1.95 eV (red 633 nm) and 2.42 eV (green 514 nm), the fits of the Raman curves with the Fano expression, Eq. (1) appears excellent on the investigated range of boron doping of the homoepitaxial films (Pruvost and Deneuville, 2001). The variations of the distortion parameter q (Fig. 6, Pruvost and Deneuville, 2001), and of Γ and $\Omega - \omega_0$ vs. the doping level exhibit abrupt transitions from the values of the undoped material to very different ones above

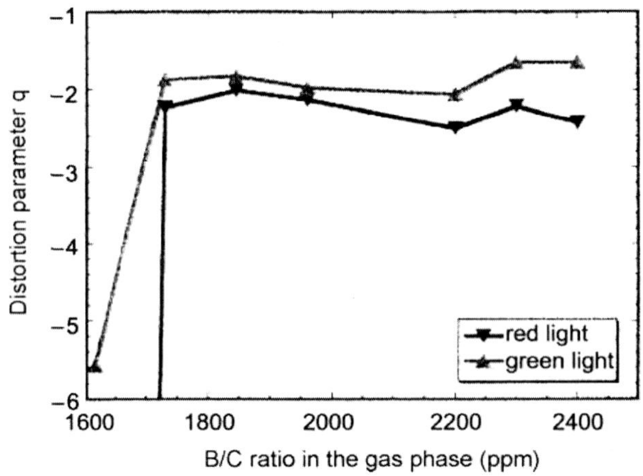

FIG. 6. Heavily boron-doped homoepitaxial films: variation of the distortion parameter q of the Fano effect vs. their B/C ratio in the gas phase with $B/C = 2400$ ppm in the gas phase. Inset: position of the peaks for 1730, 1845, 1960, 2200, 2300 and 2400 ppm.

a percolation-like threshold around $B/C = 1700$ ppm (about 3×10^{20} B-cm^{-3} corresponding to the beginning of the metallic conductivity on the boron impurity band). There is no deformation when $|q| \to \infty$. From the fit, q is always <0, with low values of $|q| \approx 2.2$, i.e., large distortion of the Lorentzian shape of the phonon above the threshold for metallic conductivity on the boron impurity band. As expected q depends on the energy of the exciting light. It is systematically larger (lower $|q|$) and a small distortion appears sooner (1640 ppm) for the green light. As the doping level increases, $|q|$ slightly increases and slightly decreases for the red and the green lights, respectively. Qualitatively, there is an abrupt threshold for the appearance of the Fano distortion on the Raman spectra of all boron-doped films (Gheeraert et al., 1993; Ager et al., 1995; Locher et al., 1995; Zhang et al., 1996; Liao et al., 1997; Pruvost and Deneuville, 2001). Such thresholds with $q < 0$ are also obtained for the (100) and (111) faces of the crystallites (Ushizawa et al., 1998), but for lower boron concentrations. This has to be contrasted with significantly larger values of $|q|$, respectively, 13 and 25 (lower distortions) and with the progressive variations of $|q|$ in n-type doped Si and p-type Ge (Abstreiter et al., 1984) when the dopant impurity band progressively merges in the Si conduction band and the Ge valence band.

Deeper interpretation needs the analysis of q. The expression of q (Eq. (1)) contains two terms and can be written as

$$q = \frac{V\dfrac{T_p}{T_e}}{\pi V^2 \rho(E)} + \frac{R(E)}{\pi \rho(E)} \tag{3}$$

The comparison of the values obtained for Γ and Ω of Eq. (2) vs. doping, gives a second term of the order of 1 in Eq. (3), as in n-type Si and p-type Ge. Therefore, the sign of q in Eq. (3) depends mainly on the first term. Their negative sign and smaller absolute values for diamond, were ascribed to a strong positive deformation potential ($d = +53.8$ to $+106$ eV are obtained from calculations (Cardona and Christensen, 1986), $d = +99.1$ eV is obtained from experiments (Reggiani et al., 1983) in undoped diamond, $d = +16$ to 19 eV is deduced from the variation of the lattice parameter in the same boron concentration range, see Section 3.2.2.2). The detailed interpretation also gives a small negative value of T_p and a relatively large positive value of T_e (in agreement with interband transitions for the electronic continuum) (Pruvost and Deneuville, 2001).

The Fano distortion of the Raman signal of the zone center optical phonon appears quite abruptly independently of the boron precursor, when [B] induces the metallic conductivity on the boron impurity band in polycrystalline and homoepitaxial diamond films. It is ascribed to quantum interference between this zone center optical phonon and a continuum of electronic transitions around the same energy involving the metallic conduction on the boron impurity band. The Fano distortion is much stronger in diamond than in Si or Ge, and involves a positive deformation potential.

5.3.2. Lower Wavenumber Raman Signals

The existence of a low wavenumber continuum with 500 and 1220 cm^{-1} wide bands have been reported in heavily doped polycrystalline MPCVD films (Gheeraert et al., 1993; Ager et al., 1995; Gonon et al., 1995c) or HFCVD films (Zhang et al., 1996; Liao et al., 1997; Wurzinger et al., 1997) and homoepitaxial MPCVD films (Pruvost et al., 2000; Pruvost and Deneuville, 2001) and on the (100) and (111) faces of isolated crystallites (Ushizawa et al., 1998) deposited by MPCVD with various precursors for boron.

In polycrystalline films and on the (100) and (111) faces of isolated crystallites, the positions of these band were found constant when various excitation energies are used (Gheeraert et al., 1993; Ushizawa et al., 1998) (which discards photoluminescence signals) and when the boron concentration increases (Zhang et al., 1996; Liao et al., 1997; Ushizawa et al., 1998). From careful examination of the results about homoepitaxial films, the position of the 1220 cm^{-1} band peak remains independent of the excitation energy and the boron concentration, while that of the band around 480 cm^{-1} shifts to slightly lower values when the excitation energy and the boron content increases (Fig. 7, Pruvost et al., 2000).

In the other materials, Fano distortions occur without these low wavenumber bands that therefore appear specific to diamond. Such bands were also reported in undoped diamond films grown by the hot-filament technique (Buckley, Moustakas, Ye and Voiron, 1989).

The total density of state of phonon has a main maximum around 1250 cm^{-1}, and a secondary maximum around 550 cm^{-1} (Tubino et al., 1972). Some relaxation of the selection rules from the disorder introduced by the boron atoms

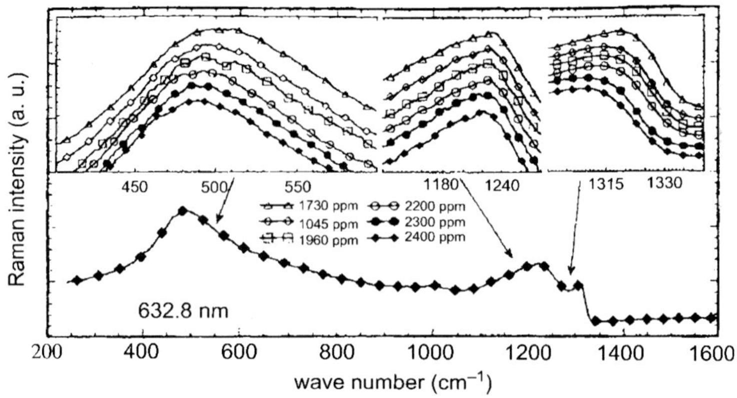

FIG. 7. Raman spectra of heavily boron-doped films.

(Gheeraert et al., 1993) or an amorphous-like spectrum (Zhang et al., 1996; Liao et al., 1997) will therefore give the Raman peak around $1220\,\text{cm}^{-1}$.

From the lack of the ω^4 relationship between the intensities of these peaks and the excitation energy, another explanation involving electronic Raman scattering caused by impurity-induced electronic states has also been proposed (Ushizawa et al., 1998).

The fast increase of the diamond lattice parameter of polycrystalline (Brunet et al., 1997) and homoepitaxial (Brunet et al., 1998) films as the boron concentration increases in this range might favor the first interpretation (Pruvost et al., 2000), but there is still no definitive conclusion.

A continuum with two wide bands around 500 *and* 1220 cm^{-1} *appears simultaneously with the Fano distortion of the zone center optical phonon. Their origins are not so clear at that time.*

5.4. Very Heavy Boron Concentrations

Finally, as reported in Section 3.1.3, additional signals around 1350 and 1580 cm^{-1} appear above about 7×10^{21} B-cm^{-3} from microcrystalline graphite in honeycomb-like structure with undoped graphite walls surrounding boron-doped diamond in both polycrystalline (Fig. 1) and homoepitaxial films.

6. Electrical Characteristics of Boron-doped Diamond Films

We get some information about the electrical characteristics of boron-doped diamond films from previous sections. The boron concentration inducing the metallic conductivity on the boron impurity band appears critical for the lattice parameter, the infrared absorption and the Raman diffusion. The homoepitaxial films and even more the polycrystalline films still contain significant concentrations of impurities and of structural defects. Their defect concentrations depend on the deposition technique and on its parameters. In the polycrystalline films, the incorporation of boron up to the metallic conductivity on the boron impurity band decreases the defect concentrations. The films are deposited in an atmosphere containing a very high concentration of atomic hydrogen, and a very high residual concentration of hydrogen [H] (around 10^{20} cm^{-3} in the homoepitaxial films) remains in the undoped films (Section 2.1.2). In the classical semiconductors post-hydrogenation resulting in high [H] neutralize some defects, but also—hopefully in a limited temperature range—induces pairing of boron with hydrogen, and therefore also the neutralization of its acceptor behavior. There is only one result about [H] in as-grown boron-doped films, but infrared absorption and electrical conductivity indicates active boron acceptors. We summarize here the main results on the influence of the boron concentration and of the temperature on the boron–deuterium pairing after post-deuteration of homoepitaxial boron-doped films. The chapter 9 of this book is devoted to this point. Then, the experimental results about the conduction in the boron-doped polycrystalline, highly oriented and homoepitaxial films will be compared with the results on IIb natural crystals and with the models for the various conduction processes. These models involve the boron concentration and its ionization energy in diamond and the temperature, as well as the nature and the concentration of the residual defects.

6.1. Summary about the Interaction of Atomic Hydrogen with Boron-doped Homoepitaxial Films (see Chapter 9)

The virgin films with 5×10^{19} B-cm^{-3} contain about 10^{19} H-cm^{-3} (Chevallier et al., 2001), a lower H concentration than that (3×10^{20} cm^{-3}) in undoped films (Samlenski et al., 1995). The H site and its effect on the electrical properties of the film remain unclear at that time. After 2 h of contact at 550 °C of the homoepitaxial boron-doped films with a deuterium plasma, SIMS measurements indicate equal concentrations of deuterium and boron down to a significant depth ('plateau'): there is deuterium–boron pairing. As the boron concentration decreases, the width of the plateau increases (Chevallier et al., 1998). The pairing neutralizes the acceptor behavior of boron in diamond (Uzan et al., 2001). Hopefully, the boron–deuterium pairing does not hold above 550 °C which explains the efficiency of boron doping in the diamond films deposited between 750 and 950 °C (Chevallier et al., 1999).

6.2. Models for the Conduction of Boron-doped Films

The conductivity of boron-doped diamond film is believed to add three components (Visser et al., 1992; Malta, von Windheim, Wynands and Fox, 1995;

Inushima et al., 2000), with the traditional subscripts 1 from holes in the valence band, 2 for conduction within a band between the valence band and the acceptor level, 3 from hopping between unionizied and ionizied boron:

$$\sigma = \sigma_1 \exp E_1/kT + \sigma_2 \exp E_2/kT + \sigma_3 \exp E_3/kT \tag{4}$$

6.2.1. *The Partially Compensated Semiconductor Model for the Hole Concentration in the Valence Band*

The first component is from $pq\mu_p$. As the mobility varies in $T^{\pm S}$, while *p*(*T*) varies as $\exp - E_A/kT$ with the reciprocal temperature, it has approximately the activation energies of the hole concentration *p*(*T*) in the valence band. Then the model focuses on the determination of *p*(*T*). It is calculated from the partially compensated semiconductor model (Collins and Lightowlers, 1979). *p*(*T*) depends on the ionization energy E_I of boron in diamond and on the density N_a (= [B]) of the acceptor atoms and that N_d of the compensating donor atoms (compensation ratio $K = N_d/N_a$). With $E_I = 0.378$ eV $>> (kT)_{300\ K}$, full thermal ionization of the boron atoms resulting in a saturation of $p = N_a - N_d$ is reached only at very high temperatures which increase as N_a and K increase.

For an ideal uncompensated p-type diamond, $N_d = 0, K = 0, N_a = [B]$ (Fig. 8) (i) the temperature for *p*(*T*) saturation remains in the measurement range only for low acceptor concentrations ($N_a = [B] = 10^{16}$–10^{17} cm^{-3}), but it reaches 2000 K for $N_a = [B] = 10^{18}$ *B*-cm^{-3}. (ii) *p*(*T*) exhibits at "medium or low" temperatures a single activation energy equal to $E_I/2$.

As the other wide band gap semiconductors, actual p-type diamond is compensated by unintentionally introduced donors with a concentration N_d. With increasing N_d, i.e., increasing K, and $N_a = 10^{18}$ cm^{-3} (Fig. 9), (i) *p*(*T*) saturate to $N_a - N_d$. (ii) there is an additional "low" temperature range increasing with K, with another activation energy equal to E_I. (iii) For $K > 0.1$, or for $K < 0.1$ when the measurement temperature is not high enough, only the E_I activation energy can be seen.

6.2.2. *Conduction on a Band Built from the Excited Levels of the Bound Hole*

There is still discussion on the existence of this second conductivity component in diamond (Visser et al., 1992; Malta et al., 1995; Inushima et al., 2000). It is ascribed to conduction on a band (formed by the interaction between the excited levels of closely spaced acceptors) between the usual boron impurity band and the valence band edge. Its activation energy would decrease from E_1 at low doping level to 0 near the metal insulator transition (Mott and Davis, 1968). It is expected to be seen for very low compensation ratio and high neutral acceptor concentration.

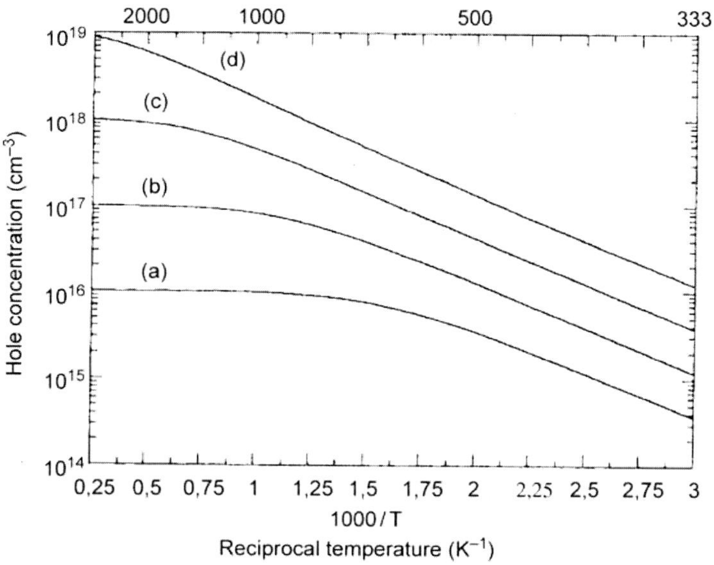

FIG. 8. Calculated hole concentration in the valence band vs. reciprocal temperature for (a) 10^{16}, (b) 10^{17}, (c) 10^{18}, (d) 10^{19} B-cm^{-3} in ideal diamond ($K = 0$).

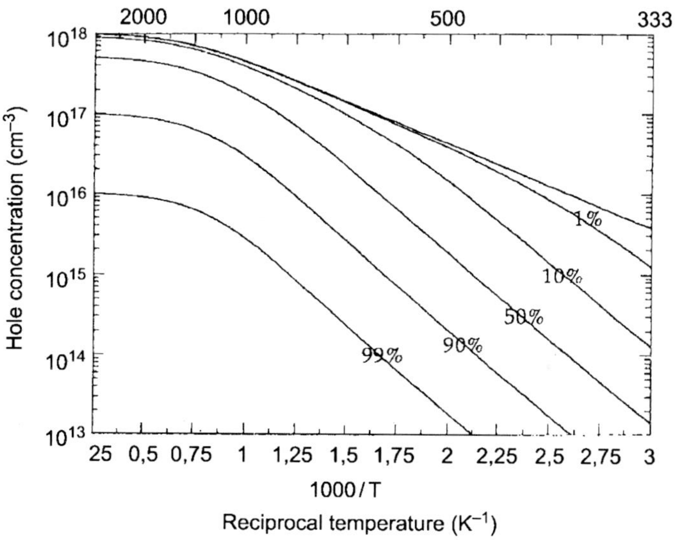

FIG. 9. Calculated hole concentration in the valence band vs. reciprocal temperature for 10^{18} B-cm^{-3} in diamond with various compensation ratios $K = N_d/N_a$ of 0, 1, 10, 50, 90 and 99%.

6.2.3. *Hopping between Ionized and Unionized Boron*

The third component is ascribed to hopping conduction between unionized and ionized nearest neighbor acceptors. Its activation energy E_3 is always lower than E_1 and is expected to decrease as K increases (Malta et al., 1995). Its pre-exponential factor has a characteristic law $\log \sigma \propto N_a^{-1/3}$ when N_a increases (Visser et al., 1992). This component is always expected to dominate the total conductivity at sufficiently low temperature. As N_a and K increase, it would dominate the conductivity up to higher temperatures.

Finally, with increasing boron concentration, long range hopping on localized levels around the Fermi level, and then metallic conductivity is expected (Mott and Davis, 1968; Visser et al., 1992; Malta et al., 1995).

6.2.4. *Model for the Hole Mobility*

The hole mobility at room temperature and its variation with temperature is also used to characterize the boron-doped diamond films.

Quite generally, a higher mobility is expected in films containing a lower concentration of defects. The experimental results are striking between homoepitaxial and highly oriented or polycrystalline films. This seems also obvious between different qualities of homoepitaxial films from different concentrations of compensating donors and of other defects.

The variation of the mobility with the temperature is expected to be a better criterion that the room temperature mobility. However, the experimental results for diamond appear significantly different from those for the classical semiconductor, while there is no specific model. Therefore, the model for conduction in the valence band (first component in Eq. (4)) in classical semiconductors is used. It holds only for monocrystalline material and ideal semiconductor (i.e., without compensating donor and other defects). It predicts a maximum of $\mu(T)$. At 'high' temperature, the scattering of the holes by the acoustical phonons decreases the mobility as the temperature increases, resulting in $\mu(T) \propto T^{-s}$ with $s = 1.5$. At 'low' temperature, the scattering of the holes by the ionized acceptors gives an increasing mobility as the temperature increases, resulting in $\mu(T) \propto T^{s}$ with $s = 1.5$.

The other conduction processes (second and third component in Eq. (4)) occur with much lower mobilities. We are not aware of model for the mobility in these conduction processes.

6.2.5. *Global Fit of $\sigma_1(T)$*

Several works fit the first conductivity component $\sigma_1(T) = pq\mu$ which includes the variation of the mobility with the temperature $\mu(T)$. For this conduction process the exponential variation of p is much faster than the power law variation of $\mu(T)$. More rigorously, the $p\mu$ product varies as $\exp - E_1/kT$ because, from its Nv factor p varies as $T^{1.5} \exp - E_1/kT$ which is in principle, perfectly compensated by the predicted $\mu \propto T^{-1.5}$ in the high temperature regime of the mobility.

6.3. Conduction in the Homoepitaxial Films

There is still a wide dispersion in the experimental results about the conductivity, the hole concentrations and mobilities in boron-doped diamond and their variations with the temperature. We will describe briefly some results for IIb crystals. We will focus on some systematic works on homoepitaxial, highly oriented and polycrystalline films. We will compare the results on IIb natural and homoepitaxial films, and will try to show at what extent the models give a good description of the experimental results. Finally, we will try to connect the information collected from the electrical and the physicochemical measurements.

6.3.1. *Conduction in the Valence Band: Experimental Results*

6.3.1.1. Natural IIb Crystals. Although they are restricted to a boron concentration in the 10^{16}–10^{17} cm^{-3} range, there is dispersion on the results about the natural IIb crystals. An activation energy for hole concentration around $E_1 = 0.368$ eV ($\cong E_I$ the ionization energy of boron) is measured up to about 300 K (Collins and Williams, 1971; Malta, von Vindheim and Fox, 1993). Because of their low [B] # $N_a \leq 5 \times 10^{16}$ cm^{-3}, at higher temperatures, the hole concentration increases more slowly and finally saturates above 1250 K to a value ascribed to the difference $N_a - N_d$ between the acceptor concentration N_a and the compensating donor concentration (Collins and Williams, 1971; Wynand, Malta, Fox, von Windheim, Fleuriat, Irvine and Vandersande, 1994). The fit of $p(T)$ in natural IIb crystal with the partially compensated semiconductor model gives $N_d \leq 5 \times 10^{15}$ cm^{-3} when $N_a = [B] \leq 5 \times 10^{16}$ cm^{-3} ($K \leq 0.1$) (Collins and Williams, 1971) and $N_d = 6 \times 10^{15}$ cm^{-3} when $N_a = [B] = 3 \times 10^{16}$ cm^{-3} ($K = 0.2$) (Malta et al., 1993). The room temperature mobilities are commonly around 520 cm^2 V^{-1} s^{-1} (Malta et al., 1993; Wynand et al., 1994), and only in the best crystals in the range 1100–1500 cm^2 V^{-1} s^{-1} (Austin, 1956; Wedepohl, 1957; Collins and Lightowlers, 1979) while an exceptional value of 2010 cm^2 V^{-1} s^{-1} has been also reported with $N_a = 5 \times 10^{16}$ cm^{-3} and $N_d = 9 \times 10^{14}$ cm^{-3} ($K = 0.018$) (Collins and Lightowlers, 1979). The mobility increases continuously as T decreases, as T^{-s}, with $s \approx 2.4$–2.8 down to 350 K even for usual samples with room temperature mobility around 520 cm^2 V^{-1} s^{-1} (Collins and Lightowlers, 1979; Malta et al., 1993), then with a lower s given as 1.5 for the best sample (Collins and Lightowlers, 1979) at the lower temperatures.

The fit of p(T) by the partially compensated semiconductor model is fully satisfying only for the best IIb crystals which have a compensation ratio below 0.1. If the mobility always decreases from 170 *K as the temperature increases, it has the* T^{-s} *variation law with* $s = 1.5$, *only for the best IIb crystals and up to* 350 *K. At higher temperature, the* $s = 2.4$–2.8 *found for the usual and best samples, is not given by the model.*

6.3.1.2. Homoepitaxial Films MPCVD Films. The boron concentration in homoepitaxial films covers a significantly wider range. We will compare from the very beginning, the results from systematic studies on homoepitaxial films in the [B] range around 10^{17} cm^{-3} to those on IIb crystals. We will compare the results obtained from various groups for higher [B].

Earlier boron-doped homoepitaxial films: 1986–1991. In the first boron-doped homoepitaxial films deposited on by MPCVD with diborane on (111) natural diamonds (Fujimori et al., 1986), the boron concentration varies around 10^{20} B cm^{-3}, the resistivity of such films varies from 10^{-3} to 1 Ω cm with a very low activation energy of 0.013 eV.

In the later works of the same group (Fujimori, Nakata and Imai, 1990; Shiomi et al., 1991), the films were deposited on the (110) and (100) surfaces of synthetic diamond with boron concentration varying from 4×10^{17} to 3×10^{20} B-cm^{-3} which correspond at room temperature to holes concentrations between 2×10^{14} and 10^{17} cm^{-3}. For films deposited on (100) surfaces, maximum mobilities from 450 ($p_{300\,K} = 9 \times 10^{13}$ cm^{-3}) and 100 cm^2 V^{-1} s^{-1} ($p_{300\,K} = 2.3 \times 10^{15}$ cm^{-3}) were obtained at temperatures of 300 and 400 K followed by a decrease with a power law T^{-s} with $2.2 < s < 3$ at higher temperatures [Fujimori et al., 1990]. In the later work, the mobility continuously decreases as the temperature increases above 300 K slightly up to 400 K, then more rapidly [Shiomi et al., 1991]. The fit of the holes concentration by the partially compensated semiconductor model gives acceptor concentrations N_a higher by a factor of 10 than the [B] SIMS values. The activation energies of $p(T)$ are equal to E_I and $E_I/2$ for $N_a = 5 \times 10^{18}$ cm^{-3} and $N_d = 5 \times 10^{17}$ cm^{-3} ($K = 0.1$) [Shiomi et al., 1991]. The activation energy of the conductivity decreases first slowly around 0.38 eV and then rapidly in the 10^{19} and 10^{20} B-cm^{-3} range [Fujimori et al., 1990] to reach metallic conductivity at 3×10^{20} B-cm^{-3} [Shiomi et al., 1991].

The activation energy for p(T) and the room temperature mobility of these earlier boron-doped films containing around 10^{17} *B-cm*$^{-3}$ *was similar to those of the usual IIb crystals. The p(T) activation energy has the* E_I *and* $E_I/2$ *components expected from the model about conduction in the valence band when* $K = 0.1$. *It decreases suddenly at very high [B] to reach zero for the metallic conductivity already given around* 3×10^{20} *cm*$^{-3}$. *The mobility decreases as p (*≅ *the ionized boron concentration) increases, which is expected. At least up to* $p = 2.3 \times 10^{15}$ *cm*$^{-3}$, *the temperature variation of the mobility has the* *same* T^{-s} *with the same* $s = 2.4–2.8$ *in the same temperature range* ($T > 350$ *K*) *than in the IIb natural crystals.*

Boron-doped homoepitaxial films: 1995–1996. In 1995, for film deposited by MPCVD with diborane, the Kobe Steel USA group obtains a single activation energy around 0.38 eV between 1.5×10^{17} and 2×10^{18} B-cm^{-3} decreasing to 0.22 eV for [B] $= 4 \times 10^{19}$ cm^{-3}, and mobilities with maximum values around 250 K decreasing from 479 to 14 cm^2 V^{-1} s^{-1} as the boron concentration increases (Malta et al., 1995). As [B] increases up to 2×10^{18} cm^{-3}, the fit of their $p(T)$ by the partially compensated model give N_a values higher than the [B] SIMS values, decreasing N_d concentrations from 12 to 3×10^{16} cm^{-3}, resulting in K values decreasing from 0.22 to 0.03.

Then this group obtains ('improved samples' (Fox, Harstell, Malta, Wynands, Tessmer and Dreyfus, 1995; Fox et al., 1996)) for the fit of $p(T)$, an acceptor concentration $N_a = 1.8 \times 10^{16}$ cm^{-3} lower than the [B] SIMS value of 4.5×10^{16} B-cm^{-3} with a compensating donor concentration of 2.2×10^{15} cm^{-3} ($K = 0.12$), and the E_I and $E_I/2$ activation energies. The hole mobility was 1590 cm^2 V^{-1} s^{-1} at room temperature increasing to 2800 cm^2 V^{-1} s^{-1} as the temperature decreases down to 160 K. Room temperature mobilities higher than 1000 cm^2 V^{-1} s^{-1} were obtained up to a boron concentration of 2×10^{18} B-cm^{-3} (Fig. 10, Fox et al., 1995). However, systematic variations of $\mu(T)$ as [B] increases were not given.

As the early boron-doped films, the first set of these boron-doped films have the performances of usual IIb natural crystals (except a maximum in $\mu(T)$ *around* 250 K) *for their lower [B]. In agreement with the physicochemical results, they*

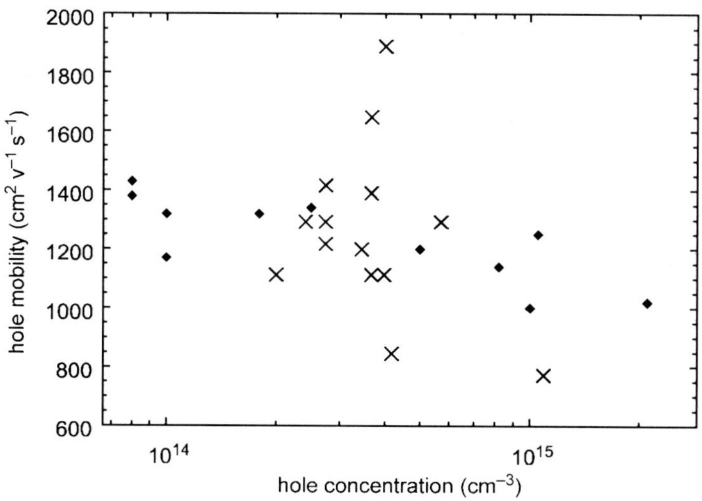

FIG. 10. Hole mobility vs. boron concentration in homoepitaxial films of diamond. From Okushi (2001) (×) and Fox et al. (1995) (◆).

show an improvement on the quality of the films when they are deposited on the (100) *surface. The fit of their p(T) indicates an improvement of their quality (a decrease of K) as [B] increases up to* 2×10^{18} cm^{-3}. *Around* 10^{17} *B-cm*$^{-3}$, *the improved samples have room temperature mobility in the range of the best IIb natural crystal. For its lower [B]* $= 4.5 \times 10^{16}$ *B-cm*$^{-3}$, $K \approx 0.1$, *p(T) exhibit the* E_I *and* $E_I/2$ *activation energy predicted by the model, and the mobility decreases continuously from* 170 K *as in IIb natural crystals. In addition,* $\mu > 1000$ cm^2 V^{-1} s^{-1} *is kept for [B] increasing up to* 2×10^{18} cm^{-3}.

Late boron-doped homoepitaxial films: *1997–2000*. In 1997, groups from Japan using MPCVD with diborane doping obtained room temperature mobilities around 1000 cm^2 V^{-1} s^{-1} with $\mu(T)$ increasing as the temperature decreases for low [B] $\approx 10^{17}$ cm^{-3}, while $\mu(T)$ exhibit maximum in the 250–350 K range for higher [B]. The fit of their $p(T)$ curves gives slowly decreasing N_d around 10^{16} cm^{-3} from $N_a = 8 \times 10^{16}$ up to 5×10^{17} cm^{-3}, then N_d increasing to 1.7×10^{17} cm^{-3} for $N_a = 1.7 \times 10^{18}$ cm^{-3} and K values always lower than 0.1, with a minimum K of 0.02 when $N_a = 4.6 \times 10^{17}$ cm^{-3} (Kiyota et al., 1997).

Recently the Electrotechnical Laboratory group (Tsukuba, Japan) reports a still higher room temperature mobility of 1840 cm^2 V^{-1} s^{-1} (Yamanaka, Watanabe, Masai, Takeuchi, Okushi and Kajimura, 1998b) for samples prepared with 0.3% CH_4 and trimethylboron, but deposited on an undoped homoepitaxial diamond buffer layer prepared with very low methane content. The fit of $p(T)$ gives an acceptor concentration $N_a = 1 \times 10^{17}$ cm^{-3} (significantly lower than the SIMS value of [B] $= 6 \times 10^{17}$ B-cm^{-3}) and a compensating donor concentration of 4×10^{14} cm^{-3} ($K = 0.004$) (Yamanaka, Takeuchi, Watanabe, Okushi and Kajimura, 1999). The mobility continuously increases up to 3370 cm^2 V^{-1} s^{-1} as the temperature decreases to 170 K.

Other laboratories reach after improvement, the room temperature mobility as well as the $\mu(T)$ *of the best IIb crystals on homoepitaxial films with low [B]. There is slight improvement of the room temperature mobility up to about [B]* $= 6 \times 10^{17}$ *cm*$^{-3}$. *Very low K value of* 0.004 *reported for the best sample of the Electrotechnical Laboratory with decreasing* μ *as T increase (*μ_{R_T} *= 1840 cm*^{2} *V*$^{-1}$ *s*$^{-1}$. *For the other ones, a maximum of* $\mu(T)$ *appears at higher and higher temperature as p increases.*

Simple σ(T) *measurements*. Simple $\sigma(T)$ measurements looking at the activation energies also give information on the films. For good quality homoepitaxial films doped from diborane and deposited by MPCVD on Ib (100) substrate, (Lagrange et al., 1998), (i) there is hopping conductivity above 300 K only when [B] $\geq 2.5 \times 10^{19}$ cm^{-3}, (ii) E_I and $E_I/2$ activation energies around 10^{18} B-cm^{-3}, (iii) the higher activation energy remains roughly constant at 0.36 eV from 5×10^{16} B-cm^{-3} to 5×10^{18} B-cm^{-3}, then drops to 0.24 eV at 10^{20} B-cm^{-3} and finally abruptly drops to 0 where the films exhibit metallic conductivity at about 3×10^{20} B-cm^{-3}. The same boron concentration for the metallic conductivity was indicated in Werner et al. (1996a).

For selected samples simple σ(T) *measurements confirm the* $E_I/2$ *activation energy, which indicates compensation* $K < 0.1$, *even for [B] around* 10^{18} *cm*$^{-3}$. *It also confirms the sudden decrease of the activation energy of* $p(T)$ *down to* 0 *for a metallic conductivity around* 3×10^{20} *B-cm*$^{-3}$.

6.3.1.3. Experimental Results on HFCVD Films. The earlier homoepitaxial films deposited by the hot-filament technique on (110) Ib substrates were doped from B_2O_3 powder diluted in acetone (Okano et al., 1989). Their conductivities exhibit at high temperature an activation energy decrease from 0.38 to 0.26 eV as the boron content increases up to a maximum value of 3×10^{20} B-cm^{-3}. Below about 300 K "very low" temperature, they have an additional conductivity component.

Later hot filament homoepitaxial films were deposited on (100) and (110) surfaces of IIa natural crystals (Visser et al., 1992) and doped from the temperature-dependant boron pressure of their BN substrate holder. The boron incorporation is systematically higher for the (110) growth (from 1.8×10^{19} to 5.1×10^{21} B-cm^{-3}) than for the (100) growth (3×10^{18} to 5.9×10^{19} B-cm^{-3}). Their conductivity activation energies at "medium-low" and "very low" temperature are ascribed, respectively, to valence band and hopping conduction. The first one decreases slowly from 0.31 to 0.26 eV up to 2×10^{19} B-cm^{-3} and then more rapidly. The second one remains nearly constant around 0.025 eV. The "low" and "medium" temperature parts of the $p(T)$ of the best film (deposited on the (100) surface with $[B] = 2.7 \times 10^{18}$ cm^{-3}) have the E_I and $E_I/2$ activation energies. Its fit by the partially compensated semiconductor model gives N_a values of $1{-}2 \times 10^{18}$ cm^{-3} and N_d around 2×10^{16} cm^{-3} ($K = 0.01{-}0.05$). The mobility of the same sample has a maximum around 590 cm^2 V^{-1} s^{-1} around 300 K and $\mu(T) \propto T^{2.2}$ and $T^{-2.8}$, respectively, below and above this maximum.

The characteristics of the HFCVD films are better for the (100) *growth direction than the other ones, which agrees with the physicochemical results. With a [B] value of* 2×10^{18} *cm*$^{-3}$, $p(T)$ *exhibits the* E_I *and* $E_I/2$ *activation energies predicted for their* $K < 0.1$, *with room temperature mobility of this best sample reaching that of the usual IIa natural crystals. Its* $\mu(T)$ *has a maximum around* 300 *K with the usual* $T^{-2.8}$ *law at higher temperature.*

6.3.2. *Comparison Between the Models and Experimental Results for the Conduction in the Valence Band*

6.3.2.1. Fit of the Experimental Results by the Partially Compensated Semiconductor Model. Fine characterization of the samples by their acceptor concentrations N_a and their compensating donor concentration N_d from electrical measurements relies on the fit of $p(T)$ by the partially compensated semiconductor model. It is expected to be fully successful only when it includes characteristic structures as saturation or variation of the activation energy in E_I and $E_I/2$, and values of N_a in agreement with those of [B] deduced from SIMS.

Saturation was almost reached only for IIb natural crystals with $[B] \leq 5 \times 10^{16}\ cm^{-3}$ (Collins and Williams, 1971; Wynand et al., 1994) as expected from Figure 8. Most of the temperature measurements of the hole concentrations do not show (Kiyota et al., 1997) or do not fit (Yamanaka et al., 1999) the high temperature part of $p(T)$ which would show the second activation energy of $E_I/2$ at low compensation ratio ($K = N_d/N_a < 0.1$). This structure appears with $K \leq 0.1$ in few MPCVD and HFCVD films on the hole concentration $p(T)$ (Shiomi et al., 1991; Visser et al., 1992; Fox et al., 1995) or on the conductivity $\sigma(T)$ (Lagrange et al., 1998) measured at sufficiently high temperature. More generally, the fits are not always fully satisfying as N_a is significantly higher or lower than [B]. If we look for the lower values of N_d, they are reported for films deposited by MPCVD, as $N_d = 4 \times 10^{14}$ B-cm^{-3} for $[B] = 1 \times 10^{17}\ cm^{-3}$ (Yamanaka et al., 1999) (lower than $N_d = 9 \times 10^{14}\ cm^{-3}$ reported in the exceptional IIb diamond (Collins and Lightowlers, 1979)) and $N_d = 2.2 \times 10^{15}\ cm^{-3}$ when $[B] = 4.5 \times 10^{16}$ B-cm^{-3} (Fox et al., 1996). For the best film deposited by HFCVD $N_d = 2 \times 10^{16}\ cm^{-3}$ for 2.7×10^{18} B-cm^{-3} (Visser et al., 1992) is obtained, with $K = 0.01$.

Systematic variations of N_d vs. [B] were reported only for few MPCVD films. The donor concentration decreases from 1×10^{17} to $3 \times 10^{16}\ cm^{-3}$ while [B] increases from 1.5×10^{17} to $2 \times 10^{18}\ cm^{-3}$ (Malta et al., 1995), slightly decreases around $10^{16}\ cm^{-3}$ from 8×10^{16} up to 5×10^{17} B-cm^{-3} and then increasing to $1.7 \times 10^{17}\ cm^{-3}$ for $[B] = 1.7 \times 10^{18}\ cm^{-3}$ (Kiyota et al., 1997).

For the good MPCVD and HFCVD homoepitaxial films, the fit of p(T) or $\sigma(T)$ by the partially compensated model gives the same qualitative trends on N_d and K when [B] increases, however, with N_a values higher than [B] from SIMS. The quality of the films improves up to around 5×10^{17}–10^{18} *B-cm*$^{-3}$.

For the best MPCVD homoepitaxial films, there is no systematic studies of N_a, N_d, and *K with [B]. The numerical values deduced from the fits are not fully consistent. This time N_a is significantly lower than the [B] value from SIMS. Neighboring values of room temperature mobilities and of $\mu(T)$ are found in films having quite different K* (0.12 (*Fox et al., 1996*) *and of* 0.004 (*Yamanaka et al., 1999*)).

6.3.2.2. Comparison of the Room Temperature Mobilities. Then, although it does not give a physical description of the samples, the room temperature mobility appears as a better criterion for the electrical quality of the boron-doped films. Figure 10 gives the room temperature mobilities vs. hole concentration for the best samples at our knowledge, the 'improved' Kobe Steel USA samples (Fox et al., 1995) using diborane (assuming $p = 10^{-3} \times [B]$) and the Electrotechnical Laboratory films (Okushi, 2001) prepared from $\approx 0.3\%$ of methane and trimethylboron. They decrease from 1590 to 1000 $cm^2\ V^{-1}\ s^{-1}$ as p increases from 4.5×10^{14} to 2×10^{15} cm^{-3} for the first ones, while with a larger spread in the μ values, they present a maximum of 1890 cm^2 V^{-1} s^{-1} around $p = 4 \times 10^{14}$ cm^{-3} in the range 2×10^{14} to 3×10^{15} cm^{-3} for the second ones. Nearly similar mean values are obtained for $2 \times 10^{14} < p < 6 \times 10^{14}$ cm^{-3}.

The room temperature mobility in the best boron-doped MPCVD diamond films is higher than those (1100–1500 cm^2 V^{-1} s^{-1}) *of the best IIb crystal* (*Austin*, 1956; *Wedepohl*, 1957). *It still do not reach the exceptional value* (2010 cm^2 V^{-1} s^{-1}) *of the exceptional IIb crystal* (*Collins and Lightowlers*, 1979).

6.3.2.3. Comparison of the Measured $\mu(T)$ with the Model for Classical Semiconductors. Although the experimental results cannot be completely understood by the classical model, the variation of the mobility with the temperature seems to be at a lesser extent a criterion for the electrical quality of the films.

The $d\mu/dT > 0$ predicted by the model at low temperature—when the mobility is limited by the concentration of the ionized doping atoms—is not found in the measurement range ($T > 170$ K) of both the best (Collins and Lightowlers, 1979) and the usual (Malta et al., 1993) IIb natural crystals. However, from their high E_I (0.378 eV) associated with their low [B] ($<5 \times 10^{16}$ cm^{-3}), the concentration of the ionized B remains very weak. On the contrary, such lack of $d\mu/dT > 0$ is only found in the best homoepitaxial diamond films when $[B] \leq 10^{17}$ cm^{-3} (Fox et al., 1995, 1996; Kiyota et al., 1997; Sonoda et al., 1997; Yamanaka et al., 1999). The T^{-s} with $s = 1.5$ of the model is found only for $[B] \leq 10^{17}$ cm^{-3} and up to about 350 K in the best natural IIb diamond (Collins and Lightowlers, 1979) as well as in the best homoepitaxial films (Kiyota et al., 1997; Yamanaka et al., 1998a,b). Above 350 K, $s = 2.4$–2.8 seems to be a quite general law from diamond, as it is found in all (including the best) IIb natural crystals (Collins and Lightowlers, 1979; Malta et al., 1993) and in all (including the best) boron-doped films (Fujimori et al., 1990; Visser et al., 1992; Kiyota et al., 1997; Sonoda et al., 1997). The $d\mu/dT > 0$ at low temperature systematically found for the higher [B] levels was ascribed to a parallel hopping conduction associated to an increasing concentration of defects (Visser et al., 1992; Malta et al., 1995; Kiyota et al., 1997). However, it might be related only to the higher [B] of the films rather than to their poorer quality. Starting from $[B] = N_a = 3 \times 10^{19}$ B-cm^{-3} with a hopping conduction involving a carrier concentration of 10^{16} cm^{-3} below 300 K, post-deuteration of the samples (which annealed the acceptor behavior of B by deuterium–boron pairing), decreases the hole concentration in the 10^{14} cm^{-3} range, removes the hopping conduction, and shifts down by 150 K the maximum of the mobility. It is not expected to decrease the concentration of other defects as annealing above 650 °C recovers the initial state (Uzan et al., 2001), because the temperatures needed to break the H–B and H–C bonds are different.

Similar $\mu(T)$ are found in the best homoepitaxial MPCVD diamond films (*with* $[B] \leq 10^{17}$ cm^{-3} *and the best IIb crystals. Similar variation laws specific to* *diamond (T^{-s} with $s \sim 2.4$–2.8) are found with all [B] in MPCVD or HFCVD films and with natural crystals when $T > 350$ K. According to the sample the $d\mu/dT > 0$ at low temperature might be connected only to high [B] or to an increasing concentration of defects. The understanding of $\mu(T)$ in diamond required additional experimental results and probably some specific adjustments for diamond of the classical model.*

6.3.3. *Conduction on a Band Built from the Excited States of the Bound Holes of Boron*

According to Inushima et al. (2000), from electrical measurements on a $p^+/p^-/p^+$ structure with a boron concentration of 10^{19} cm^{-3} in the p^+ layers, conduction also occurs between occupied and unoccupied levels on a band built from the first excited state of the bound hole of boron (0.30 eV below its ground state and 0.06 eV above the valence band edge), with an activation energy of 0.06 eV. This would be the second component of the conduction with the E_2 activation energy.

6.3.4. *Hopping Conduction Between Ionized and Unionized Boron*

It gives the third conduction component with the E_3 activation energy. Hopping conduction does not appear above 200 K in good quality MPCVD films with [B] in the range 10^{16}–10^{17} B-cm^{-3} (Fox et al., 1995, 1996; Yamanaka et al., 1999) and even around 4×10^{18} B-cm^{-3} (Inushima et al., 2000). In MPCVD doped films from diborane such component is reported below various temperatures, below 300 K for 10^{19} B-cm^{-3} (Shiomi et al., 1991), below 250 K for $[B] > 6 \times 10^{17}$ B-cm^{-3} with a low ($\leq$35 meV) nearly constant activation energy and a magnitude increasing with the boron and defects concentrations (Malta et al., 1995). For films prepared by HFCVD and doped from their BN substrate holder, below 300 K for $[B] > 2 \times 10^{18}$ B-cm^{-3}, with an activation energy around 25 meV and with the characteristic variation for the pre-exponential factor $\log \sigma \propto -(N_a)^{-1/3}$. The magnitude of the component increases as the boron concentration and as the value of K increases from increasing defects concentrations from the (100) to the (110) films (Visser et al., 1992). At still lower temperature (Visser et al., 1992) or higher boron concentration (Inushima et al., 2000), the variable range hopping between localized states around the Fermi level with $\log \sigma$ proportional to $T^{-1/4}$ is observed. Variable range hopping, and then metallic conductivity appear (Shiomi et al., 1991; Lagrange et al., 1998) without the smooth continuity expected from the model. Moreover, as we will see later, for similar doping level the hopping component increases rapidly as the concentration of structural defects increases from homoepitaxial to polycrystalline films (Malta et al., 1995).

Hopping conduction increases as [B] and/or the concentration of defect increase. While a higher concentration of donors increases the hopping conduction between unionized and ionized boron, hopping on the levels induced by structural defects and donors might be also hopping conduction paths at low temperature.

6.3.5. *Relationships Between the Physicochemical and Electrical Properties*

Only trends between the physicochemical and electrical properties have been proved.

Quite generally, (100) films with less structural defects than (110) and (111) films exhibit better electrical properties (Fujimori et al., 1990; Visser et al., 1992). The hole mobility decreases—which is mainly ascribed to an increase in the concentration of the structural defects—as the CH_4 content increases from 1 to 10% (Kiyota et al., 1997)—and as the boron content increases above about 5×10^{17} (Okushi, 2001) or $1-2 \times 10^{18}$ B-cm^{-3} (Fox et al., 1995; Kiyota et al., 1997). More precisely for good samples, from the fit of $p(T)$ by partially compensated semiconductor model, it seems that when they are in the 10^{16}–10^{17} cm^{-3} range, the concentration of the compensating donors first decreases as the boron concentration increases up to the 5×10^{17}–2×10^{18} cm^{-3} range (Malta et al., 1995; Kiyota et al., 1997). The Electrotechnical Laboratory group ascribed the higher mobility in their samples to lower methane concentration ($\approx$0.3%) in the gas phase for the growth of the boron-doped films, on a buffer undoped film grown with a very low concentration ($\leq$0.015%) of methane.

The detailed effect of the concentrations of the impurities on the film properties is not clear, in particular, because the impurity concentrations are not usually measured. However, with low boron content resulting in a room temperature hole concentration in the 10^{13}–10^{14} cm^{-3} range, improvement of the room temperature mobility from 210 to 900 cm^2 V^{-1} s^{-1}, increasing mobility as the temperature decreases to 200 K instead of a maximum around 300 K, and an increase of a factor of 10 of the hole concentration were reported as the nitrogen concentration in the feeding gas was reduced (Sonoda et al., 1997).

The compensating donor concentration in homoepitaxial films is decreased by using a quasi perfect undoped film as pseudo-substrate, low concentrations of methane for the growth of the boron-doped films and by reducing the concentration of impurities in the feeding gases.

6.4. CONDUCTION IN HIGHLY ORIENTED FILMS

The size of the diamond substrates limits that of the homoepitaxial films. Highly oriented films deposited on Si have larger areas. Boron-doped highly oriented films were deposited on undoped highly oriented films (Stoner et al., 1993; Malta et al., 1995).

For the best sample, a single activation energy of 0.346 eV for $p(T)$ without hopping conduction and a monotonic decrease in mobility was reported between 180 and 440 K with room temperature values of $p_{(300\ K)} = 5.9 \times 10^{13}\ cm^{-3}$ and $\mu_{(300\ K)} = 165\ cm^2\ V^{-1}\ s^{-1}$ (Stoner et al., 1993).

For good highly oriented and homoepitaxial films deposited in the same run (Malta et al., 1995), the highly oriented (100) film contains a three times higher boron concentration ($6 \times 10^{18}\ cm^{-3}$ instead of $2 \times 10^{18}\ cm^{-3}$) while its hole concentration $p(T)$ in the valence band is almost similar (2.4 instead of $3.1 \times 10^{14}\ cm^{-3}$ at RT), with a 0.05 eV smaller activation energy (0.305 instead of 0.355 eV). The fits of their $p(T)$ with the partially compensated semiconductor model gives similar N_a ($1.1 \times 10^{18}\ cm^{-3}$), but a five times higher N_d ($1.5 \times 10^{17}\ cm^{-3}$ instead of $3 \times 10^{16}\ cm^{-3}$) and a four times higher compensation ratio ($K = 0.13$ instead of 0.03) for the highly oriented films. The other main differences are their lower mobilities (135 instead of 427 $cm^2\ V^{-1}\ s^{-1}$ at room temperature) and their variation with temperature which exhibits a maximum at a higher temperature of 273 instead of 223 K (Malta et al., 1995).

The poorer characteristics of the highly oriented films are ascribed to their higher concentration of defects. The decrease of the activation energy of their conductivity suggests parallel conduction path through defects. The previous analysis suggests a main effect of grain boundaries, with a grain which might have the quality of the best homoepitaxial films when $[B] = 5.9 \times 10^{16}\ cm^{-3}$. *The effects of the grain boundaries remain unclear, as the variation of the mobility in* $\exp - q\Phi b/kT$ *used to describe their effect in Si does not hold for diamond* (*Fox et al.*, 1994).

6.5. Conduction in Polycrystalline Films

The early works with polycrystalline HFCVD films doped from boron trioxide (Okano et al., 1988, 1989) reported activation energies of the conductivity decreasing from 0.2 to 0.12 eV as [B] increases from about 3×10^{18} to $3 \times 10^{20}\ cm^{-3}$. Using methane, hydrogen, oxygen and diborane (Mort et al., 1989), it decreases from 0.55 to 0.1 eV from $[B] = 1 \times 10^{18}$ to 3.5×10^{19} B-cm^{-3}, then remains to this value up to 1.7×10^{20} B-cm^{-3}.

The comparisons of MPCVD polycrystalline and homoepitaxial films deposited with methane, hydrogen and diborane show (Malta et al., 1993, 1995) that their conductivities are similar at "low" temperature, but with lower activation energies (0.26 eV) below 10^{18} B-cm^{-3}, which decrease from lower [B] ($[B] > 10^{18}\ cm^{-3}$) than those of the homoepitaxial films. Fits of their $p(T)$ values give higher N_d and K than in homoepitaxial films. N_d increases slightly from 2.1 to $6.3 \times 10^{17}\ cm^{-3}$, but K decreases from 0.71 to 0.13 as [B] (lower than N_a) increases from 3×10^{17} to $7 \times 10^{18}\ cm^{-3}$.

The main difference between polycrystalline and homoepitaxial films are therefore (i) their much lower hole mobilities (1–50 $cm^2\ V^{-1}\ s^{-1}$ at room temperature) which exhibit maximum in the range 250–350 K (Malta et al., 1993, 1995). (ii) A 1–2 orders of magnitude higher hopping conduction.

Simple analysis of the conductivity $\sigma(T)$ of MPCVD films doped from BCl_3 gives still lower values of the activation energies, decreasing continuously from 0.17 to metallic conductivity as [B] increases from 10^{18} to $8 \times 10^{21}\ cm^{-3}$ (Werner et al., 1996b).

From their hole concentrations and their activation energies, and their compensation ratio which can decrease down to 0.13, the best polycrystalline films are not so far from homoepitaxial films. Their compensation ratio K, which reflects their quality, improves as [B] increases up to $7 \times 10^{18}\ cm^{-3}$*, which agrees with the results of the physicochemical measurements. This suggests a limited increase of the concentration of defects in the grain compared to that in homoepitaxial films. The main difference is in their much lower mobilities. This also suggests a much higher concentration of defects in the grains boundaries associated with the drastic increase of the hopping conduction and a drastic decrease of the mobility. These effects seem more important with the HFCVD deposition technique, and when the boron trichloride precursor is used for MPCVD deposition.*

7. The Role of Boron Doping in Diamond Thin Films Applications

Diamond thin films are used in an increasing number of applications, passive electronic devices (see Chapters 14 and 16), active electronic devices (see Chapter), and electrodes for various electrochemical applications (see Chapters 11–13). Boron doping is necessary for some of them.

7.1. Passive Electronic Devices

In passive electronic devices, high temperature (Aslam et al., 1994) and high pressure (Taher et al., 1994) detectors use highly boron-doped polycrystalline thin films.

7.2. Active Electronic Devices

Optimization of active electronic devices would include metallic ohmic contacts with low contact resistivity. This needs heavily boron-doped films in contact with an intermediate carbide obtained after deposition and annealing of a carbide forming metal. With some spread according to the nature of the metal and the post-treatments, values in the range of $10^{-5}\ \Omega\ cm^2$ are obtained, with $[B] \geq 5 \times 10^{19}\ cm^{-3}$ for HFCVD homoepitaxial films doped from B_2O_3 (Hewett, Taylor, Ziedler and Geis, 1995), and in the range of $10^{-6}\ \Omega\ cm^2$ with $[B] \geq 3 \times 10^{20}$ and $2 \times 10^{21}\ cm^{-3}$ for polycrystalline films doped, respectively, from H_3BO_3 (Nakanishi, Otsuki, Oku, Ishiwata and Murakani, 1994) and from BCl_3 (Werner et al., 1996a,b).

Some Schottky diodes use boron-doped homoepitaxial films with oxidation of the surface before the metal deposition. An increase in their boron content would decrease their serie resistance, but deteriorates their room temperature non-ideality factor from $n = 1.2$ for $[B] = 1-2 \times 10^{16}\ cm^{-3}$ to $n = 1.7$ for $[B] = 10^{17}\ cm^{-3}$ when gold is used as Schottky metal (Ebert, Vescan, Borst and Kohn, 1994; Ebert, Vescan, Gluche, Borst and Kohn, 1997). Similarly, a good non-ideality factor of $n = 1.1$ is obtained for $[B] = 4 \times 10^{16}\ cm^{-3}$ and Al contacts (Hayashi, Yamanaka, Watanabe, Sekiguchi, Okushi and Kajimura, 1997). With Au, Al, Ni, Cr and Pt very good ideality factor $n \leq 1.1$ at room temperature were recently obtained by the Electrotechnical Laboratory group with low boron incorporation ($[B] \leq 10^{17}\ cm^{-3}$) in films deposited on a buffer layer grown with a very low methane concentration (Yamanaka et al., 1999; Yamanaka, Takeuchi, Watanabe, Okushi and Kajimura, 2000).

An increase of the boron content from 10^{16} *to* $10^{17}\ cm^{-3}$ *seems to deteriorate the non-ideality factor of the Schottky diodes. This disagrees with the improvement of the quality of the good films with boron incorporation up to* 4×10^{17}–$2 \times 10^{18}\ cm^{-3}$ *derived from the transport measurement described in Section* 6.3.2.

There are few results on pn junctions. On (100) Ib diamond crystal used as n-type semiconductor at high temperature ($T \geq 610\,°C$), unspecified boron concentrations between 3×10^{17} and $8 \times 10^{19}\ cm^{-3}$ give an exponential behavior on several orders of magnitude and non-ideality factors around 2 (Borst et al., 1995). More recently, pn junctions obtained by deposition of homoepitaxial films with $[B] = 10^{17}\ cm^{-3}$ and then with a phosphorus content of $5 \times 10^{18}\ cm^{-3}$ on (111) heavily boron-doped diamond substrate, gives at room temperature ten orders of magnitude rectification at ± 10 V (Koizumi, Watanabe, Hasegawa and Kanda, 2002).

7.3. Boron-doped Electrodes for Various Electrochemical Reactions (see also chapters 11, 12, 13)

Recently, diamond electrodes appeared attractive for various electrochemical reactions. Almost all electrodes use polycrystalline films. While there are (very small) cathodic and anodic currents with non-intentionally doped films (Sakharova, Pleskov, Di Quarto, Piazza, Sunseri, Treremetskaya and Vernin, 1995; Katsuki, Wakita, Nishiki, Shimamune, Akiba and Iida, 1997; Katsuki, Takahashi, Toyoda, Kurosu, Iida, Wakita, Nishiki and Shimamune, 1998), most of the results are obtained with highly or heavily doped boron films (Xu et al., 1997; Swain, Anderson and Angus, 1998; Tenne and Lévy-Clément, 1998). The boron-doped electrodes supply not only to the solution holes for oxidation ('anodic') reactions, but also with about the same efficiency electrons for reduction ('cathodic') which is less obvious to understand.

The boron concentration of the electrodes are sometimes reported not only in the range of $10^{19}\ cm^{-3}$ (Alehashem, Chamber, Strojek and Swain, 1995; Xu et al.,

1997; Beck, Krohn, Kaiser, Fryda, Klages and Schafer, 1998; Xu and Swain, 1998), but also in the range of 10^{20} cm^{-3} (Swain and Ramesham, 1993; Ramesham and Rose, 1997; Ramesham, 1998; Swain et al., 1998; Tenne and Lévy-Clément, 1998) and even 10^{21} cm^{-3} (Li, Totir, Vinokur, Miller, Chottiner, Argoitra, Angus and Sherson, 1997; Li, Totir, Vinokur, Miller, Chottiner, Evans, Angus and Scherson, 1998; Yano, Tryk, Hashimoto and Fujishima, 1998; Sarada, Roa, Tryk and Fujishima, 1999; Yano, Popa, Tryk, Hashimoto and Fujishima, 1999). All samples are characterized by their coplanar resistivities given in the range 0.1–10 Ω cm. As we have seen in Section 6, the conductivity of the highly boron-doped films contains several components, whose relative intensities depend on the quality of the film. In particular (Section 6.5), relative intensities of these components in polycrystalline films are expected to depend on the size of the grain (concentration of the residual defects in the grain), and on the conductivity from the grain boundaries. Although easy to measure, the resistivity might not be the best criterion to characterize these heavily doped films for electrochemical applications.

On the other hand, in this boron concentration range the shapes of the Raman spectra are very sensitive to [B] as shown in Sections 5.2 and 5.3. For polycrystalline films, it contains the narrow 1332 cm^{-1} signal from the zone center optical phonon of diamond, and wide signals around 1500–1550 cm^{-1} from the parasitic graphitic phases. When [B] increases up to $\approx 3 \times 10^{20}$ cm^{-3}, the quality of the polycrystalline improves, the 1332 cm^{-1} signal becomes narrower, and the intensity of the 1500–1550 cm^{-1} signal decreases. When $[B] > 3 \times 10^{20}$ cm^{-3}, there is metallic conductivity on the boron impurity band, the intensity of the 1332 cm^{-1} peak decreases rapidly, shifts to lower wavenumber, becomes asymmetrical and widens from its interference with a continuum of electronic transitions (Fano effect). In the same time wide bands appear around 500 and 1220 cm^{-1}, and the '1332 cm^{-1}' signal more and more mixes with the 1220 cm^{-1} wide band.

From a recent systematic study (Lévy-Clément, Zenia, Ndao and Deneuville, 1999; Ndao, Zénia, Deneuville, Bernard and Lévy-Clément, 2000), the electrochemical current density depends mainly on the boron content of the 'bulk' films. For instance, at −2 V, just below the cathodic edge (about − 1.75 V for heavily doped diamond) of the potential window, it increases almost abruptly from the 10^{-5} to the 10^{-2} A-cm^{-2} range for the hydrogen evolution in a 0.1 M KCl solution.

The Raman spectrum seems to be a better criterion than the coplanar resistivity for the suitability of the diamond films as electrodes when high current densities are required in electrochemistry.

8. Summary and Conclusions

The goal of boron incorporation is p-type doping of the diamond films. However, it has also important effects on the physicochemical properties of these films. We have reviewed the effect of their boron content on some physicochemical and electrical properties of homoepitaxial, highly oriented and polycrystalline boron diamond films. We have looked for (i) the influence of the HFCVD and MPCVD deposition techniques and/or of their deposition conditions on these properties; (ii) the relationships between the physicochemical and electrical properties; (iii) the similarities or differences between IIb crystals and diamond films with [B] in the range of $10^{16}-10^{17}$ cm^{-3}.

The boron incorporation from the gas phase to the solid phase depends on the deposition technique and on the deposition conditions, on the monocrystalline or polycrystalline nature of the films.

Undoped homoepitaxial diamond films contain impurities and defects. Their concentrations are still higher in highly oriented and polycrystalline films, which contain additionally amorphous and graphitic parasitic phases.

Numerous precursors have been used to incorporate boron in the films, but more currently diborane and trimethylboron are used. The incorporation from the gas to the solid phase is not completely understood, but is efficient. The incorporation rate is about 10 times higher in the (111) than in the (100) growth directions as in crystals. It was reported around 1 for MPCVD (100) homoepitaxial films. For polycrystalline films, it varies from 0.06 to 400 according to the deposition techniques and the nature of the precursors. The size of the grain decreases continuously as the boron content increases in polycrystalline films. The deposition rates of the films are usually in the range $0.1-1$ μm. They decrease or increase as [B] increases when, respectively, MPCVD and HFCVD are used.

Independent of the deposition conditions or techniques, the boron concentration of the films modifies their structural properties, with a striking effect of the concentration inducing the metallic conductivity on the boron impurity band.

Diamond films contain from about 10^{16} to 8×10^{21} B-cm^{-3} while IIb crystals remain in the $10^{16}-10^{17}$ B-cm^{-3} range. An additional graphite phase appears in homoepitaxial and polycrystalline diamond films when [B] $\geq$ about 10^{21} cm^{-3}. The incorporation of boron expands the lattice of the polycrystalline and of the (100) homoepitaxial diamond films. For (100) homoepitaxial films up to the metallic conductivity on the boron impurity band ([B] around 3×10^{20} cm^{-3}), a simple Vegard law is found with the covalent radii of boron (0.088 nm) and of carbon (0.077 nm). The faster expansion at higher [B] is ascribed to the effect of a positive deformation potential from the metallic conductivity on the boron impurity band. Qualitatively, increasing boron incorporation up to at least 3×10^{20} B-cm^{-3}, decreases the concentration of twin planes, point defects, and even of Si impurities in polycrystalline films. This was predicted by the Bernholc model which ascribes this improvement to the higher concentrations and mobilities of charged vacancies.

Independent of the deposition conditions and techniques, the boron is on substitutional sites. Its concentration controls the characteristic infrared absorption bands of the films. In the $10^{16}-10^{17}$ B-cm^{-3} range, those of homoepitaxial films are identical to those of IIb crystals.

From the relatively high ionization energy 0.38 eV of boron in diamond, the room temperature hole concentration in the valence band is only about $10^{-3} \times$ [B]. As in natural IIb crystals, transitions of the bound hole from its fundamental to its excited states and to the valence band, respectively induce, peaks with phonons replica (shifted by about 0.16 eV) and a step (0.378 eV) in the infrared absorption spectra of the films.

These structures appear clearly in homoepitaxial films. The width of the main peaks around 0.305 and 0.347 eV increases when [B] increases from 10^{18} cm^{-3} to about 10^{20} B-cm^{-3}. This is ascribed to the coupling of the bound hole levels. In the same time, the positions of the peaks as well as that of the step remain constant. For higher [B] inducing metallic conductivity on the boron impurity band, there is a band around 0.378 eV with fine structures around 0.347 and 0.508 eV. Therefore, the boron impurity band remains distinct from the valence band.

On the other hand, there is a B-induced one-phonon band around 0.16 eV.

After calibration, the 0.16 eV band, the 0.347 eV band and the 0.378 eV step have been used to determine the concentration of boron in homoepitaxial films. The 0.16 eV band seems more reliable for such a purpose.

At a lesser extent, the structures from the electronic transitions of the bound hole and the boron-induced one phonon absorption are seen on highly oriented films. It is difficult to distinguish them in polycrystalline films.

Independent of the deposition conditions and/or techniques, an increase of the boron concentration above 10^{17} cm^{-3} generates significant evolutions of the Raman spectra of the polycrystalline films. When it induces the metallic conductivity on the boron impurity band, striking features appear on the Raman spectra of the polycrystalline and homoepitaxial films.

For polycrystalline films, the width of the 1332 cm^{-1} signal (from the zone center optical phonon of diamond) and the intensity of the 1500–1550 cm^{-1} wide band (from the parasitic 'graphitic' phases) decrease as [B] increases up to the metallic conductivity on the boron impurity band. This is ascribed to a decrease of the concentrations of the parasitic phases and of the defects in agreement with the Bernholc model. For heavy doping ($[B] > 3 \times 10^{20}$ cm^{-3}) in homoepitaxial and polycrystalline films, the zone center optical phonon signal decreases rapidly, shifts to lower wavenumbers and widens (Fano effect). In the same time, a continuum at lower wavenumbers with wide maxima around 500 and 1200 cm^{-1} appears. The Fano effect is ascribed to a quantum interference between the phonon and a continuum of electronic transitions around the same energy involving the boron impurity band. Its analysis in homoepitaxial films required a positive deformation potential. The origin of the continuum with the two wide maxima is not clear. Signals from a disorder-induced one-phonon density of state, or electronic Raman scattering induced by high boron concentrations have been proposed.

There is a limited number of systematic works on the electrical properties of boron-doped diamond films. The electrical properties depend critically on the concentrations of the boron and that of the residual defects. They are much more sensitive than the physicochemical properties to the nature and to the concentration of the defects. They depend on the deposition technique and on the details of the deposition conditions. The results are mainly on films deposited by the MPCVD technique, but the trends seem qualitatively similar with the HFCVD films. In the $10^{16}-10^{17}$ B-cm^{-3} range, the best homoepitaxial films now reach the performances of the best IIb crystals.

The conductivity of the boron-doped films depends on the concentration of boron atoms that can be ionized. Hydrogen–boron pairing reduces it. Hopefully, although the films are deposited within a high concentration of atomic hydrogen, the pairing is not stable at the usual deposition temperatures. The chapter 9 is devoted to this hydrogen (deuterium)–boron pairing.

The conductivities of the homoepitaxial and polycrystalline films increase as the concentration of boron and the temperature increase and decrease with increasing concentrations of compensating donors N_d. At room temperature with the same [B], it decreases from homoepitaxial, to highly oriented, then to polycrystalline films.

The models for the conduction of the homoepitaxial films are qualitatively extended to highly oriented and polycrystalline films. They include two main components—conduction in the valence band and by hopping between the unionized and ionized boron atoms.

At a sufficiently low ("very low") temperature, increasing with the boron and/or the defects concentration, the hopping component dominates. At sufficiently high ("high", "medium" and "low") temperature, the valence band conduction component dominates. It activation energy at "low" temperature is close to that of the hole concentration in the valence band $p(T)$. Up to about 10^{19} B-cm^{-3} for the homoepitaxial films, it is close of the ionization energy of boron. For higher [B], it decreases rapidly to zero around $[B] = 3 \times 10^{20}$ cm^{-3} (which corresponds to the metallic conductivity on the boron impurity band) for all films. As in IIb crystals, the partially compensated semiconductor model gives the holes concentration in the valence band $p(T)$. The N_d, N_a (which can be compared to [B]) and therefore the compensation ratio $K = N_d/N_a$ deduced from the fits by this model of the experimental $p(T)$ of the homoepitaxial films give good trends about the valence band conductivity component. In good homoepitaxial films, K seems to decrease below 0.1 up to $[B] = 4 \times 10^{17}$ or 2×10^{18} B-cm^{-3} according the deposition technique and the detail of deposition conditions. The best sample with $K = 0.004$ has been prepared by MPCVD with trimethylboron with $[B] = 10^{17}$ cm^{-3}. However, lower K are sometimes obtained on films with a lower room temperature mobility.

For the quality of the doped homoepitaxial films, the room temperature mobility and at a lesser extent its variation with temperature appear as better criterions than the K values deduced from the fit of the experimental results by the partially compensated semiconductor model. The best MPCVD films—doped from diborane or trimethylboron—have room temperature mobilities of 1590 or 1840 $cm^2\ V^{-1}\ s^{-1}$ for $[B] \leq 10^{17}\ cm^{-3}$. Their mobilities continuously decrease as the temperature increases above 170 K. They reach the performances of the best IIb crystals (but not yet those of the exceptional IIb crystal). Their mobility remains above 1000 $cm^2\ V^{-1}\ s^{-1}$ as [B] increases up to 4×10^{17} or 2×10^{18} B-cm^{-3} according to the deposition conditions.

The highly oriented films have a larger hopping component which dominates the conductivity up to higher temperatures and which is ascribed to their higher concentration of defects. At sufficiently high temperature the valence band conduction dominates with $p(T)$ close to those of homoepitaxial films with the same [B], but with lower mobilities. Only the best film with $[B] < 10^{17}\ cm^{-3}$, has $d\mu/dT < 0$ above 170 K as the best homoepitaxial films. This suggests grain of good quality, while the defects in the grain boundary deteriorate the mobility.

The polycrystalline films have still higher hopping components ascribed to still higher concentrations of defects in both the grains and the grain boundaries. The activation energy of their valence band conductivity component is significantly lower than the ionization energy of boron and decreases progressively as the [B] increases.

There are good qualitative correlations between some physicochemical and the electronic properties of the boron-doped films. From the high ionization energy of boron in diamond, the hole concentration at 300 K in the valence band remains very weak (10^{-3}[B]) up to the metallic conductivity on the boron impurity band which appears quite abruptly for $[B] \geq 3 \times 10^{20}\ cm^{-3}$. During the deposition of polycrystalline films, the concentration of structural defects decreases as the hole concentration increases. This agrees with the transport measurements which give a decreasing compensation ratio K. Bound and free holes interact with the lattice of the boron-doped diamond, strongly when the boron concentration induces metallic conductivity on the boron impurity band. For both the homoepitaxial and polycrystalline films: (i) for all boron concentrations, the electronic transitions (in the infrared) between the fundamental and the excited levels of the bound hole have phonon replica; (ii) for metallic conductivity on the boron impurity band, the lattice expands rapidly, which is ascribed to a positive deformation potential; (iii) for metallic conductivity on the boron impurity band, there are quantum interferences between the zone center optical phonon and a continuum of electronic transitions (Fano effect) which is also ascribed to a positive deformation potential.

Boron doping is needed for passive and active electronic devices and when the diamond films are used in electrochemistry.

The results about the electronic devices come almost exclusively from MPCVD films. They depend on the concentrations of the boron and of the residual defects. With the same qualitative trends, the quantitative results depend even more critically than the transport measurements, on the detail of the deposition conditions.

The achievement of good ohmic contacts in the resistivity range $10^{-6}\ \Omega\ cm^2$ needs heavily boron-doped films with [B] $\geq 3 \times 10^{20}\ cm^{-3}$ in contact with an intermediate carbide obtained after deposition and annealing of a carbide forming metal. The non-ideality factor of the Schottky diodes deteriorates as [B] increases. Very good ideality factor $n < 1.1$ at room temperature are only obtained when [B] $\leq 10^{17}\ cm^{-3}$, therefore with even more drastic conditions than the best transport properties.

The electrodes are deposited with both the HFCVD and the MPCVD techniques. Their properties depend mainly on their high ($> 10^{19}\ cm^{-3}$) boron concentrations.

Usually, highly boron-doped ($> 10^{19}$ B-cm^{-2}) polycrystalline films are used as electrodes. Heavy doping ($> 3 \times 10^{20}$ B-cm^{-3}) corresponding to metallic conductivity on the boron impurity band gives three order of magnitude higher electrochemical currents. Such adequate [B] are conveniently determined from the shapes of the Raman spectra that are especially sensitive to the boron concentration in this range.

After continuous improvement, the transport properties of the best boron-doped homoepitaxial films reaches that of the best IIb natural crystals. Active electronic devices have yet more stringent requirements. They are now reached with near ideal Schottky diodes. This would induce products with active electronic devices using boron-doped diamond films in the near future.

Acknowledgements

The author acknowledges all the wide band gap semiconductor group of the LEPES for their help, in particular E. Gheeraert for critical reading of the manuscript and M. Bernard of LEPES for taking care of the figures.

References

G. Abstreiter, M. Cardona, and A. Pinczuk, Light scattering in "Solids IV" (M. Cardona and G. Guntherrodt, Eds.), p. 127. Springer, Heidelberg, 1984.

J. W. Ager III, W. Walukiewicz, M. McCluskey, M. A. Plano, and M. I. Landstrass, *Appl. Phys. Lett.* 66, 616 (1995).

S. Alehashem, F. Chamber, J. W. Strojek, and G. M. Swain, *Anal. Chem.* 67, 2812 (1995).

A. E. Alexenko and B. V. Spitsyn, *Diamond Relat. Mater.* 1, 705 (1992).

E. Anastassakis, *Phys. Rev.* 186, 760 (1969).

J. C. Angus and C. C. Hayman, *Science* 241, 913 (1988).

J. C. Angus, H. A. Will, and W. S. Stanko, *J. Appl. Phys.* 39, 2915 (1968).

M. Aslam, G. S. Yang, and A. Masood, *Sensors Actuators A* 45, 131 (1994).

I. G. Austin and R. Wolfe, *Proc. Phys. Soc. B* 69, 329 (1956).

P. K. Bachmann, D. Leers, and H. Lydtin, *Diamond Relat. Mater.* 1, 1 (1991).

J. Bardeen and W. Shockley, *Phys. Rev.* 80, 72 (1950).

F. Beck, H. Krohn, W. Kaiser, M. Fryda, C. P. Klages, and L. Schafer, *Electrochem. Acta* 44, 525 (1998).

M. Bernard and A. Deneuville, unpublished.

J. Bernholc, A. Antonelli, T. M. Del Sole, Y. Bar-Yam, and S. T. Pantalides, *Phys. Rev. Lett.* 61, 2689 (1988).

T. H. Borst, P. C. Münzinger, and O. Weis, *Diamond Relat. Mater.* 3, 515 (1994).
T. H. Borst, S. Strobel, and O. Weis, *Appl. Phys. Lett.* 67, 2651 (1995).
T. H. Borst and O. Weis, *Diamond Relat. Mater.* 4, 948 (1995).
T. H. Borst and O. Weis, *Phys. Status Solidi, Appl. Res.* 154, 423 (1996).
P. Bou and L. Vandenbulcke, Proceedings of the First International Symposium on Diamond and Diamond like films (J. P. Dismuke, A. J. Purdes, B. S. Meyerson, T. D. Moustakas, K. E. Spear, K. V. Ravi, and M. Yoder, Eds.), Vol. 89-12, p. 610. The Electrochemical Society, Los Angeles, 1989.
P. R. Brosious, J. W. Corbett, and J. C. Bourgoin, *Phys. Status Solidi (a)* 121, 677 (1974).
F. Brunet, A. Deneuville, P. Germi, M. Pernet, and E. Gheeraert, *J. Appl. Phys.* 81, 1120 (1997).
F. Brunet, P. Germi, M. Pernet, A. Deneuville, E. Gheeraert, F. Laugier, M. Burdin, and G. Rolland, *J. Appl. Phys.* 83, 181 (1998).
R. G. Buckley, T. D. Moustakas, L. Ye, and J. Voiron, *J. Appl. Phys.* 66, 3595 (1989).
R. C. Burns, V. Cvetkovic, C. N. Dodge, D. J. F. Evans, M. L. T. Rooney, P. M. Spear, and C. M. Welbourn, *J. Crystal Growth* 104, 257 (1990).
M. Cardona and N. E. Christensen, *Solid State Commun.* 58, 421 (1986).
M. Cardona and N. E. Christensen, *Phys. Rev. B* 35, 6182 (1987).
G. S. Cargill III, A. Segmuller, T. F. Kuech, and T. N. Theis, *Phys. Rev. B* 46, 10078 (1992).
J. J. Charette, *Physica* 27, 1061 (1961).
D. Chateigner, F. Brunet, A. Deneuville, P. Germi, M. Pernet, E. Gheeraert, and P. Gonon, *J. Crystal Growth* 148, 110 (1995).
C.-F. Chen, S. H. Chen, T.-M. Hong, and T.-C. Wang, *Diamond Relat. Mater.* 3, 632 (1994).
C. F. Chen, Y. C. Huang, S. Hosomi, and I. Yoshida, *Mat. Res. Bull.* 24, 87 (1989).
J. Chevallier, D. Ballutaud, B. Theys, F. Jomard, A. Deneuville, E. Gheeraert, and F. Pruvost, *Phys. Status Solidi (a)* 174, 73 (1999).
J. Chevallier, A. Lusson, D. Ballutaud, B. Theys, F. Jomard, A. Deneuville, M. Bernard, E. Gheeraert, and E. Bustarret, *Diamond Relat. Mater.* 10, 399 (2001).
J. Chevallier, B. Theys, C. Grattepain, A. Deneuville, and E. Gheeraert, *Phys. Rev. B* 58, 7966 (1998).
J. Cifre, F. Lopez, J. L. Morenza, and J. Esteve, *Diamond Relat. Mater.* 1, 500 (1992).
J. Cifre, J. Puigdollers, M. C. Polo, and J. Esteve, *Diamond Relat. Mater.* 3, 628 (1994).
C. D. Clark, P. Kemmey, E. W. J. Mitchell, and B. W. Henvis, *Philos. Mag.* 5, 127 (1960).
S. E. Coe and R. S. Sussmann, *Diamond Relat. Mater.* 9, 1726 (2000).
E. Colineau, A. Deneuville, J. Mambou, and E. Gheeraert, *Appl. Phys. Lett.* 15, 2123 (1996).
E. Colineau, E. Gheeraert, A. Deneuville, J. Mambou, F. Brunet, and J. P. Lagrange, *Diamond Relat. Mater.* 6, 778 (1997).
A. T. Collins, M. Kamo, and Y. Sato, *J. Mater. Res.* 5, 2507 (1990).
A. T. Collins and E. C. Lightowlers, in "The Properties of Diamond" (J. Field, Ed.), Academic Press, London, 1979, Chapter 3.
A. T. Collins and A. W. S. Williams, *J. Phys. C: Solid State Phys.* 4, 1789 (1971).
P. A. Crowther, P. J. Dean, and W. F. Sherman, *Phys. Rev.* 154, 772 (1967).
R. Daneli, V. N. Denisov, G. Ruani, R. Zamboui, C. Taliani, A. A. Zakhidov, U. Ugawa, K. Imaeda, K. Yakushi, H. Inokuchi, K. Kikuchi, I. Ikemoto, S. Suzuki, and Y. Achiba, *Solid State Commun.* 81, 257 (1992).
B. Daudin, G. Feuillet, J. Hübner, Y. Samson, F. Widman, A. Philippe, C. Bru-Chevallier, G. Guillot, E. Bustarret, G. Bentoumi, and A. Deneuville, *J. Appl. Phys.* 84, 2295 (1998).
G. Davies, The Chemistry and Physics of Carbon, Vol. 13. Marcel Dekker, New York, 1977, p. 1.
R. F. Davis, Z. Sitar, B. E. Williams, H. S. Kong, H. J. Kim, J. W. Palmour, J. A. Edmond, J. Ryu, J. T. Glass, and C. H. Carter, *Mater. Sci. Engng B* 1, 77 (1988).
A. Deneuville and E. Gheeraert, *Il Nuovo Cimento* 52, 1 (1996).
B. V. Derjaguin, D. V. Fedosev, V. M. Lukyanovich, B. V. Spitsyn, A. Ryabov, and A. V. Lavrentyev, *J. Crystal Growth* 2, 380 (1968).
G. Dollinger, A. Bergmaier, C. M. Frey, M. Roesler, and H. Verhoeven, *Diamond Relat. Mater.* 4, 591 (1995).

O. Dorsch, K. Holzner, M. Werner, E. Obermeir, R. E. Harper, C. Johnston, P. R. Chalker, and I. M. Buckley-Golder, *Diamond Relat. Mater.* 2, 1096 (1993).

D. L. Dreyfus, A. J. Tessmer, J. S. Holmes, C. T. Kao, D. M. Malta, L. S. Plano, and B. R. Stoner, *Mater. Res. Soc. Symp. Proc.* 339, 109 (1994).

W. Ebert, A. Vescan, T. H. Borst, and E. Kohn, *IEE-EDL* 15, 189 (1994).

W. Ebert, A. Vescan, P. Gluche, T. Borst, and E. Kohn, *Diamond Relat. Mater.* 6, 329 (1997).

R. Erz, W. Dotter, K. Jung, and H. Ehrardt, *Diamond Relat. Mater.* 4, 469 (1995).

U. Fano, *Phys. Rev.* 124, 1886 (1961).

J. E. Field, Ed., "The Properties Diamond", Academic Press, London, 1979.

J. E. Field, Ed., "The Properties of Natural and Synthetic Diamond", Academic Press, London, 1992.

F. Fontaine, PhD Thesis, University of Grenoble, France, unpublished.

F. Fontaine, E. Gheeraert, A. Deneuville, Y. L. Khong, and A. T. Collins, *Diamond Relat. Mater.* 3, 737 (1994).

F. Fontaine, C. Uzan-Saguy, and R. Kalish, *Appl. Phys. Lett.* 15, 2264 (1996).

B. A. Fox, M. L. Hartsell, D. M. Malta, H. A. Wynand, C.-T. Kao, L. S. Plano, G. J. Tessmer, R. B. Henard, J. S. Holmes, A. J. Tessmer, and D. L. Dreyfus, *Diamond Relat. Mater.* 4, 622 (1995).

B. A. Fox, M. L. Harstell, D. M. Malta, H. A. Wynands, G. J. Tessmer, and D. L. Dreyfus, *Proc. Mater. Res. Soc. Symp.* 416, 319 (1996).

B. A. Fox, B. R. Stoner, D. M. Malta, J. P. Ellis, R. C. Glass, and F. R. Sivazlian, *Diamond Relat. Mater.* 3, 382 (1994).

J. A. Freitas, *Diamond Relat. Mater.* 3, 821 (1994).

N. Fujimori, T. Imai, and A. Doi, *Vacuum* 36, 99 (1986).

N. Fujimori, H. Nakata, and T. Imai, *Jpn. J. Appl. Phys.* 29, 824 (1990).

M. W. Geis, D. D. Rathman, D. J. Ehrlich, R. A. Murphy, and W. T. Lindey, *Dev. Lett.* EDL-8, 321 (1987).

E. Gheeraert, A. Deneuville, L. Brunel, and J. C. Oberlin, *Diamond Relat. Mater.* 1, 504 (1992).

E. Gheeraert, A. Deneuville, and J. Mambou, *Diamond Relat. Mater.* 7, 1509 (1998).

E. Gheeraert, P. Gonon, A. Deneuville, L. Abello, and G. Lucazeau, *Diamond Relat. Mater.* 2, 742 (1993).

E. Gheeraert, S. Koizumi, T. Teraji, H. Kanda, and M. Nesladek, *Phys. Status Solidi (a)* 174, 39 (1999).

D. Gherthsen, B. Neubauer, Ch. Dieker, R. Lantier, A. Rizzi, and H. Luth, *J. Crystal Growth* 200, 353 (1999).

G. S. Gildenblat, S. A. Grot, C. R. W. Hatfield, C. R. Wronski, A. R. Badzian, T. Badzian, and R. Messier, *Mater. Res. Bull.* 25, 129 (1990).

P. Gonon, A. Deneuville, F. Fontaine, E. Gheeraert, A. Campargue, M. Chenevier, and S. Rodolphe, *J. Appl. Phys.* 78, 7404 (1995a).

P. Gonon, E. Gheeraert, A. Deneuville, and L. Abello, *Thin Solid Films* 256, 13 (1995b).

P. Gonon, E. Gheeraert, A. Deneuville, F. Fontaine, L. Abello, and G. Lucazeau, *J. Appl. Phys.* 78, 7059 (1995c).

M. Griesser, G. Stingeder, M. Grasserbauer, H. Baumann, F. Link, P. Wurzinger, H. Lux, R. Haubner, and B. Lux, *Diamond Relat. Mater.* 3, 638 (1994).

J. R. Hardy, *Philos. Mag.* 7, 953 (1962).

R. Haubner, S. Bohr, and B. Lux, *Diamond Relat. Mater.* 8, 171 (1999).

K. Hayashi, S. Yamanaka, H. Watanabe, T. Sekiguchi, H. Okushi, and K. Kajimura, *J. Appl. Phys.* 81, 744 (1997).

C. A. Hewett, M. J. Taylor, J. R. Ziedler, and M. W. Geis, *J. Appl. Phys.* 77, 755 (1995).

H.-J. Hinneberg, M. Heck, and K. Schmidt, *Diamond Relat. Mater.* 1, 810 (1992).

S. L. Holder, L. G. Rowan, and J. J. Krebs, *Appl. Phys. Lett.* 64, 1091 (1994).

T. Inushima, T. Matsushjta, S. Ohya, and H. Shiomi, *Diamond Relat. Mater.* 9, 1066 (2000).

R. B. Jackman, B. Baral, C. R. Kingley, and J. S. Foord, *Diamond Relat. Mater.* 5, 378 (1996).

F. Jansen, M. A. Machonkin, and D. E. Kuhman, *J. Vac. Sci. Technol. A* 8, 3785 (1990).

G. Janssen, W. J. P. Enckervort, W. Vollenberg, and L. J. Gilling, *Diamond Relat. Mater.* 1, 789 (1992).

H. Jia, J. Shinar, D. P. Lang, and M. Pruski, *Phys. Rev. B* 48, 17595 (1993).

X. Jiang, C. P. Klages, R. Zachai, M. Hartweg, and H.-J. Füsser, *Appl. Phys. Lett.* 62, 3438 (1993).

S. Jin and T. D. Moustakas, *Appl. Phys. Lett.* 65, 403 (1994).

M. Kamo, Y. Sato, S. Matsumoto, and N. Setaka, *J. Crystal Growth* 62, 642 (1983).

D. R. Kania, M. I. Landstrass, A. M. Plano, L. S. Pan, and S. Han, *Diamond Relat. Mater.* 2, 1012 (1993).

N. Katsuki, E. Takahashi, M. Toyoda, T. Kurosu, M. Iida, S. Wakita, Y. Nishiki, and T. Shimamune, *J. Electrochem. Soc.* 145, 2358 (1998).

N. Katsuki, S. Wakita, Y. Nishiki, T. Shimamune, Y. Akiba, and M. Iida, *Jpn. J. Appl. Phys. Lett.* 36, 280 (1997).

Y. D. Kim, W. Choi, H. Wakimoto, S. Usami, H. Tomokage, and T. Ando, *Diamond Relat. Mater.* 9, 1096 (2000).

H. Kim, R. Vogelgesang, A. K. Ramdas, S. Rodriguez, M. Grimsditch, and T. R. Anthony, *Phys. Rev. Lett.* 79, 1706 (1997).

A. Kimura, Y. Nakatani, K. Yamada, and T. Suzuki, *Diamond Relat. Mater.* 8, 37 (1999).

H. Kiyota, E. Matshushima, K. Sato, H. Okushi, T. Ando, J. Tanaka, M. Kamo, and Y. Sato, *Diamond Relat. Mater.* 6, 1753 (1997).

D. S. Knight and W. B. White, *J. Mater. Res.* 4, 385 (1989).

K. Kobashi, K. Nishimura, K. Miyata, Y. Kawate, J. T. Glass, and B. E. Williams, SPIE Proceedings on diamond optics (A. Feldman and S. Holly, Eds.), Vol. 969, pp. 159, 1988, San Diego, CA.

S. Koizumi, M. Kamo, Y. Sato, H. Ozaki, and T. Inuzuka, *Appl. Phys. Lett.* 71, 1065 (1997).

S. Koizumi, T. Murakami, T. Inuzuka, and K. Suzuki, *Appl. Phys. Lett.* 57, 563 (1990).

S. Koizumi, K. Watanabe, M. Hasegawa, and H. Kanda, The 9th International Symposium on Advanced Materials (ISAM 2002), Tsukuba, Japan, March 3–7, 2002.

J. P. Lagrange, A. Deneuville, and E. Gheeraert, *Diamond Relat. Mater.* 7, 1390 (1998).

C. Lévy-Clément, F. Zenia, N. A. Ndao, and A. Deneuville, *New Diamond Frontier Carbon Technol.* 9(3), 189 (1999).

L. F. Li, D. A. Totir, N. Vinokur, B. Miller, G. Chottiner, A. Argoitra, J. C. Angus, and D. A. Sherson, *J. Am. Chem. Soc.* 119, 7875 (1997).

L. F. Li, D. A. Totir, N. Vinokur, B. Miller, G. Chottiner, E. A. Evans, J. C. Angus, and D. A. Scherson, *J. Electrochem. Soc.* 145, L45 (1998).

X. Z. Liao, R. J. Zhang, C. S. Lee, S. Tong Lee, and Y. W. Lam, *Diamond Relat. Mater.* 6, 521 (1997).

Y. Liou, R. Weimer, D. Knight, and R. Messier, *Appl. Phys. Lett.* 56, 437 (1990).

R. Locher, J. Wagner, F. Fuchs, M. Maier, P. Gonon, and P. Koidl, *Diamond Relat. Mater.* 4, 678 (1995).

R. Locher, C. Wild, N. Harris, D. Behr, and P. Koidl, *Appl. Phys. Lett.* 65, 34 (1994).

H. Maeda, K. Ohtsubo, M. Kameta, T. Saito, K. Kusakabe, S. Morooka, and T. Asano, *Diamond Relat. Mater.* 7, 383 (1998).

D. M. Malta, J. A. von Vindheim, and B. A. Fox, *Appl. Phys. Lett.* 62, 2926 (1993).

D. M. Malta, J. A. von Windheim, H. A. Wynands, and B. A. Fox, *J. Appl. Phys.* 77, 1536 (1995).

M. Marchywka, J. F. Hochedez, M. W. Geis, D. G. Socher, D. Moses, and R. T. Goldberg, *Appl. Opt.* 30, 5011 (1991).

S. Matsumoto, Y. Sato, M. Kamo, and N. Sekata, *Jpn. J. Appl. Phys.* 21, 183 (1982a).

S. Matsumoto, Y. Sato, M. Tsutsumi, and N. Sekata, *J. Mater. Sci.* 17, 3106 (1982b).

P. Mehta Menon, A. Edwards, C. S. Feigerle, R. W. Shaw, D. W. Coffey, L. Heatherly, R. E. Clausing, L. Robinson, and D. C. Glasgow, *Diam. Relat. Mater.* 8, 101 (1999).

K. Miyata, K. Kumagai, K. Nishimura, and K. Kobashi, *J. Mater. Res.* 8, 2845 (1993).

J. F. Morhange, R. Beserman, and J. C. Bourgoin, *Jpn. J. Appl. Phys.* 14, 544 (1975).

J. Mort, D. Kuhman, M. Machonkin, M. Morgan, F. Jansen, K. Okumura, Y. M. LeGrice, and R. J. Nemanich, *Appl. Phys. Lett.* 55, 1121 (1989).
J. Mort, M. A. Machonkin, and K. Okumura, *Appl. Phys. Lett.* 58, 1908 (1991).
N. F. Mott and E. A. Davis, *Philos. Mag.* 17, 1269 (1968).
W. Muller-Sebert, E. Wörner, F. Fuchs, C. Wild, and P. Koidl, *Appl. Phys. Lett.* 68, 759 (1996).
Y. Muranaka, H. Yamashita, K. Sato, and H. Miyadera, *J. Appl. Phys.* 67, 6247 (1990).
J. Nakanishi, A. Otsuki, T. Oku, O. Ishiwata, and M. Murakani, *J. Appl. Phys.* 76, 2293 (1994).
A. N. Ndao, F. Zénia, A. Deneuville, M. Bernard, and C. Lévy-Clément, *Diamond Relat. Mater.* 9, 1175 (2000).
K. Nishimura, K. Das, and J. T. Glass, *J. Appl. Phys.* 69, 3142 (1991).
A. Ogasawara, T. Inushima, T. Shiraishi, S. Ohya, S. Karasawa, and H. Shiomi, *Diamond Relat. Mater* 6, 835 (1997).
I. Ohana, Y. Yacoby, and D. Schmeltzer, *Solid State Commun.* 60, 839 (1986).
K. Okano, H. Nakuri, Y. Akiba, T. Kurosu, M. Iida, and Y. Hirose, *Jpn. J. Appl. Phys.* 27, L173 (1988).
K. Okano, H. Nakuri, Y. Akiba, T. Kurosu, M. Iida, Y. Hirose, and T. Nakamura, *Jpn. J. Appl. Phys.* 28, 1066 (1989).
H. Okushi, *Diamond Relat. Mater.* 10, 281 (2001).
U. Pietsch and K. Unger, *Phys. Status Solidi (a)* 80, 165 (1983).
D. F. Poferl, N. C. Gardner, and J. C. Angus, *J. Appl. Phys.* 44, 1428 (1973).
F. Pruvost, E. Bustarret, and A. Deneuville, *Diamond Relat. Mater.* 9, 295 (2000).
F. Pruvost and A. Deneuville, *Diamond Relat. Mater.* 10, 531 (2001).
R. Ramesham, *Sensors Actuators B* 50, 131 (1998).
R. Ramesham and M. F. Rose, *Diamond Relat. Mater.* 6, 17 (1997).
M. Ramsteiner and J. Wagner, *Appl. Phys. Lett.* 51, 1355 (1987).
C. J. Rauch, *Phys. Rev. Lett.* 7, 83 (1961).
L. Reggiani, D. Waechter, and S. Zukotynski, *Phys. Rev. B* 28, 3550 (1983).
J. P. Ribot, E. Gheeraert, and A. Deneuville, unpublished.
L. H. Robin, E. N. Farabaugh, and A. Feldman, *J. Mater. Res.* 5, 2456 (1990).
J. Ruan, K. Kobashi, and W. J. Choyke, *Appl. Phys. Lett.* 60, 1884 (1992).
I. Sakaguchi, M. Nishitani-Gamo, K. P. Loh, H. Haneda, and T. Ando, *Diamond Relat. Mater.* 7, 1144 (1998).
I. Sakaguchi, M. Nishitani-Gamo, K. P. Loh, H. Haneda, S. Hishita, and T. Ando, *Appl. Phys. Lett.* 71, 629 (1997).
A. Y. Sakharova, Y. V. Pleskov, F. Di Quarto, S. Piazza, C. Sunseri, I. G. Treremetskaya, and V. P. Vernin, *J. Electrochem. Soc.* 142, 2704 (1995).
R. Samlenski, J. Schmälzlin, R. Brenn, C. Wild, W. Muller-Sebert, and P. Koidl, *Diamond Relat. Mater.* 4, 503 (1995).
B. V. Sarada, T. N. Roa, D. A. Tryk, and A. Fujishima, *J. Electrochem. Soc.* 146, 1469 (1999).
Y. Sato and M. Kamo, in "The Properties of Natural and Synthetic Diamond" (J. E. Field, Ed.), p. 423. Academic Press, London, 1992, Chapter 11.
H. Shiomi, Y. Nihsibayashi, and N. Fujimori, *Jpn. J. Appl. Phys.* 28, 2153 (1989).
H. Shiomi, Y. Nishibayashi, and N. Fujimori, *Jpn. J. Appl. Phys.* 30, 1363 (1991).
H. Shiomi, K. Tanabe, Y. Nishibayashi, and N. Fujimori, *Jpn. J. Appl. Phys.* 29, 34 (1990).
Y. Show, T. Matsukawa, H. Ito, M. Iwase, and T. Izumi, *Diamond Relat. Mater.* 9, 337 (2000).
W. V. Smith, P. P. Sorokin, L. L. Gelles, and G. J. Lasher, *Phys. Rev.* 115, 1546 (1959).
S. D. Smith and W. Taylor, *Proc. Phys. Soc.* 79, 1142 (1962).
S. Sonoda, J. H. Won, H. Yagi, A. Hatta, T. Ito, and A. Hiraki, *Appl. Phys. Lett.* 70, 2574 (1997).
H. Spicka, M. Griesser, H. Hutter, M. Grasserbauer, S. Bohr, R. Haubner, and B. Lux, *Diamond Relat. Mater.* 5, 383 (1996).
B. V. Spitsyn, L. L. Bouilov, and B. V. Deryagin, *J. Crystal Growth* 52, 219 (1981).
B. R. Stoner and J. T. Glass, *Appl. Phys. Lett.* 60, 698 (1992).

B. R. Stoner, C. T. Kao, D. M. Malta, and R. C. Glass, *Appl. Phys. Lett.*, 62 (1993).
B. R. Stoner, S. Sahaida, J. P. Bade, P. Southworth, and P. Ellis, *J. Mater. Res.* 8, 1334 (1993).
T. Sugino, K. Karasutani, F. Mano, H. Kataoka, J. Shirafuji, and K. Kobashi, *Diamond Relat. Mater.* 3, 618 (1994).
G. M. Swain, A. B. Anderson, and J. C. Angus, *Mater. Res. Bull*, September, 56 (1998).
G. M. Swain and R. Ramesham, *Anal. Chem.* 65, 345 (1993).
I. Taher, M. Aslam, M. A. Tamor, T. J. Potter, and R. C. Elder, *Sensors Actuators A* 45, 35 (1994).
D. Takeuchi, H. Watanabe, S. Yamanaka, H. Okushi, and K. Kajimura, *Diamond Relat. Mater.* 9, 231 (2000).
R. Tenne and C. Lévy-Clément, *Isr. J. Chem.* 38, 57 (1998).
R. Tubino, L. Piseri, and G. Zerbi, *J. Chem. Phys.* 56, 1022 (1972).
K. Ushizawa, K. Watanabe, T. Ando, M. Nishitani-Gamo, Y. Sato, and H. Kanda, *Diamond Relat. Mater.* 7, 1719 (1998).
C. Uzan-Saguy, A. Reznik, C. Cytermann, R. Brener, R. Kalish, E. Bustarret, M. Bernard, A. Deneuville, E. Gheeraert, and J. Chevallier, *Diamond Relat. Mater.* 10, 453 (2001).
V. S. Vavilov, *Phys. Status Solidi (a)* 31, 11 (1975).
L. Vegard, *Z. Phys.* 5, 17 (1921).
E. P. Visser, G. H. Bauhis, G. Janssen, W. Vollenberg, W. J. P. van Enckevort, and L. J. Gilling, *J. Phys.: Condens. Mater.* 4, 7365 (1992).
O. A. Voronov and A. K. Rakhmanina, *Inorg. Mater.* 29, 707 (1993).
O. A. Voronov, A. V. Rakhmanina, and E. P. Khlybov, *Inorg. Mater.* 31, 851 (1995).
J. Wagner, C. Wild, and P. Koidl, *Appl. Phys. Lett.* 59, 779 (1991).
C. Wang, A. Garcia, D. C. Ingram, M. Lake, and M. E. Kordesch, *Electron. Lett.* 27, 1459 (1991).
X. H. Wang, G-H. M. Ma, W. Zhu, J. T. Glass, L. Bergman, K. F. Turner, and R. J. Nemanich, *Diamond Relat. Mater.* 1, 828 (1992).
H. Watanabe, K. Hayashi, D. Takeuchi, S. Yamanaka, H. Okushi, K. Kajimura, and T. Sekiguchi, *Appl. Phys. Lett.* 73, 981 (1998).
I. Watanabe and K. Sugata, *Jpn. J. Appl. Phys.* 27, 1808 (1988).
P. T. Wedepohl, *Proc. Phys. Soc. (London) B* 70, 177 (1957).
M. Werner, R. Job, A. Zaitsev, W. R. Fahrner, W. Sieffet, C. Johnston, and P. R. Chalker, *Phys. Status Solidi (a)* 154, 385 (1996a).
M. Werner, C. Johnston, P. R. Chalker, S. Romani, and J. M. Buckley-Golder, *J. Appl. Phys.* 79, 2535 (1996b).
C. Wild, R. Khol, N. Herres, W. Muller-Sebert, and P. Koidl, *Diamond Relat. Mater.* 3, 373 (1994).
P. Wurzinger, P. Pongratz, P. Hartmann, R. Haubner, and B. Lux, *Diamond Relat. Mater.* 6, 763 (1997).
H. A. Wynand, D. M. Malta, B. A. Fox, J. A. von Windheim, J. P. Fleuriat, D. Irvine, and J. Vandersande, *Phys. Rev. B* 49, 5745 (1994).
J. Xu, M. C. Granger, Q. Chen, J. W. Strojek, T. E. Lister, and G. M. Swain, *Anal. Chem. News Features* 1, 591A (1997).
J. Xu and G. M. Swain, *Anal. Chem.* 70, 1502 (1998).
B. G. Yacobi, A. R. Badzian, and T. Badzian, *J. Appl. Phys.* 69, 1643 (1991).
S. Yamanaka, D. Takeuchi, H. Watanabe, H. Okushi, and K. Kajimura, *Phys. Status Solidi (a)* 174, 59 (1999).
S. Yamanaka, D. Takeuchi, H. Watanabe, H. Okushi, and K. Kajimura, *Diamond Relat. Mater.* 9, 956 (2000).
S. Yamanaka, H. Watanabe, S. Masai, S. Kawata, K. Hayashi, D. Takeushi, H. Okushi, and K. Kajimura, *J. Appl. Phys.* 84, 6095 (1998a).
S. Yamanaka, H. Watanabe, S. Masai, D. Takeuchi, H. Okushi, and K. Kajimura, *Jpn. J. Appl. Phys.* 37, L1129 (1998b).
T. Yano, E. Popa, D. A. Tryk, K. Hashimoto, and A. Fujishima, *J. Electrochem. Soc.* 146, 1081 (1999).

Y. Yano, D. A. Tryk, K. Hashimoto, and A. Fujishima, *J. Electrochem. Soc.* 145, 1870 (1998).
W. A. Yarbrough and R. Messier, *Science* 247, 688 (1990).
C. Y. Yeh, Z. W. Lu, S. Froyen, and A. Zunger, *Phys. Rev. B* 46, 10086 (1992).
R. J. Zhang, S. T. Lee, and Y. W. Lam, *Diamond Relat. Mater.* 5, 1288 (1996).
W. Zhu, P. C. Yang, and J. T. Glass, *Appl. Phys. Lett.* 63, 1640 (1993).

CHAPTER 5

N-TYPE DIAMOND GROWTH

Satoshi Koizumi

ADVANCED MATERIALS LABORATORY, NATIONAL INSTITUTE FOR MATERIALS SCIENCE, 1-1 NAMIKI, TSUKUBA 305-0044, JAPAN

1. Introduction

As described before, diamond can be a favorable candidate for high-power electronics applications operated under extreme circumstances such as high temperature and serious radioactive irradiation. Diamond devices will be important candidates for space engineering that must be developed rapidly in this century. Furthermore, due to the large band gap of diamond, it is expected to form deep ultraviolet (UV) light emitting devices with the wavelength shorter than 250 nm. For these applications, we need n- and p-type semiconducting diamonds. The p-type diamond exists in nature and can be formed easily either by using high-pressure–high-temperature (HTHP) method or vapor growth technique. Boron forms acceptor level in diamond at 0.37 eV over the valence band maximum. On the other hand, n-type diamond does not exist in nature and it was very difficult to form artificially. Since 1980s, that chemical vapor deposition (CVD) technique of diamond has been established, n-type doping of diamond has been one of the most important research theme. Among the V group elements, nitrogen (N) can be the most favorable atom to be incorporated in diamond lattice because of its atomic radius being close to that of carbon. Actually, conventional HTHP diamond contains N atoms with large amount located at lattice substitutional sites in a single atomic form. As is well known, however, nitrogen forms deep donor state at 1.7 eV below the conduction band minimum due to a small displacement of N atoms from the tetrahedral coordination in the lattice. Theoretical calculation studies predicted that phosphorus (P), lithium (Li) and sodium (Na) can form shallow donor states (Kajihara, Antonelli and Bernholc, 1993). It also predicted the difficulty to dope these elements due to low solubility of them in the diamond lattice (Anderson and Mehandru, 1993). Ion implantation will be effective to solve the problem because of the non-equilibrium feature of the method. By CVD method, one can dope enough large amounts of impurity atoms as opposed to the prediction. Although a number of experimental researches were performed doping P by CVD process and Li, Na, C, O and P by ion implantation, there were no reports to convince n-type conductivity. In 1997, we have succeeded to form n-type diamond thin films by doping P during

ISBN: 0-12-752185-2
ISSN: 0080-8784

the growth of {111} epitaxial layer under well-optimized CVD experimental conditions to achieve high crystalline perfection (Koizumi, Ozaki, Kamo, Sato and Inuzuka, 1997). After the success, a study of sulfur (S) doping was reported to give n-type conductivity with much higher electron mobility (Sakaguchi, Gamo, Kikuchi, Yasu, Haneda, Suzuki and Ando, 1999). However, the p-type conductivity of the S-doped diamond by several different research groups was reported later (Kalish, Reznik, Uzan-Saguy and Cytermann, 2000; Garrido, Nebel, Stutzmann, Gheeraert, Casanova and Bustarret, 2002). These studies pointed out the possibility of unintentional boron incorporation during the growth. The results of S-doping studies are still unclear. On the other hand, P-doped diamond research revealed a lot of new features of the diamond by electrical and optical measurements. This chapter describes the technical aspects of P-doped diamond growth and the electrical properties of the diamond characterized by Hall measurements, and finally, recent results of pn junction UV light emitting diode are mentioned.

2. Phosphorus-doped Diamond Growth

2.1. Preface

Several experimental researches had already been reported about P doping before our success in 1996. In those experiments, several different doping sources such as phosphine (PH_3) (Kamo, Yurimoto, Ando and Sato, 1991; Schauer, Flemish, Wittstruck, Landstrass and Plano, 1994; Bohrs, Haubner and Lux, 1995), red phosphorus (solid) (Alexenko and Spitsyn, 1992), P_2O_5 (Okano, Kiyota, Iwasaki, Nakamura, Akiba, Kurosu, Iida and Nakamura, 1990) or trimethyl phosphite ($(CH_3O)_3P$) (Chen, Lo and Chen, 1996) were used. The doping has been performed during the growth of diamond using variety of CVD methods such as microwave plasma CVD, DC plasma CVD, hot filament CVD or chemical transport reaction method. Although some of them have succeeded obtaining conductive films, there were no results to convince n-type conductivity clearly by Hall measurements. Furthermore, there was a possibility that the conductivity is related to the defects or non-diamond precipitates due to poor crystalline perfection of the films. We have focused on the crystalline perfection of the P-doped diamond films to clarify the type of conductivity and its origin as the first step of this research.

2.2. Preliminary Experiments

In CVD diamond growth process, diamond grows under very critical surface reactions of atomic hydrogen and carbonaceous species. Therefore, even very small amount of impurity atoms induced in the process sometimes influence significantly on the growth morphology or crystalline perfection of the diamond thin films. We need to pay attention to optimize the growth condition of the doped diamond films independently adapted to each dopant to obtain better crystalline perfection of the films. In the preliminary experiments, we have observed the growth of diamond particles under P-doped condition. The diamond growth has been performed using microwave plasma assisted CVD method. The experiments have been done by old-style quartz-tube-type CVD system with the background vacuum of 10^{-7} Torr evacuated by turbo molecular pump. Methane (CH_4) and hydrogen (H_2) were used for the source gas for diamond growth and PH_3 was used as a doping source gas. P-doped diamond particles were grown on molybdenum substrates and observed by scanning electron microscopy (SEM). In Figure 1, SEM photographs of typical P-doped diamond particle are shown. The growth conditions are: methane concentration (CH_4/H_2) 0.3%, gas pressure 80 Torr, phosphine concentration (CH_4/PH_3) 8000 ppm and substrate temperature 980 °C. On the particle surface, we can identify {111} and {100} facets forming most likely a cubo-octahedron shape as shown in Figure 1(a). It is very clear that {111} facets are formed very flat in contrast to rough and complicated surface morphology of {100} facets. As shown in Figure 1(b), numerous hillocks that seem to be surrounded by {111} surface have been formed on {100}. This implies that step flow growth has not been favorably promoted and only the defective or contaminated portions may give stable nucleation sites for three-dimensional growth of diamond. On {111} facet, diamond grows rather two-dimensionally as shown in Figure 1(c). By the series of growth experiments performed under various growth conditions (CH_4/H_2: 0.15–2%, substrate temperature: 800–1000 °C), it is found that only {111} surface is formed flat and for the growth, low methane concentration and high substrate temperature were required. In cases where the methane concentration exceeds 0.3%, numerous three-dimensional nucleation appears on {111} and the grown surface is observed highly defective. Although, it is possible to grow flat {100} facet by applying high methane concentration (2%) and lower substrate temperature (800 °C) condition, the doping efficiency of P is about 1/100 compared to the case of {111}. The epitaxial diamond films grown with P-doping formed on {100} were very highly resistive. These films showed co-incorporation of hydrogen along with P atoms. We concluded that {111} is the most favorable surface orientation to form high quality P-doped layers.

2.3. Phosphorus-doped Epitaxial Diamond Thin Films

Further growth experiments have been performed using {111} single crystalline diamond substrates to obtain high crystalline perfection films. First, we started the experiments using bulky diamond substrate with hexagonal shaped {111} polished surfaces. Although we could grow n-type diamond thin films, the mobility was as low as 20–30 $cm^2 V^{-1} s^{-1}$ and the reproducibility was poor

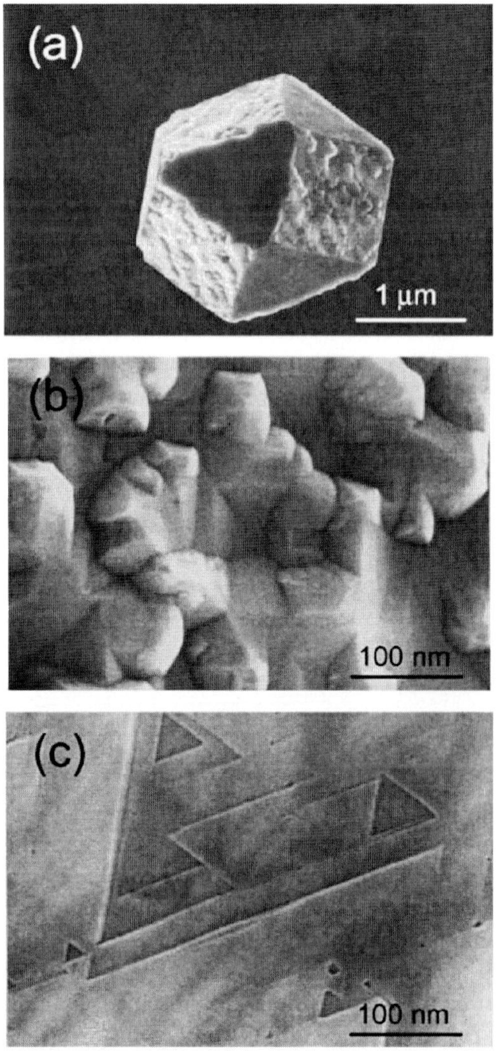

FIG. 1. SEM images of typical diamond particle grown under phosphorus doping and the magnified images of each facet of the particle: (a) overall image of the particle, (b) {100} surface of the particle and (c) {111} surface of the particle.

(Koizumi et al., 1997; Koizumi, Kamo, Sato, Mita, Sawabe, Reznik, Uzan-Saguy and Kalish, 1998). This was mostly due to poor uniformity of the shape and the color of the substrates that seriously influence the temperature measurements during the growth experiments. So, we switched to use square-shaped diamond single crystals that show better uniformity of color and shape of independent crystals. By the experiments to optimize {111} P-doped diamond growth, it was found that the growth conditions listed in Table I give highly crystallized

TABLE I

GROWTH CONDITIONS OF P-DOPED DIAMOND THIN FILMS

Source gases	PH_3, CH_4, H_2
CH_4/H_2 (%)	0.05, 0.075
PH_3/CH_4 (ppm)	50–5000
Total gas pressure (Torr)	100
Total gas flow rate (sccm)	400–1000
Substrate temperature (°C)	850–950
Substrate	Ib diamond {111} ($2 \times 2 \times 0.5$ mm^3)
Growth duration (h)	2

P-doped film growth reproducibly (Koizumi, Teraji and Kanda, 2000). For the experiments, we have used metal-chamber-type microwave plasma assisted CVD system to establish much higher purity control. Figure 2 shows schematic of the growth system. Propagation mode of the microwave is changed from rectangular to circular mode through cylindrical wave-transformer. The microwave is introduced through a quartz glass window put on the top of the chamber. In the chamber, a plasma cavity of which the size is tuned for 2.45 GHz microwave, is attached under the window. Source gases are introduced from the top of the cavity. The substrate is placed on the molybdenum holder that penetrates into the cavity through a hole at the center of reflector disk. Plasma is stably formed on the substrate. As source gases, hydrogen, methane and phosphine were used and they are independently supplied with precise control of the flow rate by mass flow controllers. Premixed phosphine/hydrogen (PH_3/H_2: 100 ppm) was used and the final purity is 6N. Hydrogen gas is supplied through Pd diffuser with the purity of 9N. The purity of methane is 6N. The growth

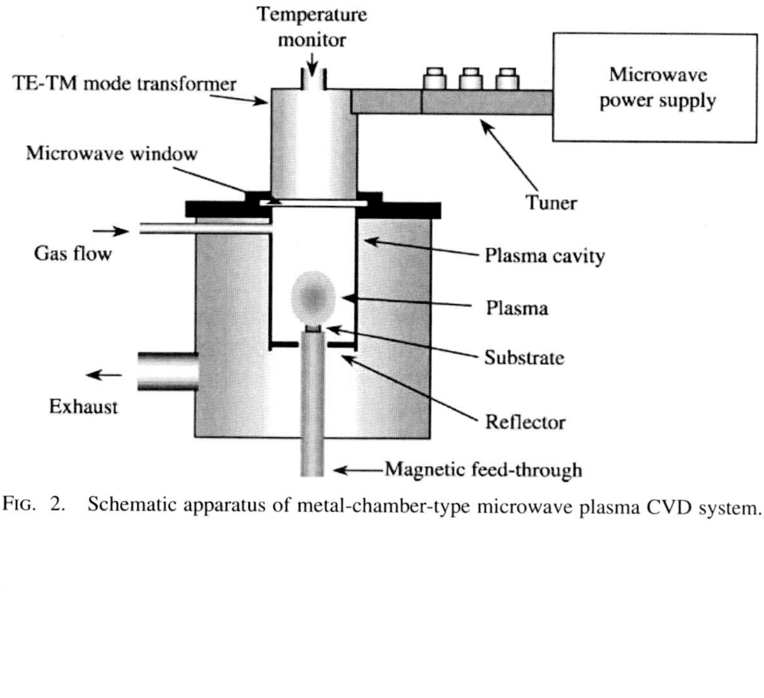

FIG. 2. Schematic apparatus of metal-chamber-type microwave plasma CVD system.

chamber is evacuated by turbo molecular pump with oil mechanical pump to its base pressure of less than 1×10^{-8} Torr. The chamber is baked out before each run of the growth experiment to reduce water partial pressure. As substrates, Ib-type high-pressure synthetic (HPHT) {111} diamond single crystals were used. The size is $2 \times 2 \times 0.5$ cm^3. The substrates are chemically oxidized using $HClO_4/H_2SO_4/HNO_3$ (1/3/4, 200 °C, 30 min) or $NaClO_3/HNO_3$ (10 wt%, 200 °C, 1.5 h) to remove organic contaminations followed by boil and rinse in ultra-pure water. This process is important to unify the initial condition of the substrate surface as oxygen terminated state. After installing the substrate into the chamber and evacuation of the chamber, the growth experiments have been performed as follows. Plasma is ignited with hydrogen flow at 10–20 Torr. The pressure is increased up to 100 Torr with keeping the plasma stable and after a few minutes, the substrate temperature rises up to 850 °C. Then methane is introduced and the substrate temperature rises up again to 900 °C and stabilized within minutes. At this stage, microwave power is normally about 600 W. Phosphine is introduced and the P-doped diamond growth is started. The temperature of the substrate is monitored through a top window by optical pyrometer. The accuracy of the absolute value of the temperature is very poor because of the optical emission from plasma. The growth rate of P-doped diamond films is 500 nm h^{-1} at a methane concentration of 0.05% and it is proportional to the methane concentration within the range of CH_4/H_2: 0.025–0.15%. The phosphine concentration in the gas phase has not influenced much to the growth rate.

2.4. Optical Microscopic Observation

In Figure 3, typical Nomarski differential interference contrast micrographs are presented. Phosphorus concentration (PH_3/CH_4) is 500 ppm during the diamond growth and the film thickness is 1 μm for all the samples in the images. In the low magnification image (Fig. 3(a)), it is clearly observed that the film is mostly very flat except for one defective portion observed in the right-top part of the photograph. The irregular growth is sometimes observed propagating from the edge of the substrate. This is considered to be due to the defective nature of roughly treated side edge of the single crystalline diamond substrates. The inhomogenity of color in the substrate related to the growth sector is observed in the image as a global contrast. Figure 3(b) shows a magnified image of flat P-doped film surface. The particles observed in the image are contaminants adsorbed accidentally during chemical oxidation process. In most part, the surface does not show any characteristic features. It is not very clear in the image, but one can see very weak contrast in the left part of the image as a band of about 10 μm width, which corresponds to the part of residual scratch of insufficient polishing of the substrate. Typically, the film surface shows some features as seen in Figure 3 (c). We can observe limited number of 'comets' on the surface. These comets are formed by very small number of local disordering such as structural defects or persistent contaminants induced accidentally on the film surface.

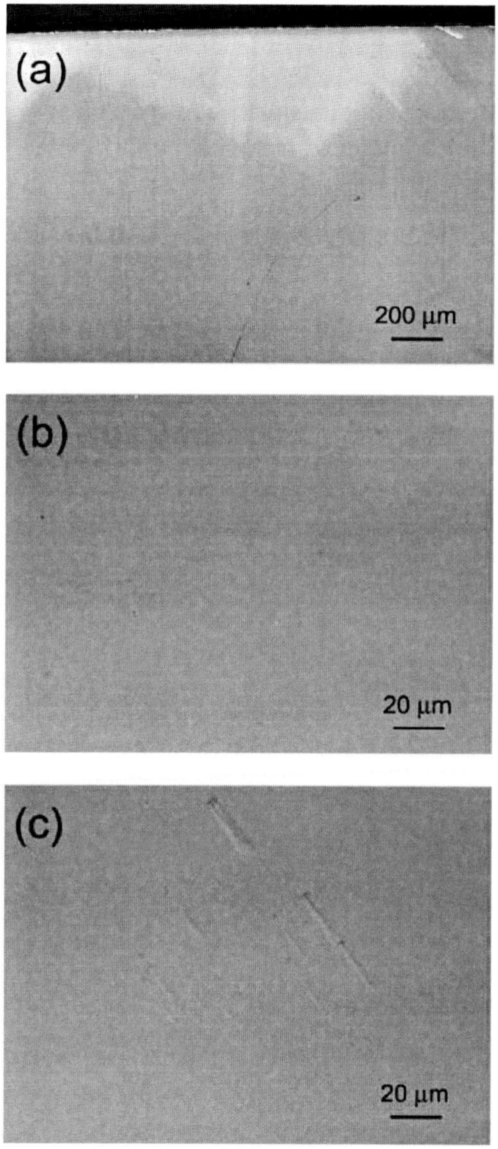

FIG. 3. Nomarski micrographs of {111} P-doped diamond surfaces.

The tail of the comet is formed by disturbing two-dimensional growth. Ideally, diamond growth proceeds with critical chemical reaction between carbon and hydrogen atoms taking place at steps and kinks. The steps capture carbon precursors and hydrogen atoms keep them in diamond form by terminating their dangling bonds. With the sequence of this process diamond grows two-dimensionally, so called 'step-flow-mode'. In the case of step-flow-mode growth, the local disordering blocks the step-flow growth and the shadows are formed at the down flow of the disordering. The comets are the evidence of high quality thin film growth. In our experiments performed with our optimized growth condition, secondary nucleation or nucleation of randomly oriented diamond particles has not been observed.

2.5. Impurity Analysis

To clarify the impurity concentration and distribution, the P-doped diamond thin films were characterized by secondary ion mass spectrometry (SIMS). An example of SIMS depth profile of P-doped sample is shown in Figure 4. This sample has been formed with three different P-concentration films stacking on the substrate to check the controllability of doping concentration. The P-concentration has been changed sequentially during diamond growth with PH_3/CH_4: 100, 500 and 1000 ppm every 1 h. The P-concentrations in the layers were 6×10^{17}, 3×10^{18} and $5 \times 10^{18}\ cm^{-3}$ for 100, 500 and 1000 ppm layers, respectively. It is confirmed that the P-concentration in the film is almost proportional to that in the gas phase during the growth. The doping efficiency was about 3–4%. The depth profile of hydrogen does not show obvious change with P-incorporation. Although this may be due to a high background signal of residual hydrogen in the analysis chamber during SIMS measurements, the hydrogen incorporation is smaller than in the case of {100} P-doped film. Silicon incorporation, which was often observed in the case of quartz-tube-type microwave CVD diamond growth, has been rarely observed in this case.

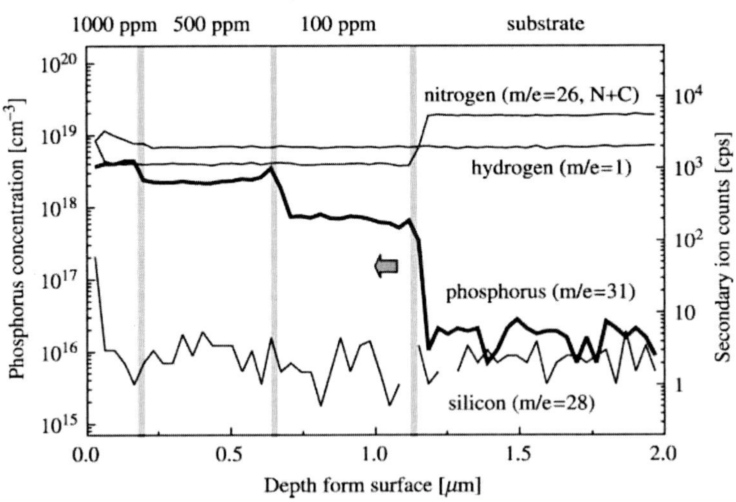

FIG. 4. SIMS depth profile of impurity atoms in the P-doped film grown at 900 °C in three steps of PH_3 concentrations in the gas phase, P/C: 100, 500 and 1000 ppm.

Nitrogen signal is observed rather strongly if we simply compare the intensity to that in the substrate. This is also due to the influence of background vacuum during the measurements and recent measurements revealed that nitrogen in the P-doped layer is at least less than 1/100 of that of type-Ib substrate. Boron signal was under the detection limit (4×10^{15} cm^{-3}) of the present SIMS measurement.

2.6. Structural Analysis

Crystalline perfection and other structural characteristics of P-doped epitaxial diamond thin films were characterized by high-energy reflection electron diffraction (RHEED), Raman spectroscopy, Rutherford back scattering (RBS) and particle induced X-ray emission (PIXE). The RHEED patterns were observed as streaks and this confirms the surfaces of the films are very flat. Twinning was rarely observed. By comparing angular dependences of channeling ions during RBS and PIXE, it was found that almost 100% of P atoms are located just at the lattice substitutional sites and the displacement of P atoms from the lattice site is negligibly small. By Raman spectroscopy, it was confirmed that the crystalline perfection of P-doped film is equivalent to that of the substrate. Figure 5 shows typical Raman spectra of P-doped film and its substrate. The measurement has been performed using a Raman spectroscopy system equipped with confocal collecting optics that can resolve 1 μm to the depth direction. The full width at half-maximum of P-doped film (a) and the substrate (b) is 1.83 and 1.73 cm^{-1}, respectively. The resolution of the monochromator is 1.69 cm^{-1} and these values are within the error of the measurement. We could not observe any obvious signature of strain in the P-doped film because the Raman peak is located at

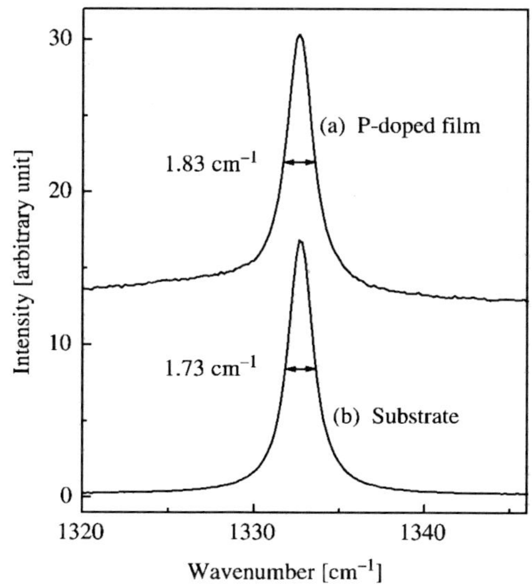

FIG. 5. Typical Raman spectra of P-doped diamond thin film and the substrate.

1332.5 cm^{-1}, i.e., the same as that observed from bulk diamond without stress. It is considered that the strain is accumulated in diamond film by P-doping because the covalent bond radius of P atom is 1.1 Å and it is larger than that of carbon (0.7 Å) in diamond lattice. The total amount of expected strain is at most 0.02% when we assume the P-concentration to be 10^{19} cm^{-3}, and the resulting shift of the Raman peak will be 0.1–0.5 cm^{-1}. The value is within the detection error of our Raman system. However, when we have grown thick P-doped films with {111} orientation, the films are often cracked along the cleavage direction ⟨110⟩ and sometimes the cracked flakes peals off from substrate. This shows that the stress is accumulated obviously in the film. At present, the maximum thickness of {111} P-doped film is limited by this phenomena and that is about 10 μm for the film doped with 10^{19} cm^{-3}.

3. Electrical Properties of Phosphorus-doped Diamond Films

The Hall effect measurement is the most convincing method to characterize the electrical conduction mechanism of semiconductors. We have performed Hall measurements for all the P-doped diamond thin films that we have grown since 1995. Since 1998, we have found an appropriate condition to grow high-quality P-doped films using metal-chamber-type microwave CVD system; the Hall measurements show n-type conductivity for all the samples.

Before Hall measurements, the samples were oxidized by acid treatment that was applied for pretreatment of diamond substrates. This process is important because as grown (hydrogen terminated) surface of P-doped films show rather higher electrical conductivity than the film itself. At present, the conduction type is not clear, but the same conduction mechanism will be taking place as that observed for intrinsic CVD diamond films. If the treatment has not been applied, it is impossible to obtain stable Hall voltage especially at a temperature lower than 200 °C. For ohmic electrode formation, we have performed systematic experiments such as changing work function of metal contacts, metal carbide formation using several different metals and annealing condition, defect inducement by ion implantation, and selective growth of heavily P-doped thin layer formation (Teraji, Koizumi, Mita, Sawabe and Kanda, 1999; Teraji, Koizumi and Kanda, 2000; Teraji and Koizumi, in preparation). Among these experiments, the implantation and heavy doped layer formation were adopted to give effectively sufficient ohmic property for van der Pauw Hall measurements. The contacts were formed at four symmetrically located points on the surface P-doped film. The ion implantation was performed through a metal shadow mask with argon ion at 40 keV with the dose of 1×10^{16} cm^{-2} to form graphitized thin layer at the film surface with the thickness of 30 nm and the diameter of 300 μm followed by rapid thermal annealing in high vacuum. The heavily doped layers were formed around four corners of the sample surface by the selective growth technique using gold (Au) and titanium (Ti) stacking film mask. Finally, Au/Ti stacking layers were formed on the modified area with the diameter of 300 μm for protection and an utility of Au wire bonding. The Hall measurements were performed under severe temperature control within the range of room temperature (RT) to 600 °C. The sample vessel was evacuated by turbo-molecular pump and helium was introduced during the measurement at the pressure of about 1 Pa. In Figure 6, temperature dependences of carrier concentrations measured from P-doped diamond thin films grown under different P-concentrations. The samples have shown stable n-type Hall signal in this temperature range. P-concentrations of the films are 4×10^{17} cm^{-3} for (a) and 5×10^{18} cm^{-3} for (b). It is clear that both the plots aligned straightly and parallel to each other within the temperature range of RT to 500 K showing single thermal activation energy level supplies electrons as conducting carriers. The carrier concentrations are almost proportional to P-concentrations. The carrier concentration (n) of partly compensated semiconductors obeys the following equation:

$$\frac{n(N_A + n)}{N_D - N_A - n} = \frac{N_C}{g_d} \exp\left(-\frac{E_a}{kT}\right) \tag{1}$$

where N_D and N_A are the donor and acceptor concentrations, E_a the activation energy of electron from donor level in the unit of eV, g_d the degeneration factor (for n-type diamond the value is 2), k the Boltzmann constant, T the temperature in Kelvin and N_C is the effective density of state and it is expressed as:

$$N_C = 12\left(\frac{2\pi m_e^* kT}{h^2}\right)^{3/2}$$

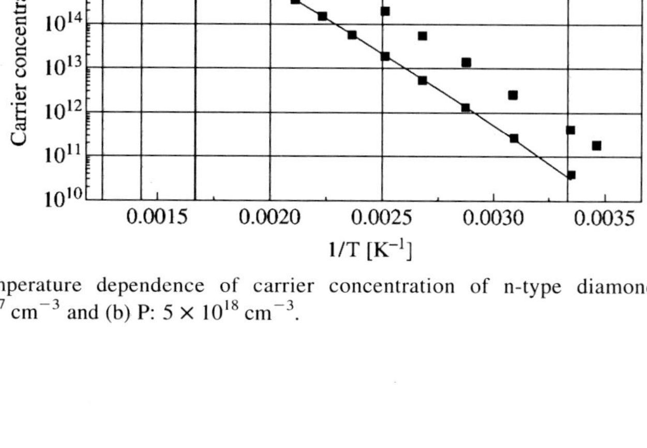

FIG. 6. Temperature dependence of carrier concentration of n-type diamond thin films. (a) P: 4×10^{17} cm^{-3} and (b) P: 5×10^{18} cm^{-3}.

where m_e^* is the effective mass of electron, i.e., $0.55m_e$ (m_e, mass of free electron), h the Plank constant.

By solving Eq. (1) and fitting the carrier concentration profile of Figure 6(a), the following values are calculated within small error: E_a, 0.59 eV; N_D, 4.5×10^{17} cm^{-3}; N_A, 6.1×10^{15} cm^{-3}. The compensation ratio (N_A/N_D) is about 1.5%. The activation energy is so large that the carrier concentration varies over six orders of magnitude with the present temperature range. At RT, the carrier concentration is at most 10^{11} cm^{-3} for both the samples in spite of large donor concentration in the films. The resistivity of these samples was 10^5–10^6 Ω cm at RT. Temperature dependences of Hall mobility of P-doped samples are shown in Figure 7. The P-concentrations are 1×10^{18} cm^{-3} (a), 8×10^{18} cm^{-3} (b) and 3×10^{19} cm^{-3} (c), respectively. At RT, the Hall mobility was 180 cm^2 V^{-1} s^{-1} for (a) and 40 cm^2 V^{-1} s^{-1} for (b). In the case of (c) the mobility dropped steeply below 400 K by the presence of hopping conductivity due to heavy doping. The maximum Hall mobility is observed at 500 K and the value was 25 cm^2 V^{-1} s^{-1} for the sample (c). For the case of (b) also, the mobility drops down at around RT and this is also considered to be the influence of hopping conductivity, but the amount of hopping carriers will be smaller than in the case of (c). In the case of (a), a small drop is observed even though it is difficult to assume the presence of hopping conductivity because of low donor concentration. The temperature dependences of carrier concentration (in between (a) and (b) in Figure 6) and resistivity (not shown here) do not show even a tendency of the hopping conductivity at RT. The deterioration mechanism is not clear at present, but may be related to deeper energy states that trap electrons. At higher temperature regime, the mobility drops rapidly with temperature increase. The relationship is proportional to $T^{-1.92}$ as shown in Figure 7 and the slightly heavier dependence than simple phonon scattering process is well known

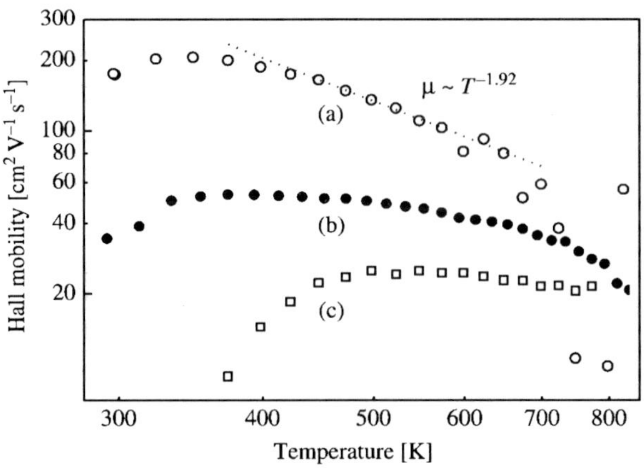

FIG. 7. Temperature dependence of carrier mobility of n-type diamond thin. (a) P: 1×10^{18} cm^{-3}, (b) P: 8×10^{18} cm^{-3} and (c) P: 3×10^{19} cm^{-3}.

to obey $T^{-1.5}$ rule. The other lightly P-doped samples also show strong temperature dependences (T^{-2}–$T^{-2.4}$) of the mobility in the temperature regime over 400 K.

4. Luminescence Characteristics of Phosphorus-doped Diamond Thin Films

Cathodoluminescence (CL) measurement is a well-known technique to characterize defects and impurity effects for diamond with high sensitivities. In the earlier stage of our P-doping experiments, the CL spectra show band-A emission and impurity band emission, or even complicated spectrum explained by donor–acceptor transition theory, as well as characteristic exciton-related emission on P-doped diamond (Sternschulte, Thonke, Sauer and Koizumi, 1999). These were mostly due to poor crystalline perfection and inferior purity control. After 1998, when we started to use metal chamber type CVD system and promoted intensive purity control of doping experiments, luminescence characteristics have been improved drastically. In Figure 8, a representative CL spectrum of the high-quality P-doped diamond thin film is shown. The spectrum was taken at 80 K with 5 keV primary electron beam. In the spectrum, the bound exciton recombination coupled with transverse optical (TO) phonon (BE_{TO}: 239 nm, 5.19 eV) is strongly observed and the other phonon replica (BE_{TO+O}: 246 nm, 5.03 eV, BE_{NP}: 232 nm, 5.34 eV) and free exciton coupled with TO phonon (FE_{TO}: 235 nm, 5.275 eV) are clearly observed. The intensity of the bound exciton recombination luminescence is almost proportional to the P-concentration in the film. With temperature increase, the intensity of BE decreases and the intensity of FE increases due to

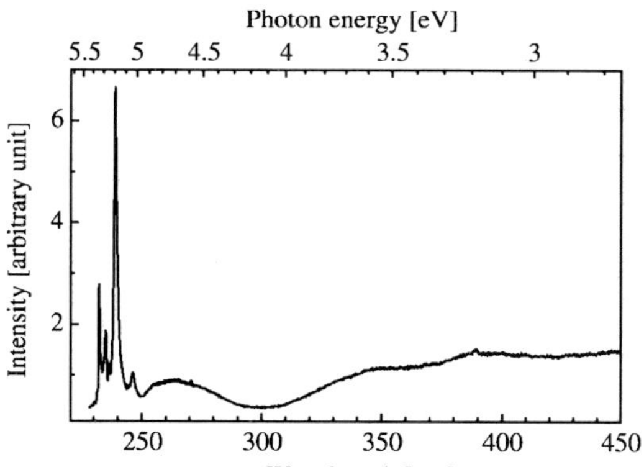

FIG. 8. Typical CL spectrum of P-doped diamond thin films. P: 4×10^{18} cm^{-3} and the carrier mobility is about 50 cm^{-2} V^{-1} s^{-1} at RT. The films thickness is 1 μm. Temperature: 80 K, primary electron beam energy: 10 keV.

dissociation of bound exciton from impurity atoms. The localization energy calculated from the fitting of the temperature dependence of BE intensity was 80 meV. The value coincides well with the energy deviation between FE_{TO} and BE_{TO}. The free exciton recombination luminescence is observed even at RT. At a temperature over 80 K, characteristic band emission sometimes appears at 275 nm (4.5 eV), though the intensity is very weak in the spectrum shown here. The origin of this band is unclear, but it must be related to the P-incorporation in diamond lattice.

5. pn Junction (Koizumi, Watanabe, Hasegawa and Kanda, 2001, 2002)

We have succeeded to grow n-type diamond by doping phosphorus (P) under well-optimized conditions for CVD of diamond thin films as mentioned previously. The wide energy band gap of diamond (5.47 eV) is attractive for the opto-electronic applications because it allows the possibility of UV light emitting devices. Although diamond has an indirect band gap, cathodoluminescence (CL) and photoluminescence (PL) studies have revealed UV luminescence. The UV luminescence is related to exciton recombination from diamond crystals that have high crystalline perfection. The main luminescence peaks appear at 5.27 eV (235 nm) by free exciton recombination with the assistance of TO phonon transition for intrinsic diamond (Dean, Lightowlers and Wight, 1965; Collins, Kamo and Sato, 1989), and at 5.21 eV (238 nm) (Dean et al., 1965; Lawson, Kanda, Kiyota, Tsutsumi and Kawarada, 1995) and at 5.19 eV (239 nm) (Sternschulte et al., 1999) by bound exciton recombination in doped diamond. The energies of these luminescence peaks are considerably higher than those observed from gallium nitride (GaN) at 3.47 eV (357 nm). Recent studies convinced unique characteristics of the free exciton recombination luminescence from high quality diamond that can be observed even at RT by CL (Watanabe et al., 1998) and electroluminescence (Horiuchi, Nakamura and Yamashita, 2000) while the other semiconducting materials exhibit the exciton related luminescence only at very low temperatures. This is considered to be due to large binding energy (90 meV) of free exciton in diamond. Our recent results show the appearance of free exciton and bound exciton recombination emission dominantly from P-doped diamond thin films by CL with very low emission intensity of the other luminescence such as A-band (2.3–2.8 eV) or 4.5 eV band which are caused by defects and impurities (Nakazawa, Tanabe, Tachiki, Kawarada and Koizumi, 2001). Thus, to have deep UV light emitting devices made of diamond is a reality.

5.1. Formation Process of pn Junction

Each B- and P-doped diamond thin film has been grown using individual microwave plasma enhanced CVD systems independently. Metal-chamber-type

CVD systems with the background pressure of 1×10^{-8} Torr were used to avoid incorporation of unexpected impurity such as silicon and nitrogen (see Section 2.3 for the detailed description). Mechanically polished $\{111\}$ surface of boron-doped single crystalline diamond that contains boron atoms over 100 ppm and shows high conductivity was used as a substrate. The size of the substrate is 2×2 mm^2 with 0.5 mm thickness. The formation process of pn structure is as follows: (i) cleaning of substrate surface, (ii) growth of B-doped layer, (iii) cleaning of the sample, (iv) growth of P-doped layer, (v) cleaning of the sample, (vi) formation of ohmic electrodes for both the layers. For steps (i), (iii) and (v), a chemical oxidation process was applied using acid mixture of $HClO_4$, H_2SO_4 and HNO_3 with the ratio of 1 : 3 : 4 at a temperature of 200 °C for 30 min, resulting in the removal of non-diamond adsorbents from the sample surface and the termination of the surface with oxygen in place of hydrogen. Each B- and P-doped layer was grown using trimethylboron (TMB: $(CH_3)_3B$) and PH_3, respectively, as impurity source gases. The experimental conditions for each B- and P-doped diamond growth are summarized in Table II. Finally, the ohmic electrodes are formed for both the p- and n-type layers. For the p-type layer, we deposited a Ti film (thickness 100 nm) with a Au evaporated film (thickness 100 nm) for protection. The films were deposited at 400 °C to make a better ohmic interface. For the n-type layer, the process is almost the same as mentioned in Section 3. The annealing process applied for ion implanted contact formation process does not affect the electrical properties of the p- and n-type layers. In addition, at such annealing temperature we do not expect significant inter-diffusion of impurities at the pn junction interface. The Au/Ti deposited contacts have been made through the mask with 150 μm in diameter on the treated surfaces. Finally, the pn sample was processed by reactive ion etching (RIE). The aluminum vacuum deposited film was formed on the sample to cover electrode area as RIE mask. RIE has been performed by conventional capacitance-coupled sputtering (etching) system operated at 13.56 MHz with the power of 200 W. As reactant gases, CF_4 and O_2 were used with mixing ratio of 1/100

TABLE II

THE EXPERIMENTAL CONDITIONS FOR B- AND P-DOPED DIAMOND GROWTH

	B-doped diamond	P-doped diamond
Source gases	TMB[a], CH_4, H_2	PH_3, CH_4, H_2
CH_4/H_2 (%)	0.1	0.05
TMB, PH_3/CH_4 (ppm)	20–100	500–2000
Total gas pressure (Torr)	80	100
Total gas flow rate (sccm)	500	400
Substrate temperature (°C)	900–950	900–920
Growth duration (h)	3	2

[a]Trimethylboron ($(CH_3)_3B$).

with the total pressure of 2–5 Pa. The etching rate is approximately 100 nm min^{-1}. Aluminum film with the thickness of 250 nm worked well as a protection mask in our etching process up to 2 μm.

5.2. Doping Characteristics

To check the impurity concentrations of each layer and inter-diffusion, SIMS analysis has been performed for the pn structure grown on insulating diamond substrate at the same growth condition (Fig. 9). The impurity concentrations of each B- and P-doped layers are $1-2 \times 10^{17}$ and $7-8 \times 10^{18}$ cm^{-3}, with the thickness of about 1 and 2 μm, respectively. The doping profile is almost uniform for each P- and B-doped layer and there is no significant intermixture region at the pn interface. Although it is not clear that the existence of impurity concentration gradient from the present SIMS result due to poor depth resolution, there is a possibility to have a gradient structure of P concentration near the interface due to the growth process. In the initial growth period of P-doped layer on the B-doped layer by CVD, it takes a few minutes for the impurity concentration to saturate in the gas phase. It will give a gradient structure of P concentration within the thickness of 10–50 nm depending on the gas flow rate during CVD of P-doped layer. The electrical properties of each layer have been characterized by Hall measurements using the layers grown simultaneously on insulating diamond substrates. p- and n-type conductivity have been confirmed for B- and P-doped layers. The activation energies of carrier concentration were 0.37 and 0.59 eV for B- and P-doped layers, respectively. These values are in good agreement with the published data for those impurity-doped diamonds (Yamanaka et al., 1998; Koizumi et al., 2000). Resistivity of B- and P-doped layer is 500 and 10^5 Ω cm, respectively. In case of the present diamond pn junction, most of the operating voltage is applied to n-type layer and it will be consumed for the resistive heating of the layer during forward current operation.

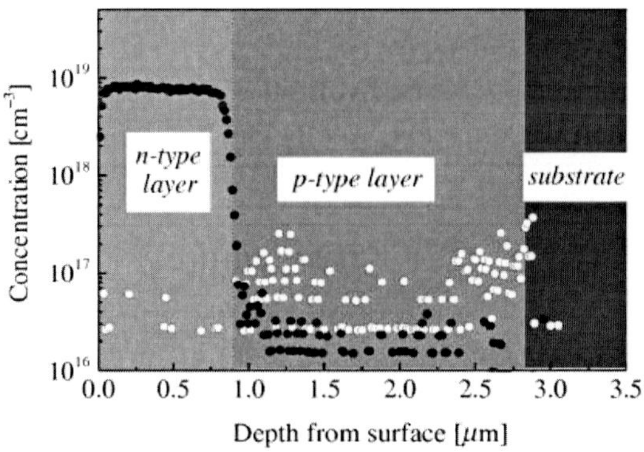

Fig. 9. Typical SIMS depth profile of diamond pn junction: solid circles represent phosphorus (^{31}P) and open circles represent boron (^{11}B). The substrate is non-doped diamond.

Due to the high resistivity, it is unfavorable to operate present pn junction under high-power high-frequency conditions. The carrier mobility of the B- and P-doped layer is 300 and 60 $cm^2\ V^{-1}\ s^{-1}$, respectively, at RT. For the n-type layer with P-concentration of $7-8 \times 10^{18}\ cm^{-3}$, the present value of the mobility is quite reasonable. The mobility observed for the B-doped layer would be the highest value as a {111} epitaxial layer ever reported. These characteristic values are summarized in Table II.

5.3. Electrical Properties of Diamond pn Junction

Current–voltage ($I-V$) characteristics of the pn junction show clear rectifying property (Fig. 10(a)). The $I-V$ data has been acquired over the voltage applied on the electrode of p-type layer. Breakdown was not observed for the reverse voltage over 100 V. The turn-on voltage is about 5 V. The rectification ratio was over eight orders of magnitude when forward (n-type negative) and reverse (n-type positive) voltage of ± 10 V was applied (Fig. 10(b)). The ideal factor was about 3.8 deduced from the slope of a linear portion at around 4 V in Figure 10(b), and that implies poor perfection of the pn junction. A strong deviation of the $I-V$ profile from the exponential relationship over a current of 10^{-6} A and an applied voltage of 5–6 V is due to high series resistance of the sample (mainly of the P-doped layer) of about $10^5\ \Omega$. The resistance related to B-doped layer and the substrate is negligibly small as compared to that of P-doped layer. Capacitance–voltage measurement showed a clear voltage dependent increase of the capacitance in reverse voltage region, convincing the existence of a depletion region and its narrowing with the voltage increase to the forward

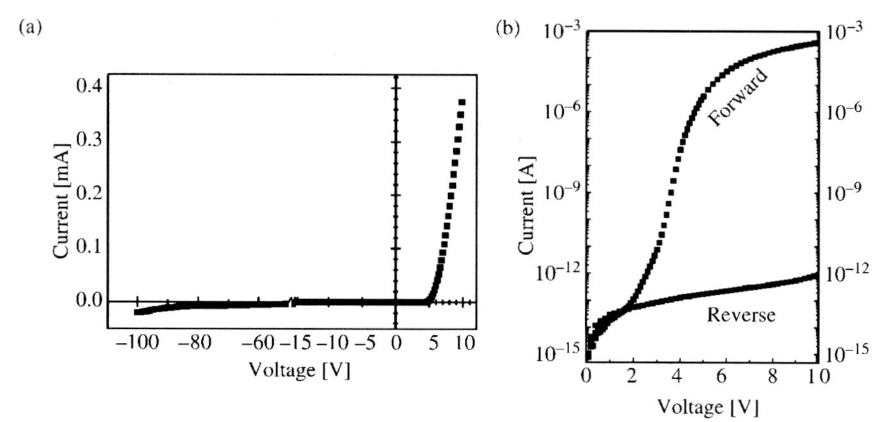

FIG. 10. Representative current–voltage ($I-V$) characteristics of diamond pn junction. In linear plots (a), the 'Voltage [V]' shows the applied voltage to p-type diamond. In semi-logarithmic plots (b), the 'forward' direction corresponds to the case when the positive voltage was applied to p-type diamond.

voltage. The $1/C^2-V$ plots shows linear relationship in most cases and abrupt junction interface is formed. However, occasionally, for specific samples, better linear relationship has been observed for $1/C^3-V$ profile than $1/C^2-V$ profile and this implies the impurity-graded junction has been formed as discussed previously. The built-in potential was about 4.7 V estimated from the profiles. By electron beam induced current (EBIC) analysis, a clear current response has been observed strongly at the pn junction confirming the existence of depletion region at the pn junction (Sekiguchi and Koizumi, accepted for publication). These results convince that the diode characteristics observed from the present sample originated from the pn junction.

5.4. Characterization of Light Emission

For forward current exceeding 100 μA, light emission was observed. In the present structure of samples, light emission was observed around the electrode of the sample because the pn junction and the light emitting area exist behind the electrode and the driving voltage is applied perpendicular to the sample surface (Fig. 11). The gold wires are bonded to each electrode on the n-type layer and the forward bias is applied for several electrodes. The color of the light was blue-violet. Increasing the forward current, we could observe the UV light emission in the spectrum (Fig. 12(a)). The driving voltage and current are 28 V and 10 mA, respectively. For the present sample, above 5 mA, we can see sharp emission peak located at 235 nm (5.27 eV) that is attributed to free exciton recombination luminescence coupled with TO phonon (FE_{TO}). The small peak (shoulder) located at around 243 nm (5.10 eV) is attributed to the phonon replica of FE_{TO}

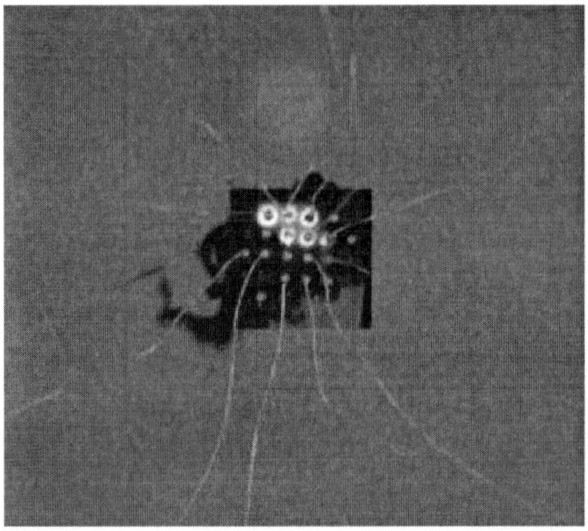

Fig. 11. Visible image of diamond LED under forward biasing. The circular shaped electrodes with their diameters of 150 μm are formed with the separation of 150 μm from each other. Light emission can be seen around the electrodes.

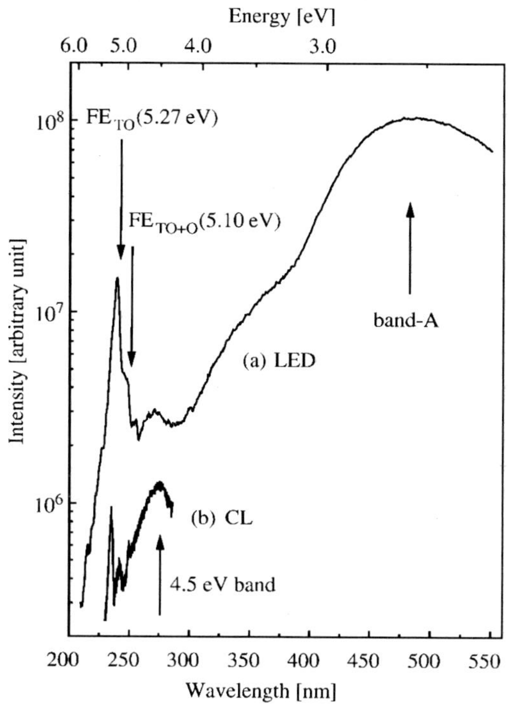

FIG. 12. Optical emission spectra of the pn junction operated at forward current of 10 mA (a). A representative CL spectrum of P-doped diamond thin film taken at RT is shown in (b) for comparison.

peak. Also observed are broadband emissions at an UV region with peak energy located at 4.5 eV and a broadband emission in the visible region. The origin of a 4.5 eV band emission is not very clear yet, but it is often observed for B- and P-doped CVD diamond thin films by CL and PL analysis (Watanabe et al., 1998; Sternschulte et al., 1999). The band emission in the visible region is attributed to band-A emission that is often observed by CL and PL from CVD diamond thin films. The blue-violet color of the light is due to the band-A emission. The intensity ratio of the free exciton peak against the band-A was 1 : 7. The appearances of the free exciton and the 4.5 eV band emissions are very well consistent with RT CL spectrum observed from P-doped diamond thin film (Fig. 12(b)). This result supports high crystalline perfection of the P-doped layer prepared in the present study. In our series of B-doping experiments for {111} diamond thin films, the quality of the B-doped layer was not so good that exciton peak bound to boron was negligibly weak even at 110 K. The visible lights such as band-A luminescence observed in the spectrum are most likely emitted from B-doped layer. It is required to improve the crystalline perfection of the B-doped layer to get the diamond LED that emits monochromatic UV light of 235 nm.

6. Summary

n-Type diamond growth was one of the most difficult theme in the diamond research field. We have succeeded to obtain n-type diamond by doping phosphorus during CVD diamond growth on {111} diamond surface under well-optimized growth condition for high crystalline perfection diamond thin films. The surface of P-doped diamond surface is smooth with a characteristic morphology that convinces step-flow-mode growth is proceeding. Doping efficiency of P atoms was 3–4%. From RT to over 600 °C, the P-doped diamond thin films shown clear n-type characteristics as confirmed by Hall measurements. At present, the maximum doping concentration is about $5 \times 10^{19}\ \mathrm{cm}^{-3}$, and over the concentration of $10^{19}\ \mathrm{cm}^{-3}$, hopping conductivity is observed at RT as dominant carrier transportation mechanism. For lower P-concentration films, clear thermal activation of the carriers has been observed and the activation energy was 0.6 eV which is consistent with the other characterization results obtained from FTIR and photocurrent measurements. Even for the heavily doped films, the thermally activated characteristic has been observed at high temperatures nominally over 400 K. The compensation ratio is as low as 1.5%. Characteristic luminescence spectrum has been obtained by CL analysis which attributes bound exciton recombination related to P atoms. At RT, the free-exciton recombination has been observed that convinces high crystalline perfection of P-doped diamond thin films grown in the present study. A clear diode characteristic has been observed from pn junction formed with the present P-doped diamond thin film grown on B-doped p-type diamond thin film. The rectification ratio was 10^8–10^{10} at RT. The light emission was observed with the diode operated with forward biasing of around 25 V. The spectrum showed the appearance of sharp UV light emission at 235 nm (5.27 eV) that attributes free exciton recombination coupled with TO phonon. The spectrum was also involved with broadband emissions caused by impurity and crystalline defects. To get monochromatic UV light emission from diamond pn junction, it is required to improve the crystalline perfection of boron-doped films.

References

A. E. Alexenko and B. V. Spitsyn, *Diamond Relat. Mater.* 1, 705 (1992).

A. B. Anderson and S. P. Mehandru, *Phys. Rev. B* 48(7), 4423–4427 (1993).

S. Bohrs, R. Haubner, and B. Lux, *Diamond Relat. Mater.* 4(2), 133–144 (1995).

C. F. Chen, S. F. Lo, and S. H. Chen, *Diamond Relat. Mater.* 5(6–8), 766–770 (1996).

A. T. Collins, M. Kamo, and Y. Sato, *J. Phys. Condens. Matter.* 1, 4029 (1989).

P. J. Dean, E. C. Lightowlers, and D. R. Wight, *Phys. Rev.* 140, A352 (1965).

J. A. Garrido, C. E. Nebel, M. Stutzmann, E. Gheeraert, N. Casanova, and E. Bustarret, *Phys. Rev. B* 65(16) (2002) (art. no. 165409).

H. Horiuchi, K. Nakamura, and S. Yamashita, *Jpn. J. Appl. Phys.* 39, L604 (2000).

S. A. Kajihara, A. Antonelli, and J. Bernholc, *Physica B* 185(1–4), 144–149 (1993).

R. Kalish, A. Reznik, C. Uzan-Saguy, and C. Cytermann, *Appl. Phys. Lett.* 76(6), 757–759 (2000).

M. Kamo, H. Yurimoto, T. Ando, and Y. Sato, in "Proceedings of the Second International Conference on New Diamond Science and Technology" (R. Messier, J. T. Glass, J. E. Butler, and R. Roy, Eds.), p. 637. Materials Research Society, Pittsburgh, 1991.

S. Koizumi, M. Kamo, Y. Sato, S. Mita, A. Sawabe, A. Reznik, C. Uzan-Saguy, and R. Kalish, *Diamond Relat. Mater.* 7(2–5), 540–544 (1998).

S. Koizumi, H. Ozaki, M. Kamo, Y. Sato, and T. Inuzuka, *Appl. Phys. Lett.* 71, 1064 (1997).

S. Koizumi, T. Teraji, and H. Kanda, *Diamond Relat. Mater.* 9, 935 (2000).

S. Koizumi, K. Watanabe, M. Hasegawa, and H. Kanda, *Science* 292, 1899 (2001).

S. Koizumi, K. Watanabe, M. Hasegawa, and H. Kanda, *Diamond Relat. Mater.* 11(3–6), 307–311 (2002).

S. C. Lawson, H. Kanda, H. Kiyota, T. Tsutsumi, and H. Kawarada, *J. Appl. Phys.* 77, 1729 (1995).

K. Nakazawa, K. Tanabe, M. Tachiki, H. Kawarada, and S. Koizumi, *Phys. Rev. B* 64(23) (2001) (art no. 235203).

K. Okano, H. Kiyota, T. Iwasaki, Y. Nakamura, Y. Akiba, T. Kurosu, M. Iida, and T. Nakamura, *Appl. Phys. A* 51, 344 (1990).

I. Sakaguchi, M. N. Gamo, Y. Kikuchi, E. Yasu, H. Haneda, T. Suzuki, and T. Ando, *Phys. Rev. B* 60, R2139 (1999).

S. N. Schauer, J. R. Flemish, R. Wittstruck, M. I. Landstrass, and M. A. Plano, *Appl. Phys. Lett.* 64(9), 1094–1096 (1994).

T. Sekiguchi and S. Koizumi, *Appl. Phys. Lett.* 81(11), 1987–1989 (2002).

H. Sternschulte, K. Thonke, R. Sauer, and S. Koizumi, *Phys. Rev. B* 59, 12924 (1999).

T. Teraji, M. Katagiri, S. Koizumi, T. Ho, and H. Kanda, *Jpn. J. Appl. Phys.* 42(8A), L882–L884 (2003).

T. Teraji, S. Koizumi, and H. Kanda, *Appl. Phys. Lett.* 76(10), 1303–1305 (2000).

T. Teraji, S. Koizumi, S. Mita, A. Sawabe, and H. Kanda, *Jpn. J. Appl. Phys.* 38, L1096–L1098 (1999).

H. Watanabe, K. Hayashi, D. Takeuchi, S. Yamanaka, H. Okushi, K. Kajimura, and T. Sekiguchi, *Appl. Phys. Lett.* 73(7), 981–983 (1998).

S. Yamanaka, H. Watanabe, S. Masai, D. Takeuchi, H. Okushi, and K. Kajimura, *Jpn. J. Appl. Phys.* 37(10A), L1129–L1131 (1998).

This page is intentionally left blank

Chapter 6

TRANSPORT AND DEFECT PROPERTIES OF INTRINSIC AND BORON-DOPED DIAMOND

C. E. Nebel

Walter Schottky Institut, TU-München, Am Coulombwall, D-85748 Garching, Germany

1. Introduction

The electronic properties of diamond are determined by intrinsic defects, and extrinsic contaminants or dopants, most prominently nitrogen, silicon, boron and phosphorus, as well as different structural morphologies. Natural and high pressure high temperature (HPHT) synthetic diamonds are single crystalline wide band-gap semiconductors with a 5.48 eV indirect band gap at 300 K (Collins, Lawson, Davies and Kanda, 1990; Mainwood 1994a,b) and well understood electronic properties such as high electron ($4500\ \text{cm}^2\ \text{V}^{-1}\ \text{s}^{-1}$) and hole mobilities ($3800\ \text{cm}^2\ \text{V}^{-1}\ \text{s}^{-1}$) (Isberg, Hammersberg, Johansson, Wikström, Twitchen, Whitehead, Coe and Scarsbro, 2002), high carrier saturation drift velocity ($2.7 \times 10^7\ \text{cm s}^{-1}$) (Ferry, 1975; Nebel and Stutzmann, 2001; Nebel, 2003), high electric breakdown field ($10^7\ \text{V}^{-1}\ \text{cm}^{-1}$) (Collins, 1994a,b), small dielectricity (5.7) (Mainwood 1994a,b), and high surface acoustic wave velocity ($1.2 \times 10^4\ \text{m}^{-1}\ \text{s}^{-1}$) (Fox and Dreifus, 2001). Diamond is mostly grown hetero-epitaxially on lattice-mismatched substrates such as silicon (Jiang and Jia, 1995), SiC (Kawarada, Suesada and Nagasawa, 1995), boron nitride (Polo, Sanchez, Wang, Esteve and Andujar, 1997) and a variety of metals by chemical vapor deposition (CVD) techniques that results in polycrystalline morphologies. Under certain optimized nucleation and growth conditions the grains grow highly (100)-oriented (HOD) (Jiang, Klages, Zachai, Hartweg and Füßer, 1993; Locher, Wild, Herres, Behr and Koidl, 1994; Jiang et al., 1998), polycrystalline (PCD) with statistically oriented grains (Wild, Kohl, Herres, Müller-Sebert and Koidl, 1994) on silicon or even single-crystalline (Schreck, Hörmann, Roll, Lindner and Stritzker, 2001). In these films, bulk diamond properties and grain boundary effects have to be taken into account for an understanding of the electronic properties. In some cases, homoepitaxy is applied to grow high quality intrinsic or doped monocrystalline CVD layers (Hayashi,Yamanaka, Watanabe, Sekiguchi, Okishi and Kajimura, 1998). The properties of these films resemble characteristics of natural or HPHT synthetic diamond, as no grain boundaries are present (Isberg et al., 2002).

ISBN: 0-12-752185-2
ISSN: 0080-8784

To characterize CVD diamond we have applied a variety of experiments. Primary techniques for transport properties are temperature-dependent dark conductivity, Hall effect, transient- and spectrally resolved photoconductivity (SPC) measurements. Optical investigations have been applied using transmission/reflectivity experiments and photo-thermal deflection spectroscopy (PDS) (Jackson and Amer, 1982). To identify nitrogen and carbon defects, electron paramagnetic resonance (EPR) techniques are used. On boron-doped diamond, capacitance–voltage (*CV*) measurements, deep-level transient spectroscopy (DLTS), optically excited DLTS techniques and SPC experiments have been used to characterize the acceptor properties of boron and defects as well as the interaction of boron with deuterium.

2. Electronic Properties of CVD Diamond

2.1. Experimental Details

The diamond films investigated and discussed in this chapter were grown hetero-epitaxially on silicon wafers by microwave plasma assisted CVD, using methane and hydrogen as source gases. Polycrystalline diamond (PCD) films with statistical orientation of grains are grown in pure CH_4/H_2 gas mixtures whereas highly oriented diamond (HOD) layers with (100) faceted columnar texture are deposited by adding well-defined amounts of nitrogen to the CH_4/H_2 gas mixtures (Füßer, Rösler, Hartweg, Zachai, Jiang and Klages, 1993; Jiang et al., 1993). From Raman experiments on high quality CVD diamond films we know the volume fraction non-diamond (N_{ND}) relative to diamond (N_D) components which is typically in the range of $N_{ND}/N_D \approx (2–8) \times 10^{-3}$.

After deposition, the silicon substrates have been removed by etching in hydrofluoric acids. The freestanding diamond films were cleaned in a solution of ($CrO_3 + H_2SO_4$) at 150 °C for 30 min or in oxygen plasma to remove graphitic carbon and to obtain oxygen termination. For transport experiments we evaporated coplanar interdigital Ti (200 Å)/Pt (100 Å)/Au (3000 Å) contacts on the growth side of the films. To achieve ohmic properties the films were annealed at 500 °C for 10 min. For transient photoconductivity experiments where Schottky contact properties are required semitransparent (25% at 220 nm) Cr/Au layers were evaporated in sandwich geometry on the substrate and growth side of the films.

2.2. Dark Conductivity and Majority Carriers

No truly intrinsic diamond with conductivity activation energies in the range of 2.5 eV expected for the large band gap of $\cong$ 5.48 eV of diamond is available. At room temperature such activation energy causes a carrier density much less than one electron per 10^{15} cm^3 diamond. In real diamond defects, extrinsic contaminants, grain boundary or surface effects govern the electronic properties and generate activation energies significantly smaller than 2.5 eV.

Type Ib diamonds, that containing up to 300 ppm single substitutional nitrogen, have an optical absorption threshold of about 1.7 eV (Farrer, 1969), which is attributed to the photo-ionization of substitutional nitrogen, acting as a deep donor. It is, therefore, assumed that the dark conductivity in Ib-diamond is n-type. Typical dark resistivities measured in single crystal Ib-(Borst and Weis, 1996), IIa-(Collins, 1994a,b, p. 284) and PCD-diamond are shown in Figure 1. At sufficiently high temperatures ($T > 400$ K), the resistivity in Ib and IIa is activated with 1.7 eV, the thermal ionization energy of substitutional nitrogen. The resistivity in type IIa diamond ('non-semiconducting diamond') is about four orders of magnitude higher, but shows the same activation energy as type Ib (Fig. 1). Obviously, the small amount of single substitutional nitrogen, even present in IIa diamond, is sufficient to dominate conductivity. However, this result cannot be generalized for type IIa diamond. Redfield (1954), for example, detected by photo-Hall experiments in several highly resistive layers (comparable to type IIa diamond) a positive Hall-voltage, which indicates that at least the photo-response of such diamonds can be dominated by holes. Figure 1 also shows typical dark resistivities of high quality intrinsic CVD PCDs. The conductivity is activated with increasing energies towards the higher temperatures, approaching 1–1.7 eV at the highest applied temperatures, where the resistivity is in between type IIa and Ib diamond. At lower temperatures the resistivity of PCD diamond is

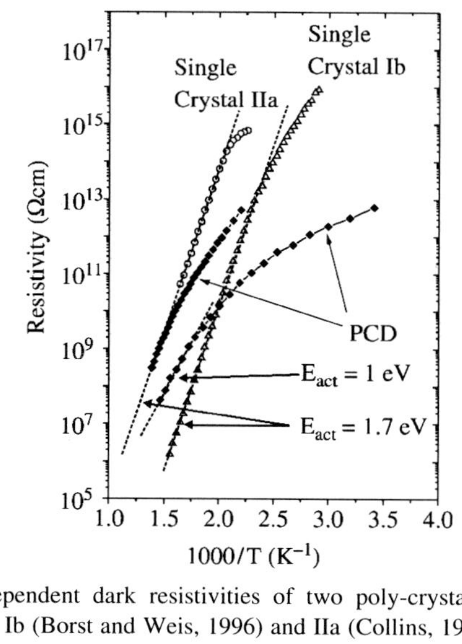

FIG. 1. Temperature dependent dark resistivities of two poly-crystalline CVD diamond films (PCD), of a single crystal Ib (Borst and Weis, 1996) and IIa (Collins, 1994) diamond.

several orders of magnitude lower than in single crystal diamond. This is attributed to conduction in grain boundaries.

To our knowledge, no Hall- or thermo-power data on undoped CVD diamond are available in the literature as these films are too resistive and too thermally conductive (20 W cm^{-1} K^{-1}, Hartmann and Reichling, 2001) for such experiments. Alternatively, we have characterized contact properties by means of transient photoconductivity experiments (Nebel, Münz, Rohrer, Graeff, Stutzmann, Zachai and Güttler, 1996; Nebel, Münz and Stutzmann, 1997), to determine if n- or p-type conductivity governs the properties of intrinsic CVD diamond. The set-up used for these experiments is schematically shown in Figure 2.

The band bending at a metal/semiconductor interface arises from space charges that are present at the interface. The potential barrier in the space charge layer extends over a fraction of the sample thickness. Its spatial width depends on the type of doping (holes or electrons), on the doping density and defect properties of the semiconductor. A Schottky barrier in a n-type semiconductor gives rise to electron depletion and in a p-type semiconductor to hole depletion, respectively.

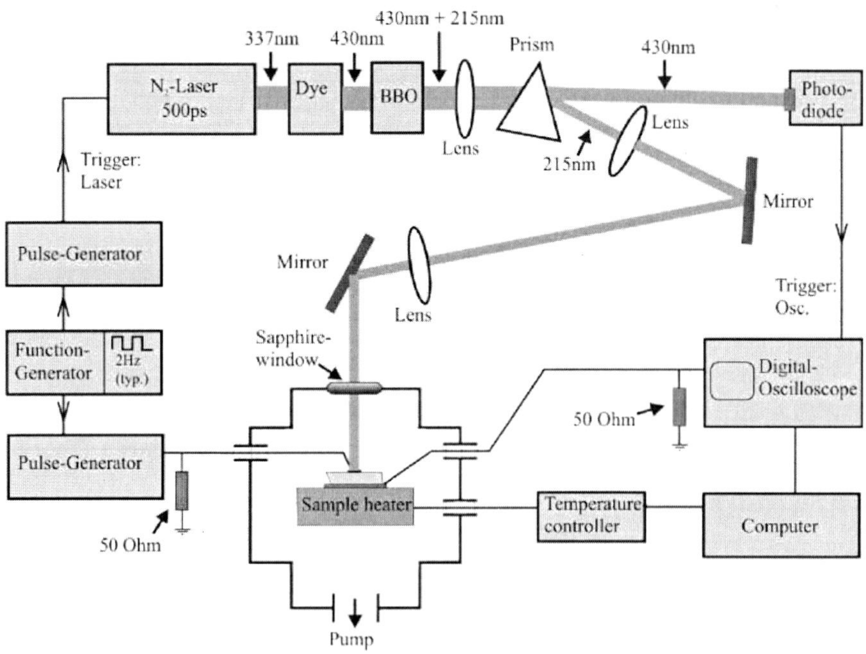

FIG. 2. Time-of-flight set-up. A N_2-pumped dye laser with a frequency-doubling BBO crystal is used as the light source. The sample is placed in a cryostat and the field is applied by a pulse generator. The laser emission and pulse generator are triggered by a function generator. The signal is recorded by a fast digital oscilloscope. The set-up is controlled by a computer.

Transient photoconductivity experiments can be applied to explore such space charge properties (Street, 1983). Here, a short strongly absorbed light pulse of 500 ps duration excites a number of mobile carriers in the depletion layer where carriers drift in the built-in electric field. A charge, Q, is induced on the opposite electrodes of magnitude:

$$Q = N_0 q \frac{\bar{x}}{d} \tag{1}$$

where $\bar{x}$ is the average distance photo-generated carriers drift in the space-charge field, d is the sample thickness, q is the elementary charge and N_0 the number of photo-excited carriers. In practice, the photocurrent is the measured quantity and the collected charge Q is obtained by numerical integration. At zero bias, the drift of carriers is due to the built-in field, $\vec{F}_0(x)$, in the depletion layer. Application of an external DC voltage modifies the built-in field, thus establishing an effective field $\vec{F}_{\text{eff}}$. The photocurrent generated in a non-uniform field is

$$\vec{j}(t) = q(n\mu_n + p\mu_\text{p})\vec{F}_{\text{eff}}(t) \tag{2}$$

where $\vec{F}_{\text{eff}}(t)$ is the field experienced by the charge at time t after the excitation light pulse, and μ is the mobility of electrons (n) and holes (p). From the sign of the current (charge) one obtains the type of band bending that is a depletion layer either for electrons (n-type semiconductor) or for holes (p-type semiconductor).

Figure 3 shows typical charge collection data (Q) measured on a Cr/Au-poly-crystalline CVD diamond interface. Q is negative at zero bias, indicating an electron depletion layer. This is detected on all our HOD and PCD films. We, therefore, state, that high quality poly-crystalline CVD diamond is n-type. We assume that the small amount of substitutional nitrogen is enough to shift the Fermi level towards the conduction band, giving rise to n-type properties.

Application of forward voltages (U) reduces the built-in potential at the interface. Q decreases and finally becomes zero for $U = +11.3$ V. Such a high

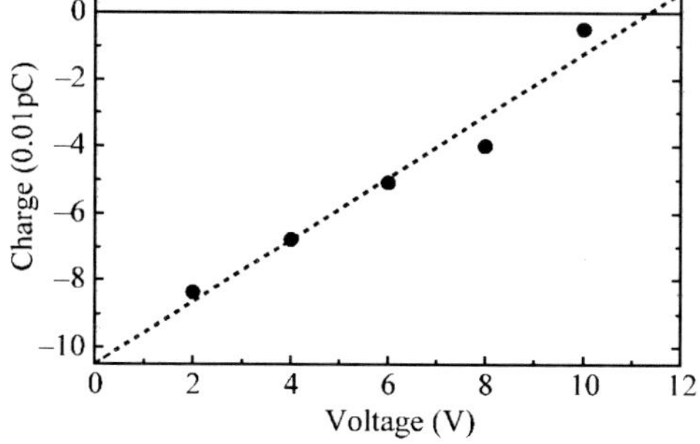

FIG. 3. Charge collection data measured in the depletion layer of a Cr contact evaporated on CVD PCD diamond.

voltage is required to empty deep traps by electric field enhanced emission to establish flat band conditions.

2.3. Transport Properties of Electrons and Holes

2.3.1. *Hole Mobility*

In the literature, the mobility of holes in boron-doped CVD diamond films has been determined by a combination of conductivity and Hall effect data. Figure 4 summarizes the results measured at room temperature on boron-doped natural and synthetic (Grot, Hatfield, Gildenblat, Bazian and Badzian, 1991) (open squares), homo-epitaxial grown single crystal CVD diamond (Takeuchi, Yamanaka, Watanabe and Okushi, 2001) (open circles), hetero-epitaxially grown HOD (Stoner, Kao, Malta and Glass, 1993; Boettger, 1995) (full diamonds and dots) and poly-crystalline diamond (Huang, Guo, Hwang and Chang, 1996) (full squares) films. The single crystalline diamond films were deposited on HPHT synthetic Ib (001) diamond substrates by means of microwave-plasma CVD (Takeushi et al., 2001). Mobilities of up to $1800\,\mathrm{cm}^2\,\mathrm{V}^{-1}\,\mathrm{s}^{-1}$ are reproducibly obtained for hole densities smaller than $10^{15}\,\mathrm{cm}^{-3}$ (open circles in Figure 4). There seems to be an upper limit to μ_h in boron-doped diamond. The scattering of data is attributed to residual impurities in these films, such as Fe and Mo. At hole densities larger than $10^{15}\,\mathrm{cm}^{-3}$ ionized impurity scattering at boron acceptors becomes dominant, causing a decrease in mobility that follows a $\mu_\mathrm{h} \approx 1/p$ law (dashed line in Figure 4) (Seeger, 1999), where p is the density of holes that is proportional to the density of ionized acceptors (N_A).

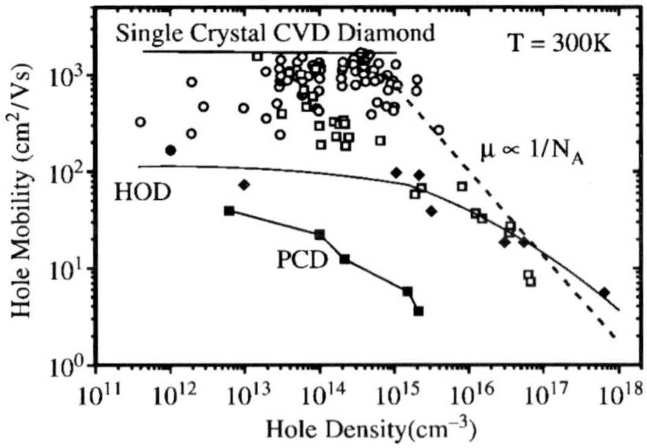

FIG. 4. Hole mobilities measured as a function of hole densities at room temperature on natural and synthetic (□ Grot et al., 1991), on homo-epitaxially grown single crystal (○ Takeuchi et al., 2001), on poly-crystalline CVD diamond grown hetero-epitaxially on silicon either highly (HOD) (◆ Boettger, 1995, ● Stoner et al., 1993]), or randomly oriented (PCD) (■ Huang et al., 1996). The full lines are guide to the eyes and the dashed line shows the expected variation assuming scattering at ionized impurities.

In PCD the mobilities are 10–50 times smaller, varying between 3 and 40 $cm^2 V^{-1} s^{-1}$, and in HOD diamond μ_h is in the range of 70–165 $cm^2 V^{-1} s^{-1}$ at $T = 300$ K. These significantly lower mobilities are generated by grain boundaries, where the effective mobility μ_{eff} can be calculated as demonstrated by Orton and Powell (1980)

$$\mu_{eff} = \mu_B \exp\left[-\frac{E_{GB}}{kT}\right] \tag{3}$$

where μ_B is the mobility in the grains, E_{GB} the energy barrier at grain boundaries, k the Boltzmann constant and T the temperature. Taking into account typical bulk mobilities of single crystal diamond ($\mu_B = 200-1800$ $cm^2 V^{-1} s^{-1}$) and the effective mobility in PCD ($\mu_{eff} \cong 40-165$ $cm^2 V^{-1} s^{-1}$), results in grain boundary energy barriers E_{GB} of 5–100 meV.

Temperature dependent variations of μ_h in a natural type IIb (Dean, Lightowlers and Wight, 1965) and a homo-epitaxially grown CVD diamond ($N_A = 6 \times 10^{17}$ cm^{-3}) (Yamanaka, Watanabe, Masai, Takeuchi, Okushi and Kajimura, 1998) are compared in Figure 5. The mobility in the CVD film increases from 1200 $cm^2 V^{-1} s^{-1}$ at 340 K to 3400 at 170 K. This is in good overall agreement with the data of the natural type IIb diamond. Below 400 K, μ_h exhibits the $T^{-3/2}$ temperature dependence expected for acoustic phonon scattering (Seeger, 1999). Above 400 K, the slope becomes steeper, showing a T^{-s} dependence with $s = (2.7-2.9)$, which is a result of optical phonon scattering (Reggiani, Bosi, Canali, Nava and Kozlov, 1981). Recently, it has been shown that holes in intrinsic diamond propagate with even

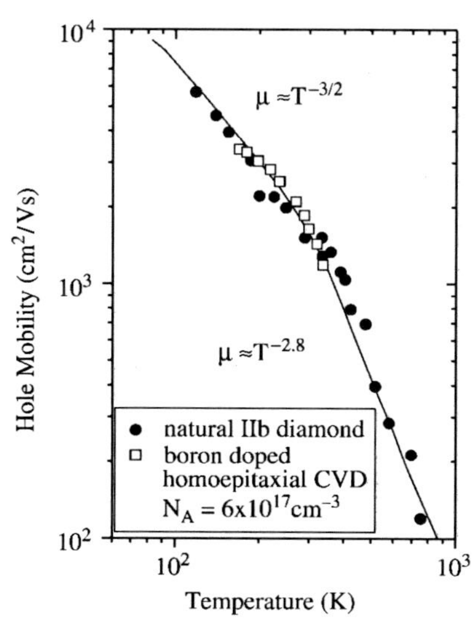

FIG. 5. Temperature dependent variation of the hole mobility in natural IIb-diamond (● Dean et al., 1965) and in homo-epitaxially grown single crystal CVD diamond (□ Yamanaka et al., 1998).

$3800\ cm^2\ V^{-1}\ s^{-1}$ at room temperature (Isberg et al., 2002). Up to now, this is the highest mobility of holes in a semiconductor (see, for example: Si: $450\ cm^2\ V^{-1}\ s^{-1}$, Ge: $1900\ cm^2\ V^{-1}\ s^{-1}$, GaAs: $400\ cm^2\ V^{-1}\ s^{-1}$, GaN: $10-30\ cm^2\ V^{-1}\ s^{-1}$ (Sze, 1981)).

2.3.2. *Electron Mobility*

In 1997, n-type doping of diamond by phosphorus was achieved (Koizumi, Kamo, Sato, Ozaki and Inuzuka, 1997; Koizumi, Kamo, Sato, Mita, Sawabe, Reznik, Uzan-Saguy and Kalish, 1998). The donor level is about 590 meV below the conduction band minimum. Typical electron mobilities measured by Hall effect are shown in Figure 6. At room temperature the mobility is in the range of $70\ cm^2\ V^{-1}\ s^{-1}$ for $6 \times 10^{18}\ cm^{-3}$ donors and $240\ cm^2\ V^{-1}\ s^{-1}$ for $3 \times 10^{18}\ cm^{-3}$ phosphorus, respectively. The temperature dependence above 300 K varies according to $\mu \propto T^{-1}$. At lower temperatures a steep decrease is detected. For comparison, scattering at ionized impurities ($\mu \propto T^{1.5}$) is also drawn in (dashed line), which does not fit the experimentally detected data at all. Further experiments and growth optimizations are required to elucidate the electron transport properties in phosphorus-doped diamond.

Carrier mobilities in nominally undoped polycrystalline CVD diamond have been characterized by transient photoconductivity experiments in coplanar or sandwich contact geometry. Typical results are shown in Figure 7 for three polycrystalline CVD layers (Plano, Landstrass, Pan, Han, Kania, McWilliams and Ager, 1993). In these experiments, both electrons and holes contribute to the transient photocurrent, so that the result is a combined mobility. With increasing

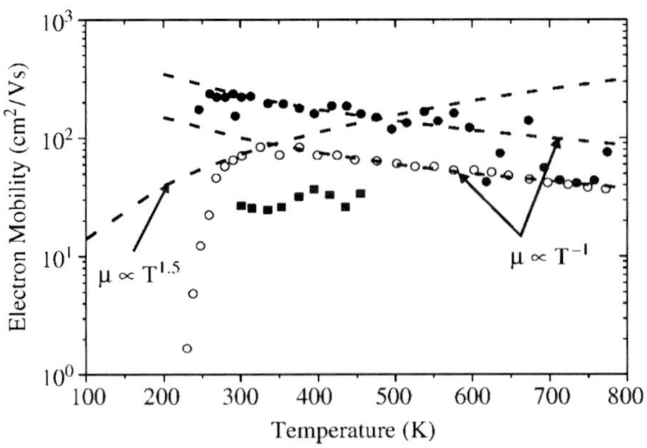

FIG. 6. Temperature-dependent electron mobilities in phosphorus doped homo-epitaxially grown CVD diamond with phosphorus to carbon ratios: ■ P/C = 100 ppm, ○ P/C = 1000 ppm, ● P/C = 500 ppm (data from Koizumi et al., 1997, 1998). The dashed lines represent (i) scattering at ionized impurities ($\mu \propto T^{1.5}$) and (ii) a best fit to the data ($\mu \propto T^{-1}$).

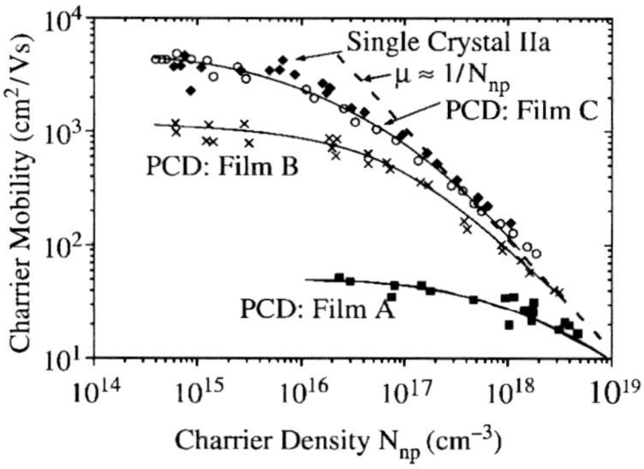

FIG. 7. Combined carrier mobilities measured by transient photoconductivity experiments on different quality PCD films and in a single crystal IIa-diamond film (data from Plano et al., 1993). The full lines are guide to the eyes, the dashed line shows the variation expected for scattering at ionized impurities.

film quality the transport properties of CVD diamond approaches those of natural type IIa diamond. At low carrier density the combined mobility is about 4000 $cm^2 V^{-1} s^{-1}$, which is comparable to single crystal IIa natural diamond. The improvement in electrical behavior from film A to C correlates with increased grain size (ranging from 3 to 6 μm (A), 10 to 12 μm (B) and 14 to 16 μm (C)), higher growth temperatures, and improved purity of the diamond films. Details can be found in Pan, Kania, Han, Ager, Landstrass, Landen and Pianetta (1992) and Plano et al. (1993). With increasing carrier density, N_{np}, the mobility decreases, approaching a $\mu \propto 1/N_{np}$ characteristic. At low densities the mobility saturates and becomes independent of N_{np}. The values for the mobility at low excitation densities ($\cong$ 4000 $cm^2 V^{-1} s^{-1}$) are close to recently published results of 4500 $cm^2 V^{-1} s^{-1}$ for electrons, measured by time-of-flight experiments in intrinsic monocrystalline CVD diamond (Isberg et al., 2002). Obviously, electrons propagate in diamond with unmatched mobilities (see, for example, Si: 1500 $cm^2 V^{-1} s^{-1}$, Ge: 3900 $cm^2 V^{-1} s^{-1}$, GaN: 900 $cm^2 V^{-1} s^{-1}$ (Sze, 1981)). Only GaAs with 8500 $cm^2 V^{-1} s^{-1}$ is better.

2.3.3. *Carrier Trapping and Lifetimes*

The lifetime of photo-generated carriers in the conduction and valence band is surprisingly short as shown in Figure 8 measured in two high quality polycrystalline CVD layers (Bergonzo, Tromson, Mer, Guizard, Foulon and Brambilla, 2001). Ultra fast detections of transient currents reveal that the initial decay is well described by an exponential time dependence with time constants in the range 340–550 ps. In some samples (upper curve in Figure 8) the decay turns into a slower relaxation. This is shown in Figure 9, where a $I_{ph}(t) \propto t^{-1}$ variation governs the long time decay (Nebel, Stutzmann, Lacher, Koidl and Zachai, 1998).

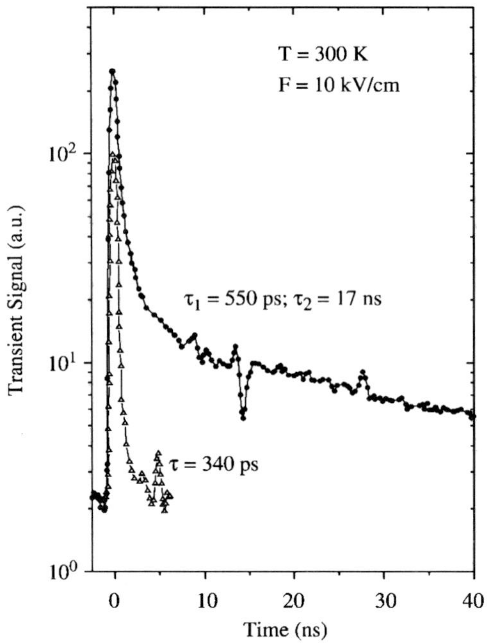

FIG. 8. Typical transient current decays in high quality PCD films (from Bergonzo et al., 2001). The initial decays follow an exponential time dependence with constants of 340 and 550 ps. In one of these films (●) the decay turns into a slow relaxation with a time constant of 17 ns.

We have calculated the density of traps, N_T, that generates the fast initial decay by (Street, 1982)

$$N_T = \frac{1}{\tau v_{th} \beta_{cap}} \tag{4}$$

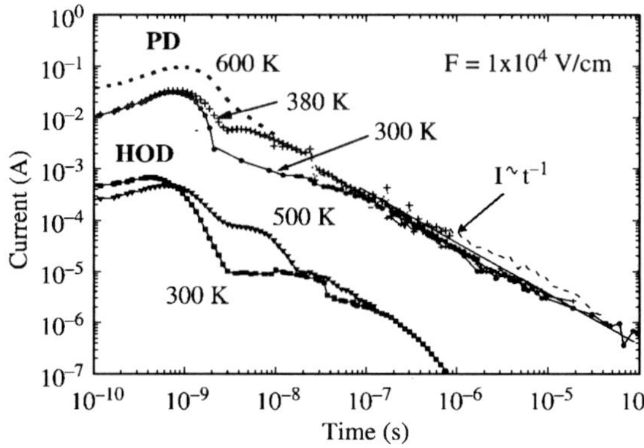

FIG. 9. Transient photocurrents measured in PCD and HOD films at different temperatures and the indicated electric field. The full line represents the slope of a t^{-1} decay.

where τ is the deep trapping lifetime ($\cong$ 340–550 ps), v_{th} the thermal velocity of carriers ($\cong 10^7$ cm s^{-1}) and β_{cap} the capture cross-section. For neutral traps we assume $\beta_{cap} \cong 10^{-15}$ cm^2 (Street, 1982). Based on these parameters a trap density of about 10^{17} cm^{-3} is calculated. This is the density of trapping centers most likely present at grain boundaries.

The decay in the time regime 1 ns to 100 μs is investigated by standard transient photo-excitation experiments where a pulsed laser illumination of 500 ps FWHM generates the electron–hole pairs (Nebel et al., 1997). As shown in Figure 9, photocurrents in PD turn into a long lasting decay that is proportional t^{-1}.

This characteristic time dependence indicates a hopping-down current decay in a continuum of localized states as proposed by Robertson and O'Reilly (1987) and Robertson (1992) for amorphous carbon. Such states are present at grain boundaries, which will be shown later by PDS and persistent photo current data also. For this case, the transient current can be described by Nebel, Street, Johnson and Tsai (1992) and Nebel and Street (1992)):

$$I_{ph}(t) = n(t)\mu(t)qFA \quad (5)$$

where $n(t)$ is the carrier density, $\mu(t)$ the mobility of the center of charge, q the elementary charge, F the applied electric field and A the illuminated area. Both, carrier density and mobility can be time dependent. After generation of carriers by a pulsed laser illumination, electrons and holes rapidly thermalize into localized states as shown in Figure 10. The center of charge follows the demarcation energy, E_D, around which carriers are distributed within a narrow

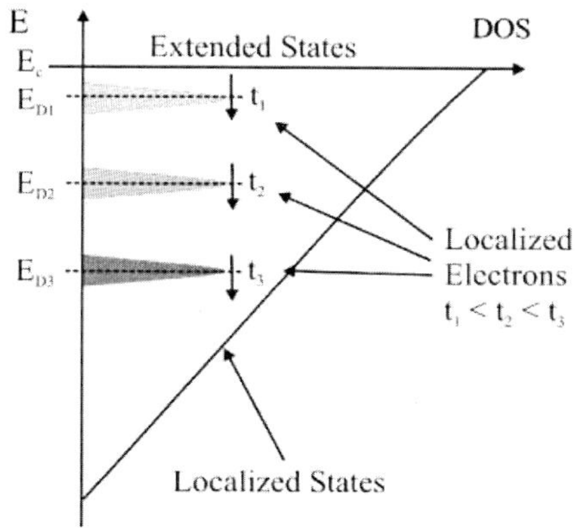

FIG. 10. Schematics of carrier thermalization in the hopping-down regime. After carrier generation by a short laser pulse, they get trapped fast in the density of localized states. In a large density of localized state distribution, carriers propagate by hopping. The center of charge shifts deeper into the DOS with increasing time. Carriers are distributed within a narrow width around the demarcation energy E_D.

energy range ($\cong kT$). E_D is obtained by equating the hopping rate, $\nu(E_D)$, to the measurement time, t by

$$\nu(E_D)t = 1 \tag{6}$$

with

$$\nu(E_D) = \nu_0 \exp\left\{-\frac{2r(E_D)}{\alpha(E_D)}\right\} \tag{7}$$

where ν_0 is the Raman or attempt-to-escape frequency, which is typically in the range of 10^{12}–$10^{13}\ \mathrm{s}^{-1}$, $r(E_D)$ is the hopping or tunneling distance at the demarcation energy and $\alpha(E_D)$ the localization length at E_D.

The localization length α is energy dependent and may be described by

$$\alpha(E) = \frac{\hbar}{[2m_{\mathrm{eff}}(E_c - E)]^{1/2}} \tag{8}$$

where m_{eff} is the effective mass of a carrier and E_c is the energy which separates delocalized from localized states. For carriers, trapped deep in the distribution of localized states ($E_c - E > 200$ meV), α is a weakly varying function of energy and typically in the range of (2–10) Å, dependent on the effective mass, m_{eff}. In the following, we assume that α is constant. The propagation of carriers at E_D is governed by a diffusion-like mechanism. From the Einstein relation, given by

$$\mu = \frac{qD}{kT} \tag{9}$$

where k is the Boltzmann constant, and T the temperature, the mobility is related to the diffusion constant, D, by:

$$D = \frac{\nu(E_D)r^2(E_D)}{6} \tag{10}$$

The hopping mobility, μ_{hop}, is, therefore:

$$\mu_{\mathrm{hop}}(E_D(t)) = \frac{1}{6}\frac{q}{kT}\{r(E_D(t))\}^2\frac{1}{t} \tag{11}$$

The time dependent variation of the tunneling distance $r(E_D)$, calculated from Eqs. (6) and (7) is:

$$r(E_D) = \frac{\alpha(E_D)}{2}\ln(\nu_0 t) \tag{12}$$

Combining Eqs. (5)–(12) and assuming a constant carrier density, N_0, provides the time dependent photocurrent decay, given by:

$$I_{\mathrm{ph}}(t) = N_0 FA\frac{q^2}{6kT}\left\{\frac{\alpha(E_D)}{2}\ln(\nu_0 t)\right\}^2 t^{-1} \tag{13}$$

For times $t > 10^{-9}$ s the decay is dominated by the last term in Eq. (13). The transient photo-response is consequently well described by:

$$I_{ph}(t) \propto t^{-1} \tag{14}$$

This supports our experimental data, where $I_{ph}(t) \propto t^{-1}$ characteristics have been detected in PD in the time regime 10^{-9}–10^{-4} s and in HOD in the regime 2×10^{-8}–2×10^{-7} s. The $I_{ph}(t) \propto t^{-1}$ feature is universal for 'hopping-down thermalization' of carriers in localized densities-of-states like in amorphous hydrogenated silicon (Nebel et al., 1992; Nebel and Street, 1993). Our experimental data measured in PD and HOD diamond indicate that localized electronic states are present and distributed in the band gap of diamond. Most likely these states are at or in the vicinity of grain boundaries. Unfortunately, the $I_{ph}(t) \propto t^{-1}$ characteristic cannot be used to calculate specific properties of the DOS, like energy distribution and density of states (DOSs).

The interaction of photo-generated carriers with traps and defects in the band gap of diamond has been characterized by charge collection measurements. Here, the ratio of the electrical charge, Q_{IN}, induced in the external circuit to the charge generated by a photon or a charged particle, Q_0, is evaluated. Hecht (1932) discussed this problem in detail and he came up with the famous 'Hecht equation' for charge collection, given by

$$\eta = \frac{Q_{IN}(F)}{Q_0} = \frac{\mu\tau F}{d}\left\{1 - \exp\left(-\frac{d}{\mu\tau F}\right)\right\} \tag{15}$$

where d is the sample thickness, μ the carrier mobility, τ the deep trapping lifetime and F the applied electric field. $\mu\tau F$ is the 'Schubweg', d_D, of carriers, which is the drift length carriers can move before deep trapping or recombination terminates the propagation. In the event that $\mu\tau F << d$ one can neglect the exponential term, which leads to:

$$\eta = \frac{Q_{ind}(F)}{Q_0} = \frac{d_D}{d} \tag{16}$$

If both, electrons and holes, propagate in a charge collection experiment, the Schubweg is given by:

$$d_D = (\mu_e \tau_e + \mu_h \tau_h)F \tag{17}$$

where 'e' stands for electrons and 'h' for holes.

Figure 11 shows three efficiency spectra recorded on a single crystal type IIa and two poly-crystalline CVD layers of 25 μm (sample A) and 300 μm (sample B) thickness. The measurements were made in a vacuum using a constant 10 kV cm^{-1} bias field (Bergonzo et al., 2001). Only in single crystal IIa-diamond, η reaches 100% which means that carriers propagate through 200 μm of diamond. In CVD poly-diamond the efficiency peaks at about 40% in A and 2.5% in B which corresponds to a drift length of 10 μm in A and 7.5 μm in B. As in these experiments electron–hole pairs were generated by alpha particles

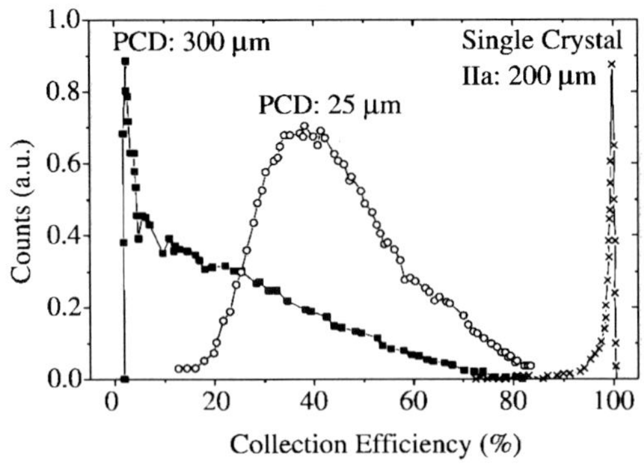

FIG. 11. Charge collection spectra recorded on two PCD films of 25 and 300 μm thickness and in a single crystal IIa-diamond of 200 μm thickness (from Bergonzo et al. (2001)).

impinging different parts of the detector, the spread of collection efficiencies represents locally varying electronic properties of the detector, corresponding to crystallites with varying electronic quality. The spread in collection efficiencies ranges between 20 and 80% (5–20 μm) in A, and 2 and 70% (6–210 μm) in B. In addition, the collection efficiency varies with film thickness (Manfredotti, Fizzotti, Polesello, Vittone and Wang, 1996). Low values are observed on the substrate side and high values on the growth side of the CVD layers. This is due to growth properties of PCD, where crystallites are small at the beginning of the deposition (substrate side) and become larger as the film thickness increases. Note, the collection efficiency is a function of thermal and optical pre-treatments ('sample priming') as well as of irradiation time of the films, as traps and defects get charged during the experiments that cause electric field and recombination variations (Viatkus, Inushima and Yamazaki, 1993; Bergonzo et al., 2001).

2.4. Optical Detected Sub-Band Gap Absorption

Sub-band gap absorption properties of CVD diamond have been measured by optical transmission/reflectivity and PDS experiments (Jackson and Amer, 1982; Nesladek, Vanecek and Stals, 1996). Typical results are shown in Figure 12 including data obtained on type Ib, IIa and IIb synthetic diamond (Rohrer, Graeff, Janssen, Nebel, Stutzmann, Güttler and Zachai, 1996). The absorption spectra of amorphous carbon (a-C) and hydrogenated amorphous carbon (a-C : H) from Hauser (1977) and Smith (1984) are also depicted for comparison. In a-C the absorption coefficient α is in the range of 10^4–10^5 cm^{-1}. Hydrogen induced defect compensation in a-C:H gives rise to a decrease in α of one to two orders of magnitude. In highly (100) oriented CVD diamond (HOD) α is strongly dependent on growth conditions and can be as low as in good poly-crystalline

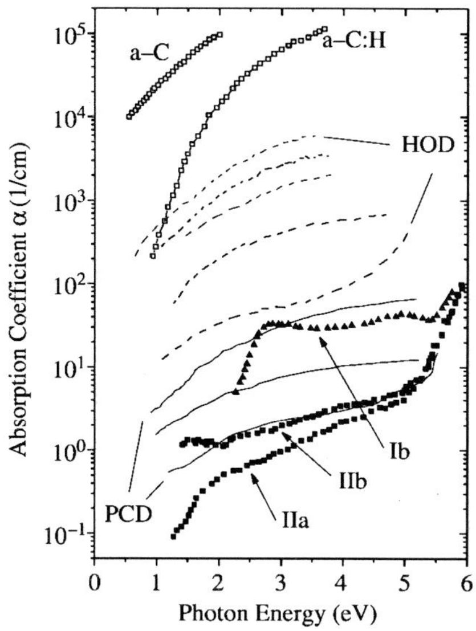

FIG. 12. Optical absorption coefficients of a-C and a-C : H (Hauser, 1977; Smith, 1984), of typical HOD films (dashed lines), of high quality PCD layers (full lines) and of single crystal Ib, IIa and IIb-diamond.

CVD layers (PCD). Up to now, however, the PCD films show the lowest sub-band gap absorption, especially if they are grown rather thick to achieve large grains and polished on the substrate side to remove small grains and graphitic inclusions. The sub-band gap absorption coefficient in these films varies between 0.5 cm^{-1} at 1 eV and 10 cm^{-1} at the onset of band-to-band transitions (5.48 eV).

It is interesting to note that the sub-band gap absorption of CVD diamond (HOD, PCD) does not resemble the properties of single crystalline type Ib and IIa diamond. The monotonous increase of $\alpha(h\nu)$ in PCD and HOD is discussed based on a sub-band gap DOS distribution that is schematically shown in Figure 13 (Nesladek et al., 1996 and Nesladek et al., this issue). The model is based on a theoretical approximation of amorphous carbon films, suggested by Bredas and Street (1985) and Robertson and O'Reilly (1987). According to the model of Nesladek et al. (1996) amorphous graphite is present at grain boundaries. In graphite, three electrons form sp^2-bonds to neighboring carbon atoms (σ-bonds) and one electron forms a π-orbital perpendicular to the σ-bonds. The π-bonds are weaker than the σ-bonds and are, therefore, closer to the Fermi level than the σ-bonds. Due to disorder, the π-and σ-bonds are energetically broadened. The filled π-(bonding) and the empty π^*-states (anti-bonding) are assumed to be Gaussian shaped. The distribution of filled σ-(bonding) and empty σ^*-states (anti-bonding) is supposed to be exponentially decreasing towards the center of the band gap. Σ (valence band) and Σ^* (conduction band) represent the bands

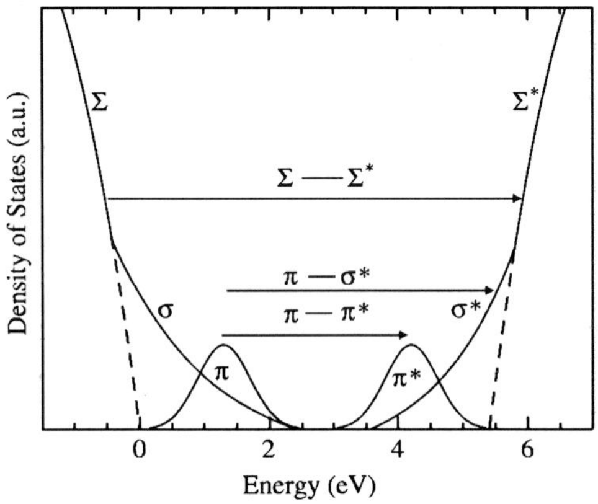

FIG. 13. DOS model for amorphous graphite present at grain boundaries of poly-crystalline diamond. Also shown are the dominant optical transitions (arrows).

of diamond. Figure 13 shows the dominant optical transitions, which have been taken into account to calculate the optical absorption (Rohrer, 1998) based on

$$\alpha(h\nu) \propto \frac{1}{h\nu}\int_{E_F-h\nu}^{E_F} g_{oc}(E)g_{em}(E+h\nu)\mathrm{d}E \tag{18}$$

where g_{oc} is the occupied and g_{em} the empty DOS distribution, $h\nu$ is the photon energy, E_F the Fermi level and E the energy. The transition matrix element is assumed to be constant. For $\pi \rightarrow \pi^*$ transitions we use

$$g_{oc}(E) = \frac{A}{w\sqrt{2\pi}}\exp\left[-\frac{(E-E_\pi)^2}{2w^2}\right] \tag{19}$$

and

$$g_{em}(E+h\nu) = \frac{A}{w\sqrt{2\pi}}\exp\left[-\frac{(E+h\nu-E_{\pi^*})^2}{2w^2}\right] \tag{20}$$

where A is a constant which is proportional to the defect density, E_π and E_{π^*} are the maxima and $2w$ the FWHM of the Gaussian distributed π- and the π^*-states.

For $\pi \rightarrow \sigma^*$ transitions we use

$$g_{oc}(E) = \frac{A}{w\sqrt{2\pi}}\exp\left[-\frac{(E-E_\pi)^2}{2w^2}\right] \tag{21}$$

and

$$g_{em}(E+h\nu) = B\exp\left[\frac{(E+h\nu-E_{G\sigma})}{E_{t\sigma^*}}\right] \tag{22}$$

where $E_{G\sigma}$ is the distance between the σ-and σ*-states in a-C, $E_{t\sigma^*}$ is the characteristic energy of the exponential tail of the σ*-band and B a constant which is proportional to the σ* density.

For transitions from the valence band maximum to the conduction band minimum ($\Sigma \rightarrow \Sigma^*$) we use exponentially distributed states:

$$g_{\mathrm{em}}(E) = N_{\mathrm{v}} \exp\left(-\frac{E}{E_{t\Sigma}}\right) \tag{23}$$

and

$$g_{\mathrm{em}}(E + h\nu) = N_c \exp\left(\frac{E + h\nu - E_{\mathrm{G}}}{E_{t\Sigma}}\right) \tag{24}$$

where E_{G} is the band gap of diamond (= 5.48 eV), $E_{t\Sigma}$ the characteristic energy of the exponentially distributed conduction and valence band states and N_{c} and N_{v} are the effective DOSs of the conduction and valence band.

The results of these calculations are shown in Figure 14. The dashed lines represent $\pi\pi^*$, $\pi\sigma^*$ and $\Sigma\Sigma^*$ transitions and the full lines the resulting overall optical absorptions that takes into account all contributions. The calculations reveal that the peak of the π^* distribution is about 3.4–3.8 eV above the peak of the π-distribution. The FWHM of the Gaussians is about 1.45–1.6 eV. The energy gap between σ–σ* states is about 5 eV and the characteristic energies of

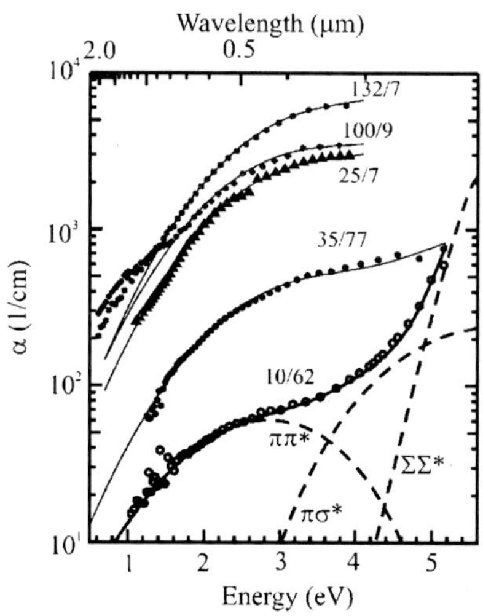

FIG. 14. Absorption spectra recorded on PCD films with different nitrogen contents. The numbers represent the nitrogen density in ppm and the film thickness (for example, 132/7 is: 132 ppm N, 7 μm thickness). The dashed lines are contributions of the $\pi\pi^*$, $\pi\sigma^*$ and $\Sigma\Sigma^*$ transitions and the full lines the resulting overall absorption.

their tails vary between 180 and 250 meV. These data are in reasonable agreement with data published in the literature (Nesladek et al., 1996).

The empty π^*-and σ^*-states at grain boundaries are traps for electrons and the filled π and σ states traps for holes. As they are present all over the gap of polycrystalline CVD diamond they dominate the dark conductivity, especially at low temperatures, as shown in Figure 1, and give rise to the non-exponential photocurrent decay, as shown in Figure 9, and to grain size limited drift lengths of carriers as measured by charge collection experiments (Fig. 11).

2.5. Paramagnetic Defects in CVD Diamond

From studies on natural and synthetic diamond it is known that nitrogen is the dominant impurity in diamond, with many possible configurations, from isolated substitutional to aggregates (Smith, Sorokin, Gelles and Lasher, 1995). The substitutional dispersed form is a deep donor with a level at 1.7 eV below the bottom of the conduction band, giving rise to a well-established electron spin resonance (EPR) signal, known as the P1 center. On the other hand very little is known about nitrogen in CVD diamond films, and the data available in the literature are not consistent. For example, despite the fact that nitrogen is a deep donor, it has been argued that 'intrinsic' diamond, which contains typically 10–80 ppm nitrogen, is p-type.

We have thus investigated a series of samples, which have been deposited on (100) silicon wafers by microwave plasma assisted CVD from gas mixtures of 1.5% CH_4 in H_2 and a nitrogen admixture varying from 25 up to 5000 ppm N_2/H_2. The deposition conditions were similarly to those published by Füßer et al. (1993). The absolute nitrogen content has been detected by elastic recoil detection, and varied from 6 to 100 ppm (Rohrer et al., 1996).

For the EPR and light induced (L)EPR experiments free standing samples were placed in the TE_{102} cavity of a standard X-band EPR spectrometer. All experiments have been performed at room temperature using 0.07–100 kHz magnetic field modulation, with phase sensitive detection. Samples are illuminated with a Hg/Xe arc lamp for the LEPR experiments. Low-pass cut-off optical filters were used to control the photon energy in the LEPR experiments.

The nitrogen related P1 resonance ($g = 2.0024$) can be detected in high quality PCD and HOD films (Graeff, Rohrer, Nebel, Stutzmann, Güttler and Zachai, 1996; Rohrer et al., 1996; Graeff, Nebel, Stutzmann, Flöter and Zachai, 1997) at low microwave power of $<10\ \mu W$. The typical density is about 1–5% ($\cong 5 \times 10^{16}\ cm^{-3}$) of the total nitrogen content as detected by elastic recoil detection analysis (ERDA). In Figure 15 typical LEPR signals from a highly oriented, (100) textured CVD diamond with a nitrogen content <15 ppm are shown for two different orientations of the sample with respect to the magnetic field. The signals consist of a central line with $g = 2.0024$ and a peak-to-peak width ΔH_{pp} of 0.18 G, surrounded by two satellite peaks coming from the hyperfine interaction of the electron spin ($S = 1/2$) with the ^{14}N nuclei

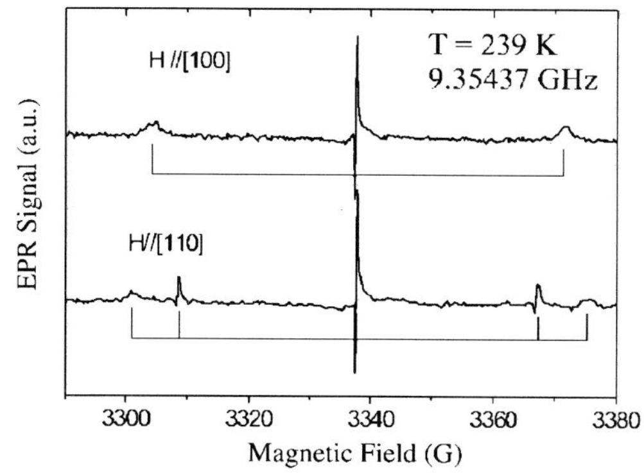

FIG. 15. Typical LESR spectra of the P1 center ($g = 2.0024$) measured in a highly oriented, (100) textured CVD diamond film, for two different orientations of the sample with respect to the magnetic field. Indicated in the figure are the expected positions for the satellite lines.

(nuclear spin $I = 1$). The unpaired electron of the nitrogen can be situated on any one of the four C–N bonds. The symmetry axis of the hyperfine interaction with the ^{14}N is parallel to any of the four $\langle 111 \rangle$ bond directions. It is found, that crystallite disorientation causes a linewidth broadening of the satellite lines, which can be used to quantitatively characterize the diamond film texture (Graeff et al., 1997). The position of the satellite peaks (ΔH) in relation to the central line is given by

$$\Delta H = \sqrt{A_{\parallel}^2 \cos^2 \theta + A_{\perp}^2 \sin^2 \theta}$$

where $A_{\parallel} = 40.93$ G and $A_{\perp} = 29.20$ G are the parallel and perpendicular components of the hyperfine tensor, respectively, and θ is the angle between the external magnetic field H_0 and the symmetry axis of the hyperfine interaction, the four $\langle 111 \rangle$ bond directions.

For samples with randomly oriented crystallites, which contain more than 20 ppm of nitrogen (ERDA), this line can be detected only after UV illumination with photons of energy $h\nu > 4$ eV. It indicates, that in these films substitutional nitrogen is fully compensated by deep defects or by electron transfer to grain boundaries. The UV illumination counteracts the compensation, so that P1 can be detected by EPR. An isotropic set of two satellite peaks is found at ± 29.2 G relative to the central line.

For samples with less than 20 ppm nitrogen (ERDA), the $g = 2.0024$ signal can be detected without illumination. The density of P1 centers in such films is about 1% of the total amount of nitrogen as detected by ERDA (about 5×10^{16} spins per cm^3 for a sample with 7×10^{18} cm^{-3} nitrogen). Illuminating such a sample with UV light ($h\nu > 4$ eV), rises the density of P1 centers for a factor three. Obviously, in these films the nitrogen donor concentration is only partially

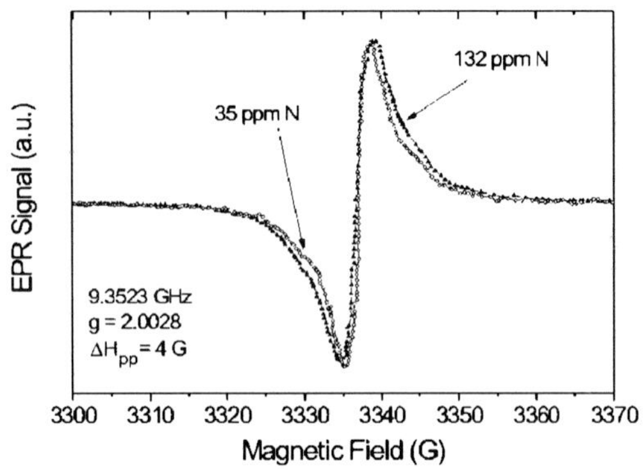

FIG. 16. EPR signal $g = 2.0028$, which consists of a single central line with unresolved shoulders (indicated by arrows). The signals arise from unpaired electrons in carbon dangling bond orbitals, which are hyperfine-coupled to hydrogen atoms about 2 Å away.

compensated. As a significant density of P1 centers is detected by EPR, the Fermi-level in PCD and HOD films must be in close vicinity of substitutional nitrogen. CVD diamond, therefore, is an n-type semiconductor. The majority of nitrogen as detected by ERDA, however, does not resemble P1 characteristics. Nitrogen contamination of CVD diamonds is a consequence of nitrogen contamination of the growth chamber or contamination of the CH_4/H_2 gas mixtures, used in the CVD process.

The broad line $g = 2.0028$ with a ΔH_{pp} of 4 G, as shown in Figure 16, is present in all samples, independent of nitrogen content, or crystal orientation, both in the dark (EPR) and under illumination (LEPR). The signal $g = 2.0028$ consists of a single central line with unresolved shoulders (indicated by arrows in Figure 16). This specific shape has been described by Zhou, Watkins, MacNamara-Rutledge, Messmer and Chawla (1996) as a central line, arising from an unpaired electron in a carbon dangling bond orbital, which is hyperfine-coupled to a hydrogen atom about 2 Å away. The defect is generated by hydrogen, entering stretched C–C bonds at grain boundaries, allowing the carbon atoms to relax back. Hydrogen forms a bond with one of the carbon atoms, and leaves the other bond with an unpaired electron. The density of $g = 2.0028$ paramagnetic states can be raised after UV illumination to about 1% of the equilibrium density of defects, which corresponds to 2.3×10^{18} spins per cm^3. We attribute this to electron trapping in grain boundaries, where positively charged carbon dangling bonds are present, which can capture electrons and, therefore, rise the density of paramagnetic $g = 2.0028$ centers. Figure 17 shows a graph of the $g = 2.0028$ carbon defect density versus the absorption coefficient at 3 eV of films shown in Figure 14. The linear relation between the increase in sub-band gap absorption at 3 eV (α_{PDS} (3 eV)) and carbon dangling bond density supports the model that $g = 2.0028$ centers are present at grain boundaries.

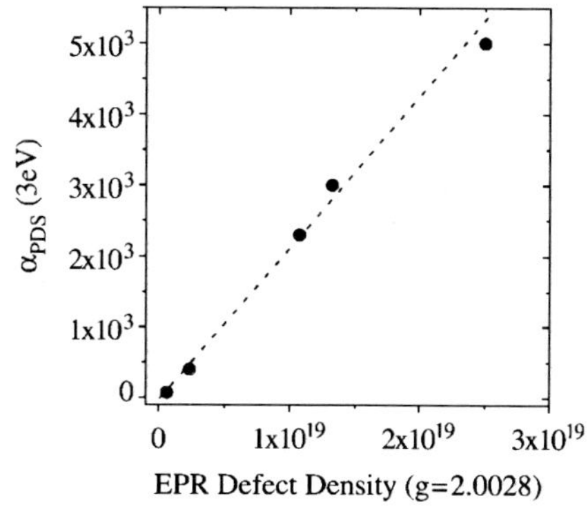

FIG. 17. Optical absorption at 3 eV photon energy vs. EPR carbon dangling bond density ($g = 2.0028$) (dashed line for guiding the eyes).

2.6. Spectral Resolved Photoconductivity

Sensitive methods to detect sub-band gap properties are SPC and PDS. PDS experiments detect all states that absorb light, while SPC is sensitive only to absorption events that generate mobile carriers. In the following, we briefly summarize the basic theory of SPC.

The current density, $\vec{j}$, in a semiconductor is given by the conductivity, σ, and the applied electric field, $\vec{F}$:

$$\vec{j} = \sigma\vec{F} \tag{25}$$

The conductivity is a function of

$$\sigma_D = q(n_D\mu_n + p_D\mu_p) \tag{26}$$

where q is the elementary charge, n_D and p_D are electron and hole densities in the dark and μ_n and μ_p are electron and hole mobilities, respectively. Illuminating a semiconductor generates additional carriers (n_{ph} and p_{ph}) by absorption of photons, so that the conductivity is enhanced

$$\sigma = q\{(n_D + n_{ph})\mu_n + (p_D + p_{ph})\mu_p\}$$

$$= q(n_D\mu_n + p_D\mu_p) + qn_{ph}\mu_n + qp_{ph}\mu_p = \sigma_D + \Delta\sigma_n + \Delta\sigma_p \tag{27}$$

$\Delta\sigma_n$ and $\Delta\sigma_p$ are photo-conductivities of electrons and holes, respectively. In thermodynamic equilibrium we assume that generation rate, $G(x)$, and recombination rate, $R(x)$, are equal

$$G(x) - R(x) = 0 \tag{28}$$

The generation rate can be calculated by

$$G(x) = -\eta\frac{d\Phi(x)}{dx} \tag{29}$$

where η is the quantum efficiency of photo-generation of carriers, and $\Phi(x)$ is the photon flux, given by

$$\Phi(x) = \Phi_0(1 - \rho)\exp(-\alpha x) \tag{30}$$

where α is the absorption coefficient, ρ the reflectivity loss at the surface of the semiconductor, and Φ_0 the photon flux of the light source. The generation rate G is, therefore:

$$G(x) = \eta\Phi_0(1 - \rho)\alpha \exp(-\alpha x) \tag{31}$$

Taking into account the lifetime τ_n of electrons, the recombination rate $R(x)$ is:

$$R(x) = \frac{n_{\mathrm{ph}}(x)}{\tau_n} \tag{32}$$

Using Eqs. (28), (31) and (32), we can calculate the photo-generated electron density, which is:

$$n_{\mathrm{ph}}(x) = \eta\tau_{\mathrm{n}}\Phi_0(1 - \rho)\alpha \exp(-\alpha x) \tag{33}$$

Taking into account Eq. (27), the photoconductivity of electrons in a layer of thickness d is:

$$\begin{aligned} \Delta\sigma_{\mathrm{n}} &= q\eta\mu_{\mathrm{n}}\tau_{\mathrm{n}}\Phi_0(1 - \rho)\alpha \int_0^d \exp(-\alpha d) \\ &= q\eta\mu_{\mathrm{n}}\tau_{\mathrm{n}}\Phi_0(1 - \rho)[1 - \exp(-\alpha d)] \end{aligned} \tag{34}$$

For weak absorption, where $\alpha d << 1$, we replace the exponential function e^x by $(1 + x)$ and get for electrons

$$\Delta\sigma_{\mathrm{n}} = q\eta\mu_{\mathrm{n}}\tau_{\mathrm{n}}\Phi_0(1 - \rho)\alpha d \tag{35}$$

and for holes:

$$\Delta\sigma_{\mathrm{p}} = q\eta\mu_{\mathrm{p}}\tau_{\mathrm{p}}\Phi_0(1 - \rho)\alpha d \tag{36}$$

The lifetime of carriers strongly depends on defects, Fermi-level and light excitation density. For weak excitation, where photo-generated carrier densities are smaller than dark-carrier densities, the lifetime of photo-generated electrons is given by

$$\tau_{\mathrm{n}} = \frac{1}{\beta_{\mathrm{n}}\bar{v}_{\mathrm{n}}N_{\mathrm{n}}} \tag{37}$$

where β_{n} is the capture cross-section of N_{n} recombination centers for electrons that are neutrally or positively charge centers. $\bar{v}_{\mathrm{n}}$ is the mean thermal velocity of electrons. For holes, the same formalism can be applied

$$\tau_{\mathrm{h}} = \frac{1}{\beta_{\mathrm{h}}\bar{v}_{\mathrm{h}}N_{\mathrm{p}}} \tag{38}$$

where β_h is the capture cross-section of N_p recombination centers for holes that are neutrally or negatively charged centers. $\bar{v}_h$ the mean thermal velocity of holes.

In case of non-intentionally doped CVD diamond the Fermi level is about 1–1.7 eV below the conduction band. Diamond is, therefore, an n-type semiconductor where below the Fermi level all states are occupied with electrons. As a consequence, the density of recombination centers for holes, N_p, is significantly larger than for electrons, N_n. Due to this asymmetry, the lifetime of minority carriers (holes in diamond) is much shorter than the life-time of electrons (majority carriers). Photoconductivity measurements in non-intentionally doped CVD diamond films are thus governed by electron properties. The normalized spectrally resolved photocurrent density $j_{ph}(h\nu)$ is

$$\frac{j_{ph}(h\nu)}{\Phi_0(h\nu)} \approx \tau_n \alpha(h\nu) \tag{39}$$

where j_{ph} and Φ_0 are measured as a function of photon energy ($h\nu$). Applying SPC experiments, consequently reveal the 'lifetime-absorption coefficient product' ($\tau\alpha$). Under certain conditions, it is possible to keep τ constant ('constant photocurrent measurements, CPM' (Grimmeiss and Ledebo, 1975)), subsequently, the normalized photocurrent is proportional to the absorption coefficient $\alpha(h\nu)$.

SPC experiments were performed using coplanar contact configurations where two bar-like Ti/Pt/Au contacts 1 mm apart were evaporated on top of the diamond films. The set-up is shown in Figure 18. As light sources, a halogen and a xenon lamp were used. The light was dispersed in a Spex 340E monochromator using three different gratings, blazed at 1000, 500 and 250 nm, respectively, and was mechanically chopped at 2 Hz. All optical parts employed, including the vacuum chamber window, have an optical transparency up to 200 nm. The signal was detected via a lock-in amplifier. The light intensity was measured with a pyroelectric detector. For transport measurements, the samples were mounted in a vacuum chamber with pressures lower than 10^{-6} mbar. To remove water and other contaminants from the surface, a heat treatment at 500 K for 30 min was applied.

Normalized SPC data measured on a high quality PCD film and on a single crystal type Ib HPHT synthetic diamond are shown in Figure 19 and compared to optical absorption results (photoconductivity and PDS data on PCD from Nesladek, Stals, Stesmans, Iakoubovskij and Adriaenssens, 1998). While the optical absorption is a monotonously increasing function of photon energy, the photocurrent shows distinct features. In PCD two dominant absorption shoulders with onsets around 1.2 and 2.2 eV are detected. By comparison with the results deduced from the Ib-diamond where the photocurrent response is due to substitutional nitrogen, it is obvious that the broad shoulder arising at 2.2 eV is due to single substitutional nitrogen (P1 center) which is also detected by EPR experiments ($\cong 10^{18}$ cm^{-3}) in this film (Graeff et al., 1996). The absorption at 1.2 eV is detected on the as-grown sample or after hydrogen plasma treatment

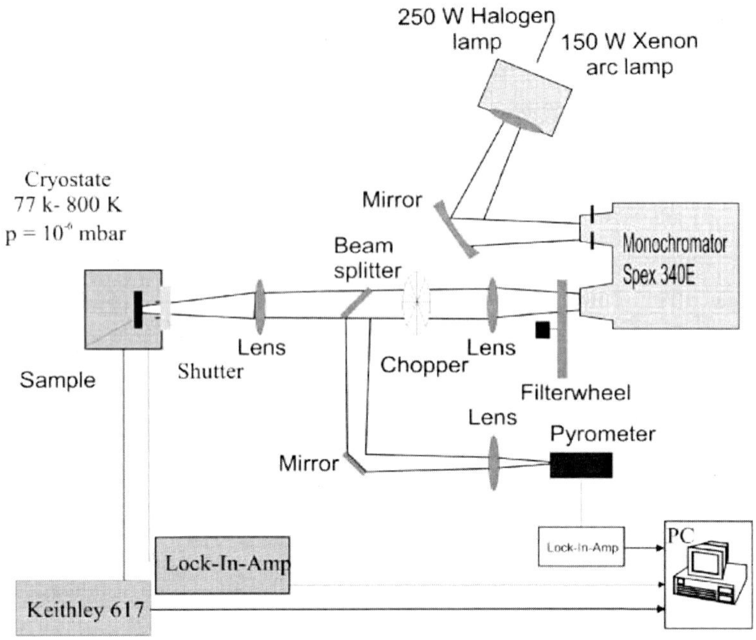

FIG. 18. SPC set-up. As light sources, a halogen and xenon lamp were used. The light is dispersed in a Spex 340E monochromator using three different gratings and mechanically chopped at 2 Hz. The signal is detected via lock-in amplifier. The light intensity is measured with a pyroelectric detector. Samples are mounted in a vacuum chamber with pressures $\leq 10^{-6}$ mbar.

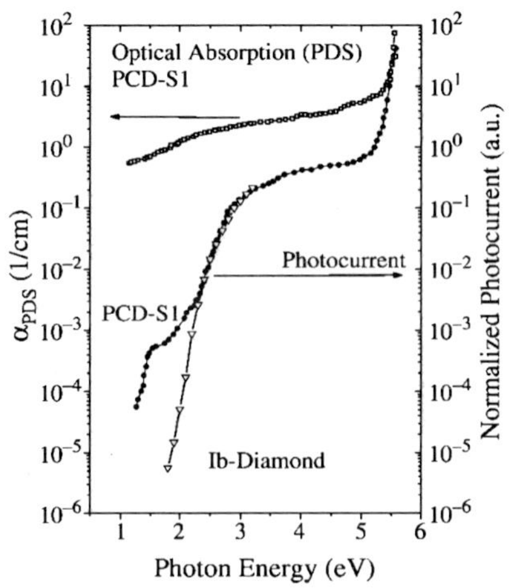

FIG. 19. Comparison of optical absorption (□) and normalized SPC (●) measured on a PCD film (data from Nesladek et al., 1998). Also shown is the photocurrent detected in a single crystal Ib-diamond (▽).

and diminishes after oxidation. It is, therefore, attributed to surface defects of hydrogen terminated diamond.

Obviously, photocurrent and PDS measurements reveal very different absorption features. While PDS is sensitive to absorption in general, photocurrent experiments (SPC) detect only those absorption events that generate propagating carriers. SPC is, therefore, dominated by absorption events in high quality volume fractions. In the case of large grain PCD, these are comparable to single crystalline type Ib diamond.

However, applied on small grain polycrystalline films, reveals grain boundary properties. In these films the absorption is dominated by amorphous graphite. Spectrally resolved photocurrents of different quality PCD films are shown in Figure 20 (data of i2 and S1 from Nesladek et al., 1998). The onset detected in all films occurs at about 1 eV. The photocurrents increase orders of magnitudes in the regime 1–3 eV, turn into a shoulder and finally increase again towards higher photon energies in the regime $h\nu > 4.5$ eV. The good PCD films are approximately 'solar blind' and fulfill the requirements for UV detectors. On the best samples measured, the SPC in the regime $h\nu > 5.5$ eV is about three orders of magnitude larger than in the regime $h\nu \leq 4.5$ eV. In the literature even six orders of magnitude have been reported (Whitfield, Lansley, Gaudin, McKeag, Rizvi and Jackman, 2001).

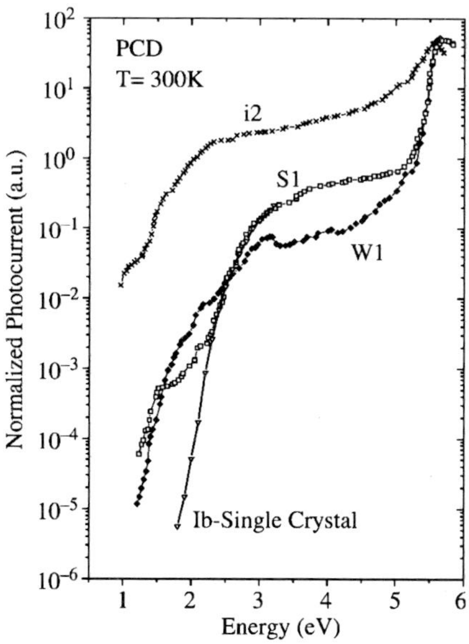

FIG. 20. Normalized spectrally resolved photocurrents in PCD diamond with decreasing quality from S1 (□, data from Nesladek et al., 1998), W1 (◆) to 2i (X, data from Nesladek et al., 1998). Also shown is the photocurrent in single crystal Ib-diamond (∇).

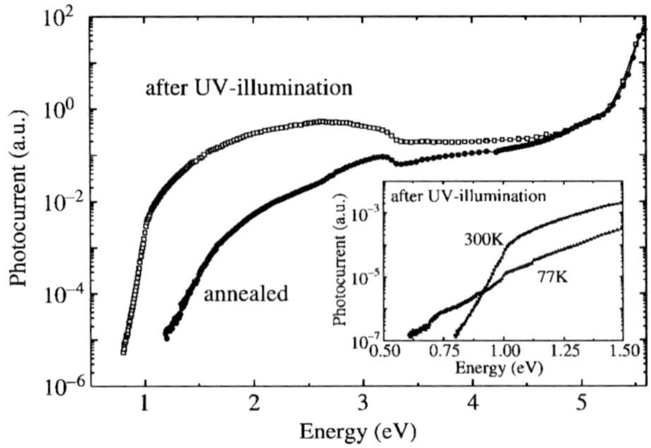

FIG. 21. Spectrally resolved normalized photocurrents measured in the annealed state (●) and after UV illumination (□) in PCD. Inset: shift of the photocurrent onset from 0.8 eV at 300 K to 0.6 eV at 77 K.

Around 1 eV, photocurrents can be significantly enhanced by pre-illumination with UV-light ($h\nu \geq 5.5$ eV) (Rohrer, Graeff, Nebel, Stutzmann, Güttler and Zachai, 1997; Nebel, Waltenspiel, Stutzmann, Paul and Schäfer, 2000). A typical example is shown in Figure 21 where a PCD film has been pre-illuminated for 10 min at a flux of 10^{13} photons/(s cm^2). The onset of the SPC shifts to energies as small as 0.8 eV at 300 K, and to 0.6 eV at 77 K (see insert in Figure 21). This is due to a filling of the DOS distribution of amorphous graphite, which is present at grain boundaries, with photo-generated carriers by UV-bias pre-illumination, where carriers are trapped in states, which are not in thermal contact with the conduction- and valence-band. Spectrally resolved illumination during the SPC experiments excites these carriers, they propagate due to the applied electric field and eventually recombine.

2.7. Persistent Photocurrents

PCD shows persistent photocurrent (PPC) effects which are illustrated in Figure 22 as a function of temperature (Nebel et al., 2000). The dark current is recorded after 30 min annealing at 750 K. It is activated with $\cong 0.85$ eV in the low and 1.47 eV in the high temperature regime. The photocurrent is generated by illumination with an UV light source emitting photons of 5.5 eV energy (flux: 10^{13} photons/s cm^2). The measurements are performed from low to high temperatures. At $T = 300$ K, the $\eta\mu\tau$-product is in the range of 10^{-6} cm^2 V^{-1}. Illuminating the PCD layer at $T = 300$ K for 10 min with the light and dose as described above, and then switching off the light and leaving the sample in the dark for 30 min gives rise to a dark current which is only slightly smaller than the photocurrent ($\equiv$ PPC).

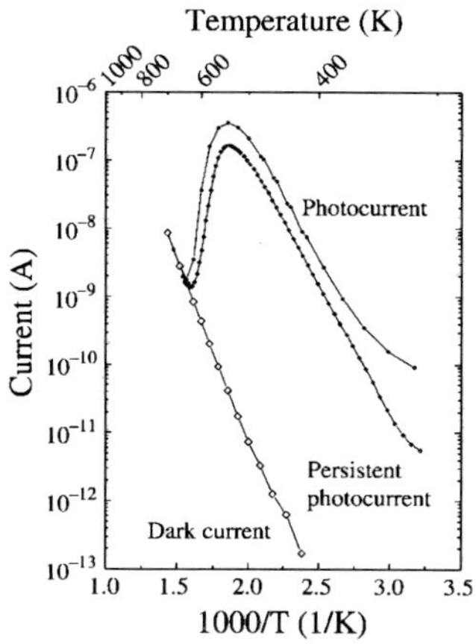

FIG. 22. Dark curren (◇), Photocurrent (●) and persistent photocurrent (■) in PCD.

Annealing characteristics of PPCs are shown in Figure 23 (symbols). The currents have been normalized by

$$I_{\text{PPCN}}(t) = \frac{I_{\text{PPC}}(t) - I_{\text{D}}}{I_{\text{PPC}}(t=0) - I_{\text{D}}} \tag{40}$$

where t represents the time relative to the beginning of the relaxation experiment ($t = 0$). The decay is stretched exponentially following a time dependence well described by (Street, Kakalios and Hayes, 1986)

$$I_{\text{PPCN}}(t) \propto \exp\left[-\frac{t}{\tau}\right]^{\beta} \tag{41}$$

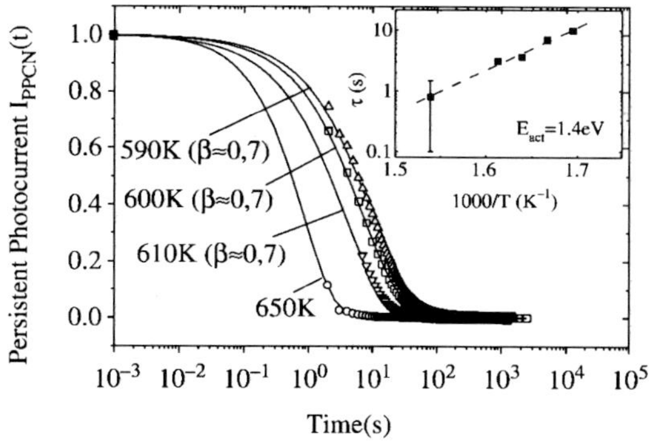

FIG. 23. Normalized PPC decays (symbols) and calculated fits (full lines) based on Eq. (41). Inset: PPC annealing time constants (squares) and calculated fit (full line) based on Eq. (42).

where β is the dispersion parameter and τ the relaxation time constant. The fits (full lines in Figure 23) calculated for $\beta = 0.7$ and τ, as shown in the inset Figure 23, are in reasonable agreement with the experimental data. τ is activated

$$\tau(T) = \tau_0 \exp\left[\frac{E_{\mathrm{act}}}{kT}\right] \tag{42}$$

with $E_{\mathrm{act}} \cong 1.4$ eV. The data indicate a relaxation time constant at $T = 300$ K of 6000 years. The stretched exponential characteristic ($\beta = 0.7$) reflects disorder induced broadening of the 1.4 eV trap center.

Combining bulk properties of Ib-diamond and amorphous carbon in grain boundaries results in a DOS which is schematically shown in Figure 24. UV-light of $h\nu = 5.5$ eV generates electron–hole pairs in the grains. As PCD diamond is an n-type semicondutor the lifetime of holes is significantly shorter than of electrons. The probability of electrons to reaching grain boundaries is, therefore, much higher than for holes. In grain boundaries electrons are captured in the states above the Fermi level, shifting E_F towards the conduction band. E_{act} of the conductivity decreases after UV-illumination as experimentally detected by PPC. The increased occupation after UV-illumination is also detected by spectrally

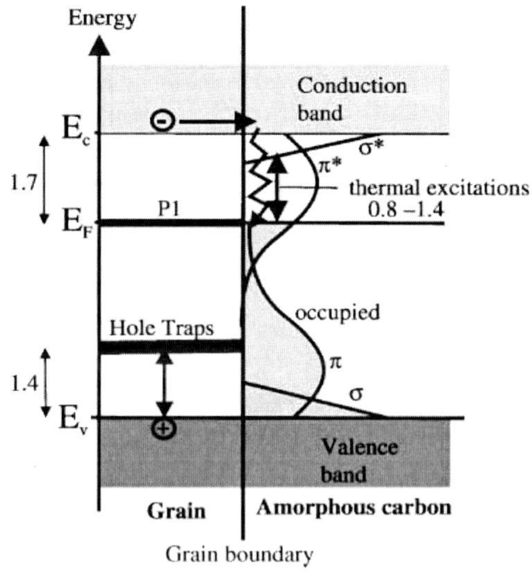

FIG. 24. Hybrid DOS which combines the properties of single crystal diamond with amorphous carbon at grain boundaries. Photo-generated electrons are trapped in anti-bonding π^* and σ^* states where transport takes place by thermal activated hopping. Holes are trapped in the bulk of grains. To establish thermodynamic equilibrium, electrons and holes have to overcome an energy barrier of about 1.4 eV.

resolved PC (Fig. 21). In grain boundaries the conductivity is governed by thermally activated hopping in antibonding π^*- and σ^*-states (Sklovskii, Levin, Fritzsche and Baranovskii, 1990). The temperature dependent shift of the transport level relative to the Fermi level gives rise to the experimentally detected variation of the activation energy (for example: dark current is activated with 0.84 eV at 450 K and with 1.45 eV at 725 K). The origin of PPC is, however, the separation of electrons and holes due to hole trapping and recombination in grains and electron trapping in grain boundaries. At sufficiently high temperatures of $T > 600$ K electrons are thermally re-emitted into the grains and thermodynamic equilibrium is established again. The energy barrier to equilibrate is about 1.4 eV (Nebel et al., 2000).

2.8. Summary of Electronic Properties of n-Type CVD Diamond

The electronic properties of CVD diamond are a combination of bulk and grain boundary effects. In large grain CVD diamond they approach the properties of single crystal Ib or IIa-diamond. In small grain CVD diamond the optical and electronic properties are governed by amorphous carbon like DOSs present at grain boundaries. Therefore, the properties are widely spread: the dark conductivity at high temperature is activated with energies between 1 and 1.7 eV. The activation energy decreases significantly towards lower temperatures. From charge collection experiments it is concluded that nominally undoped-CVD diamond is an n-type semiconductor. The mobility of holes and electrons in CVD diamond can be as high as in natural or HPHT synthetic diamond but generally μ_e and μ_h are significantly smaller in PCD and HOD due to energy barriers at grain boundaries. The typical drift length measured by charge collection experiments is in the range of the grain size. The sub-band gap absorption of poly-CVD diamond is well described by an amorphous graphite DOS distribution present at grain boundaries. EPR experiments detect carbon defects (H1) at grain boundaries and nitrogen (P1) in the range of 10^{17}–10^{18} cm^{-3} which is the origin of the n-type character of CVD diamond. Finally, spectrally resolved photocurrents show very different properties compared to optical absorption measurements. The onset of photocurrent is around 1 eV, increasing several orders of magnitude towards a shoulder in the range 2–3 eV. Only in high quality PCD layers are the PC characteristics comparable to Ib single crystalline diamond. As photo-generated electrons and holes are separated by trapping in grains (holes) and in grain boundaries (electrons) a PPC effect is generated in these films.

If progress in hetero-or homo-epitaxial growth of single crystal CVD diamond continues, it is reasonable to assume that the electronic properties of CVD diamond will approach those of single crystal natural or HPHT synthetic diamond or maybe become even better.

3. Defect and Acceptor Properties of Boron-doped Diamond

3.1. Defects in Boron-doped Diamond

The optical absorption of boron acceptors in diamond dominates the spectral response in the regime 0.36–2.5 eV. Techniques like SPC and PDS are thus not helpful to investigate defect properties of boron-doped diamond. To overcome this problem, we have applied CV and thermally (DLTS) or optically (ODLTS) excited deep-level-transient-spectroscopy, which belong to the most sensitive characterization methods to explore defect and doping properties (Lang, 1974). As depletion layers of Schottky contacts on diamonds are characterized, boron acceptors are ionized and, therefore, do not interfere with optical excitations of deep defects.

These techniques are used to investigate boron acceptor densities in diamonds, contact properties of metals on diamonds, defects in the energy range 0.6–2.4 eV above the valence band, and hydrogen interactions with boron. Experiments have been performed both on HPHT IIb-diamond and on boron-doped homoepitaxially grown CVD diamond films.

3.1.1. Experimental Details

In CV, DLTS and ODLTS experiments the capacitance, C, is calculated by the detection of the phase shift, φ, between the applied sinusoidal voltage $V(t)$ and the measured current $I(t)$, given by

$$tg(\varphi) = -\frac{1}{\omega RC} \tag{43}$$

where R is the resistivity of the semiconductor and ω the applied 'wobble frequency' ($= 2\pi f$). Due to the high resistivity even of boron-doped diamond, it is necessary to use a frequency generator which allows the adjustment of the frequency within typical 10 Hz to 100 kHz at a fixed amplitude of 100 mV. The resulting phase shift φ is monitored by a lock-in amplifier (Zeisel, 2001). The depletion layer width of the Schottky contact can be varied by application of forward or reverse voltages. Switching the diode after a filling pulse into reverse, gives rise to a transient capacitance which is recorded via a digitizing oscilloscope. The time dependent capacitance relaxation of a depletion layer, where an acceptor and a trap level is present, is given by:

$$C(t) = C_0[1 - \Delta C(t)] = C_0\left[1 - \frac{N_T^+}{2N_A^-}\exp\left(-\frac{t}{\tau}\right)\right] \tag{44}$$

We assume that traps can be occupied by single holes. C_0 is the equilibrium capacitance at a certain reverse voltage, N_T^+ is the density of occupied (holes) trap levels above the ionized acceptor density N_A^- (we refer to the valence band edge), τ is the reemission time constant and t the time. From the relaxation dynamics we can calculate the energy of the trap following the procedure of Lang (1974), ('boxcar method') where an Arrhenius plot of $\log(\tau T^2)$ versus $1000/T$ is used to calculate the energy of the trap (T is the temperature). The density of traps can be calculated from variations of C by:

$$\Delta C_t = C(t=0) - C(t \rightarrow \infty) \tag{45}$$

Combining Eqs. (44) and (45) results in:

$$N_T = 2N_A \frac{\Delta C_t}{C_0} \tag{46}$$

where C_0 is known from CV measurements. Thermal excitation of defects can be explored within reasonable time windows for defect levels between 400 and 900 meV away from the nearest band edge by DLTS experiments. For even deeper traps we have applied optically-excited DLTS (ODLTS), where, instead of phonons, photons are used to excite carriers out of defects. In this case, the absorption coefficient, α, can be calculated by:

$$\alpha = \beta_{opt} N_T \propto \frac{1}{\Phi} \frac{dC(t)}{dt}\bigg|_{t \geq 0} \tag{47}$$

Here, β_{opt} is the optical cross-section of the traps, N_T is the trap density and Φ is the photon flux ($cm^{-2}\ s^{-1}$). In the case of ODLTS, at $t = 0$ monochromatic light is switched on and the resulting change in capacitance is detected via lock-in technique and the photon flux via a pyroelectric sensor. The time derivative of $C(t)$ at the beginning of the illumination is proportional to the absorption coefficient (Eq. (47)). Like in conventional DLTS the absolute change of capacitance is used to calculate the trap density (Eq. (46)).

The CV-, DLTS, ODLTS set-up is schematically shown in Figure 25. As a light source we use a tungsten halogen lamp in combination with a monochromator and filters to absorb second harmonic excitation. The sample is mounted in a cryostat which allows temperature variations between 4 and 500 K. The intensity of the illumination is measured in situ by a pyroelectric detector. To fill traps we apply a -3 V forward voltage for 10 s. Then the junction is reversed (in most cases to $+2$ V) and the transient capacitance is recorded. In case of ODLTS the light is switched on after a delay time of about 60 s. This time delay is required to achieve a constant capacitance, meaning a well defined starting condition.

We used a boron-doped HPHT IIb-diamond, in the following named sample A, and a homo-epitaxially grown boron-doped CVD diamond of 3 μm thickness on a Ib-diamond substrate, named B (Jiang, Willich, Paul and Klages, 1999). In case of sample A, we evaporated sandwich contacts with Al as Schottky contact and Ti/Pt/Au for Ohmic contacts. The Ti(100 Å)/Pt(150 Å)/Au(1500 Å) contacts were annealed at 500 °C for 10 min to form Titanium-Carbide which gives rise to

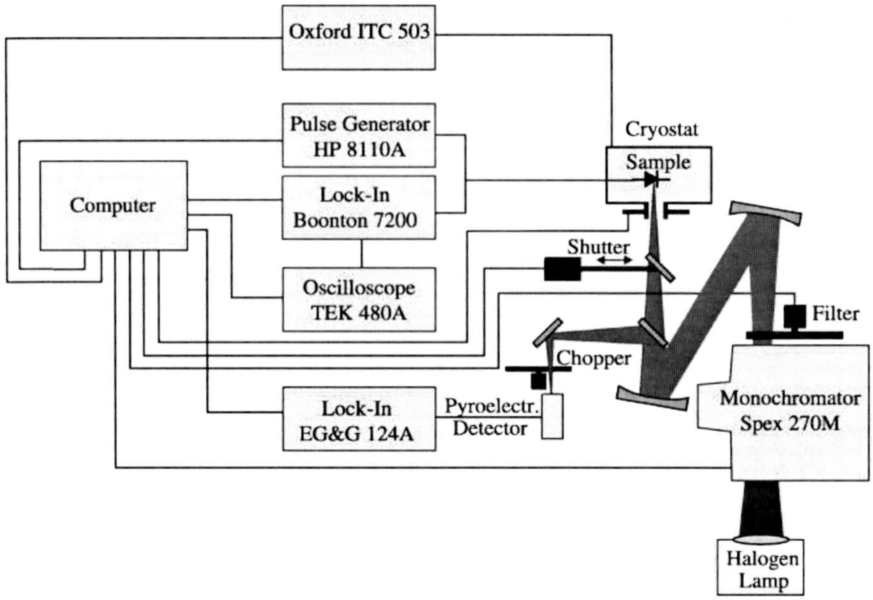

FIG. 25. CV, DLTS and ODLTS set-up. Samples are placed in an Oxford cryostat and can be illuminated by monochromatic light. The capacitance is measured by a lock-in amplifier or a Boonton CV-bridge and is digitized by an oscilloscope. Forward or reverse voltages are applied via a pulse generator. A pyro-electric sensor detects the photon flux.

Ohmic properties. For sample B, coplanar configurations were applied due to the insolating Ib-diamond substrate. In this case the distance between the Al and Ti/Pt/Au contacts was 50 μm. The size of the contact pads was 1 mm × 1 mm.

The ODLTS experiments on sample A were performed by shining light through the diamond into the depletion layer of the Al contact (back side illumination). For sample B with coplanar contacts the gap between the Al and Ti/Pt/Au contacts was illuminated.

3.1.2. *Aluminum Schottky Contact Properties*

Figure 26 shows a typical CV-plot measured on sample A. The capacitance C is given by

$$C = A\frac{\partial Q}{\partial V} = A\sqrt{\frac{q\varepsilon_0\varepsilon N_A}{2(V_{bi} + V_{rev})}} \tag{48}$$

where A is the diode area, ε_0 the dielectric constant, ε is the relative dielectric number of the semiconductor, N_A the acceptor density, V_{bi} the built-in-potential and V_{rev} the externally applied voltage in reverse direction.

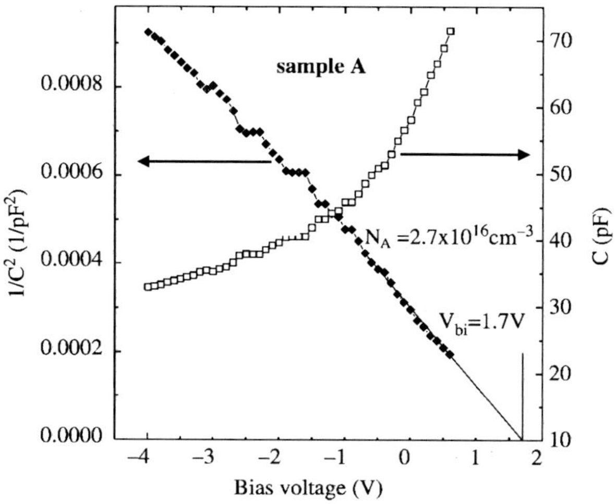

FIG. 26. CV plot measured in sample A. The acceptor density is 2.7×10^{16} cm^{-3} and the built-in voltage 1.7 V.

From the CV characteristic of an ideal diode it follows that $1/C^2$ is a linear function of V_{rev}:

$$\frac{1}{C^2} = \frac{2(V_{\text{bi}} + V_{\text{rev}})}{A^2 q \varepsilon_0 \varepsilon N_{\text{A}}} \tag{49}$$

The condition $1/C^2 = 0$ provides the built-in voltage, V_{bi}, and from the slope one can calculate the acceptor density, N_{A}, which gives an acceptor density of 2.7×10^{16} cm^{-3} and a built-in potential V_{bi} of 1.7 V.

To further investigate the Schottky contact barrier we have performed 'internal photoemission spectroscopy experiments' in short circuit conditions. In the case of internal photoemission the normalized photocurrent follows the Fowler law (Fowler, 1931):

$$\frac{I_{\text{ph}}}{\Phi} \propto [h\nu - \Theta_{\text{B}}]^2 \tag{50}$$

where I_{ph} is the photocurrent, Φ the photon flux, $h\nu$ the photon energy, q the elementary charge and Θ_{B} the contact barrier.

Figure 27 shows the results of these measurements carried out at $T = 296$, 270 and 380 K. The linear dependence for $h\nu \geq 2.4$ eV is consistent with internal photoemission of holes from Al into diamond. For $h\nu < 2.4$ eV the photo-response is due to absorption in the depletion layer of the diamond Schottky contact. The barrier height Θ_{B} is calculated from the extrapolation of the linear part to zero as shown in Figure 27, which gives $\Theta_{\text{B}} = 2.35$ eV.

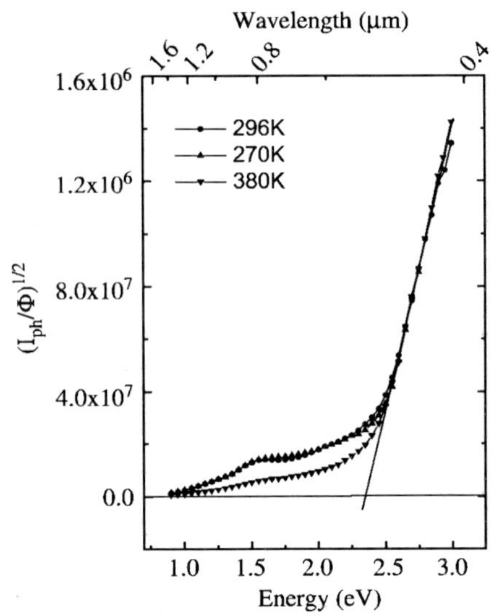

FIG. 27. Fowler plot of the spectral response of the Schottky junction on sample A. The contact barrier is 2.35 eV. The photocurrent in the regime $h\nu < 2.4$ eV is due to absorption in the depletion region of the Schottky junction. The measurements have been performed at different temperatures (296, 270 and 380 K).

A schematic view of the aluminum–diamond Schottky junction is given in Figure 28. Taking into account the built-in potential (qV_{bi}) of 1.7 eV as detected by CV-measurements and the relation:

$$\Theta_{\mathrm{B}} = qV_{\mathrm{bi}} + (E_{\mathrm{F}} - E_{\mathrm{v}}) \tag{51}$$

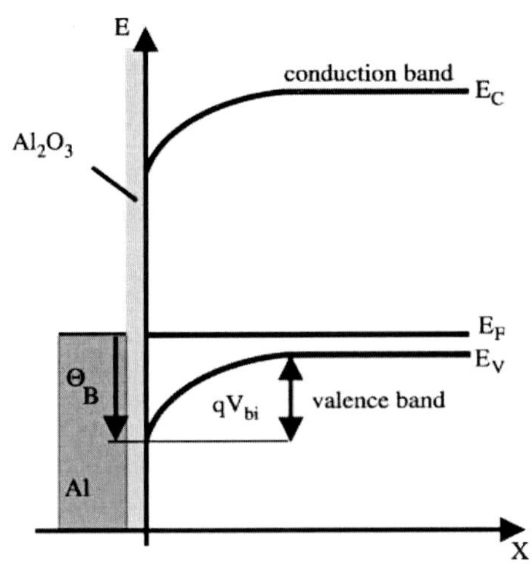

FIG. 28. Schottky contact of Al on p-type diamond. There is a thin Al_2O_3 oxide layer between Al and the diamond.

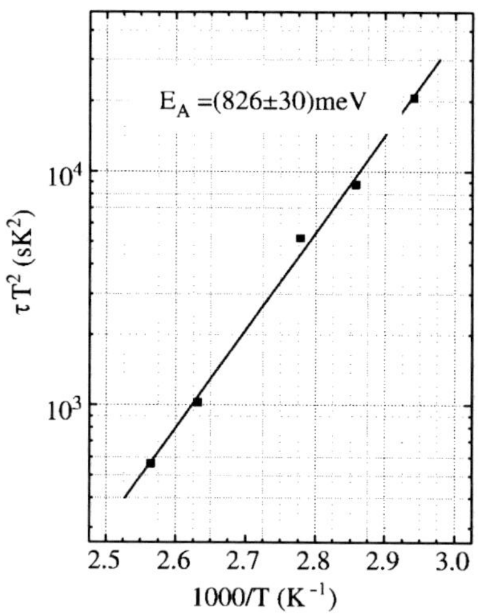

FIG. 29. Activation energy of the DLTS experiments on sample A. A transition at 0.826 eV is detected.

where $E_F - E_v$ is the Fermi energy, we calculated $E_F - E_v = 0.6$ eV. In comparison with Hall results, this is a very high activation energy as Hall effect measurements yield $E_F - E_v = 0.36$ eV. From the literature it is known that the barrier height, Θ_B, of Al on p-type diamond is 1.95 eV (Mönch, 1995). We assume that the 0.4 eV higher barrier in our case is caused by a thin Al_2O_3 oxide layer at the interface as shown in Figure 28.

3.1.3. Deep Defects

DLTS experiments performed on sample A reveal a trap with 826 meV activation energy as shown in Figure 29 (Zeisel, Nebel and Stutzmann, 1998). The density is about 10^{15} cm^{-3}. In sample B, this defect cannot be detected.

The corresponding ODLTS results are shown in Figure 30 (sample A). The onset of absorption occurs at about 0.85 eV, steeply increasing towards higher photon energies and turning over into a plateau at about 1.5 eV. The rise of the ODLTS signal for $h\nu > 2.4$ eV is a result of internal photoemission as shown in Figure 27.

To determine the trap level energy from optical absorption we use theoretically calculated photo-ionization cross-sections from Inkson (1981), given by:

$$\sigma_0 \propto \frac{\rho(h\nu - E_T)^{1/2}}{h\nu} \frac{(h\nu - E_T)^{\delta}}{(h\nu)^{2\gamma}} \tag{52}$$

where $h\nu$ is the photon energy, E_T is the trap level energy and $\rho(E)$ the density-of-states in the valence band. δ is 0 for indirect and 1 for a direct optical transitions.

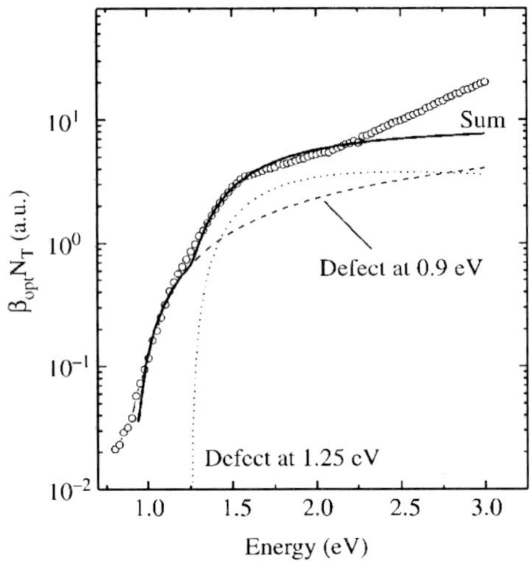

FIG. 30. ODLTS data of sample A (open circles). The absorption in the regime $h\nu < 2.4$ eV can be discussed based on two absorption centers (dashed lines), one at 0.9 eV and one at 1.25 eV. The full line is the resulting fit taking into account both centers. At $h\nu > 2.4$ eV the increase may be related to internal photoemission.

To fit this data, we used two defects as shown in Figure 30. One is 0.9 eV above the valence band edge, E_v, and a second 1.25 eV above E_v. The comparison of DLTS and ODLTS data indicates that the 0.826 eV thermal and the 0.9 eV optical excitation energies are most likely related to the same transition. The superposition of the 0.9 and 1.25 eV absorptions in the regime 1.25–1.5 eV limits the accuracy of the calculated threshold energy of the defect at 1.25 eV (see below).

Investigations of other boron-doped diamond layers using ODLTS experiments reveal spectra which are summarized in Figure 31. Here we present data from two homo-epitaxially grown CVD samples on Ib-diamond substrates and of a hetero-epitaxially CVD film, which all contain about 5×10^{17} cm^{-3} boron, and of a synthetic type IIb layer with 2×10^{16} cm^{-3} boron.

All spectra show an absorption shoulder in the spectral regime 1.3–1.8 eV. Calculating optical cross-section fits to the data indicate that a defect with optical ionization energy of 1.25 eV is present as discussed in sample A. In the IIb-diamond and in one of the CVD films, the ODLTS spectra show also thresholds around 0.9 eV. This is in agreement with DLTS and ODLTS data on sample A (Fig. 30). Different spectral shapes in the regime 0.9–2 eV are due to superposition of absorption arising from the 0.9 and 1.25 eV transitions.

To characterize the origin of the excitations at 0.9 and 1.25 eV, we have performed carbon implantation experiments in combination with ODLTS experiments, which is discussed in Section 3.1.4.

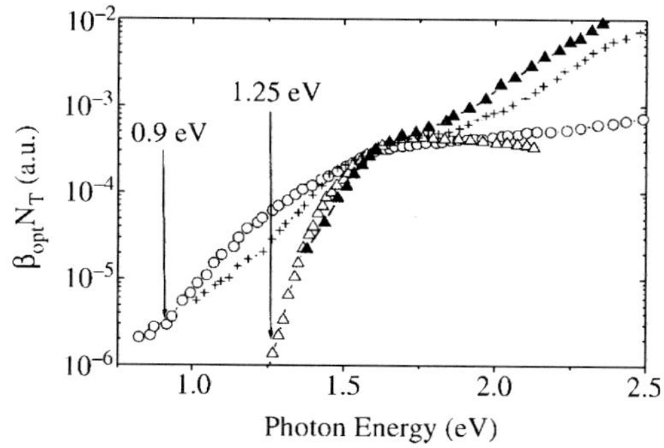

FIG. 31. ODLTS spectra of several boron-doped diamond films. Spectra x and Δ are measured in samples grown by CVD on Ib substrates, spectrum ○ is detected in IIb-diamond with $2 \times 10^{16}\ cm^{-3}$ boron, and spectrum ● is measured in a boron-doped polycrystalline CVD diamond film grown on Si. The CVD films have boron concentrations of $5 \times 10^{17}\ cm^{-3}$. Arrows indicate the threshold of excitation for the two energies 0.9 and 1.25 eV.

3.1.4. Carbon Implantation

To illuminate the nature and origin of defects in boron-doped diamond we have performed carbon implantation experiments on IIb-diamond containing $2.5 \times 10^{16}\ cm^{-3}$ uncompensated acceptors. C^+-implantation generates vacancies and interstitials in diamond (Kalish, Uzan-Saguy, Philosoph, Richter and Prawer, 1997), which can be detected by CV and ODLTS experiments. The implantation dose of C^+-ions of 160 keV energy was kept as low as possible and was $10^{10}\ cm^{-2}$ (dose I) and $10^{11}\ cm^{-2}$ (dose II). TRIM calculations give average vacancy densities of $5 \times 10^{16}\ cm^{-3}$ for dose I and $5 \times 10^{17}\ cm^{-3}$ for dose II within a shallow region extending 230 nm into the bulk (Fig. 32) (Ziegler, Biersack and Littmark, 1985).

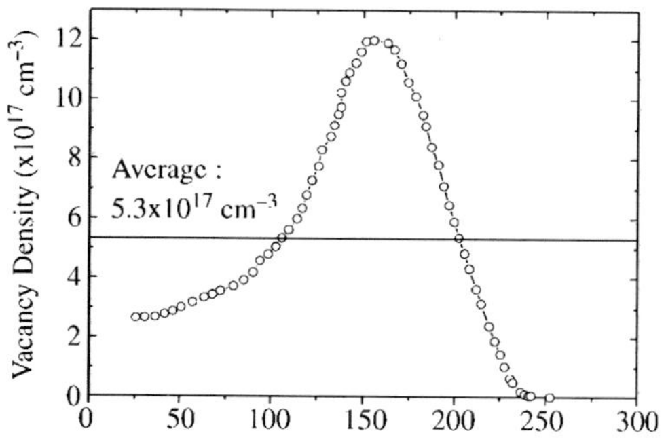

FIG. 32. TRIM computer simulation of vacancies generated by 160 keV C^+ ion implantation with a dose of 10^{11} ions/cm^2.

The measured defect density is about $7 \times 10^{15}\,\text{cm}^{-3}$ (dose I) and $2 \times 10^{16}\,\text{cm}^{-3}$ (dose II) which is significantly smaller than calculated by TRIM, however, much larger than the initial density of $2 \times 10^{14}\,\text{cm}^{-3}$ before implantation.

To characterize carbon implantation damage in diamond we applied ODLTS experiments, where the capacitance ΔC_{SS} is measured after a defined delay time, t, in-between the start of the illumination ($t = 0$) and the detection of ΔC_{SS}. In this case ΔC_{SS} approaches a steady state $\Delta C_{\text{SS}}(t \rightarrow \infty)$ since all traps are ionized. The amplitude of the capacitance $\Delta C_{\text{SS}}(t \rightarrow \infty)$ is given by

$$\Delta C_{\text{SS}} = \frac{C_0 N_{\text{D}}}{2 N_{\text{A}}} \tag{53}$$

where N_{D} is the defect density and N_{A} the boron acceptor density. Since N_{A} is known from CV profiling, the defect density N_{D} can be directly calculated using this equation. In practice, the capacitance is measured 200 s ($= t_s$) after the light has been switched on. ΔC_{SS} is then calculated by

$$\Delta C_{SS} = C(t_S) - C(t = 0) \tag{54}$$

and normalized to C_0, given by

$$C_0 = A \sqrt{\frac{q \varepsilon_0 \varepsilon N_{\text{A}}}{2(V_{\text{bi}} + V_{rev})}} \tag{55}$$

Figure 33 shows the normalized ODLTS spectra ($\Delta C_{SS}/C_0$) as a function of photon energy for different implantation doses. The untreated (not-implanted) diode shows an increase in signal up to 1.15 eV, and remains approximately constant for higher photon energies.

C^+-ion implantation gives rise to a significant modification of this characteristic. As shown in Figure 33, the low energy regime (regime A) is not affected by implantation. However, at about 1.3 eV a steep increase of the spectrum is

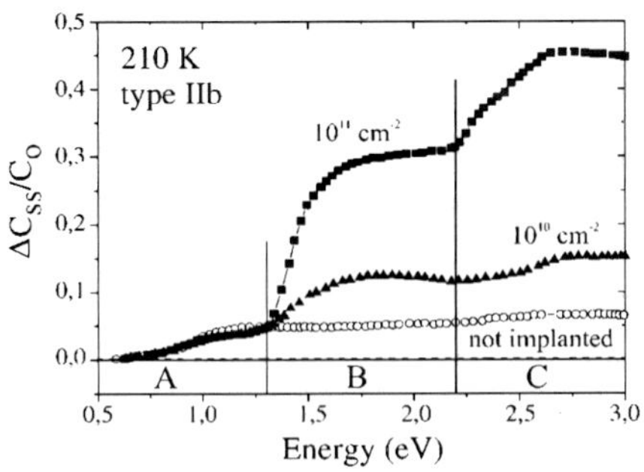

FIG. 33. Normalized ODLTS data measured on IIb-diamond with (a) no implantation damage (open dots) and after carbon implantation with a dose of 10^{10} ions/cm^{-2} (full triangles) and 10^{11} ions/cm^{-2} (full dots), respectively.

observed, saturating in the energy regime 1.8–2.2 eV (regime B). The amplitude of the plateau increases with increasing implantation dose. At photon energies larger than 2.2 eV (regime C) the spectrum increases again, approaching another plateau in the regime 2.6–3 eV. The amplitude of this second plateau is also a function of implantation dose. A discussion of this feature is, however, beyond the scope of this chapter, as internal photoemission (Fig. 27) and defect absorption have to be considered. In the following, we thus discuss only data measured in the regime 0.8–2.2 eV.

The C^+-implantation obviously generates defects with an optical threshold at 1.25 eV. This is in good agreement with data detected on sample A, where also a defect level is detected in non-implanted boron-doped diamond at 1.25 eV.

We have applied isochronal annealing experiments in combination with ODLTS to characterize the dynamics of defect annealing. ODLTS spectra measured after 1 h of isochronal annealing at selected temperatures are shown in Figure 34. The defect densities in region B are constant up to about 550 K then decrease rapidly between 550 and 650 K and approach a constant density at temperatures $\geq$773 K.

The annealing kinetics are shown in Figure 35 and are well described by

$$N(t, T) = N_0 \exp[-r(T)\tau_A] + N_{res} \tag{56}$$

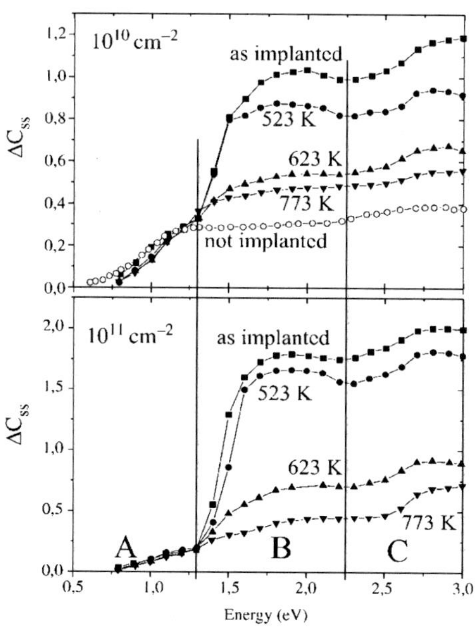

FIG. 34. ODLTS spectra (DCSS) measured after 1 h isochronal annealing at several temperatures on samples with doses: 10^{10} (upper figure) and 10^{11} ions/cm^2 (lower figure). Also shown is the spectrum of the non-implanted diode (open circles, upper graph).

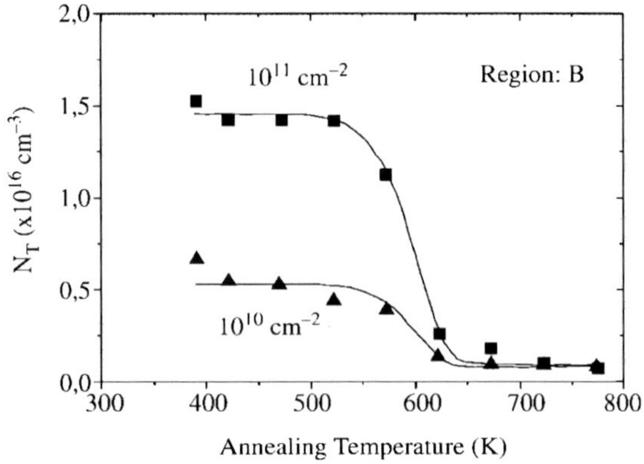

FIG. 35. Annealing kinetics of defects (squares and triangles) measured in the temperature regime 400–780 K. The kinetics are well described by exponential relaxations (full lines) with an activation energy of 1.25 eV. Defect densities are calculated from region B (Fig. 34) after 1 h isochronal annealing at different temperatures.

where r is a thermally activated annealing rate:

$$r(T) = r_0 \exp\left[-\frac{E_{\text{act}}}{kT}\right] \tag{57}$$

Here, E_{act} is the activation energy of diffusing specimens and r_0 the rate pre-factor. r_0 can be calculated by

$$r_0 = \frac{\nu}{n_i} \tag{58}$$

where ν is the Raman frequency $4 \times 10^{13}\ \text{s}^{-1}$ (Raman energy: $\hbar\varpi_{\text{LO}} = 163$ meV (Dean et al., 1965)) and n_{i} the number of jumps needed to find a trapping site. To estimate the number of jumps we assume that this number corresponds to the ration of lattice sites to trapping sites. The density of trapping centers is given by the detected defect density $\approx 2 \times 10^{16}\ \text{cm}^{-3}$ and the density of carbon atoms is $1.77 \times 10^{23}\ \text{cm}^{-3}$ (Kaiser and Bond, 1959). This results in $r_0 \approx 10^7\ \text{s}^{-1}$. The calculated fits to the data (full lines) are shown in Figure 35 and reveal an annealing energy of 1.25 eV for the defect detected by ODLTS at the same energy level.

Obviously, the defect densities in region B increase with implantation dose. We thus conclude that defects detected by ODLTS are most likely vacancies. A comparison with TRIM calculations shows, however, that only 5% of the calculated vacancy density is detected by our ODLTS experiments. From annealing experiments we find a surprisingly small activation energy of 1.25 eV for defect B. For neutral vacancies migration energies of 2.4 eV have been published (Collins, 1992) and typical annealing temperatures are above 1000 K. By characterizing the decay of neutral vacancies using photoluminescence (GR1) Lawson, Davies, Collins and Mainwood (1992) and Collins, Allers and Hiscock (1998) observe a recombination of vacancies after annealing at 460 °C for 130 min. This recombination is attributed to the migration of carbon interstitials to vacancies. The calculated migration energy is 1.6 eV. Massarani and Bourgaoin (1976) published activation energies for the migration of carbon interstitials of even 1.3 eV.

The annealing energy of 1.25 eV detected by us may be related to the same process, namely that carbon interstitials become mobile at temperatures as low as 600 K, diffuse, and finally recombine with vacancies nearby. As our experiments were performed in boron-doped diamond, we assume that the vacancies are positively charged (V^+). The energy of V^+ is most likely different to energies of neutral vacancies, V^0. We, therefore, do not expect too much agreement with the data available in the literature. In addition, the migration of V^+ may be affected by electric fields present in depletion layers where our CV and ODLTS experiments are performed.

3.1.5. Defect Properties

Firstly, we discuss the nature of the defect detected at 0.9 eV. An interesting and plausible model was introduced by Collins and Rafique (1978) who applied photo capacitance measurements. They detected an onset of absorption at $h\nu \geq 1$ eV. By taking into account the boron acceptor level at 0.36 eV and a defect located about 1.25 eV above the valence band edge it is reasonable to assume that the 0.9 eV transition is related to optically induced hole excitations from defects into ionized acceptors, followed by thermal excitation into the valence band as shown in Figure 36. This model is supported by the reasonably good match of transition energies detected by Collins and us.

ODLTS experiments show, that in all investigated boron-doped samples a defect level, about 1.25 eV above the valence band maximum is detected.

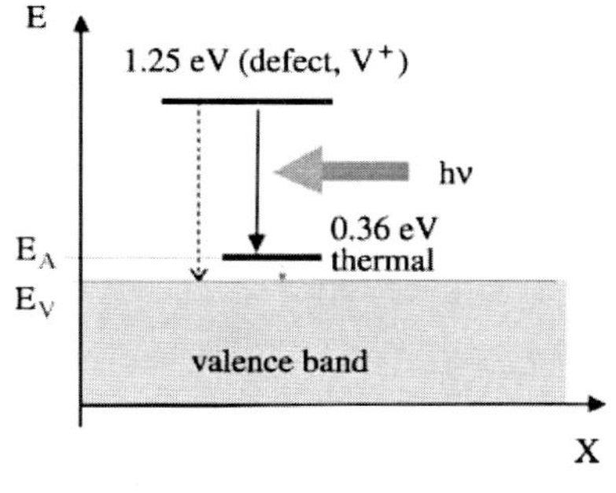

FIG. 36. Schematic defect acceptor model used to discuss the DLTS and ODLTS data. For $h\nu \geq 0.9$ eV holes are excited into the boron acceptor level from which they are thermally excited into the valence band. At higher photon energies ($h\nu \geq 1.25$ eV) direct transitions into the valence band take place.

This defect increases with C^+-implantation and may, therefore, be attributed to carbon vacancies (V^+). On most of the samples, the superposition of the 0.9 and 1.25 eV excitations prevents a detailed characterization of optical and thermal properties of the level at 1.25 eV as the dynamic ranges of the spectra are too low.

Only in one of our boron-doped samples, was the 0.9 eV transition absent, so that we could characterize the deep defect in detail. The results of ODLTS experiments, performed on a homo-epitaxial boron-doped CVD-diamond (sample B), are shown in Figure 37. The absorption profile increases over five orders of magnitude and is temperature dependent. The optical thresholds are at about 1 eV, and the density of this defect is about 10^{16} cm^{-3}. None of the four theoretical cross-sections (Eq. (52)) calculated from Inkson (1981) can be used to calculate reasonable good fits to the data. The measured energy dependence is broader than Inkson's predictions, which is due to thermal broadening of optical transitions.

To take thermal effects into account, we used a model which is schematically shown in Figure 38. Here, the lattice energy plus the electronic energy of the initial (lower parabola) and the final (upper parabola) state is plotted versus a configurational coordinate. After photo-ionization, holes are delocalized in the valence band. The optical transition, therefore, has to include vibronic states of the defect and the DOSs of the valence band. We have shown that such an optical cross-section is well described by (Zeisel, Nebel, Stutzmann, Gheeraert and Deneuville (1999)):

$$\sigma(h\nu) \propto \frac{1}{h\nu} \sum_p W(p) \frac{(h\nu - E_T - p\hbar\Omega)^{\delta+1/2}}{(h\nu - p\hbar\Omega)^{2\gamma}} \tag{59}$$

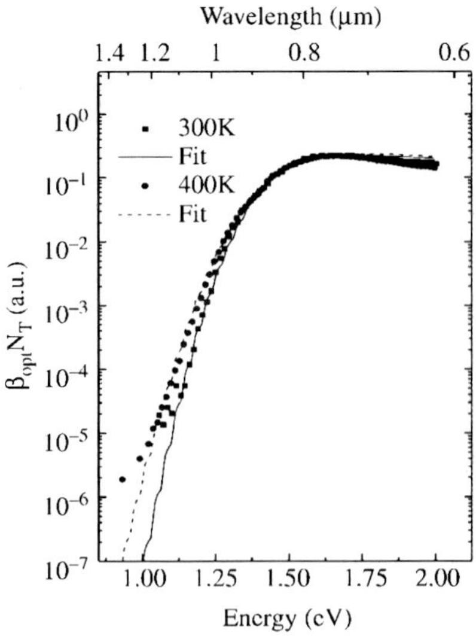

FIG. 37. ODLTS data (full circles, full squares) measured at $T = 300$ and 400 K in sample B and the calculated fits (dashed line, full line).

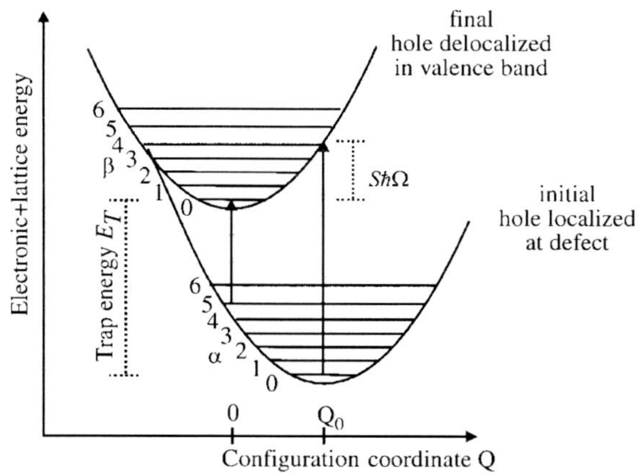

FIG. 38. Configurational coordinate diagram for deep levels. The two parabolas are the initial and final defect state with vibrational levels. The two parabolas are shifted vertically by the trap energy E_T, and horizontally by the change in configuration coordinate Q. Optical transition between initial and final defect states are marked with arrows. The Franck–Condon $S\hbar\Omega$ is also drawn in.

where γ and δ are the same parameters as in Eq. (52), $W(p)$ are weighting functions given by:

$$W(p) = \exp\left[\frac{p\hbar\Omega}{2kT} - S\coth\left(\frac{\hbar\Omega}{2kT}\right)\right] I_p\left[S\cos\mathrm{ech}\left(\frac{\hbar\Omega}{2kT}\right)\right] \tag{60}$$

where I_p is the Bessel function, $\hbar\Omega$ is the energy of phonons, S is the Huang–Rhys factor and $h\nu$, the energy of photons. p is defined to be

$$p = \alpha - \beta \tag{61}$$

where α and β are quantum numbers of the vibrational states of the ground and of the ionized state of the defect (Figure 38), respectively.

Eq. (59) contains five free parameters of interest, namely the trap energy E_T, the two parameters for the photo-ionization cross-section γ and δ, the Huang–Rhys factor S, which is a part of $W(p)$ of Eq. (60), and the phonon energy $\hbar\Omega$. We assume 'allowed' ($\delta = 0$) and 'direct' ($\gamma = 1$) optical transitions from defects into the valence band. Up to now, nothing is known about the phonon energies $\hbar\Omega$ of this defect. In our calculations we assume energies in the range of 30–60 meV which is reasonable as in absorption and luminescence experiments on diamond these energies are commonly detected. S and E_T have been used as fitting parameters.

The best fits are calculated for a defect energy E_T of 1.28 eV (thermal excitation energy), a Huang–Rhys factor S of 4.5, a phonon energy of $\hbar\Omega = 35$ meV, and a constant Franck–Condon shift of 0.16 eV. For $S = 4.5$, the defect lattice coupling is in-between 'weak' and 'strong'. The optical ionization energy can be calculated by:

$$E_{\text{opt}} = S\hbar\Omega + E_{\text{T}} \tag{62}$$

and results in 1.44 eV.

Based on these results, the two excitation thresholds at 0.83–0.9 and 1.25 eV, detected by ODLTS in a variety of samples, are attributed to only one defect at 1.28 eV (thermal excitation energy), from which holes are either directly excited into the valence band (optical excitation energy: 1.44 eV) or into boron acceptor levels at 0.36 eV above E_{v} (optical excitation energy: ≈ 0.9 eV) from where they are thermally excited into the valence band. We attribute the 1.28 eV defect to positively charged vacancies, generated either by carbon implantation (post growth treatment) or carbon ion bombardment during the CVD growth of the films in RF or microwave plasmas.

3.2. Long-living Excited States in Boron-doped Diamond

3.2.1. Introduction

The investigation of boron impurities in diamond has been extremely fruitful over the last 30 years. Lyman spectra in the infrared, electronic Raman effect (Kim, Vogelgesang, Ramdas, Rodriguez, Grimsditch and Anthony, 1997, 1998), and cathodoluminescence (Sharp, Collins, Davies and Joyce, 1997) in the ultraviolet have revealed most of the electronic structure of this 'shallow acceptor'.

The electronic properties of boron in diamond are, however, still not fully understood. Pokrovskii, Smirnova and Khvalkovskii (1995), Pokrovskii, Altukhov, Smirnova and Khvalkovskii (1998) and Pokrovskii, Smirnova and Khvalkovskii (1996) have shown that the lifetime of photo-excited carriers in excited states in some group III and group V impurities in diamond, silicon and germanium is a few orders of magnitude longer than that of free carriers. They applied DC- and modulated photoconductivity experiments in the infrared, which showed that in boron-doped diamond with concentrations $>10^{18}$ cm^{-3} and a high compensation degree (>0.1) at low temperatures hopping photoconductivity (HPC) takes place in two excited states of boron, having ionization energies of 130 and 170 meV. The lifetimes of holes in these states at low temperatures ($T < 90$ K) is 4×10^{-8} and 2×10^{-7} s, respectively, which is about 3–4 orders of magnitude larger than the lifetime of holes in the valence band ($\approx 10^{-11}$ s). These long-living excited states are much deeper in energy than well-known excited states of the $p_{3/2}$-Lyman series. The prolonged lifetime is a result of the forbidden transition to the 1s ground states. 'Forbidden' refers to the fact that optical transitions between even parity states are not allowed by the electric-dipole selection rule (Kim et al., 1998). Smith and Taylor (1962) calculated that there must be a forbidden transition at about 266 meV to explain the anomalously large width of the absorption peak at 304 meV. Collins, Lightowlers and Dean

(1969) found three forbidden transitions at 240, 266 and 289 meV, applying photoconductivity in the IR where the capture of holes from the valence band to these states together with a cascade emission of one or more optical phonons generates a series of minima in the spectral response. The forbidden transition at 268 meV has been detected by optical absorption in a heavily doped (5×10^{18} cm^{-3}) synthetic diamond (Collins, Dean, Lightowlers and Sherman, 1965) where the transition becomes partially allowed due to some mixing of electronic states in the heavily doped specimen (Kohn, 1957). It is assumed that the ground and excited levels are of even parity and, therefore, may be Raman active. Up to now, however, electronic Raman experiments could not detect these transitions as the Raman cross-section for transitions to higher lying even parity states are significantly reduced and have thus escaped experimental observations (Kim et al., 1998).

In this chapter we discuss long-living excited boron states measured by SPC in the energy range 3–3.6 eV (Nebel, Rohrer, and Stutzmann, 2001). Three well-defined photoconductivity maxima are detected at 3.35, 3.4 and 3.44 eV in highly boron-doped diamond ($(4-8) \times 10^{18}$ cm^{-3}) which invert to minima with decreasing boron density. These features are interpreted as being due to the rapid capture of photo-excited holes from the valence band to the long-living excited states of boron, together with the emission of one or more LO phonons. At high doping densities the wave-function overlap to neighboring excited states is sufficient to allow tunneling transitions which gives rise to the hopping photocurrent (HPC) maxima at well-defined excitation energies whereas at low doping concentrations with negligible wave-function overlap well pronounced minima occur at the same photon energies.

3.2.2. Experimental Details

Boron-doped CVD-diamond specimens have been grown homo-epitaxially on insulating highly oriented (100 texture) CVD-diamond (Jiang et al., 1999) at 770 °C, using 2 sccm CH_4, 491 sccm H_2 and a mixture of tri-propyl-borate in H_2. Three samples with 0.6 μm thickness have been grown with doping densities of 8×10^{18} cm^{-3} (sample A), $(5-7) \times 10^{18}$ cm^{-3} (sample B) and $\approx 10^{17}$ cm^{-3} (sample C) as analyzed by ERDA. The nitrogen content in these layers is in the range of $(1-4) \times 10^{17}$ cm^{-3}. To extend these experiments to lower doping concentrations, a HPHT synthetic IIb-diamond has also been investigated (sample D), with an uncompensated acceptor density of 5×10^{16} cm^{-3} as measured by Hall experiments.

The photoconductivity experiments were carried out between 77 and 500 K, using a liquid nitrogen cryostat. An Xe-arc lamp illuminates a 0.34 m monochromator equipped with a 1200 cm^{-1} grating, blazed at 250 nm. The spectral resolution of the monochromator in the energy range 3–3.7 eV is about 10 meV. The monochromatic light is focused on the coplanar Ti/Pt/Au contacts on top of the diamond. The contacts, separated by 1 mm apart, show ohmic characteristics in the applied temperature and electric field regime ($<2000\ V^{-1}\ cm^{-1}$). The photocurrent is detected by a lock-in amplifier. In all cases, a light modulation frequency of 3 Hz was used. The photocurrent spectra are normalized by the photon flux measured by a pyroelectric sensor. Bias light illumination was applied using a halogen lamp ('white light') or a second monochromator as a light source.

3.2.3. Results

Figure 39 shows normalized photocurrent spectra measured at 85 K with and without white bias light illumination (note that the bias light is not chopped). In sample A, pronounce peaks at 3.34, 3.4 and 3.44 eV can be detected, in sample B peaks at 3.36 and 3.4 eV are revealed, using white bias light illumination (the highly doped specimens). In sample C, just a small peak at about 3.35 eV is measured, whereas in sample D a photocurrent quenching is observed in the same spectral regime. Obviously, peaks invert with decreasing doping density. Without white bias illumination only the dominant peak at 3.35 eV can be seen in A and B.

The large peak (P1) at 3.34 eV in sample A is broadened on the low energy side resulting in a FWHM of about 50 meV. The other two peaks at 3.4 (P2) and 3.44 eV (P3) are significantly smaller in amplitude and just slightly larger than the emerging photocurrent continuum. P1 and P2 are separated by 60 meV and P2 and P3 by 40 meV. Such discrete structures in photoconductivity are generally not expected for optical transitions from a deep defect level into the valence band.

The peak energies and amplitudes are strongly temperature dependent as shown in Figure 40. The maxima are largest at 150 K decreasing towards higher and lower temperatures. Above 190 K they disappear completely. A comparison with conductivity data shown in Figure 41 indicates that the spectral structure emerges in the hopping regime of the conductivity ($T < 200$ K). Above 190 K the conductivity is singly activated with 300 meV, here transport in the valence band dominates, the acceptors are thermally ionized. The sharp rise of the photocurrent and the onset of hopping transport in the dark conductivity (Fig. 41) below 200 K indicate that hopping transport also dominates the photoconductivity below 190 K. The temperature dependent amplitude of the dominant peak at about 3.34 eV was evaluated by subtracting the offset at 3.1 eV from the maxima. Finally, the difference was normalized by the largest amplitude calculated at 150 K.

In samples A and B the photocurrent emerged at low temperatures if the specimen was simultaneously illuminated with bias light. Therefore, we applied photocurrent excitation spectroscopy experiments to investigate the effects of bias illumination. Here, the photocurrent generated by the absorption of the three peaks is detected as a function of monochromatic bias light

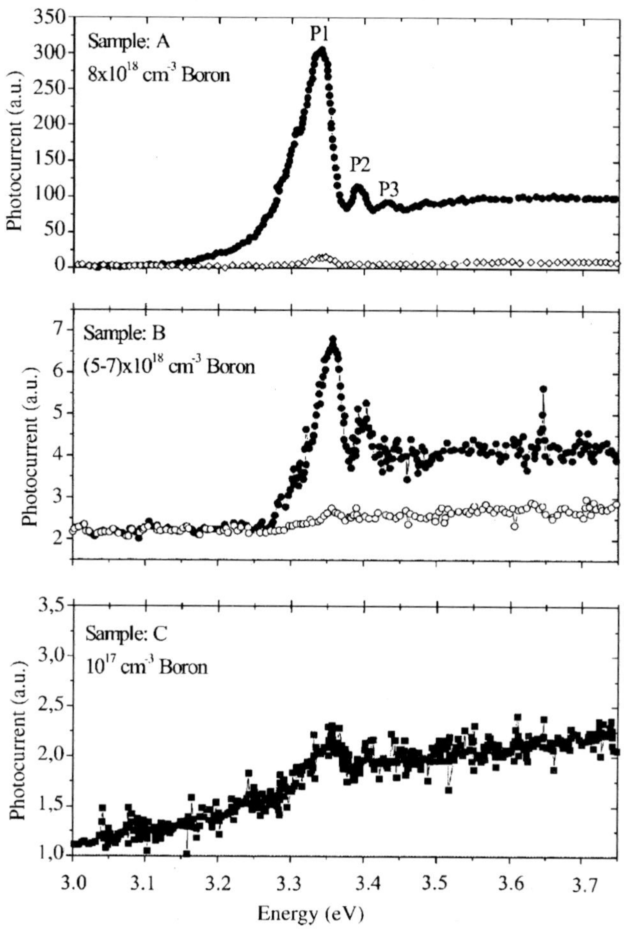

FIG. 39. Spectrally resolved photocurrent experiments measured at $T = 85$ K on boron-doped diamond layers containing different doping concentrations as indicated. Full dots and full squares represent data measured with CW white bias illumination. Open dots represent data detected without CW-bias.

illumination. We used a band pass filter which is transparent between 3.1 and 3.6 eV to excite the peaks, this light is chopped. The bias light (not chopped) was varied between 0.8 and 4 eV. We found that the only radiation effective in producing these enhancements was one, with quantum energies larger than 3.05 eV as shown in Figure 42. The amplitude increases approximately linearly showing a well-defined dip at 3.35 eV. This is clearly a 'spectral hole burning' effect where the CW illumination decreases the occupation of the ground level significantly so that the excitation via the weak chopped light becomes too small to be detected. A weak second minimum about 165 meV higher in energy (Fig. 42) is also detected, indicating phonon assisted features.

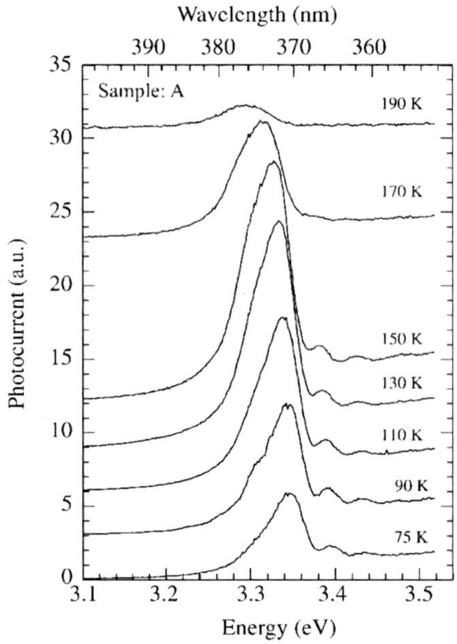

FIG. 40. Temperature dependent spectrally resolved photocurrent experiments measured with CW white bias illumination in sample A containing 8×10^{18} cm^{-3} boron. With increasing temperature the peaks shift towards lower energies and vanish above 190 K. Note, the photocurrents are shifted vertically for better comparison.

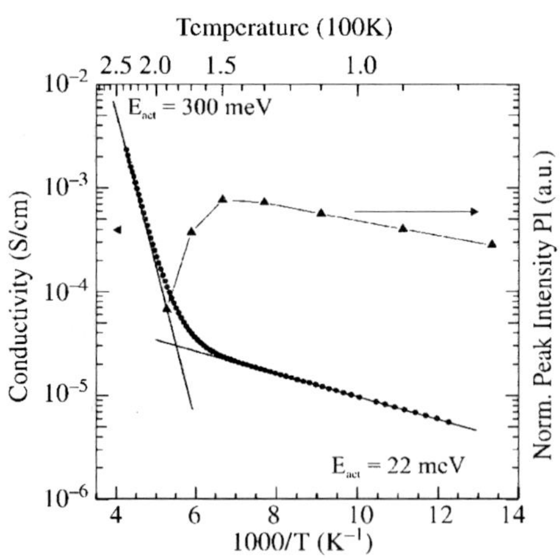

FIG. 41. Dark conductivity of sample A and normalized P1 amplitude plotted as function of reciprocal temperature. The P1 maximum emerges in the hopping regime of the conductivity where $E_{act} = 22$ meV.

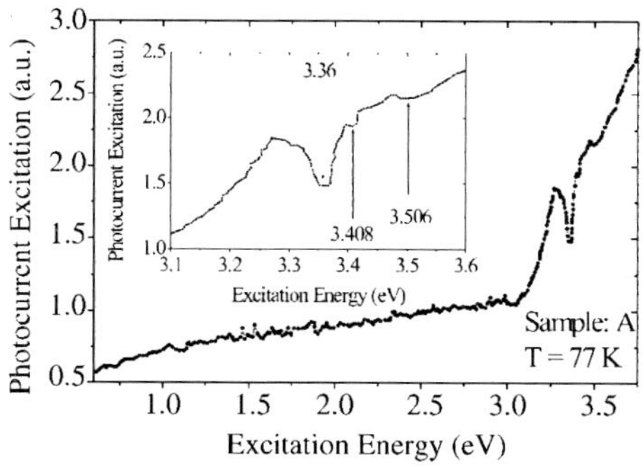

FIG. 42. PES in sample A measured at 77 K. A threshold is detected at 3.05 eV and a dip at 3.36 eV. The insert shows the expanded regime 3.1–3.6 eV where three dips are detected at 3.36, 3.408 and 3.506 eV.

3.2.4. *Discussion*

Based on the facts that photoconductivities: (a) show distinct maxima, (b) are related to hopping transport in boron acceptor states, and (c) are stimulated by bias light, we would now like to discuss the model shown in Figure 43.

The acceptors are partially compensated by deep defects from which holes can be optically excited into the valence band. Holes in the valence band are captured very fast into excited levels of boron from which they rapidly thermalize into the 1s ground state. Several excited states with forbidden optical transitions to the ground level (long-living states) accumulate holes. At such high doping densities ($N_B \geq 5 \times 10^{18}$ cm^{-3}), the average distance r_A between boron atoms can be calculated by:

$$r_A = \left(\frac{4\pi}{3}N_B\right)^{-1/3} \tag{63}$$

Inserting N_B results in an average distance $r_A < 36$ Å. The Bohr radius r_{Bohr} of the ground state is about (3–8) Å (Borst and Weis, 1996), of excited states r_{Bohr} is increasing, following (Böer, 1990)

$$r_{Bohr} = r_{Bohr,n=0}n^2 \tag{64}$$

where n is the principle quantum number. For $n = 2$ the Bohr radius $r_{Bohr} = (12–32)$ Å, indicating that the wave function overlap to neighboring states is high, therefore, hopping photoconductivity takes place in these states. In sample C with 10^{17} cm^{-3} boron $r_A = 133$ Å, a very small peak at the detection limit can be measured while in sample D with about 5×10^{16} cm^{-3} and $r_A = 168$ Å a photocurrent quenching takes place.

According to Collins et al. (1969) and Pokrovskii et al. (1995, 1996, 1998) the long-living excited states are 200, 240, 266 and 289 meV above the 1s ground level. They capture holes if the excess energy can be emitted by a single or

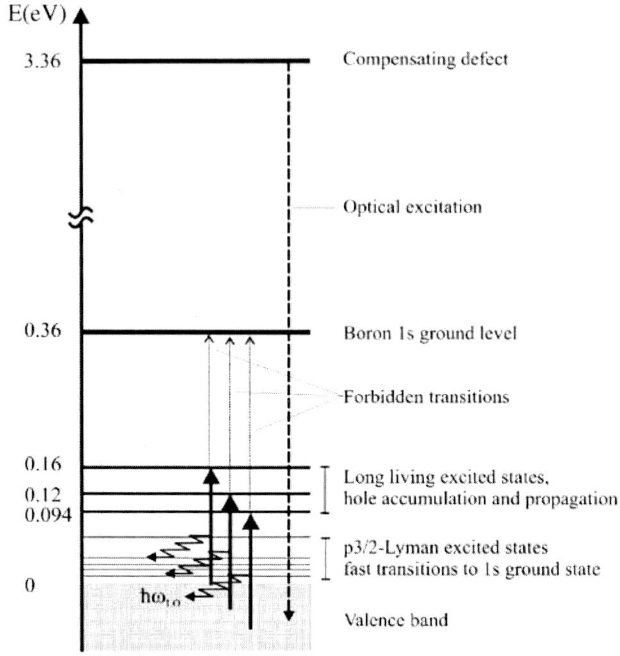

FIG. 43. Schematic energy diagram of a deep defect and boron acceptor states in diamond. Holes are optically excited from a compensating deep defect 3.36 eV above E_V into the valence band. Long-living excited states capture holes, if the excess energy can be emitted by a single or multiple LO phonon emission. The propagation of holes in long-living excited states gives rise to the detected photocurrent maxima. Excited states of boron with allowed dipole transitions to the 1s ground state have a too short lifetime to accumulate holes and to generate a photoresponse maximum.

multiple emission of LO phonons of 165 meV energy, like in oscillatory photocurrent experiments.

Assuming single LO emission, the three levels are 3.185, 3.235 and 3.275 eV above the compensating defect and should relate to the 200, 240 and 266 meV forbidden transitions of the acceptor. This is in reasonable agreement with the energy gap between the maxima which is 60 meV between P1 and P2 and 30 meV between P2 and P3. The population of these states should be dependent on the exciting photon energy and, therefore, should vary with periods of 165 meV. In photoconductivity excitation spectroscopy (PES) (Fig. 42, arrow) we detect a second minimum about 165 meV higher in energy which supports this requirement. In Figure 39, there are some modulations, they are too small to be used as a convincing argument. Collins et al. (1969) showed that the two phonon emission process is hardly detectable. They demonstrated in the case of two or more phonon emissions that the phonon energy becomes less well defined and coupling to fine structure in the absorption spectrum is less apparent. Thus, these transitions have hardly been resolved.

Finally, we want to address the bias light effect. The onset energy is 3.05 eV indicating that the excitation of deep trapped holes increases the population of the long-living states according to the establishment of quasi Fermi levels. While the high doping concentration is required to allow hopping photoconductivity in the long-living states, it partially destroys the forbidden nature of the transition. The bias illumination causes an occupation of the acceptor ground level, which prevents the fast transitions from the long-living states to the 1s ground state in the case of resonant capture by single LO phonon emission, and therefore, carrier accumulation is achieved. For decreasing doping densities the mixing of electronic states decreases and the transitions become truly forbidden. However, the hopping propagation also vanishes due to the increasing distance between neighboring sites. In this case, carriers accumulate in long-living states, and cause a quenching of the photoresponse.

3.3. Deuterium Passivation of Boron

Hydrogen rapidly diffuses in boron-doped diamond (Chevallier, Theys, Lusson, Grattepain, Deneuville and Gheeraert, 1998; Chevallier, Ballutaud, Theys, Jomard, Deneuville, Gheeraert and Pruvost, 1999), as shown in Figure 44. To explore the electronic variations caused by hydrogen in diamond, we have combined deuteration and CV-experiments (Nebel, Zeisel and Stutzmann, 1999; Zeisel et al., 1999)). In our experiments deuterium radicals are generated in a DC plasma discharge, which is about 10 cm upstream from the sample and electronically confined (Fig. 45). The HPHT IIb-diamond layers are heated to 400 °C on a substrate holder and exposed to deuterium radicals at pressures of typically 0.8 mbar for a given time. Then, the layers are cleaned in a $CrO_3:H_2SO_4$ acid to remove hydrogen from the surface. Subsequently, Ag is evaporated to form Schottky contacts, for Ohmic contacts we used Ti/Pt/Au.

In Figure 46 CV-plots measured before and after such a deuteration are shown. The capacitance is drastically decreased after deuteration. Annealing at 900 °C for 1 h recovers the initial *CV*-properties. This is in agreement with recent results of Chevallier et al. (1999).

The decrease of capacitance is due to a corresponding increase in the depletion layer width. We assume that this is a result of hydrogen interaction with boron. Chevallier et al. (1998, 1999) proposed that H^+ is the diffusing species in p-type diamond. H^+ and B^- acceptors subsequently form neutral BH-pairs. Theoretical considerations by Mehandru and Anderson (1994) support this interpretation. These authors predict the formation of a mid-gap donor level by hydrogen which traps the electron from neutral hydrogen ($B^0 + H^0 \rightarrow H^+ + B^-$), leaving a mobile H^+ which neutralizes the B^- states (formation of a neutral BH-complex). As a result a compensated semiconductor is generated within the depth of deuterium diffusion.

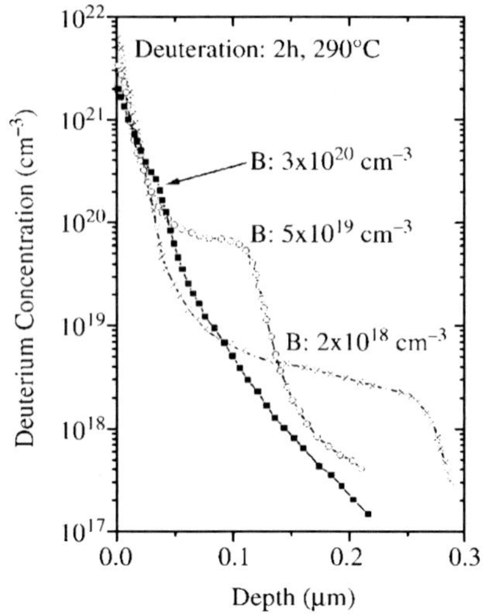

FIG. 44. Deuterium diffusion profiles in homo-epitaxial boron-doped diamond as a function of boron concentration. Samples were exposed to a deuterium RF-plasma at 290 °C for 2 h (data from Chevallier et al., 1998).

To discuss the decrease in capacitance after deuteration we adopted this model and assumed a metal/compensated diamond/p-diamond structure as shown in Figure 47. The capacitance $C(V)$ can be calculated by

$$C(V) = \frac{\varepsilon_0 \varepsilon_r A}{\sqrt{\frac{2\varepsilon}{qN_A}(V_{bi} + V) + L_D^2\left(1 - \frac{N_{AP}}{N_A}\right)}} \tag{65}$$

where ε_0 is the dielectric constant, $\varepsilon_r = 5.7$, A is the contact area, N_A is the acceptor density, V_{bi} the built-in potential, V is the applied external voltage, L_D is the depth of boron passivation (= deuterium diffusion length) and N_{AP} the remaining acceptor density in the passivated region. The best fit to our data is shown in Figure 46 (full line) and has been calculated assuming $L_D = 700$ nm,

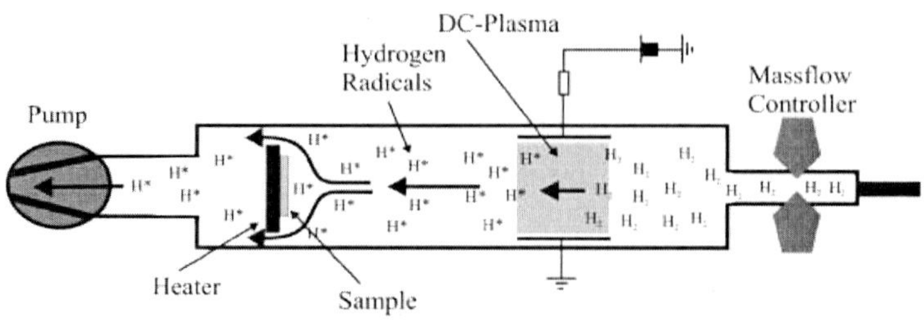

FIG. 45. Experimental set-up of the DC deuterium plasma excitation for boron passivation. Typical parameters used: pressure: 0.8 mbar, DC voltage: 1.3 kV, hydrogen flow: 20 sccm.

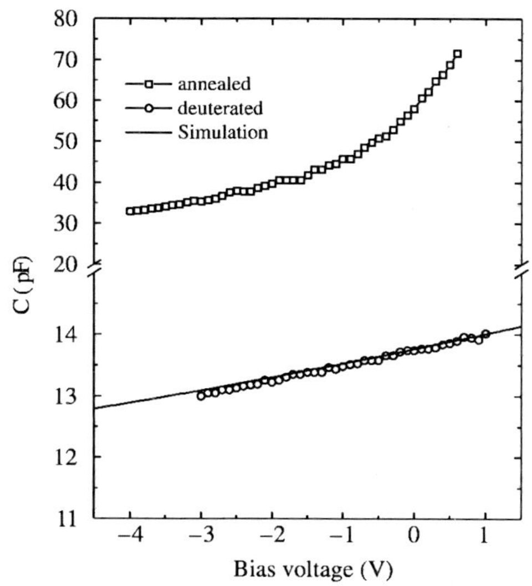

FIG. 46. CV-plot measured before and after a deuteration at 400 °C in sample A. Full line is the calculated fit based on Eq. (65).

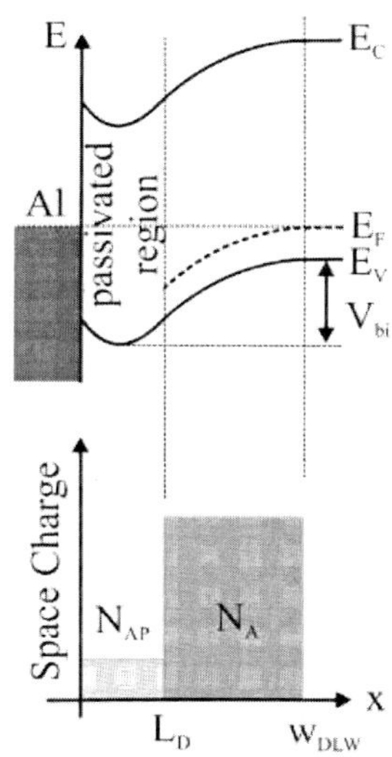

FIG. 47. Schematic model of the hydrogen compensation of boron-doped diamond. Within the diffusion length of deuterium, L_D, the acceptors are passivated. This causes a decrease in C as shown in Figure 46. N_{AP} is the remaining acceptor density in the deuterated layer, N_A is the acceptor density and w_{DLW} the depletion layer width.

$N_{AP} < 10^{14}\ \mathrm{cm}^{-3}$ and $V_{\mathrm{bi}} = 2.1$ V. The calculated fit is reasonably good, indicating that within the diffusion depth of deuterium most of the boron is compensated. Taking into account the compensation depth of deuterium ($L_{\mathrm{D}} = 700$ nm) the deuteration temperature ($T = 400\,°\mathrm{C}$) and duration ($t_0 = 120$ min), we have calculated the effective diffusion coefficient $D = L_{\mathrm{D}}^2/t_0$ which is $4.7 \times 10^{-13}\ \mathrm{cm}^2\ \mathrm{s}^{-1}$. This is about a factor five faster than data reported by Chevallier et al. (Fig. 48). The origin for this discrepancy is up to now not yet clear. One significant difference to the experiments performed by Chevalier et al. (1998, 1999) is the much lower boron density in our case. Our samples typically contain $10^{16}\ \mathrm{cm}^{-3}$ boron whereas Chevallier et al. used layers with typically $10^{19}\ \mathrm{cm}^{-3}$ or more.

As pointed out by Mehandru and Anderson (1994) the hydrogen release from a boron–hydrogen complex in diamond is either positively charged or neutral, corresponding to the two possible reactions:

$$\mathrm{B{:}H} \rightarrow \mathrm{B}^- + \mathrm{H}^+$$

or

$$\mathrm{B{:}H} \rightarrow \mathrm{B}^- + \mathrm{H}^0 + \mathrm{H}^+$$

To investigate whether hydrogen is positively charge or not, we have applied bias annealing experiments on deuterated samples (Zeisel, Nebel and Stutzmann, 2000). We used Ag Schottky contacts which show *CV* characteristics as shown in Figure 49a (open symbols). This *CV* characteristic has been explained above with a two layer model (Fig. 47). The slight voltage dependence of the *CV* data is due to the highly resistive passivated layer, where most of the applied bias

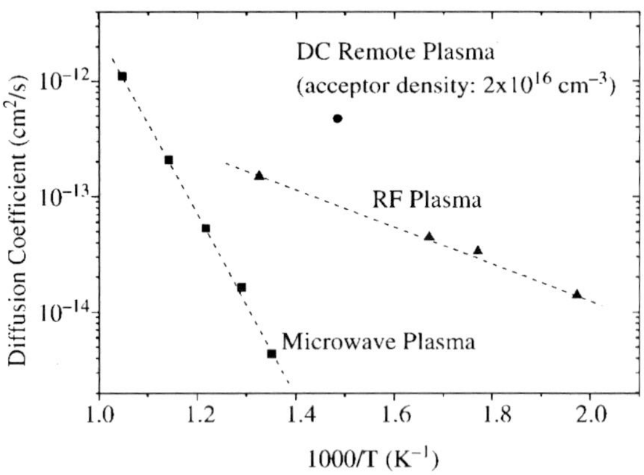

FIG. 48. Deuterium diffusion coefficients as measured by several passivation techniques. The microwave and RF plasma passivation data are taken from Chevallier et al. (1998, 1999). These diffusion experiments are performed on layers with $10^{19}\ \mathrm{cm}^{-3}$ boron or more. The full dot is the result of our calculation for the remote deuterium passivated boron-doped layer with an acceptor density of $10^{16}\ \mathrm{cm}^{-3}$.

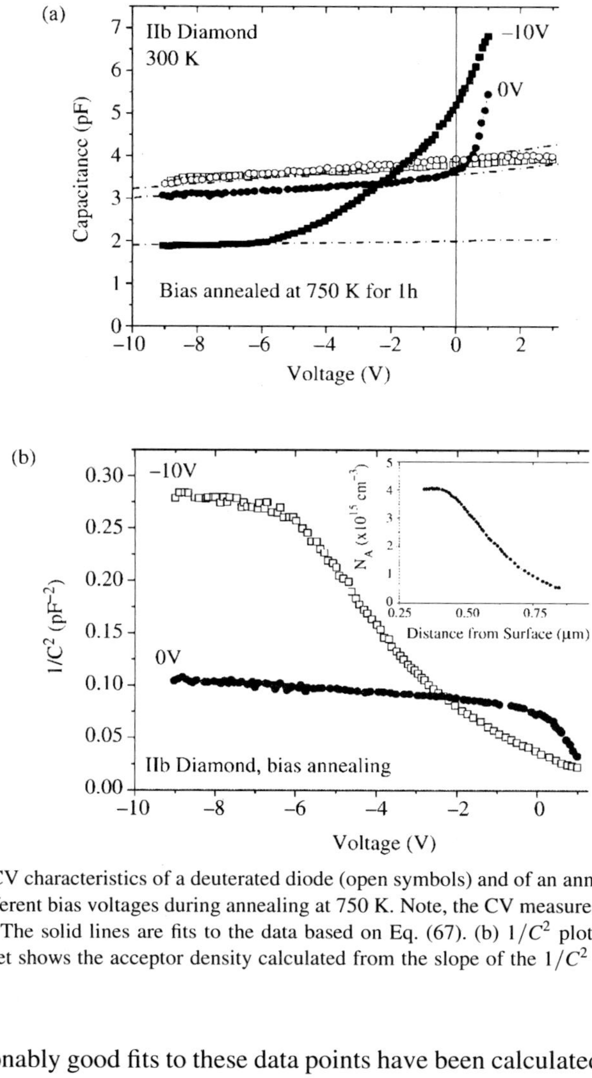

FIG. 49. (a) CV characteristics of a deuterated diode (open symbols) and of an annealed diode (full symbol) for different bias voltages during annealing at 750 K. Note, the CV measurements have been done at 300 K. The solid lines are fits to the data based on Eq. (67). (b) $1/C^2$ plot of the annealed diodes. The inset shows the acceptor density calculated from the slope of the $1/C^2$ data.

drops. Reasonably good fits to these data points have been calculated by Eq. (65). The diodes were then heated for 30 min at 630 K in the presence of an applied electric field. No significant change of the *CV* characteristic could be detected. This is in agreement with the results of Chevallier et al. (1999).

Annealing at 750 K for 60 min, however, is changing the *CV* characteristic significantly, as can be seen in Figure 49 (full symbols). We have performed these annealing experiments with zero bias field (0 V) and with − 10 V reverse bias. The results show that at 750 K or above deuterium is diffusing and, therefore, changing the *CV* characteristic. Especially the − 10 V reverse bias experiment results in a significantly changed *CV* dependence (Fig. 49a), approaching properties of a diode, which has not been deuterium treated. The diode annealed without bias (0 V), varies weakly in the regime -9 to 0 V, turning into a strong increase of C for forward voltages above 0 V. This is also shown in Figure 49b where a $1/C^2$ plot is displayed.

These features are discussed, based on a field enhanced deuterium diffusion model, as shown in Figure 50. During annealing, the boron–hydrogen complexes

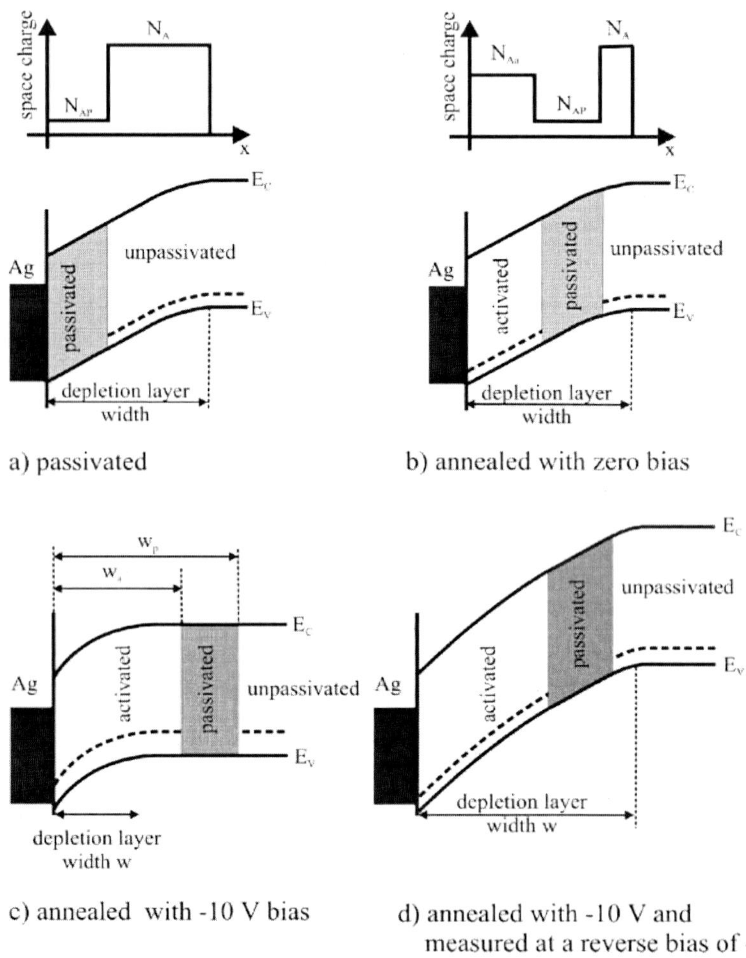

FIG. 50. Schematic hydrogen diffusion model in the depletion layer of a Schottky diode (a) after hydrogen in-diffusion a passivated surface layer is generated; (b) after annealing at 750 K for 60 min at zero applied external bias. Hydrogen propagates deeper into the bulk. The motion is partially diffusion and partially drift in the depletion field of the Schottky diode. (c) After annealing at 750 K for 60 min in the presence of a bias field (-10 V). Hydrogen has propagated out of the depletion layer of the diode and passivates boron deep in the bulk. (d) Applying a reverse voltage to the diode shown in (c) extends the depletion layer deep into diamond. The passivated regime becomes part of it again and affects the CV characteristics as shown in Figure 49.

are broken and hydrogen is released. Positively charged hydrogen will drift in the electric field of the diode depletion layer. The 'diffusion' of hydrogen is, therefore, accelerated by a drift, and hydrogen will be swept out of the depletion layer. Sooner or later hydrogen will be trapped again, forming a passivated layer, which has moved deeper into the bulk of diamond. The depletion layer of the Ag-Schottky junction recovers properties of a diode, which has not been exposed to hydrogen. This is schematically shown in Figure 50(a–d). It is interesting to note, that similar annealing experiments have been applied on hydrogen treated p-doped silicon (Tavendale, Alexiev and Williams, 1985), with comparable results.

For the diode annealed at 750 K for 60 min at − 10 V bias, we have calculated the doping profile, by (Sze, 1981):

$$N_\mathrm{A} = \frac{2}{A^2 q\varepsilon_0\varepsilon_\mathrm{r}}\frac{1}{\mathrm{d}(1/C^2)/\mathrm{d}V} \tag{66}$$

The result is plotted in Figure 49b as inset, and shows clearly that the acceptor density is decreasing in the range 400–800 nm from the metal contact.

For a better discussion of the CV data shown in Figure 49, we have solved the Poisson equation for partially passivated diodes, as shown in Figure 50b and d (Zeisel, 2001). The total width $w(V)$ of the depletion layer is given by

$$w(V) = \sqrt{\frac{2\varepsilon_0\varepsilon_r}{qN_\mathrm{A}}(V_\mathrm{bi} + V) + (w_\mathrm{p})^2 + \frac{N_{AP}}{N_\mathrm{A}}(w_a^2 - w_\mathrm{p}^2) + \frac{N_{Aa}}{N_\mathrm{A}}w_\mathrm{A}^2} \tag{67}$$

where the w_a and w_p are defined in Figure 50c. N_{Aa}, N_{AP} and N_A represent the activated, passivated and bulk acceptor densities, respectively (Fig. 50b). Based on Eq. (67) we have calculated fits to the data, shown in Figure 49a. Note, that this model simulates properties, where the deuterium-passivated regime is part of the depletion layer. In case of the diode annealed at − 10 V reverse bias, this is the case only for strong reverse voltages, namely in the regime − 6 to − 10 V (Figure 50d). For voltages in the range − 6 to + 1 V, the deuterium-passivated regime is not anymore part of the depletion layer, and Eq. (67) cannot be used. This regime can be discussed based on

$$w(V) = \sqrt{\frac{2\varepsilon_0\varepsilon_r(V_\mathrm{bi} + V_{rev})}{qN_{Aa}(x)}} \tag{68}$$

where $N_{Aa}(x)$ is the activated boron acceptor density.

In the following, we discuss application of Eq. (67) to fit the data shown in Figure 49a. The results (dashed lines) are in a reasonably good agreement with the experimentally data. Parameters used are $N_\mathrm{A} = 2.2 \times 10^{16}\ \mathrm{cm}^{-3}$, $N_{Aa} = 4 \times 10^{15}\ \mathrm{cm}^{-3}$, $N_{AP} = 5 \times 10^{13}\ \mathrm{cm}^{-3}$. For the diode annealed at zero voltage we get: $w_\mathrm{A} = 500$ nm and $w_\mathrm{p} = 750$ nm, and for the diode annealed at − 10 V bias: $w_a = 800$ nm and $w_\mathrm{p} = 1.4\ \mu$m.

Obviously, deuteration experiments in the presence of electric fields cause significant variations. The first deuteration experiment, performed without electric field at 673 K for 120 min showed that hydrogen diffused into diamond. The passivated width was calculated to extend 700 nm deep into the bulk. Subsequent annealing of the diamond at 750 K for 60 min in the presence of electric fields of a diode, gives rise to a change in hydrogen distribution. In case of zero bias annealing of the Ag/diamond Schottky junction, a layer of about 500 nm width recovers to its original doping properties. In case of a − 10 V bias annealing, the width of the recovered layer increases to even 800 nm.

These calculations and measurements show that hydrogen propagates as a positively charged particle (H^+) in boron-doped diamond. The diffusion can be accelerated by electric fields (drift enhancements). The evaluation of diffusion experiments has to take this into account, as especially in DC, AC or microwave plasma deuteration treatments electric fields are present which may cause drift enhanced diffusion of hydrogen in p-type diamond.

3.4. Summary of Defect and Acceptor Properties

CV-, DLTS/ODLTS-and photoconductivity-experiments provide us with a new insight into the properties of boron-doped diamond. We detect a new defect at 1.28 eV above the valence band maximum (thermal activation energy). The defect density can be enhanced by carbon implantation and partially annealed at rather low temperatures between 500 and 700 K, with activation energy of 1.25 eV. The defect lattice coupling is in-between weak and strong (Huang–Rhys factor $S = 4.5$), the Franck–Condon shift is 160 meV, so that the optical excitation energy is 1.44 eV. It is very likely, that these defects are positively charged vacancies (V^+); however, further investigations are required to confirm this interpretation. This defect is present in all the investigated boron-doped CVD films, which may be due to carbon ion bombardment during the growth of diamond.

In addition, in most of the investigated films a shallow transition was detected by DLTS at 0.83 eV (thermal excitation) and by ODLTS at about 0.9 eV (optical transitions). We attribute this feature to holes, excited from defects at 1.28 eV into boron acceptor 1s ground states at 360 meV above E_v, from where they are thermally excited into the extended states of the valence band.

The interactions of holes with excited levels of boron in diamond have been characterized by SPC experiments. Three long-living excited states in the spectrum of the hydrogen like boron acceptor have been detected at 200, 240 and 266 meV above the 1s ground state, which trap holes by resonant LO-phonon emission. This gives rise to pronounced SPC maxima or even to quenching effects at distinct photon energies. The discussion of the results is based on theories developed for boron as a hydrogen-like center in diamond, where besides the 1s ground state a series of excited states are existent. A few of them are electric dipole-transition forbidden, giving rise to lifetime enhancements of trapped holes.

Finally, we have shown that deuterium passivates boron in diamond. Deuterium, diffused into diamond at temperatures in the range of 400 °C, causes the formation of a passivated layer, which gives rise to variations of the *CV* dependence. This has been discussed based on a boron–deuterium-passivation model. Applying bias-voltages during annealing experiments accelerates the deuterium diffusion, which supports the model, that deuterium propagates as a positively charged particle in boron-doped diamond.

These experiments demonstrate that the electronic properties of diamond are still either fully characterized nor fully understood. Although, research on diamond is now nearly half a century old, new and exciting features being detected. The understanding of the electronic properties of diamond is required to further optimize growth and doping qualities, to produce a genuinely new wide-band-gap semiconductor for applications in electronics, ranging from electrochemistry, UV-sensors and LEDs, to bio-electronic devices.

Acknowledgements

The author is greatly indebted to C.F.O. Graeff (EPR), J. Münz (transient photoconductivity), E. Rohrer (SPC and PDS), A. Waltenspiel (spectrally resolved PC and persistent PC) and R. Zeisel (CV, DLTS and ODLTS experiments) who performed most of the experiments either as Diploma or PhD students, or as post-docs, to M. Stutzmann for stimulating discussions and to A. Flöter, P. Koidl, C.-P. Klages, X. Jiang and R. Zachai for growing CVD diamond films. We gratefully acknowledge the support of samples by E. Gheeraert and A. Deneuville, from Laboratoire d'Etudes des Proprietes Electroniques des Solides, and CNRS Grenoble for their help with samples. We thank H. Sternschulte from the Universität of Augsburg who did the carbon implantation and TRIM calculations. The author would also like to thank the Institute of Physics, the American Physical Society, Elsevier and Wiley-VCH for the permission to reproduce figures. The work was partially funded by the Bundesministerium für Bildung, Wissenschaft, Forschung und Technologie under contract number 03N1001G4.

References

P. Bergonzo, D. Tromson, C. Mer, B. Guizard, F. Foulon, and A. Brambilla, Particle and radiation detectors based on diamond, *Phys. Status Solidi A* 185, 167–181 (2001).

K. W. Böer, "Survey of Semiconductor Physics". Van Nostrand Reinhold, New York, 1990, p. 492.

E. Boettger, "Polykristalline Diamantschichten für elektronische Anwendungen". PhD Thesis. Universität Hamburg, p. 95 (1995).

T. H. Borst and O. Weis, Boron-doped homoepitaxial diamond layers: fabrication, characterization and electronic applications, *Phys. Status Solidi A* 154, 423–444 (1996).

J. L. Bredas and G. B. Street, Electronic properties of amorphous carbon films, *J. Phys. C: Solid State Phys.* 18(21), L651–L655 (1985).

J. Chevallier, D. Ballutaud, B. Theys, F. Jomard, A. Deneuville, E. Gheeraert, and F. Pruvost, Hydrogen in Monocrystalline CVD Boron Doped Diamond, *Phys. Status Solidi A* 174, 73–81 (1999).

J. Chevallier, B. Theys, A. Lusson, C. Grattepain, A. Deneuville, and E. Gheeraert, Hydrogen-boron interactions in p-type diamond, *Phys. Rev. B* 58, 7966–7969 (1998).

A. T. Collins, The characterisation of point defects in diamond by luminescence spectroscopy, *Diamond Relat. Mater.* 1(5/6), 457–469 (1992).

A. T. Collins, Breakdown field and saturation carrier velocity in diamond in "Properties and Growth of Diamond" Emis Data Reviews Series No. 9 (G. Davies, Ed.), pp. 288–289. INSPEC Publications, 1994a.

A. T. Collins, n-Type conduction and electron mobility in diamond in "Properties and Growth of Diamond" Emis Data Reviews Series No. 9 (G. Davies, Ed.), pp. 284–287. INSPEC Publications, 1994b.

A. T. Collins, L. Allers, and J. Hiscock, The annealing of interstitial-related optical centres in type II natural and CVD diamond, *Diamond Relat. Mater.* 7(2–5), 228–232 (1998).

A. T. Collins, P. J. Dean, E. C. Lightowlers, and W. F. Sherman, Acceptor-impurity infrared absorption in semiconducting synthetic diamond, *Phys. Rev.* 140, A1272–A1274 (1965).

A. T. Collins, S. C. Lawson, G. Davies, and H. Kanda, Indirect energy gap of C^{13} diamond, *Phys. Rev. Lett.* 65, 891–894 (1990).

A. T. Collins, E. C. Lightowlers, and P. J. Dean, Role of phonons in the oscillatory photoconductivity spectrum of semiconducting diamond, *Phys. Rev.* 183, 725–730 (1969).

A. T. Collins and S. Rafique, Photocapacity measurements on natural semiconducting diamond, *J. Phys. C: Solid State Phys.* 11(7), 1375–1380 (1978).

P. J. Dean, E. C. Lightowlers, and D. R. Wight, Intrinsic and Extrinsic Recombination Radiation from Natural and Synthetic Aluminum-Doped Diamond, *Phys. Rev.* 140, A352–A368 (1965).

R. G. Farrer, On the substitutional nitrogen in diamond, *Solid State Commun.* 7, 685–688 (1969).

D. K. Ferry, High-field transport in wide-band-gap semiconductors, *Phys. Rev. B* 12, 2361–2369 (1975).

R. H. Fowler, The analysis of photoelectric sensitivity curves for clean metals at various temperatures, *Phys. Rev.* 38, 45–56 (1931).

A. B. Fox and D. L. Dreifus, Diamond SAW devices in "Properties, Growth and Applications of Diamond" Emis Data Reviews Series No. 26 (M. H. Nazare and A. J. Neves, Eds.), pp. 375–385. INSPEC Publications, 2001.

H. J. Füßer, M. Rösler, M. Hartweg, R. Zachai, X. Jiang, and C.-P. Klages, Some aspects of heteroepitaxy for diamond film growth, *Proceedings of the Third International Symposium on Diamond Materials, Electrochemical Society* 93-17, 102–110 (1993).

C. F. O. Graeff, C. E. Nebel, M. Stutzmann, A. Flöter, and R. Zachai, Characterization of textured polycrystalline diamond by electron spin resonance spectroscopy, *J. Appl. Phys.* 81, 234–237 (1997).

C. F. O. Graeff, E. Rohrer, C. E. Nebel, M. Stutzmann, H. Güttler, and R. Zachai, Optical excitation of paramagnetic nitrogen in chemical vapor deposited diamond, *Appl. Phys. Lett.* 69, 3215–3217 (1996).

H. G. Grimmeiss and L.-A. Ledebo, Spectral distribution of photoionization cross sections by photoconductivity measurements, *J. Appl. Phys.* 46, 2155–2162 (1975).

S. A. Grot, C. W. Hatfield, G. Sh. Gildenblat, A. R. Bazian, and T. Badzian, Electrical properties of selectively grown homoepitaxial diamond films, *Appl. Phys. Lett.* 58, 1542–1544 (1991).

J. Hartmann and M. Reichling, Thermal transport in diamond in "Properties, Growth and Applications of Diamond" Emis Data Reviews Series No. 26 (M. H. Nazare and A. J. Neves, Eds.), pp. 32–39. INSPEC Publications, 2001.

J. J. Hauser, Electrical, structural and optical properties of amorphous carbon, *J. Non-Cryst. Solids* 23, 21–41 (1977).

K. Hayashi, S. Yamanaka, H. Watanabe, T. Sekiguchi, H. Okishi, and K. Kajimura, Diamond films epitaxial growth by step-flow mode, *J. Cryst. Growth* 183, 338–346 (1998).

K. Hecht, Zum Mechanismus des lichtelektrischen Primärstromes in isolierenden Kristallen, *Zeitschrift für Physik* 77, 235–245 (1932).

J. T. Huang, W. H. Guo, J. Hwang, and H. Chang, Reliability of Hall effect measurements on chemical vapor deposited polycrystalline B-doped diamond films, *Appl. Phys. Lett.* 68, 3784–3786 (1996).

C. Inkson, Deep impurities in semiconductors: II. The optical cross section, *J. Phys. C* 14, 1093–1101 (1981).

J. Isberg, J. Hammersberg, E. Johansson, T. Wikström, D. J. Twitchen, A. J. Whitehead, S. E. Coe, and G. A. Scarsbro, High carrier mobility in single-crystal plasma-deposited diamond, *Science* 297, 1670–1672 (2002).

W. B. Jackson and N. M. Amer, Direct measurement of gap-state absorption in hydrogenated amorphous silicon by photothermal deflection spectroscopy, *Phys. Rev. B.* 25, 5559–5562 (1982).

X. Jiang and C. L. Jia, Diamond epitaxy on (001) silicon: an interface investigation, *Appl. Phys. Lett.* 67, 1197–1199 (1995).

X. Jiang, C.-B. Klages, R. Zachai, M. Hartweg, and H.-J. Füßer, Epitaxial diamond thin films on (001) silicon substrates, *Appl. Phys. Lett.* 62, 3438–3440 (1993).

X. Jiang, K. Schiffmann, C.-P. Klages, D. Wittorf, C. L. Jia, and K. Urban, Coalescence and overgrowth of diamond grains from improved heteroepitaxy on silicon (001), *J. Appl. Phys.* 83, 2511–2518 (1998).

X. Jiang, P. Willich, M. Paul, and C.-P. Klages, In situ boron doping of chemical vapor-deposited diamond films, *J. Mater. Res.* 14, 3211–3220 (1999).

W. Kaiser and W. L. Bond, *Phys. Rev.* 115, 857–863 (1959).

R. Kalish, C. Uzan-Saguy, B. Philosoph, V. Richter, and S. Prawer, Loss of electrical conductivity in boron-doped diamond due to ion-induced damage, *Appl. Phys. Lett.* 70, 999–1001 (1997).

H. Kawarada, T. Suesada, and H. Nagasawa, Heteroepitaxial growth of smooth and continuous diamond thin films on silicon substrates via high quality silicon carbide buffer layers, *Appl. Phys. Lett.* 66, 583–585 (1995).

H. Kim, R. Vogelgesang, A. K. Ramdas, S. Rodriguez, M. Grimsditch, and T. R. Anthony, Electronic Raman and infrared spectra of isotopically controlled blue diamonds, *Phys. Rev. Lett.* 79, 1706–1709 (1997).

H. Kim, R. Vogelgesang, A. K. Ramdas, S. Rodriguez, M. Grimsditch, and T. R. Anthony, Electronic Raman and infrared spectra of acceptors in isotopically controlled diamonds, *Phys. Rev. B* 57, 15315–15327 (1998).

W. Kohn, Solid State Physics (F. Seitz and D. Turnbull, Eds.), Vol. 5, p. 257. Academic Press, New York, 1957.

S. Koizumi, M. Kamo, Y. Sato, S. Mita, A. Sawabe, A. Reznik, C. Uzan-Saguy, and R. Kalish, Growth and characterization of phosphorus doped n-type diamond thin films, *Diamond Relat. Mater.* 7, 540–544 (1998).

S. Koizumi, M. Kamo, Y. Sato, H. Ozaki, and T. Inuzuka, Growth and characterization of phosphorous doped {111} homoepitaxial diamond thin films, *Appl. Phys. Lett.* 71, 1065–1067 (1997).

D. V. Lang, Deep-level transient spectrocopy: a new method to characterize traps in semiconductors, *J. Appl. Phys.* 45, 3023–3032 (1974).

S. Lawson, G. Davies, A. T. Collins, and A. Mainwood, Migration energy of the neutral vacancy in diamond, *J. Phys.: Condens. Matter* 4, L125–L148 (1992).

R. Locher, C. Wild, N. Herres, D. Behr, and P. Koidl, Nitrogen stabilized ⟨100⟩ texture in chemical vapor deposited diamond films, *Appl. Phys. Lett.* 65, 34–36 (1994).

A. Mainwood, Electronic band structure of diamond in "Properties and Growth of Diamond" Emis Data Reviews Series No. 9 (G. Davies, Ed.), pp. 3–8. INSPEC Publications, 1994a.

A. Mainwood, Optical constants of diamond in "Properties and Growth of Diamond" Emis Data Reviews Series No. 9 (G. Davies, Ed.), pp. 9–12. INSPEC Publications, 1994b.

C. Manfredotti, F. Fizzotti, P. Polesello, E. Vittone, and F. Wang, A study of polycrystalline CVD diamone by nuclear techniques, *Phys. Status Solidi A* 154, 327–350 (1996).

B. Massarani and J. C. Bourgaoin, Defects at low temperature in electron-irradiated diamond, *Phys. Rev. B* 14, 3682–3689 (1976).

S. P. Mehandru and A. B. Anderson, The migration of interstitial H in diamond and its pairing with substitutional B and N: molecular orbital theory, *J. Mater. Res.* 9, 383–386 (1994).

W. Mönch, "Semiconductor Surfaces and Interfaces" Surface Sciences. Springer, Berlin, 1995, p. 368.

C. E. Nebel, Electronic properties of CVD diamond in "Semiconductor Science and Technology" Special Issue: Diamond Electronics (R. Jackman, Ed.), Vol. 18(3), 2003, S1-S11.

C. E. Nebel, J. Münz, E. Rohrer, C. F. O. Graeff, M. Stutzmann, R. Zachai, H. Güttler, in "Transport Properties of CVD and Synthetic Diamond" The Physics of Semiconductors, Berlin, Germany (M. Scheffler and R. Zimmermann, Eds.), Vol. 1, pp. 95–98. World Scientific, Singapore, 1996.

C. E. Nebel, J. Münz, and M. Stutzmann, Electronic properties of CVD and synthetic diamond, *Phys. Rev.* 55, 9786–9791 (1997).

C. E. Nebel, J. Rohrer, and M. Stutzmann, Long living excited states in boron doped diamond. *J. Appl. Phys.* 89 (4), 2237–2240 (2001).

C. E. Nebel and R. A. Street, Electron and hole transport in a-Si : H at high field and low temperature, *Mater. Res. Soc. Symp. Proc.* 258, 830–852 (1992).

C. E. Nebel and R. A. Street, High field transport in a-Si : H, *Int. J. Mod. Phys. B* 7, 1207–1258 (1993).

C. E. Nebel, R. A. Street, N. M. Johnson, and C. C. Tsai, High-electric-field transport in a-Si : H. I. Transient photoconductivity, *Phys. Rev. B.* 46, 6789–6802 (1992).

C. E. Nebel and M. Stutzmann, Transport properties of diamond: carrier mobility and resistivity in "Properties, Growth and Applications of Diamond" Emis Data Reviews Series No. 26 (M. H. Nazare and A. J. Neves, Eds.), pp. 40–52. INSPEC Publications, 2001.

C. E. Nebel, M. Stutzmann, F. Lacher, P. Koidl, and R. Zachai, Carrier trapping and release in CVD diamond films, *Diamond Relat. Mater.* 7, 556–559 (1998).

C. E. Nebel, A. Waltenspiel, M. Stutzmann, M. Paul, and L. Schäfer, Persistent photocurrents in CVD diamond, *Diamond Relat. Mater.* 9, 404–407 (2000).

C. E. Nebel, R. Zeisel, and M. Stutzmann, CV and DLTS experiments in boron-doped diamond, *Phys. Status Solidi A* 174, 117–127 (1999).

M. Nesladek, L. M. Stals, A. Stesmans, K. Iakoubovskij, and G. J. Adriaenssens, Dominant defect levels in diamond thin films: a photocurrent and electron paramagnetic resonance study, *Appl. Phys. Lett.* 72, 3306–3308 (1998).

M. Nesladek, M. Vanecek, and M. L. Stals, Defect induced optical absorption in CVD diamond films, *Phys. Status Solidi A* 154, 283–303 (1996).

J. W. Orton and M. J. Powell, The Hall effect in polycrystalline and powdered semiconductors, *Rep. Prog. Phys.* 43(11), 1263–1307 (1980).

L. S. Pan, D. R. Kania, S. Han, J. W. Ager III, M. Landstrass, O. L. Landen, and P. Pianetta, Electrical transport properties of undoped CVD diamond films, *Science* 255, 830–833 (1992).

M. A. Plano, M. I. Landstrass, L. S. Pan, S. Han, D. R. Kania, S. McWilliams, and W. J. Ager III, Polycrystalline CVD diamond films with high electrical mobility, *Science* 260, 1310–1312 (1993).

Y. E. Pokrovskii, I. V. Altukhov, O. I. Smirnova, and N. A. Khvalkovskii, Competition between long-living and $D^-(A^+)$ states of donors and acceptors in Silicon, *Phys. Status Solidi B* 210, 815–819 (1998).

Y. E. Pokrovskii, O. I. Smirnova, and N. A. Khvalkovskii, Long-living excited states of impurities in Si, *Solid State Commun.* 93, 405–408 (1995).

Y. E. Pokrovskii, O. I. Smirnova, and N. A. Khvalkovskii, Long-living excited states of impurities in diamond-like semiconductors in "Proceedings of the 23rd International Conference on the Physics of Semiconductors, Berlin" (M. Scheffler and R. Zimmermann, Eds.), pp. 2689–2692. World Scientific, Singapore, 1996.

M. C. Polo, G. Sanchez, W. L. Wang, J. Esteve, and J. L. Andujar, Growth of diamond films on boron nitride thin films by bias-assisted hot filament chemical vapour deposition, *Appl. Phys. Lett.* 70, 1682–1684 (1997).

A. G. Redfield, Electronic Hall effect in diamond, *Phys. Rev.* 94, 526–537 (1954).

L. Reggiani, S. Bosi, C. Canali, F. Nava, and S. F. Kozlov, Hole-drift velocità in natural diamone, *Phys. Rev. B* 23, 3050–3057 (1981).

J. Robertson, π-Bonded clusters in amorphous carbon materials, *Philos. Mag. B* 66, 199–209 (1992).

J. Robertson and E. P. O'Reilly, Electronic and atomic structure of amorphous carbon, *Phys. Rev. B* 35, 2946–2957 (1987).

E. Rohrer, in "Photoleitungs-Spektroskopie von Diamant" Selected Topics of Semiconductor Physics and Technology (G. Abstreiter, M. Stutzmann and P. Vogl, Eds.), Vol. 10. Walter Schottky Institute, Technische Universität München, 1998.

R. Rohrer, C. F. O. Graeff, R. Janssen, C. E. Nebel, M. Stutzmann, H. Güttler, and R. Zachai, Nitrogen-related dopant and defect states in CVD diamond, *Phys. Rev. B* 54, 7874–7880 (1996).

E. Rohrer, C. F. O. Graeff, C. E. Nebel, M. Stutzmann, H. Güttler, and R. Zachai, Sub-bandgap spectroscopy of chemical vapor deposition diamond, *Mater. Sci. Engng* B46, 115–118 (1997).

M. Schreck, F. Hörmann, H. Roll, J. K. N. Lindner, and B. Stritzker, Diamond nucleation on iridium buffer layers and subsequent textured growth: a route for the realization of single-crystalline diamond films, *Appl. Phys. Lett.* 78, 192–194 (2001).

K. Seeger, "Semiconductor Physics". Springer, Berlin, 1999.

S. J. Sharp, A. T. Collins, G. Davies, and G. S. Joyce, Higher resolution studies of shallow bound exciton luminescence in diamond, *J. Phys.: Condens. Matter* 9(33), L451–L455 (1997).

B. I. Sklovskii, E. I. Levin, H. Fritzsche, and S. D. Baranovskii, Advances in Disordered Semiconductors (H. Fritzsche, Ed.), Vol. 3, p. 161. World Scientific, Singapore, 1990.

F. Smith, Optical constants of a hydrogenated amorphous carbon film, *J. Appl. Phys.* 55, 764–771 (1984).

W. V. Smith, P. P. Sorokin, I. L. Gelles, and G. J. Lasher, Electron-spin resonance of nitrogen donors in diamond, *Phys. Rev.* 115, 1546–1552 (1959).

S. D. Smith and W. Taylor, Optical phonon effects in the infra-red spectrum of acceptor centres in semiconducting diamond, *Proc. Phys. Soc.* 79, 1142–1153 (1962).

B. R. Stoner, C.-T. Kao, D. M. Malta, and R. C. Glass, Hall effect measurements on boron-doped, highly oriented diamond films grown on silicon via microwave plasma chemical vapor deposition, *Appl. Phys. Lett.* 62, 2347–2349 (1993).

R. A. Street, Trapping parameters of dangling bonds in hydrogenated amorphous silicon, *Appl. Phys. Lett.* 41, 1060–1062 (1982).

R. A. Street, Measurements of depletion layers in hydrogenated amorphous silicon, *Phys. Rev. B* 27, 4924–4932 (1983).

R. A. Street, J. Kakalios, and T. M. Hayes, Thermal equilibration in doped amorphous silicon, *Phys. Rev. B* 34, 3030–3033 (1986).

S. M. Sze, "Physics of Semiconductor Devices". Wiley, New York, 1981, pp. 245.

T. Takeuchi, Y. Yamanaka, H. Watanabe, and H. Okushi, Device grade B-doped homoepitaxial diamond thin films, *Phys. Status Solidi A* 186, 269–280 (2001).

A. J. Tavendale, D. Alexiev, and A. A. Williams, Field drift of the hydrogen-related, acceptor-neutralizing defect in diodes from hydrogenated silicon, *Appl. Phys. Lett.* 47, 316 (1985).

R. Viatkus, T. Inushima, and S. Yamazaki, Enhancement of photosensitivity by ultraviolet irradiation and photoconductivity spectra of diamond thin films, *Appl. Phys. Lett.* 62, 2384–2386 (1993).

M. D. Whitfield, S. P. Lansley, O. Gaudin, R. D. McKeag, N. Rizvi, and R. B. Jackman, High-speed diamond photoconductors: a solution for high rep-rate deep UV–laser applications, *Diamond Relat. Mater.* 10, 650–656 (2001).

C. Wild, R. Kohl, N. Herres, W. Müller-Sebert, and P. Koidl, Oriented CVD diamond films: twin formation, structure and morphology, *Diamond Relat. Mater.* 3, 373–381 (1994).

S. Yamanaka, H. Watanabe, S. Masai, D. Takeuchi, H. Okushi, and K. Kajimura, High-quality B-doped homoepitaxial diamond films using trimethylboron, *Jpn. J. Appl. Phys.* 37, L1129–LL113 (1998).

R. Zeisel, in "Optoelectronic Properties of Defects in Diamond and AlGaN Alloys" Selected Topics of Semiconductor Physics and Technology (G. Abstreiter, M.-C. Amann, M. Stutzmann, and P. Vogl, Eds.), Vol. 36, 2001.

R. Zeisel, C. E. Nebel, and M. Stutzmann, Deep level transient spectroscopy of synthetic type IIb diamond, *J. Appl. Phys.* 84, 6105–6108 (1998).

R. Zeisel, C. E. Nebel, and M. Stutzmann, Passivation of boron in diamond by deuterium, *Appl. Phys. Lett.* 74, 1875–1876 (1999).

R. Zeisel, C. E. Nebel, and M. Stutzmann, Capacitance–voltage profiling of deuterium passivation and diffusion in diamond Schottky diodes, *Diamond Relat. Mater.* 9, 413–416 (2000).

R. Zeisel, C. E. Nebel, M. Stutzmann, E. Gheeraert, and A. Deneuville, Photocapacitance study of boron-doped chemical vapour deposited diamond, *Phys. Rev. B* 60, 2476–2479 (1999).

X. Zhou, G. Watkins, K. M. MacNamara-Rutledge, R. P. Messmer, and S. Chawla, Hydrogen-related defects in polycrystalline CVD diamond, *Phys. Rev. B* 54, 7881–7890 (1996).

J. F. Ziegler, J. P. Biersack, and U. Littmark, "The Stopping and Range of Ions in Solids". Pergamon, New York, 1985.

CHAPTER 7

OPTICAL PROPERTIES OF CVD DIAMOND

Miloš Nesládek and Ken Haenen

LIMBURGS UNIVERSITAIR CENTRUM, INSTITUTE FOR MATERIALS RESEARCH (IMO), WETENSCHAPSPARK 1, B-3590 DIEPENBEEK, BELGIUM

Milan Vaněček

ACADEMY OF SCIENCES OF THE CZECH REPUBLIC, INSTITUTE OF PHYSICS, CUKROVARNICKA 10, CZ-16200 PRAGUE 6, CZECH REPUBLIC

1. Introduction

Diamond optical properties have always been associated with the attraction of diamond gems. This fascination will certainly remain, but new interest in diamond as a novel optical and electronic material comes forward. This quest, due to diamond's extreme properties is currently one of the largest drives for the diamond thin film research and especially the development of Chemical Vapor Deposition (CVD) technology for diamond preparation. The CVD diamond technology has recently had a great success in the preparation of epitaxial pn junctions as well as growth of extremely high mobility epitaxial electronic grade layers (Koizumi, Watanabe, Hasegawa and Kanda, 2001; Isberg, Hammersberg, Johansson, Wikström, Twitchen, Whitehead, Coe and Scarsbrook, 2002). Diamond, as the novel electronic and optical material has to be characterized to the sub *part per million impurity* (ppm) levels and this has to be done for thin, very low absorbing CVD diamond films. This chapter deals in detail with the recent developments in the optical characterization techniques, allowing the thin diamond film characterization, which is complementary to classical optical techniques. The reader can also find very comprehensive reviews on the classical optical techniques elsewhere (Pankove, 1975; Field, 1992; Böer, 1990).

Looking at the history of understanding of diamond's extraordinary properties the optical-based techniques were always of central importance for acquiring direct information about diamond purity, presence of color centers, the optical lattice vibrations and related IR properties. Optical techniques such as the optical absorption, luminescence and variety of related techniques and also the magnetic resonance techniques enabled an important development of diamond industry and gave it very powerful tools for controlling diamond quality. It also allowed tracing up the gemstone geographical origin, the difference between natural and synthetic diamond, the presence of lattice defects and impurities upon various treatments such as recent HPHT treatment of a natural brownish diamond making it gem-transparent

ISBN: 0-12-752185-2
ISSN: 0080-8784

(Kiflawi, Fisher, Kanda and Sittas, 1996; De Weerdt and Kupriyanov, 2002). Nowadays the research on the optical properties has got another dimension; it has amended the characterization of diamond as new optical and wide bandgap electronic material. As the optical material, the theoretical limits have been achieved as discussed by Blank (2003) and high-quality optical diamond is available for various applications such as IR windows and high-power laser optics and others. In the electronic area, just recently as described by Koizumi (2003), first diamond bipolar devices have been realized, based on optimization of P-doped n-type diamond, which includes the characterization by photocurrent-based (PC) (Bube, 1992) optical techniques such as the Photothermal Ionization Spectroscopy (PTIS) (Lifshits, 1993) and Constant Photocurrent Method (CPM) (Vanecek, Kocka, Stuchlik, Kozisek, Stika and Triska, 1983) and the Fourier Transform Photocurrent Spectroscopy (FTPS) (Vanecek and Poruba, 2002). These techniques, in parallel with the Photothermal Deflection Spectroscopy (PDS) (Boccara, Fournier, Jackson and Amer, 1980b) and more classical techniques such as FTIR, laser calorimetry (LC) (Pickles, Madgwick, Sussman and Wort, 2000; Ruf, Cardona, Pickles and Sussmann, 2000) and Raman spectroscopy became basics of the novel diamond history. Based on these techniques one was able to trace very low concentration of defects and impurities in CVD diamond films and reveal their electronic structure or to describe their influence on the electronic transport properties. To summarize these developments, this chapter deals with the following topics: defect-induced and IR absorption in undoped optical quality CVD diamond, studied by PDS and the use of PC methods applied to extrinsic defects and the electronic structures of phosphorus and boron in doped CVD diamond. Figure 1 shows one of the novel CVD

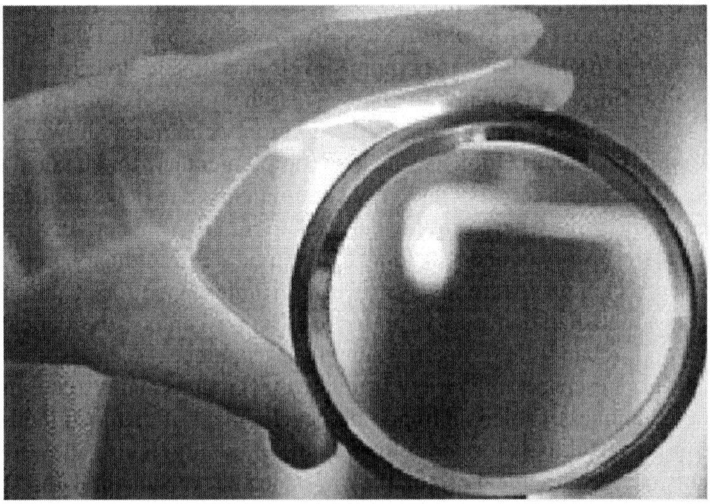

FIG. 1. Optical CVD diamond window produced by the PE MW CVD technique. With the courtesy of Element Six Ltd, UK.

diamond applications: a large CVD diamond IR window prepared by the Element Six Ltd.

2. Thin Film Optical Absorption Techniques

The principle of the defect-induced optical absorption is to translate the spectral (e.g., energy E of incident photons) dependent information of the measured optical absorption coefficient $\alpha(E)$ or the reflection $r(E)$ to the photoionization (optical) cross-section of defect states $\sigma(E)$, which codes in as a fingerprint the quantum-mechanical information about the defects and their electronic structure and their interactions with the host lattice atoms. The classical optical transmission techniques work typically in the range between the full absorption ($\alpha d \sim 5$ corresponds to light transmission of 1%, where d is the sample thickness), and the detection sensitivity limit (which is typically $A \sim \alpha d \sim 0.005$, with A the absorptance), i.e. a dynamic range of three decades in magnitude. This dynamic range limit is also due to the fact that the reflection measurements are done with a precision of about 0.1% on the measured reflectance value, implementing about 0.1% error margin as the limit for the determination of the optical absorption calculated from the expression $1 = T + R + A$, where T, R and A are the transmittance, reflectance and absorptance respectively. There are also numerous other limiting factors, such as the light scattering which besides the reduction of the transmitted light in the specular configuration, induces a superposing spectral dependence over the optical absorption coefficient $\alpha(E)$. As an example, if we take a 1 μm thick diamond film at the wavelength of the full absorption ($\alpha d = 5$), it implies that we can determine α just from about 50,000 cm^{-1} to the value of 50 cm^{-1}, e.g., at the indirect diamond gap of diamond (Fig. 2) and we are not able to see the subgap region at all. If we are now interested in the subgap absorption caused by 1 ppm defects of impurity atoms, we have to measure the absorption coefficient α down to say six decades in magnitude lower on the dynamic range, for which we have to still find a spectral dependence of $\sigma(E)$ with a reasonable resolution and sensitivity. It is obvious that classical transmission measurements with the dynamic range of three decades in magnitude fall well short of this requirement. On the other hand, for example, by photoluminescence (PL), for some impurities, we are able to see even *part per billion* (ppb) concentration of defects. Of course, to do this we need a radiative transition, which is not always the case; therefore our thin film absorption techniques such as PDS and CPM are very powerful to monitor the photoionization cross-section of defects in the gap. In the past years we have applied PDS for monitoring the optical absorption in thin diamond films (Nesládek, Vanecek and Stals, 1996a). This technique which has its roots in photacoustics has been carefully studied for measurements of the optical absorption in various thin film materials.

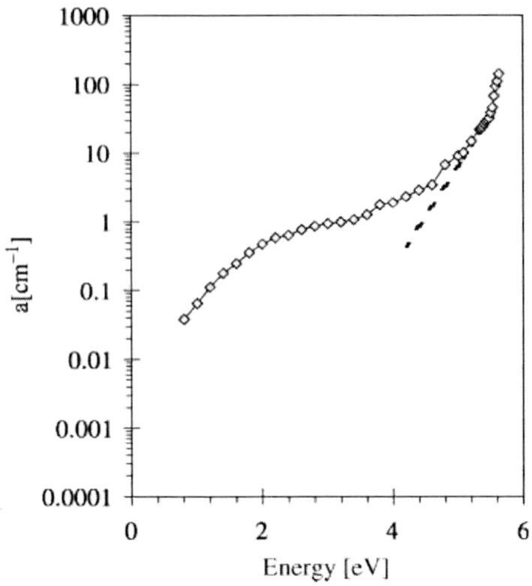

FIG. 2. The spectral dependence of the optical absorption coefficient α for an optical quality CVD diamond film, measured by PDS. The dotted line represents a fit of the indirect diamond gap.

2.1. Photothermal Deflection Spectroscopy

In the beginning of the 1980s, the theory of the mirage effect or of the so-called PDS measurements has been developed for solids by Danièle Fournier (Fournier, Boccara and Badoz, 1979). PDS is a contact less and non-destructive technique. PDS was originally used to investigate thermal properties of materials (such as the thermal diffusivity), because of the very high sensitivity of this technique, temperature changes smaller as mK can be detected. The PDS technique relies on a minute change of the index of refraction, due to the optically generated heat in the solid matter, caused by light absorption of monochromatic light (e.g., the pump beam). When a laser (probe) beam is passing through a region of material with a heat-induced gradient of the refractory index n, the probe beam is deflected. By pulsing the pump beam periodically one gets periodic changes of the gradient of the refractory index and thus induces periodic deflection of the probe beam detected by a lock-in amplifier through the use of position sensitive detectors. This is very basic principle of the PDS method, which can be found in more detail in various publications (Jackson, Amer, Boccara and Fournier, 1981; Jackson and Amer, 1982; Jackson, Amer, Biegelsen, Nemanich and Knights, 1983; Nesládek, Meykens, Stals, Vanecek and Rosa, 1996b). Two important configurations of the PDS setups are mainly used, as explained in Figure 3. In the transversal setup the pump beam and the probe beam are perpendicular to each other. While in the collinear mode, both beams are parallel with each other interacting along all the optical length of the studied sample. The schematic drawing of a transversal and the collinear setups is plotted in Figure 3.

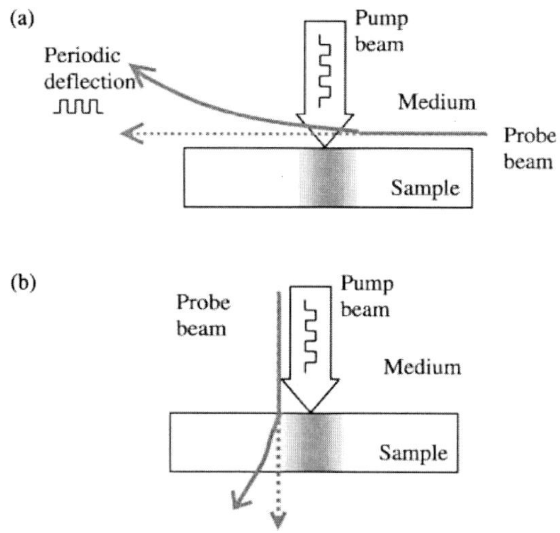

FIG. 3. A schematic drawing of a PDS setup. (a) Corresponds to the transversal and (b) to a collinear configuration.

2.2. Transversal Photothermal Deflection Spectroscopy

In the transversal PDS (t-PDS) method, the pump beam is perpendicular to the sample surface, while the probe laser beam is directed parallel to the sample just grazing its surface, as shown in Figure 3(a). In this case the heat diffuses out the sample to the surrounding medium, such as CCl_4 (with a large change of the refractory index with the temperature). It is then the thermal gradient in this medium that is probed (Boccara, Fournier and Badoz, 1980a; Boccara et al., 1980b). In the case of collinear PDS (c-PDS) measurements, the probe beam propagates perpendicular to the sample surface, quasi-parallel with the pump beam, as shown in Figure 3(b) and the changes of the refractory index of the measured material itself cause the periodic deflection of the probe beam. Both configurations allow, in principle, a local testing of the absorption of a given sample. The collinear method, however, can only be used for samples that neither absorb nor scatter the probe beam. Spatial resolution in the surface plane ((x,y)-plane) is, for both methods, determined by the width of the pump beam. In the z-direction, the resolution of the transverse method depends strongly on the thermal diffusion length inside the sample. For thermally thin samples, t-PDS measures the total (surface and bulk) accumulated absorption over the whole thickness of the sample. For the collinear method, the resolution and the locality of the measurement are determined by the cross-sectional area of the pump beam and the probe beam. By changing the wavelength λ of the pump beam, the PDS technique yields the spectral dependence of the optical absorption $\alpha(E) = \alpha(\lambda)$ of the material under investigation (Fig. 4).

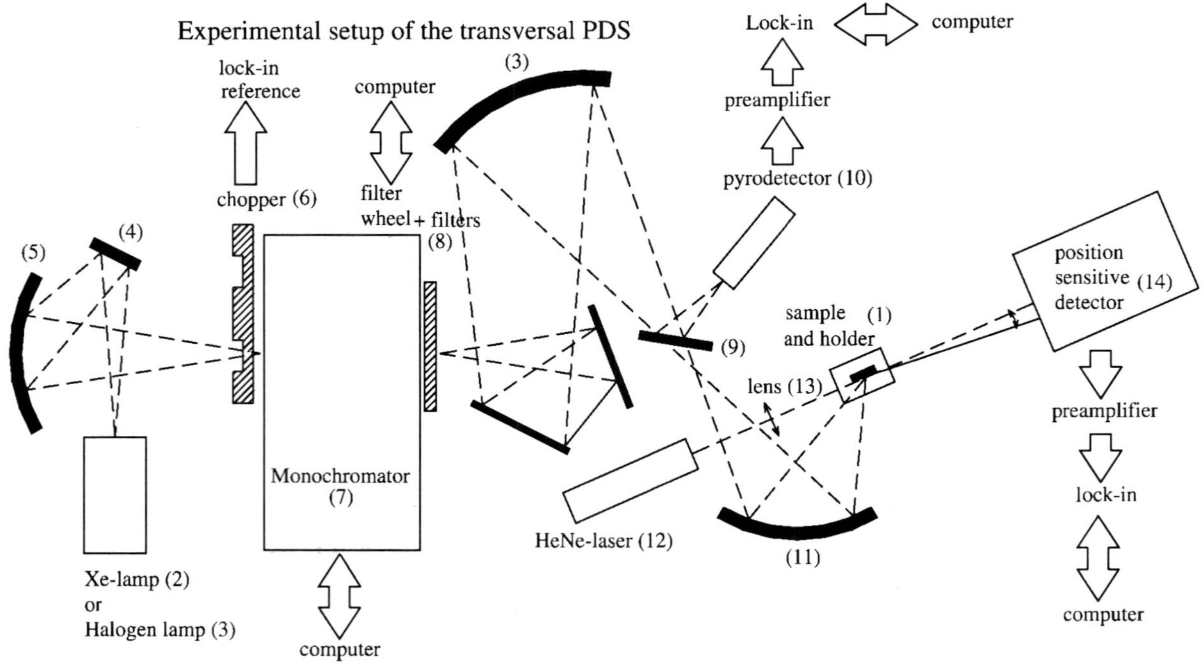

FIG. 4. Schematic of a transversal Photothermal Deflection Spectroscopy (t-PDS) setup, using W-halogen and Xe-arc lamp sources and a three-grating monochromator for measurements in the range from 0.3 to 6 eV.

In the incoherent limit and when light scattering is negligible, the normalized PDS-signal S is proportional to the sample absorbance and for both experimental arrangements—transversal and collinear PDS—given by

$$S = S(E) = S_{sat}[1 - \exp(-\alpha(E)d)] \quad (1)$$

where d is the effective optical thickness of the sample and S_{sat} stands for a proportionality constant called the saturation value. This constant includes all apparatus conversion and amplification factors (deflection-voltage/current), the distance of the probe beam to the sample surface, the dimensions of the probe beam, the reflection coefficient, the width of the sample and the material parameters (thermal properties) of the sample and the deflecting liquid. The saturation signal S_{sat} is the PDS-signal when the incident power is completely absorbed in the film (strong absorption with $\alpha d > 1$).

When d is known and the PDS signal is saturating (S_{sat}), the absolute optical absorption $\alpha(E)$ can be calculated (for energies where no saturation occurs) in a very straightforward way:

$$\alpha(E) = -1/d \ln(1 - S(E)/S_{sat} \quad (2)$$

In practice, this procedure is not as straightforward as it looks. The calculation of absolute absorption values requires the knowledge of the saturation value S_{sat}. In the investigated energy range, the PDS-signal of most thin CVD diamond films does not saturate because diamonds are very transparent and the full absorption is reached above 5.6 eV for typically mm-thick diamond plates. Thus, for optically polished samples, the matching of $\alpha(E)$ calculated from the optical transmission and reflection measurements is the standard way to scale absolutely the absorption PDS spectra. An example of the scaling of an optical absorption spectrum recorded by t-PDS and matched with the corresponding integrating sphere transmission measurements is shown in Figure 2. Another possibility for the absolute scaling is using laser calorimetry (LC) (Remes, Neslédek and Pickles, 2001). In the ideal case the c-PDS measurements can be combined with LC in one setup, allowing a direct calibration of the PDS results on the absolute scale. It should be remarked that this is only possible for a collinear configuration (Section 2.3) where the sample is in the air and the heat exchange with the environment is controlled. Also, LC is limited by light scattering excluding measurement on unpolished samples. Thus $\alpha(E)$ can only be absolutely scaled by a LC measurement in the case of polished CVD windows or epitaxial layers grown on polished substrates.

2.3. Collinear Photothermal Deflection Spectroscopy

The collinear configuration is especially interesting for studying the IR absorption of CVD diamond plates which can be used for numerous IR applications such as high power laser optics, e.g., CO_2 laser opto-couplers.

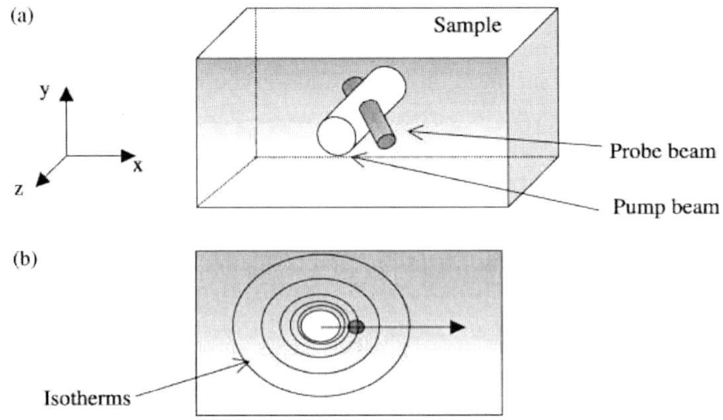

FIG. 5. (a) Beam positions in the crossing c-PDS setup. (b) Schematic representation of the created isotherms.

The schematic collinear setup with the principle described above is shown in Figures 5 and 6.

In our PDS configuration (Fig. 6) we use just one ZnSe lens for focusing both the CO_2 and HeNe lasers. Due to the difference of wavelengths of the two lasers (He–Ne 633 nm and CO_2 laser 10.6 μm) the focused beam radius a_p fulfils the theoretical condition of $a_{probe} < a_{pump}$. The experimental details such as the lens parameters and the detection equipment can be found in the works of Meykens, Haenen, Nesládek, Stals, Pickles and Sussman (2000), Remes, Nesládek, Meykens, Pickles and Sussman (2000) and Remes et al. (2001). The corresponding laser beam interaction length L is set by the angle of crossing laser beams (in the diamond sample); for our particular situation, $L \sim 1951$ μm, is less than twice the Rayleigh length of the pump and probe beams. This means the radius (waist) of the laser beam can be assumed to be constant over the

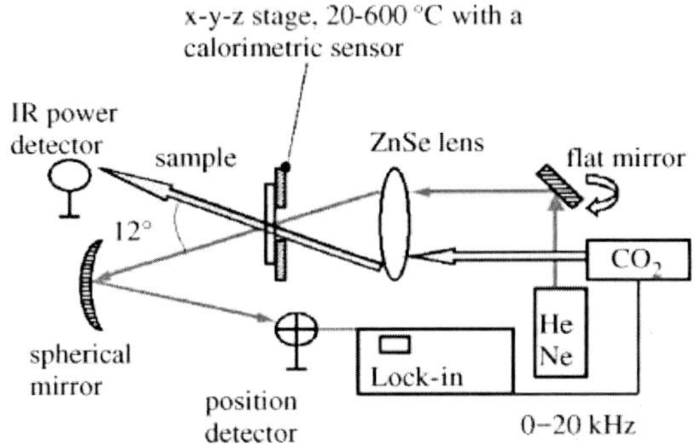

FIG. 6. A schematic setup of a collinear Photothermal Deflection Spectroscopy (c-PDS) setup used in this work, in combination with LC for the absolute scaling of the optical absorption coefficient $\alpha(E)$.

interaction region. Moreover, the interaction region is larger than the sample thickness. Therefore, the advantage of such a c-PDS arrangement is that an integrated absorption over the full sample thickness (in the z-direction) is measured. In the opposite case when L is much smaller than the thickness of the sample (this can be done by crossing the beams under a wide angle), a resolution in the z-direction can be obtained. As mentioned above our setup (Fig. 6) is additionally equipped with LC allowing direct calorimetric measurement of the absolute value of the absorption coefficient α. In the LC experiments, the temperature T rises from its starting value T_0 upon continuous CO_2 laser heating in time t and when the laser heating is interrupted, the subsequent temperature decay as the sample is cooled by the heat radiation is monitored. Absorbance is then calculated according to the equations

$$A = [1 - \exp(-\alpha d)]/[1 - R\exp(-\alpha d)] \tag{3}$$

$$\mathrm{d}T/\mathrm{d}t = H - g(T - T_0) \tag{4}$$

where the heat generation rate $H = AP_0/m/C$ describes the heating (P_0, the laser power, C the specific heat capacity, m the sample mass) and the heat loss G described by a linear temperature loss coefficient g are $G = g(T - T_0)$.

2.4. Theoretical Considerations for PDS

A complete theoretical treatment of the c-PDS measurements can be found in Jackson et al. (1981). There are two important limiting cases for the collinear PDS arrangement (Boccara et al., 1980a; Jackson et al., 1981; Zimering and Boccara, 1997). For high-modulation frequencies, when the thermal diffusion length μ of the deflecting medium is much smaller than the pump beam radius $\mu << a_p$, the amplitude of the deflection, θ_A, is given by:

$$\theta_A = 2\frac{r}{a_p^2}\frac{1}{n}\frac{\partial n}{\partial T}\frac{P_0}{\omega\rho C\pi^2 a_p^2}(1 - \exp(-\alpha d))\exp\left(-\frac{r^2}{a_p^2}\right) \tag{5}$$

where P_0 is the power of the pump beam, ω the angular frequency ($\omega = 2\pi\upsilon$), υ the laser modulation frequency, ρ the density of the deflecting medium ($\rho_{diamond} = 3.51 \times 10^3$ kg m^{-3}), (C is 0.51 J g^{-1} K^{-1} for IIa diamond at 20 °C, $\mathrm{d}n/\mathrm{d}T$ is the refraction index gradient $((1/n)\,\mathrm{d}n/\mathrm{d}T)_{diamond} = 4.04 \times 10^{-6}$ K^{-1}, $\mathrm{d}n/\mathrm{d}T = 1.6 \times 10^{-5}$ K^{-1} at 10.6 μm) (Tropf, Thomas and Linevsky, 1998). The deflection of the probe beam is at its greatest when the distance r between the probe beam and the pump beam is $a_p/\sqrt{2}$.

In the second case, for low modulation frequencies, the thermal diffusion length is much larger than the pump beam radius $\mu >> a_p$, and the Photothermal Deflection (PTD) signal θ_A depends only on the thermal diffusivity and is independent of the modulation frequency. In this case the signal is much higher than in the first case. The amplitude of the PDS signal θ_A is given by:

$$\theta_A = \frac{1}{n_0}\frac{\partial n}{\partial T}\frac{P_0}{\kappa\pi^2 r}(1 - \exp(-\alpha l))\left(1 - \exp\left(-\frac{r^2}{a_p^2}\right)\right) \tag{6}$$

with κ the thermal conductivity of the deflecting medium ($\kappa_{\text{diamond}} = 2000\ \text{W m}^{-1}\ \text{K}^{-1}$). In this case, the deflection amplitude is at its greatest for $r \sim 1.12a_p$.

By changing the position of the probe beam relative to the pump beam (Fig. 7(a)), the temperature field in the sample can be measured for the above two cases (Fig. 5(b)), also allowing an experimental verification of the pump beam radius and the determination of the thermal diffusivity. This has been done in Figure 7(a); the theoretical fitting shows a perfect agreement with experiment. Both cases described by Eqs. (5) and (6) are quite important for the measurement of the absorption coefficient in diamond. For very low f, $d < \mu$, the PTD signal is

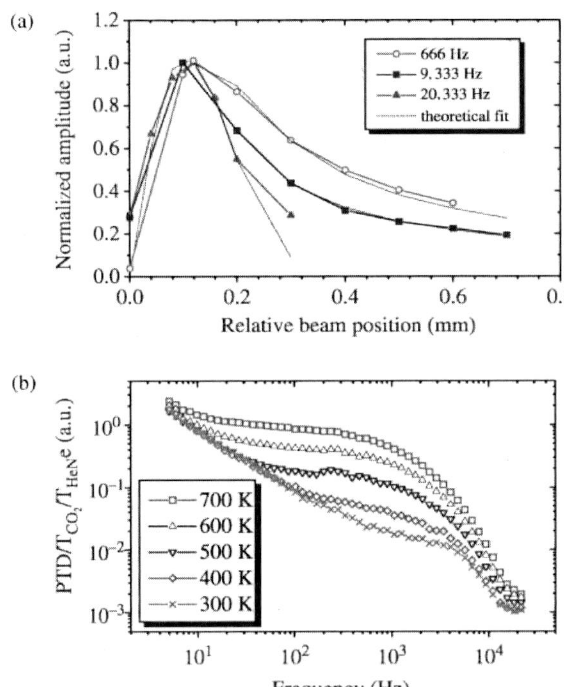

FIG. 7. (a) The amplitude S of the c-PDS signal as a function of the relative distance between the probe and the pump beam, measured for various modulation frequencies. The experimental data are fitted with a combination of the theoretical expressions (5) and (6). (b) Frequency and temperature dependence of a normalized PTD signal in high quality ($\alpha = 0.04\ \text{cm}^{-1}$ at 300 K) CVD diamond optical window at the CO_2 laser wavelength 10.6 μm.

high and then it starts to decrease with increasing frequency. The PTD is thus insensitive to f when $a < \mu < d$ and inversely proportional to f for $\mu < a < d$ which is the second limiting case (Boccara et al., 1980b). The most sensitive PTD measurements can be done when $a < \mu < d$. In this case, the PTD amplitude is proportional to the absorption coefficient α, the thermo-optic coefficient $(\mathrm{d}n/\mathrm{d}T)/n$, where n is the index of refraction, and the inverse thermal diffusivity κ, but we need to know the thermal conductivity. If the sample is fully homogeneous and of a high optical quality, it is only a normalization problem. But a polycrystalline sample with heterogeneous (non-diamond) inclusions can have different thermal conductivity in different regions; thus the knowledge of κ is important for mapping of the optical absorption coefficient.

If we are interested in the measurement of $\alpha(T)$ at various temperatures, $\kappa(T)$ is needed for the right evaluation. Due to the fact that $\kappa(T)$ and therefore the thermal diffusion length is temperature dependent, the low frequency region (Eq. (6)) where the PDS signal starts to be independent of frequency is shifting with temperature (Fig. 7(b)). This is an important problem for a correct measurement of $\alpha(T)$ because for the entire temperature range, for the signal linearity, we need to be in the region described either by Eq. (5) or (6), otherwise a full complicated expression must be used (Boccara et al., 1980b). Also, at low frequency and low temperatures, the sample dimensions start to be comparable with the thermal diffusion length and this can influence the PTD signal; thus the optimal frequency is in the range of a few hundred Hz for CVD diamond plates of 1 mm thick in our focusing geometry.

2.5. Sub Bandgap Optical Absorption in Diamond

In the UV region, the ideal pure natural diamond crystal will start to absorb strongly at 5.46 eV due to the indirect gap absorption involving phonons close to the zone boundary (Fig. 2) (Field, 1992; Madelung, 1996). On the other hand in the IR region, due to diamond's excellent transparency, it is a highly attractive material to replace ZnSe optics, as for example, output window material for CO_2 lasers. A breakthrough in laser performance could be realized when using diamond, since the effect of thermal lensing can be practically fully eliminated (Muys, 1999). Nowadays, the state-of-the-art CVD diamonds almost reach the thermo-optical properties of the best natural type IIa diamonds (Willamowski, Ristau and Welsch, 1998). Moreover, CVD diamond windows of a few mm thick and several cm large (10–15 cm) can be produced (Fig. 1).

In the IR region the diamond will absorb due to various phonon processes (Klein, Hartnett and Robinson, 1992; Thomas, Tropf and Szpak, 1995). In the purest diamond, due to lattice symmetry, the one-phonon processes are forbidden. Due to the fact that the incident photon momentum is very small compared to Brillouin zone phonons, allowed 2-phonons and multiple-phonon processes summates in total to zero momentum in high-purity material. On the other hand, in materials containing impurities the lattice symmetry is disturbed in the close vicinity of an impurity allowing 1-phonon absorption modes. In this case, the energy of local vibrations of strongly bound or light (low Z) impurities is well above the energy of the lattice optical phonon (zone-center phonon of 165 meV). For example, localized phonon vibrations of C–H are in the range of 341–349 meV. On the contrary heavy (high Z) or weakly bound impurities (like Si, clustered Ni or interstitial C) induce quasi-local phonons with energies comparable with intrinisic phonon energies. However, the harmonic-oscillator perturbation models do not predict any temperature dependence of such optical absorption on phonons (Remes et al., 2001).

Well, what news the IR-PDS spectroscopy in CVD diamond films can tell us? First thing to remark in CVD diamond is the characteristic continuum absorption appending to the absorption edge and decaying towards the IR region as shown first in our work (Nesládek et al., 1996b). In some cases a continuous absorption also has been observed in natural IIa diamond where it was never satisfactorily explained. This absorption is of course quite low and only reachable by a sensitive technique such as PDS. In our work (Nesládek, Vanecek, Rosa, Quaeyhaegens and Stals, 1995; Nesládek et al., 1996b) we were able to attribute clearly this absorption to non-diamond carbon present mainly at the grain boundaries even in the highest quality CVD diamond films. Also in some cases we have observed a similar absorption in a homoepitaxial CVD diamond layers.

The characteristic absorption is discussed further, and in the last section of this chapter the c-PDS technique is used to map the spatial dependence of the optical absorption coefficient at the CO_2 laser wavelength (10.6 μm) for a set of free standing CVD diamond windows and to study the fundamental limits of IR absorption.

2.6. Optical Absorption in Diamond: the sp^2 Carbon Absorption Model

Various models of optical absorption, such as those based on the theory of Tauc (1974), and others were developed to describe the optical absorption in short-range ordered materials such as amorphous silicon (a-Si) and amorphous hydrogenated silicon (a-Si:H). For amorphous carbon, the situation is more complex because of the mixed π–σ bonding environment. In the case of a-C and a-C:H the sp^3 sites form four σ bonds and the sp^2 sites form three σ bonds and one weaker π bond. The electronic structure of a-C and diamond-like carbon is reviewed by Robertson (1992, 1997). In a brief summary, in Tauc's model only the conservation of energy and no conservation of momentum is assumed. The momentum matrix elements T_{ij} are then replaced by their average value T_0 (Cody, 1984) and the optical absorption coefficient α can be written as

$$\alpha(E) = \frac{C}{E}\int g_{\mathrm{i}}(\varepsilon)f(\varepsilon)g_{\mathrm{f}}(\varepsilon + E)[1 - f(\varepsilon + E)]\mathrm{d}\varepsilon \qquad (7)$$

where g_{i} and g_{f} are the initial and final density of states, respectively, $f(\varepsilon)$ and $f(\varepsilon + E)$ the Fermi-Dirac occupation functions for the initial and final states and C a constant proportional to the square of the momentum matrix elements T_0^2.

We used the same approximation for a-C defects in diamond and assumed constant matrix elements. Specific for low-pressure CVD diamond is its synthesis in the pressure-temperature region in which graphite is thermodynamically stable over diamond. Thermodynamic growth models (Fedoseev, Varshavskaya, Lavrentev and Deryagin, 1977; Piekarczyk, 1992; Spear and Frenklach, 1994) based on the preferential etching of the non-diamond phase and/or diamond surface stabilization by atomic hydrogen, always include a certain probability for the incorporation of sp^2 carbon in the diamond films (Knight and White, 1989; Shroder, Nemanich and Glass, 1990). Amorphous carbon is located mainly in grain boundaries and at other defects, such as disclocations and other strain-induced defects. Nowadays it is known that, for example, in novel materials such as Ultra Nano-Crystalline Diamond (UNCD) both phases—amorphous or sp^2 bonded carbon and sp^3 bonded rings—actually interpenetrate (Gruen, 2001). It is then plausible for such materials to introduce bonding disorder. In Figure 8 the comparison of the Raman scattering factor and the PDS absorption is plotted showing CVD diamond films containing rather large amount of sp^2 carbon according to Vorlicek, Rosa, Vanecek, Nesládek and Stals (1997). Since then CVD diamond materials have been optimized greatly and the Raman spectra of high-purity material do not show any sp^2 phase. In such case the PDS subgap absorption measurement is necessary to characterize these very pure films.

In our previous work (Nesládek et al., 1996b) we have modeled polycrystalline CVD diamond as a two-phase system consisting of pure diamond and grain boundaries, containing non-diamond (sp^2) carbon. We distinguish two different phases: a diamond phase where the conduction and valence bands of diamond are modeled by two parabolic shape densities of states (DOS) with band edges E_{C} and E_{V}. At the boundary with a second sp^2 phase, both the valence and the conduction band edges are modified to include distorted sp^3 sites with differences in bond-angle and bond-length. Even in the highest quality CVD diamond sp^2 sites principally exist at specific types of grain boundaries even just as bonds terminating reconstructed surfaces of grain boundaries (Mora, Steeds and Butler, 2002). These modifications are implemented in our model by introducing an exponential tail of localized σ-, σ^*-states extending in the gap. The energies at which the parabolic bands transform into the localized exponential tails are noted by E_m and $E_{m'}$. Furthermore, a pseudogap is formed by the localized π- and π^*-gaussian bands centered around E_1 and E_2 in Figure 9(a).

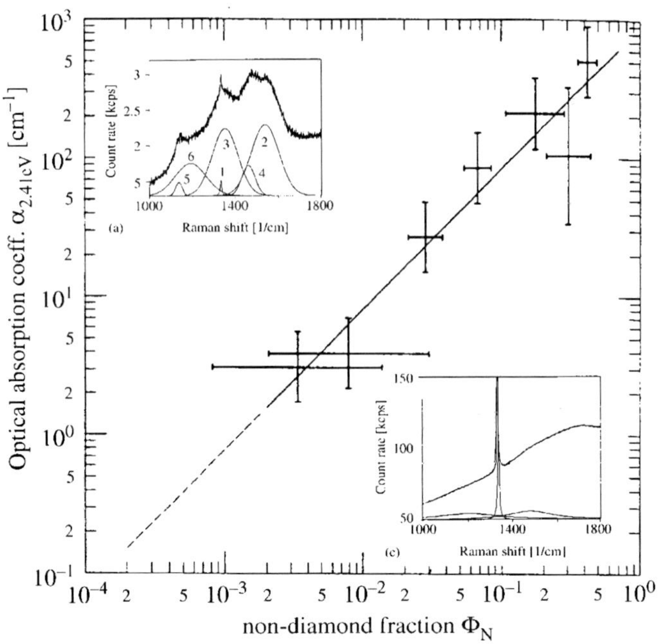

FIG. 8. Comparison of the optical absorption at Ar laser line 2.44 eV and the Raman factor obtained from Raman spectroscopy, excited by the Ar line. The inset shows the deconvolution procedures for the Raman intensity for various non-diamond component, according to Vorlicek et al. (1997).

All the different possible optical transitions we considered are described in Figure 9(a).

1. $\pi-\pi^*$ transitions in amorphous carbon regions.
2. $\pi-\sigma^*$ transitions in amorphous carbon–diamond phase boundary regions, where σ^* indicates the disorder tailing of valence or conduction band, respectively.
3. $\Sigma-\Sigma^*$ indirect transitions between the valence and conduction band of diamond.

In addition to these transitions, based on c-PDS results, we introduce a spectrally independent absorption on large non-diamond carbon inclusions (Nesládek et al., 1996a,b; Remes et al., 2000).

As an example, Figure 9(b) shows optical absorption coefficients measured by PDS for three CVD diamond films with various non-diamond carbon concentrations according to Nesládek et al. (1996b).

2.7. Study of IR Absorption by Collinear PDS

The IR absorption properties of natural diamond as well as their infrared spectra have been thoroughly measured by various authors. To stress the

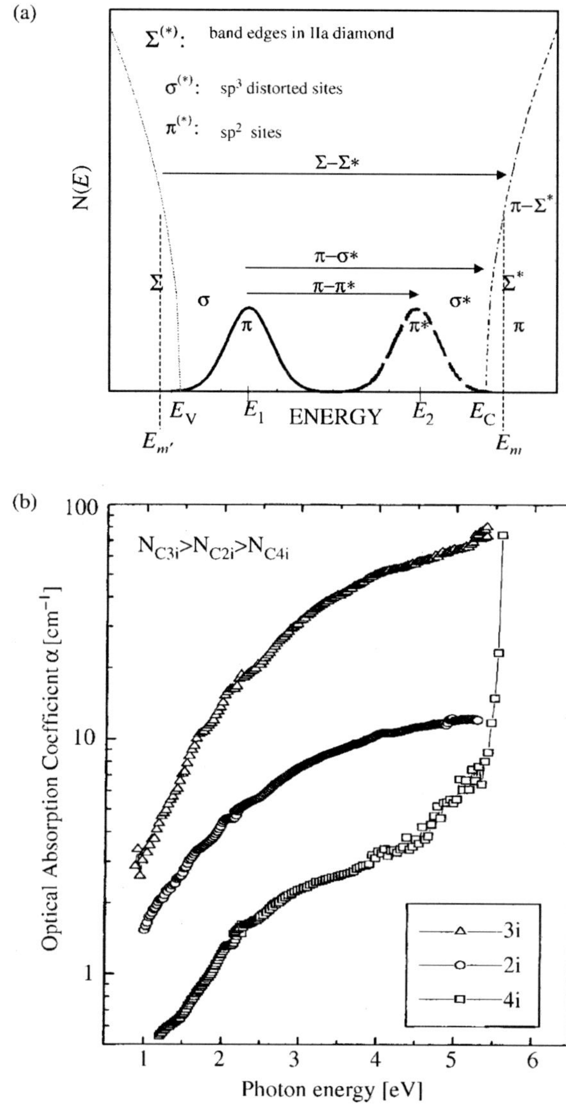

FIG. 9. (a) Model of the density of states in the gap of CVD diamond. The π and π^* bands appear due to the presence of amorphous carbon regions. Shown are also two parabolic bands Σ, Σ^* for diamond and the exponentially broadened tails σ, σ^* (starting at energy $E_{m'}$ and E_m). (b) Experimental data of CVD diamond absorption measured by PDS for samples 2i, 3i, 4i showing various concentration of non-diamond carbon N_C ($N_{C4i} < N_{C2i} < N_{C3i}$) according to Nesládek et al. (1996b).

importance, for example, in 1977 an Element Six Ltd natural diamond window was used in the Pioneer Venus spacecraft for infrared radiometry experiments to withstand the very hostile atmosphere on Venus. Nowadays, large CVD diamond wafers close to such high-purity natural diamond are available up to 4 in. in diameter. Since then many other optical applications requiring the highest technical standards followed. The majority of novel applications is based on

- Exceptional broad band infrared transparency and transparency in the optical till ultraviolet part of light spectrum.
- Excellent thermal conductivity and a very low thermal expansion coefficient.
- High fracture strength and Young modulus, high laser damage threshold.

Due to these properties, diamond is for many applications an ultimate replacement for standard IR materials such as ZnSe, Ge, sapphire and others. As an illustration, a problem encountered in high-power optics is the thermal lensing, i.e., thermal distortion of the laser beam caused by thermally induced refractive index gradients, as the window is heated by the transmitted laser beam. This issue can be minimized by using CVD diamond as the window material due to its high thermal conductivity, a low absorption coefficient and a small change of the refractive index with temperature. This enables use of diamond for construction of components like exit windows, output couplers, beam splitters and lenses for high power CO_2 lasers, including components in synchrotron beam lines and optics for space. The extreme high laser damage threshold allows more compact systems with diamond optics and reduces maintenance of optics. Table I compares the values of these quantities for diamond and ZnSe at 10.6 μm wavelength.

The IR absorption in the purest CVD diamond windows using techniques such as PDS and calorimetry has been studied (Remes et al., 2001; Davies, Mainwood, Piccirillo, Lewis, Mollart, Nesládek and Remes, 2002; Piccirillo, Davies, Mainwood, Scarle, Penchina, Mollart, Lewis, Neslédek, Remes and Pickles,

TABLE I

COMPARISON BETWEEN DIAMOND AND ZnSe MATERIAL PROPERTIES THAT ARE OF IMPORTANCE WHEN THESE MATERIALS ARE USED FOR HIGH-POWER OPTICS

	CVD Diamond (optical)	ZnSe
Thermal conductivity (W m-K^{-1})	2000	17
Absorption at 10.6 μm (cm^{-1})	0.1–0.03 (Typical: 0.05)	0.0005
Absorption of AR coating per surface (%)	0.1	0.1
dn/dT (10^{-6} K^{-1})	10	57

2002); specifically we were interested in the theoretical limit and the absorption at the CO_2 laser wavelength and how this absorption behaves with temperature.

In the IR part of the spectra for the pure diamond, containing very little impurities, we have only a well-documented two- and multiple-absorption region. Further IR diamond absorption drops to values of about 0.03 cm^{-1} at 10.6 μm, found from the previous investigations (Serrano, Cardona and Ruf, 2000). The detailed description of all absorption features in the spectra can be found, for example, in the excellent article of Klein and co-workers (1992) who mapped in detail all possible phonon combinations leading to two- and multiple-phonon absorption. The PDS technique in collinear configuration, combined with LC is an ideal instrument for performing the IR mapping. Thus not only average absorption, but also local variations of optical absorption coefficient, which can be due to impurities or local growth imperfection, or grain boundaries as discussed in Section 2.6 can be studied. To do so, PDS measurements have been carried out in the region of low-chopping frequency i.e., where $a_p < \mu < d$ is the pump beam radius (waist), μ the diffusion length and d the sample thickness. According to Eq. (6) the PDS signal S is proportional to the absorption coefficient ($\alpha \sim S \sim \mathrm{d}n/\mathrm{d}T)\alpha/\kappa$, where T is the temperature, n the index of refraction and κ the thermal conductivity. Using the configuration from Figure 6 a 2D scanning could be performed generating absorption maps with a precision determined by the focus of the HeNe beam (20 μm) and the CO_2 laser beam (100 μm). Figure 10 was performed at 633 Hz at 300 K. The optical quality CVD material provided by Element Six Ltd (DeBeers) has been studied and compared with a normal grade

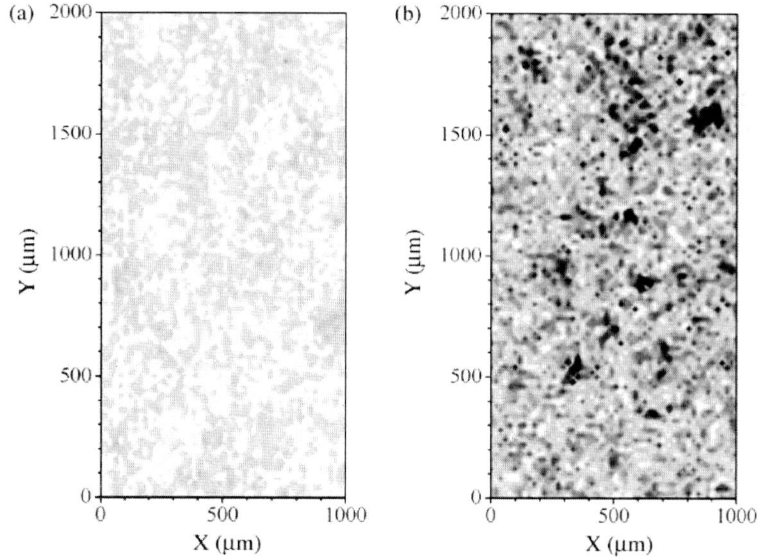

FIG. 10. The normalized PTD maps of (a) high ($\alpha = 0.04$ cm^{-1}) and (b) moderate ($\alpha = 0.1$ cm^{-1}) optical quality CVD diamond windows measured at 300 K, 633 Hz and 10.6 μm.

CVD diamond with higher absorption. It should be noted that all measured samples are of excellent quality and on-sight fully transparent with negligible concentrations of impurities such as N with concentrations well below 100 ppb; so the extrinsic absorption and consequently 1-phonon absorption are out of the C-PDS measurement sensitivity reach. The PDS mapping of this sample shows rather uniform absorption, though we can distinguish different regions.

The scan of sample shown in Figure 10(b) shows regions of very low absorption as well regions of higher absorption (dark color) which could be clearly attributed, by comparing with a microscopic image, to non-diamond carbon inclusions, similar to absorption measured by transversal PDS in some diamond plates. These inclusions so-called 'black spots' are generated during growth at the grain boundaries. If the growth conditions are not optimal, a large anisotropical strain can develop at the grain boundaries of adjacent crystals with different crystallographic orientation (non-coherent grain boundaries) generating such macroscopic carbon inclusions. It is the mastering of the deposition process to develop low-strain high-optical transparency films such as those presented in Figure 1. Such layers have the lowest measured absorption of about 0.025–0.03 cm^{-1}. We believe that only the macroscopic inclusions, but not sp^2 carbon at grain boundaries contributes to enhanced absorption in low-impurity CVD diamond, as discussed below.

2.8. Temperature Dependence of the IR Absorption

For IR components working at higher temperature the knowledge about the temperature dependence of the optical absorption coefficient is essential. The temperature dependence of the optical absorption is also important for tracing the physical mechanism of the absorption. As discussed, at 10.6 μm at room temperature (RT) we typically always measure absorption of ~0.03–0.05 cm^{-1}; by measuring the temperature dependence of α we hoped to clarify the absorption mechanism. In classical transmission measurements it is very difficult to measure at enhanced temperatures where the black body irradiation of the substrate is quite intense, thus giving a false contribution to the absorption spectra. These measurements are not reliable without special precautions. Mollart and Lewis (2001) have measured this absorption (Fig. 11) in a classical transmission spectrophotometer by using a clever trick and placing the heating element on side of the diamond crystal, off side the light beam, using the very high diamond thermal conduction to equalize the temperature with the heater stage.

In Figure 12 we plot a normalized PTD signal, measured for various temperatures (T). The figure also shows the interference fringes of both the CO_2-pump and HeNe-probe laser beams as a function of temperature. The interferences are caused by the fact that the sample undergoes a thermal expansion with a consequent increase of the sample thickness. From the

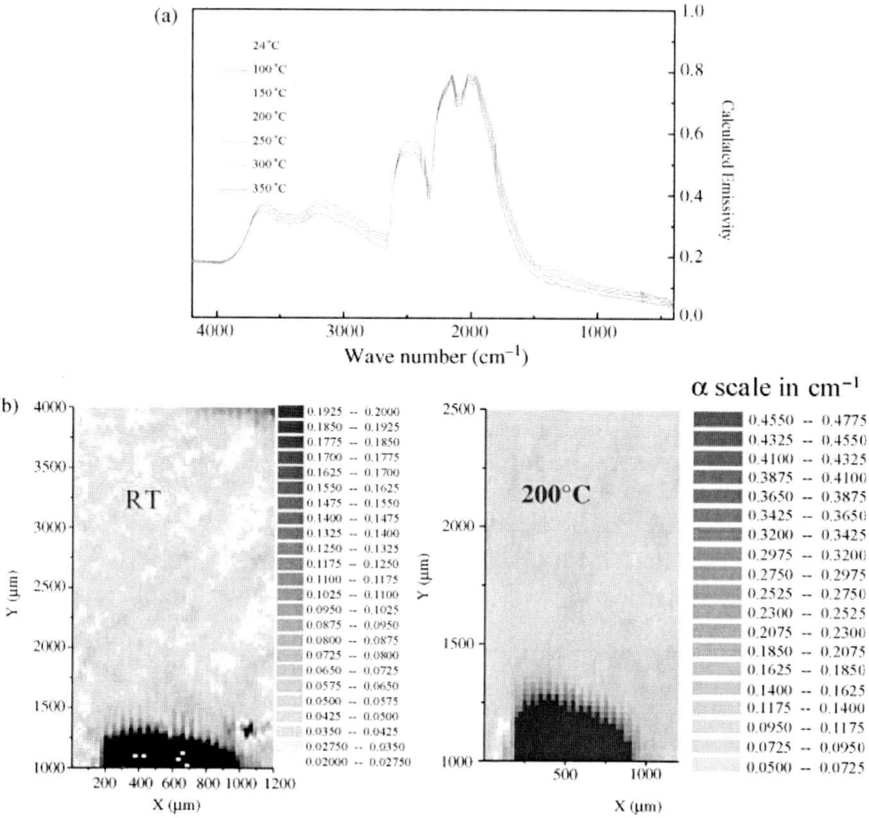

FIG. 11. (a) Optical transmission according to Mollart and Lewis (2001) measured at elevated temperatures, with the courtesy of Tim Mollart and (b) optical absorption maps measured by PTD at room temperature (RT) and temperature of 200 °C.

temperature dependence of the interference fringes the linear expansion rate with temperature and eventually the factor $(1/n)\mathrm{d}n/\mathrm{d}T$ can be calculated. It should be noted that due to our experimental setup the interference is actually composed of a few sinus-waves induced by the thermal dilatation and the temperature induced changes in the interaction area of both laser beams. When using the $\kappa(T)$ data measured from Ruf et al. (2000) we were able to evaluate $\alpha(T)$ plotted in Figure 12. For comparison calorimetric data were measured at the same time. The calorimetric measurements are difficult to perform at high temperatures due to the fact that the small temperature changes of the sample heated by a laser are modulated on a large sample temperature. We were able to measure α by calorimetry till temperatures of 150 °C at maximum. The dependence measured by calorimetry compared with the PTD data gave us an excellent agreement.

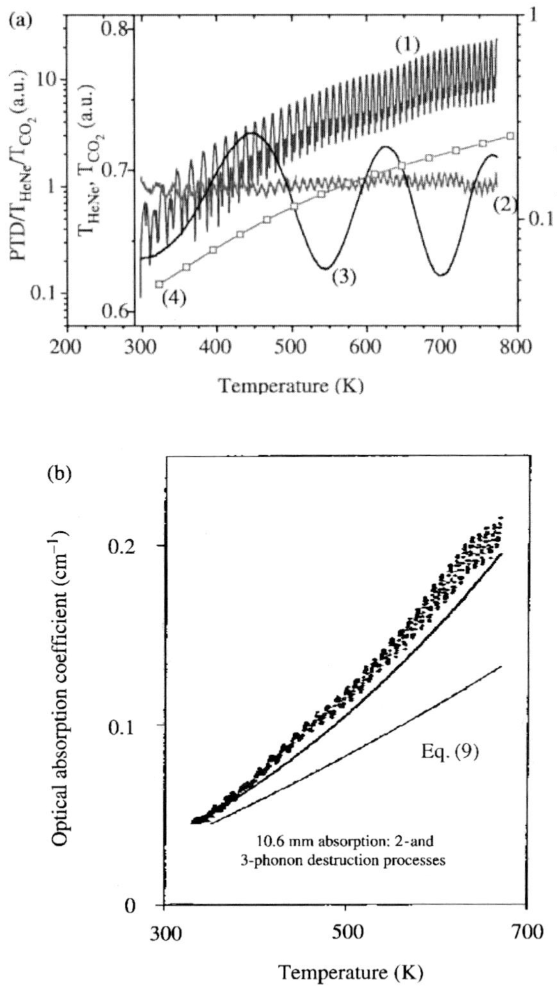

FIG. 12. (a) The normalized PTD signal (1), the transmittance of HeNe (2) and CO_2 (3) laser beams, as well as $\alpha(T)$ (4) measured in temperature region 300–800 K on a natural type IIa single crystal diamond. (b) The PDS measured optical absorption coefficient in the range 300–700 K compared with a theoretical calculation by Piccirillo et al. (2002), involving processes described by Eq. (9), and summating all possible 2 and 3 phonon contributions including a temperature shift, leading to a good agreement of experimental PDS (T) absorption at 10.6 μm.

From the measurement of the optical absorption coefficient in the wide temperature range there is strong evidence of the thermally activated nature of the intrinsic absorption at 10.6 μm. The absorption compared to IIA natural and synthetic diamond is identical reaching the value of about 0.3 cm^{-1} at about 500 °C, thus leading to an increase of about 10 times compared to the RT value. When comparing with the absorption of normal grade CVD diamond which contains some amount of non-diamond carbon, interestingly, though the absorption coefficient starts at higher value, it levels off with the highest purity CVD or IIa diamond at higher temperatures, confirming clearly the intrinsic origin of the 10.6 μm absorption.

The optical absorption at the same region has been recently measured by other authors also, though the problems of the black body radiation had to be addressed (Thomas et al., 1995). Thomas et al. have suggested a model based on the multi-phonon absorption including two-phonon acoustic–acoustic interactions to explain the temperature increase of α. Recently in a joint collaboration with Element Six Ltd (deBeers), King's College and QinetiQ a new model has been developed (Davies et al., 2002; Piccirillo et al., 2002). Based on the probability of creating two phonons by a temperature increase, through the Bose–Einstein population terms, the absorption coefficient at temperature T and the phonon frequency ν relates to phonon population n_p as:

$$I(\nu, T) = I(\nu, 0)(1 + n_p[E_1, T])(1 + n[E_2, T]) \tag{8}$$

where $n_p(E, T) = \exp((E/k_B T) - 1)^{-1}$.

The phonon energies from the Brillouin zone can be found from the phonon dispersion curves (Klein et al., 1992). In the two-phonon absorption, due to summation rules and the crystal symmetry, the transitions at $h\nu$ can involve, for example, a phonon of energy say 1300 cm^{-1} and the second one at $E_2 = h\nu - E_1$. It is obvious that based on the phonon dispersion curves the absorption at 10.6 μm (i.e., 943 cm^{-1}) cannot be due to such combination of two optical phonons and only the possibility of an acoustic–acoustic interaction as discussed by Thomas et al. (1995) is possible. According to Davies et al. (2002) based on the Debye approximation at these low energies the number of phonons is proportional to E^2. The temperature dependence of the absorption can thus be calculated by including the term $(1 + n(E))$ for each phonon at energy E contributing to the sum in Eq. (8). Figure 12(b) shows a possible contribution to the optical absorption at 10.6 μm, as calculated by Davies and compared with PDS measurements on a CVD diamond sample. A simple mechanism based on the acoustic interactions outlined above fails to explain experimental data. On the other hand Davies suggested, when taking into account processes with a destruction of phonons, a zero total momentum after the summation of the **k** vectors of phonons were involved. Based on this assumption e.g., $h\nu = E_2 - E_1$ Davies could derive combinations leading to the $E_2 - E_1 = 943$ cm^{-1} absorption, with a constraint of k being identical with $|k| = (0.25 \pm 0.1)\pi/a$, obtained from the phonon dispersion curves, E_2 being an optical phonon mode and E_1 the acoustic one. Because the optical phonon modes are relatively flat one easily gets directly $E_2 = 1275$ cm^{-1} and $E_1 = 335$ cm^{-1} for the absorption at 10.6 μm.

The absorption coefficient calculated is expressed as a sum of all processes leading to $h\nu = E_2 + E_1$, because all processes involve the coupling to the same phonons leading to the term, Piccirillo et al. (2002):

$$I(v_{-},T) = I(v_{+},0)n(E_1)(1 + n[E_2,T]v_{-}/v_{+} \quad (9)$$

Due to increase in temperature there is a red shift in the energy of all two-phonon absorption features related to the same sum of two phonons. Thus in the final calculations this change should be taken into account. One can also include higher order processes such as 3-phonon contributions where two phonons are generated and one anihilated. The reader can find details about these calculations in the article of Piccirillo et al. (2002). When summating the processes based on the two phonons and three phonons interactions with annihilating the phonon one obtains the empirical dependence depicted in Figure 12(b). This behavior agrees very well with the measured data. It should be noted that according to the measurements of Thomas et al. (1995) and also Mollart and Lewis (2001) or our measurements, one not only sees increased absorption at 10.6 μm, but also a wing of the increased absorption appears in this IR range starting say, from about 1500 to 1200 cm^{-1}. Thus for its description one needs to do the above summation for all phonons leading to the absorption in the measured range.

2.9. Summary

Section 2 has been devoted to intrinsic absorption of CVD diamond films measured by using PDS. PDS has proved to be a very powerful technique to trace up the very low optical absorption in undoped high-purity CVD diamond films. Various experimental configurations of these techniques, specifically the transversal mode using monochromatic light produced by a W-halogen or Xe lamps or a collinear configuration using two laser beams have been successfully employed, allowing to reach a sensitivity of the measurements close to $\alpha d = 10^{-6}$; enabling to study defects on ppm scale. The dominant absorption in the optical spectra is associated with a presence of non-diamond carbon at the grain boundaries. The unanimous correlation to non-diamond carbon was possible due to the use of Raman spectroscopy. Recently, existence of π-bonded carbon at grain boundaries has been studied by means of PL mapping (Mora et al., 2002) together with high-resolution electron microscopy.

The total non-diamond carbon absorption can be considered in a two-phase approximation; i.e., intrinsic absorption of the pure diamond, the extrinsic or defect absorption due to lattice imperfection and impurities and an absorption in a second sp^2 containing phase region at the grain boundaries. This continuous absorption is decaying with increasing wavelength with cutoff at IR energies, when it is taken over by multi-phonon absorption region and eventually the absorption on C–H vibrational modes. Still at lower energies a fundamental limit of 0.025–0.03 cm^{-1} is reached.

By using the PTD mapping we were able to distinguish the higher absorbing regions in the optical quality CVD diamond. It was found that due to anisotropy strain at grain boundaries of crystallites with different orientation, larger non-diamond carbon inclusions appear to be leading to an enhanced absorption region. When the deposition process is mastered the overall absorption can be as low as 0.025 cm^{-1} inside the grains in optical grade CVD diamond. The c-PDS configuration not only allows to measure at room temperature but, but also at high temperatures allowing tracing up the mechanism of optical absorption. Recently, based on our experimental data an empirical model including a simultaneous phonon emission and absorption suggested the 10.6 μm absorption limit which is of crucial importance for use of diamond in CO_2 high power laser components.

3. The Extrinsic Absorption in CVD Diamond based on Photoelectrical Techniques

The recent developments in preparation of CVD diamond for novel high-power, high-frequency, high-temperature electronics and UV optoelectronics (Koizumi et al., 2001; Isberg et al., 2002) quests ppb-sensitive detection methods for determining electronic-active defects and dopants to optimize electronic grade material CVD diamond. As discussed in Nebel (2003) a significant progress has been reached in the synthesis of optical quality polycrystalline diamond deposited by the microwave plasma enhanced chemical vapor deposition (MWPECVD) techniques where the optical absorption around 10 μm wavelength reaches its theoretical limits given by intrinsic multi-phonon absorption/emission processes (Ruf et al., 2000; Remes et al., 2001; Davies et al., 2002; Piccirillo et al., 2002). In diamond for electronics one has to go further. We would like to prepare such material where the defects in the gap, i.e., in the near IR and the visible part of the spectra will be on ppb levels. Further, for some applications, one needs to do that on a large size as the electronic industry with developed large area processing requires. Well how far are we to that dream? Significant progress in the material technology allowed to prepare material with mobilities in the range of 4000 $cm^2 V^{-1} s^{-1}$ on 0.5×0.5 cm^2 area and thickness of a few hundred microns (Isberg et al., 2002). Another example is n-type diamond which seemed to be an undefeatable hurdle and now we have single crystal n-type P-doped diamond as discussed by Koizumi (2003).

Contrary to optical grade material which has been intensively studied by optical characterization techniques such as IR, the investigation of CVD diamond for electronic applications is non-sufficient and frequently contradictory. As discussed in Section 2.1 standard optical spectroscopy is not sensitive enough to detect shallow and deep-gap states with ppm concentrations in thin layers. Amongst the most important characterization techniques have always been magnetic resonance methods such as EPR allowing detection of very low spin-active materials (Twitchen, Baker, Newton and Johnston, 2000). But the impurity is not always spin active. The photoluminescence (PL) is probably one of the ultimately sensitive techniques, but we know that it is not quantitative and not always is the radiation recombination probability sufficient; an example is the direct transition on P and B impurities which are not directly observable. In this respect photoelectrical techniques such as photocurrent in various configurations are very interesting due to their sensitivity and universality. In this chapter we will discuss various photocurrent configurations including Fourier Transform Photocurrent Spectroscopy, for detection of B, P and H-treatment related defects (Nesládek, Stals, Stesmans, Iakoubovskij, Adriaenssens, Rosa and Vanecek, 1998; Nesládek et al., 1999b).

3.1. The Principle of Photoelectric Methods

The photocurrent technique can offer still more sensitivity than the optical absorption for high-quality photosensitive crystals. Photocurrent is an indirect technique for measuring the optical absorption based on monitoring the carriers generated by incoming photons N_{ph} (i.e., proportional to $\alpha(E)$). It is indirect in the sense that carrier mobility μ as well as the lifetime τ can be wavelength dependent whose dependence is necessary to eliminate. Photocurrent spectroscopy is based on the fundamental equation, relating the secondary photocurrent (i.e., photocurrent on structures with ohmic contacts) and the optical absorption coefficient as:

$$I \sim e(1-R)N_{\mathrm{ph}}\mu\tau(1-\exp(-\alpha d))F \tag{10}$$

where e is the electric charge, R the reflection coefficient and F the electric field applied. The full derivation of the expression can be found in the work of Bube (1992). If we want to obtain $\alpha(E)$ and further on the photoionization cross-section $\alpha(E) = N_{\mathrm{i}}f(E)\sigma(E)$; where $N_{\mathrm{i}}f(E)$ is the number of occupied impurities we have to simplify the experimental measurement conditions by such way that in Eq. (10) either all parameters can be set independent on the energy of the incident light or they are known. This implies specific configurations of the PC setup, as discussed further.

Based on the way the defect interacts with the host lattice, impurities can be divided into two different classes, shallow-level defects and deep-level defects (Böer, 1990). The photoionization cross-section of deep-level impurities, is often fitted with the Inkson formula (Inkson, 1981) convoluted with a Gaussian phonon broadening term (van Enckevort and Versteegen, 1992); there are various transitions possible and one of the models for a direct transition is:

$$\sigma_{\mathrm{Inkson}}(E) = A\int_{-\infty}^{+\infty} \frac{(\varepsilon - E_{\mathrm{I}})^{3/2}}{\varepsilon(\varepsilon - B)^2} \frac{\exp\left(\dfrac{-(E-\varepsilon-\Delta E_{\mathrm{ph}})^2}{2w^2}\right)}{\sqrt{2\pi w^2}}\,\mathrm{d}\varepsilon \tag{11}$$

where E is the energy, E_{I} the photoionization offset, ε the integration constant,

A and B are constants, and $2w$ the FWHM of the Gaussian term. ΔE_{ph} and w are related through the so-called Huang-Rhys factor, which gives the electron–phonon coupling of the defect. A modification of this formula has been used by Haenen, Meykens, Neslédek, Knuyt, Stals, Teraji, Koizumi and Gheeraert (2000a).

$$\sigma(E) = A \int_{-\infty}^{+\infty} \frac{\sqrt{\varepsilon - E_{\mathrm{I}}}}{\varepsilon(\varepsilon - B)^{b}} \frac{\exp\left(\dfrac{-(E-\varepsilon)^2}{2w^2}\right)}{\sqrt{2\pi w^2}} \mathrm{d}\varepsilon \qquad (12)$$

with b a numerical factor between 1 and 3.

3.2. The Photocurrent Setup

Our work uses two basic setups. The first one is derived by a rather simple way from the PDS setup presented in Figure 4 (Section 2.2), by using just the monochromatic pump-beam, i.e., without the probe beam and detector, numbers (12) and (14) in Figure 4. The photocurrent is first preamplified by a low-noise ac preamplifier and measured further by the lock-in amplifier technique. In addition, we measure the phase shift between the ac-chopped light and the recorded photocurrent. By keeping the photocurrent constant the quasi-Fermi levels are fixed and consequently the $\mu\tau$ product in Eq. (10), leading to a simple relation $1/N_{\mathrm{ph}} \sim (1 - \exp(-\alpha d))$. This approach, known as the Constant Photocurrent Method (CPM), is valid for a known type of majority transport. In reality the situation is much more complex as it can be found, for example, in Bube (1992); this can occur when both donor- and acceptor-like centres are present in the material, implying the necessity of working with quasi-Fermi levels.

In some spectroscopic configuration we are not interested directly in $\alpha(E)$, but rather in changes of the lifetime, like for example in the oscillatory photoconductivity measurements, where—as discussed later—we search at which energy a capture of carriers moving in the band occurs into a sharp electronic levels, reflected by a decrease of the lifetime, and allowing to find exactly the energy of such sharp defect energy level. In such a case it is not necessary to work in the CPM mode.

Here we summarize briefly the technique used:

1. Constant Photocurrent Method (CPM): photocurrent and $\mu\tau$ product are kept constant, the absorption $\alpha(E)$ is proportional to $\alpha(E) = 1/N_{\mathrm{ph}}(E)$; where N_{ph} is the number of carriers needed to keep the PC at a constant chosen value (Vanecek et al., 1983).
2. Photocurrent method and oscillatory photoconductivity (PC): measurement of the photocurrent spectra can be influenced by the recombination lifetime changes (Fig. 13(a)).

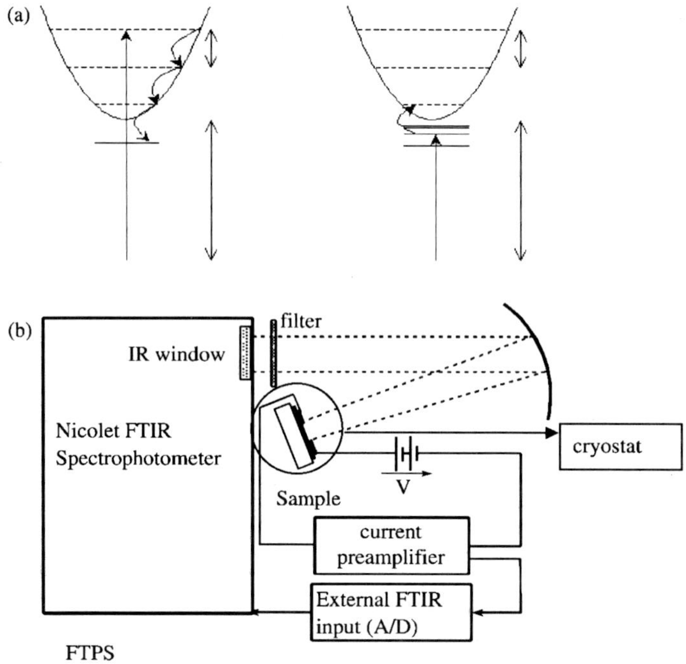

FIG. 13. (a) Model of optical transitions showing the capture of photoexcited carriers on collection levels (oscillatory conductivity) and photo-thermal induced transition from excited states of a defect level to the conduction band. (b) FTPS is based on a traditional FTIR spectroscopy system, measuring photocurrent on a diamond sample used directly as an external detector.

3. Photothermal Ionization Spectroscopy (PTIS): The carrier is optically excited from a ground state of a defect level to defect excited electronic states close to the conduction (E_C) band minimum (or valence band maximum), from which it is further photoionized (i.e., transferred to the conduction band minimum or valence band maximum) by an additional thermal energy proportional to k_BT, where k_B is the Boltzmann constant. The electronic structure of the impurity can be obtained (Fig. 13(a)).
4. Fourier Transform Photocurrent Spectroscopy (FTPS): The photocurrent spectra are obtained with Fourier Transform principles, e.g., Inverse Fourier Transform of an interferogram, obtained in photocurrent mode. This method combines principles of PC and PTIS as discussed further.

The techniques 1–3 are discussed in detail in various publications (Vanecek et al., 1983; Bube, 1992; Lifshits, 1993; Rohrer, Graeff, Janssen, Nebel, Stutzmann, Guttler and Zachai, 1996; Rohrer, Nebel, Stutzmann, Floter, Zachai, Jiang and Klages, 1998); here we deal in detail only with the FTPS method applied very recently to diamond (Kravets, Ogorodniks, Poruba, Moravec, Neslédek, Rosa and Vanecek, 2002; Vanecek and Poruba, 2002).

3.3. Fourier Transform Photocurrent Spectroscopy

Fourier transform photocurrent spectroscopy (FTPS) is based on traditional FTIR spectroscopy (i.e., a system using transmittance/reflectance measurements), having its important advantages such as quite short acquisition time, high resolution and high light throughput giving rise to a good signal/noise ratio. FTPS is not as simple as ac photocurrent spectroscopy (measured with the help of Fourier transform technique), but it can be combined with PTIS (Kravets et al., 2002; Vanecek and Poruba, 2002) by choosing the proper measurement temperature. Also important is the white-light illumination, i.e., not only thermally induced transitions as for PTIS, but also transitions induced by a bias light (which is always present in FTIR spectroscopy) are monitored. The FTPS setup is explained in Figure 13(b).

The basis is a Fourier-transform infrared (FTIR) spectrometer equipped for the near infrared region (KBr and CaF_2 beamsplitter, globar or white-light source) with an external light beam output for use of an external detector. A diamond sample is placed as an external detector, i.e., when normalized to N_{ph}, we directly get the spectroscopic signal related to the absorption of the light in the sample. The FTIR signal from the sample is thus normalized to the baseline FTIR signal from a spectrally independent detector—either a pyrodetector of deuterated triglycine sulfate (DTGS), or a Si photodiode. The signal from the pyrodetector is normalized by the number of photons by multiplying with $1/E$, E being the photon energy, in order to conserve the proportionality to the number of photons hitting the detector, as it is the case for the quantum detector (Si photodiode). Unlike the previous ac setups, where the photocurrent, modulated at a fixed frequency is detected by an lock-in amplifier, here, the photocurrent has the form of an interferogram, dependent on the scanning velocity of the mirror. The electronic detection circuit consists of a stable voltage source (e.g., a Keithley) and preamplifier (e.g., an EG&G). For the FTPS measurements, the sample is placed onto a cold finger of an Oxford Instruments liquid nitrogen cryostat with CaF_2 optical windows. Temperature regulation sets the sample temperature within the 77–450 K measurement range. Similarly as in the previous cases, we measure thin films with coplanar or interdigitated contacts. The photocurrent is amplified by a preamplifier, digitized and fed into the external interface of the FTIR spectrometer (the external detector option of the Nicolet FTIR spectrometer). Various scan-velocity can be used depending on the character of the signal. Typical scanning mirror velocity 0.158 cm s^{-1} is used, number of scans ranged from 64 to 4096, spectral resolution 4 to 32 cm^{-1}. FTPS spectra measured have to be corrected for the signal linearity and frequency dependence of the photocurrent signal, as described in the work of Griffiths and deHaseth (1986). The necessity to provide such a correction comes from the principle of FTIR detection, i.e., the signal at each wavenumber ν is modulated at different frequency $f = 2\nu$, where v is the scan velocity. The frequency dependence is determined for each sample or detected by measuring the FTIR spectrum at several different scan velocities. In order to increase the dynamic range of our measurement in the IR region, we divide the spectrum into 2 regions with the help of a thick silicon wafer polished on both faces. It acts as a filter removing the high-energy photons from the incident light beam thus pumping up the precision of measurement of the IR spectral part. The resulting measurement extends down to 3000 cm^{-1} (Vanecek and Poruba, 2002). The normalized FTPS photocurrent is proportional to the absorptance A and in a low-absorption region directly to the optical absorption coefficient α.

The necessary condition for this interpretation is the signal linearity, i.e., the constantness of the $\mu\tau$ product. The linearity of the FTPS signal can be checked with light intensity, scan velocity and applied voltages to find the behavior of spectrally dependent parameters. In Section 3.4 we discuss the spectroscopy of main defects in undoped CVD diamond films as well as the photocurrent characterization of B and P-doped films, allowing to find the photoionization onset E_I (e.g., the energy position of dopants in the gap) and the electronic structure of defects.

3.4. Photocurrent Spectra in Undoped CVD Diamond Films

In Figure 14, the optical absorption coefficient $\alpha(E)$ is calculated from a CPM spectrum and compared to a PDS measurement for an optical quality transparent CVD diamond window of a thickness 1 mm, polished from both sides.

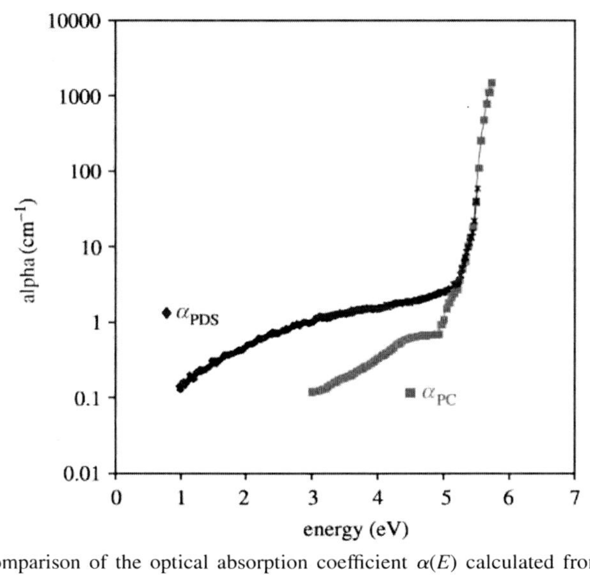

Fig. 14. A comparison of the optical absorption coefficient $\alpha(E)$ calculated from PDS (optical-upper trace) and CPM (photocurrent-lower trace) spectroscopy, measured on 1 mm thick optical quality CVD diamond plate, prepared by MW-PECDV method.

We can clearly see that the optical absorption calculated from the photocurrent copies the absorption from PDS in the region of the indirect absorption edge, showing further the optical absorption in the diamond subgap. In a closer look we can see that while in PDS the indirect edge absorption is already masked by non-diamond carbon absorption, in CPM we can still measure lower absorption towards the gap.

This difference is based on the fact that PDS is monitoring all possible optical transitions, i.e., including also those which do not lead to any free carrier and eventually thus showing the transitions on non-diamond carbon inclusions.

While in this case absorption measured by PDS the optical absorption coefficient will level off at about 0.1 eV at about 0.03–0.05 cm^{-1} for a good quality CVD diamond material, the optical absorption calculated from CPM is decaying quite strongly for materials with low extrinsic defect concentration. This behavior is very characteristic for every undoped polycrystalline CVD diamond measured so far, also by other groups (Gheeraert, Deneuville, Bustarret and Fontaine, 1995; Rohrer et al., 1996; Rossi, Salvatori, Scotti, Conte and Cappelli, 2000). As an example, we show characteristic photocurrent spectra in Figure 15, measured for hydrogenated and annealed CVD diamond films. Similar to the case of a deep level in semiconductors described by a photoionization

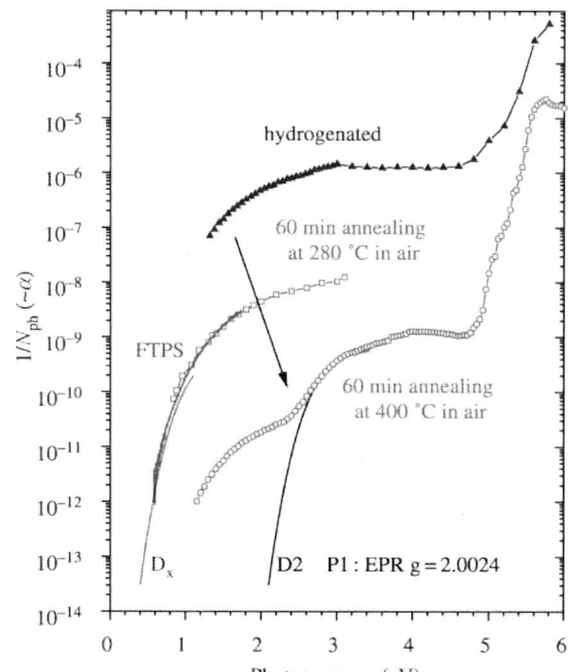

FIG. 15. Optical absorption spectra measured by CPM where $\alpha(E)$ is proportional to $1/N_{ph}$. The polished CVD diamond sample is measured after hydrogenation and upon further annealing.

cross-section by Eq. (11) we attempted to fit the IR part of the photocurrent spectra by this formula, yielding E_I from IR defects at about 0.9 eV. We have denoted this characteristic defect as D_x (Nesládek et al., 1998).

It has been realized by the group of Jackman (Gaudin, Whitfield, Foord and Jackman, 2001), that the subgap photocurrent, relative to the photocurrent at the absorption edge decreases drastically upon specific annealing treatments, including oxidation and CH_4 treatment at high temperatures, which has allowed to prepare sensitive and solar-blind detectors. It was our goal to study how much changes of the subgap absorption could be attributed to the changes of $\alpha(E)$ (i.e., changes in the defect concentration) and how much to the $\mu\tau$ product changes due to surface changes upon treatments such as the plasma hydrogenation or chemical oxidation.

First of all it is important to remark that, when measuring the photocurrent in a coplanar configuration on high-quality CVD diamond, if the sample surface is hydrogenated, such sample is extremely photosensitive. The $\mu\tau$ product easily reaches 10^{-4} cm^2 V^{-1} s^{-1}. Such hydrogenation is carried out in H plasma at temperatures typically 800 °C. On the other hand, when an undoped CVD diamond sample surface is oxidized (for example in sulfochromic acid at 200 °C) $\mu\tau$ drops significantly, typically 10^{-7} cm^2 V^{-1} s^{-1} or lower.

In our previous work we have thoroughly studied the secondary photocurrent upon hydrogenation, oxidation and annealing (Nesládek et al., 1998; Rosa, Pangrac, Vanecek, Vorlicek, Neslédek, Meykens, Quaeyhaegens and Stals, 1998; Nesládek, Meykens, Haenen, Stals, Teraji and Koizumi, 1999a). In Figure 15 two distinct features are present for an annealed sample. First, a shoulder with an onset at about 2.25 eV, attributed to substitutional nitrogen (Nesládek et al., 1998).The EPR spectroscopy revealed the P1 defect in concentrations ranging from far below 1 ppm to 100 ppm in N-doped films (Rosa et al., 1998) with g factor ($g = 2.0024$). This defect in photocurrent spectra is denoted as D2. The fitting by the Inkson formula yields the optical excitation energy of E_I being 2.25 eV. This value is about 0.55 eV higher than the thermal ionization energy of the nitrogen (1.7 eV) donor ion CVD diamond (van Enckevort and Versteegen, 1992). Nitrogen defect has a trigonal symmetry, with two resting bonds forming a lonely pair. N is slightly off-side of the substitutional position leading to a local distortion and consequently the C-dangling bond opposite to N-lonely pair induces the free electron spin, observed by EPR. Due to the large lattice distortion the Huang-Rhys factor is about 0.45 eV, leading to the shift between the optical and thermal ionization energy (Koppitz, 1986; van Enckevort and Versteegen, 1992).

The second-IR feature, denoted as discussed above, the D_x defect, appears upon the hydrogenation of the sample. Fitting of the D_x photoionization cross-section places this defect at about 0.9–1.2 eV for various films measured (Rosa et al., 1998; Kravets et al., 2002). The data in Figure 15 show that a slightly annealed sample at 280 °C for 60 min exhibits an intermediate behavior. To find if the photo-carriers generated by light absorption at this D_x energy are holes or electrons, a photo-Hall effect has been measured using an illumination from an IR laser at 825 nm, schematically shown in Figure 16.

The photo-Hall measurements shown in Figure 17 have proved unanimously p-type conductivity of photo carriers with a concentration of $10^{12}\,cm^{-3}$ and a mobility of $57\,cm^2\,V^{-1}\,s^{-1}$ on a mild annealed sample at 200 °C. Taking into account that the resistivity of the sample is 1.3 MΩ this leads to a dark carrier concentration of carriers of $8.4^{10}\,cm^{-3}$. After the hydrogenation and subsequent annealing at 280 °C, the surface induced H-conductivity has thus dropped from 7.7×10^{-5} S (i.e., the mobility of $45\,cm^2\,V^{-1}\,s^{-1}$ and a carrier concentration of 1.1×10^{13}) to the value of 7.6×10^{-7} S. Upon further annealing the dark conductivity further drops with a dark activation energy of 1.5 eV. This behavior is depicted in detail in Figure 17.

The sheet resistance is rising upon annealing of the sample in air. The mechanism of the surface conductivity has been discussed in Hayashi,

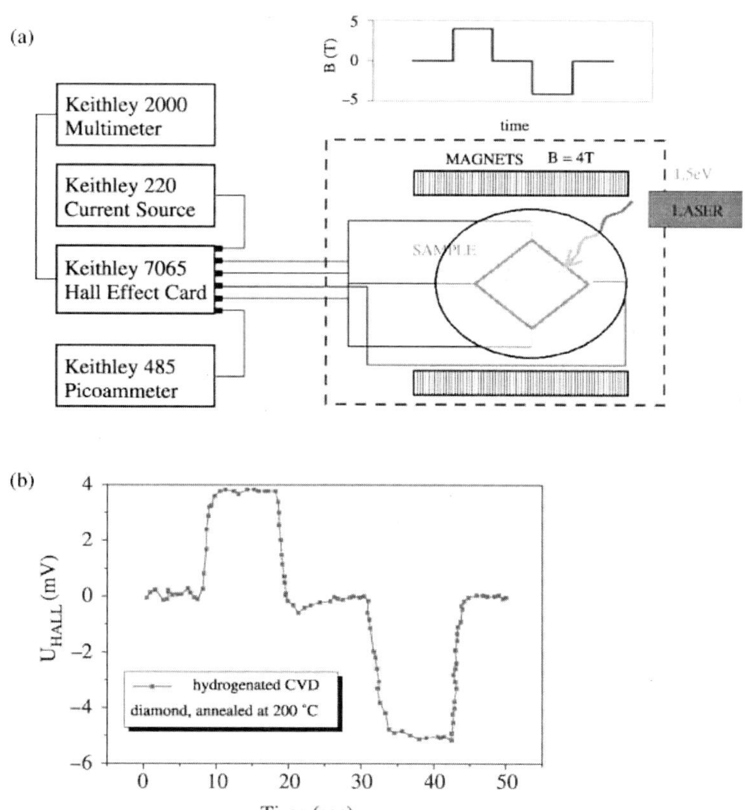

FIG. 16. The photo-Hall setup (a) and the Hall voltage response (b) measured on application of a 1 T pulsed magnetic field at RT for the sample from Figure 15, annealed at 280 °C for 60 min.

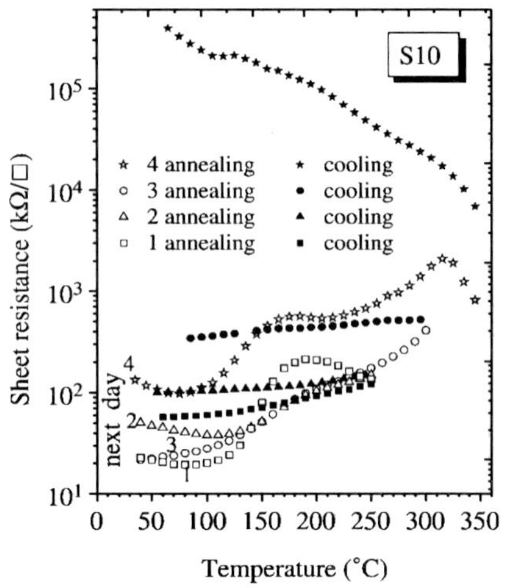

FIG. 17. The sheet resistance of a plasma hydrogenated sample as a function of annealing and cooling in the air.

Yamanaka, Watanabe, Sekiguchi, Okushi and Kajimura (1997), Looi, Jackman and Foord (1998) and Maier, Riedel, Mantel, Ristein and Ley (2000). Upon annealing at high-vacuum conditions at temperatures below 800 °C only the surface adsorbates are removed. On the other hand upon annealing in air, the surface is partially oxidized and the surface conductivity reduced. In Figure 17 at still higher temperatures of about 330 °C, an emission of the captured carriers starts to contribute to the conductivity, as shown for example by TSC measurements (Gheeraert, Deneuville, Gonon, Benabdesselam and Iacconi, 1998; Tromson, Bergonzo, Brambilla, Mer, Foulon and Amosov, 1999).

Using the picture of the transversal doping model suggested by Riestein and Ley (Blank, 2003) (Fig. 18) the surface adsorbants pin the E_F at the surface thus making the CVD diamond p-type. In this case the D_x level is partially empty and photocurrent is generated by an excitation of electrons from the valence band maximum into empty states at the D_x level. It should be noted that the E_F is pinned at E_V at the surface in a few nm thickness as pointed out by Nebel, Sauerer, Ertl, Stutzmann, Graeff, Bergonzo, Williams and Jackman (2001). In the bulk E_F is the determined by the concentration of equilibrium carriers. But because this is negligible for undoped samples, surface band bending induced carriers makes the material p-type and very photosensitive. Upon annealing, the total concentration of adsorbates is decreasing; so the upwards band-bending is reduced, leading to a shift of the E_F towards the midgap and to a decreased photosensitivity. This is due to the fact that nitrogen (deep donor) is the dominating bulk defect and the majority photocarriers are electrons for the annealed samples, as confirmed by Hall measurements.

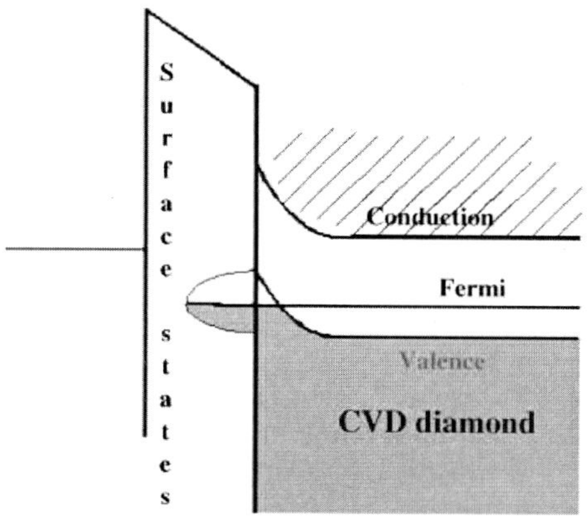

FIG. 18. The transfer doping model of Ristein depicting the pinning of E_F by surface adsorbates, complemented by a deep level D_x, seen in photocurrent spectroscopy.

In Figure 19 FTPS spectra for a PE-CVD sample grown in our ASTeX MW system (sample S20, N below $1 \times 10^{16}\,\mathrm{cm}^{-3}$ from EPR) for the temperature range 77–300 K are shown. This sample has been previously annealed by the same procedure as above. There are a few features to be discussed. In the temperature range 200–300 K we can observe photoionization of free carriers from the level D_x as discussed above with the photoionization threshold at RT

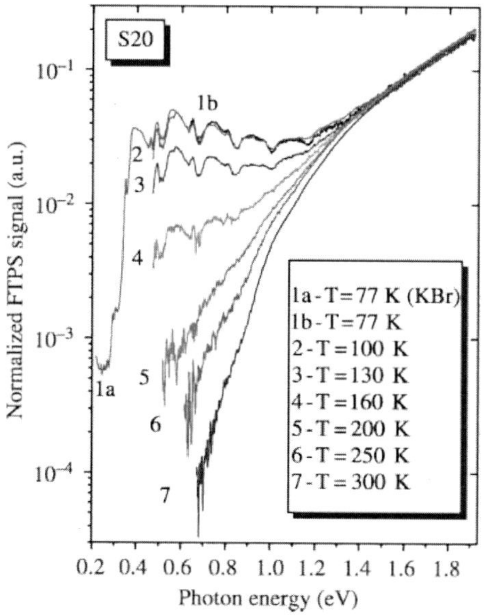

FIG. 19. The photocurrent spectra of a nominally undoped sample S20 ($N < 1 \times 10^{16}\,\mathrm{cm}^{-3}$), showing the deep defect D_x and a shallow level ascribed to residual boron.

around 0.9 eV (Vanecek, Kravets, Poruba, Rosa, Nesládek and Koizumi, 2002). With sample cooling, from a noise below 0.9 eV a shallow level is emerging, clearly identified below 160 K and saturating below 100 K. We can see the peaks in the spectrum at 0.303 eV and 0.347 eV and full photoionization starting around 0.37 eV. The photocurrent signal exhibits two series of equidistant minima, with period around 165 meV. When we compare the detailed spectrum taken at 77 K with 77 K optical absorption data and photocurrent data presented by Gheeraert, Koizumi, Teraji, Kanda and Nesládek (1999), we can see a very good agreement in position of photocurrent peaks at 303 and 347 meV and the optical absorption peaks of the first two excited states of the boron acceptor. These are the collection levels for the effect of oscillatory photoconductivity, already observed in boron-doped diamond films, period being the LO photon energy (Rohrer et al., 1998). This points to the presence of boron in some of our nominally intrinsic CVD diamond layers. We can just speculate how boron got into the deposition apparatus never used for boron doping. This experimental sample (S20) was deposited at reduced pressure and it caused an expansion of the plasma closer to the walls of the Al top-cup of the reactor vessel. Because Al and also Cu alloys sometimes contain traces of B we speculate that B could be of this origin. The concentration of boron should be quite low and we take this as a demonstration of the high sensitivity of FTPS. Because of this unintentional B-doping, together with a low N-concentration, the sample shows high photosensitivity even in mid-annealed oxidized sample; thus B takes the role of the transfer doping but shifting the E_F by a bulk doping rather than pinning at the surface.

As shown in Rosa et al. (1999) for the substitutional nitrogen donor level in diamond there is a small shift of threshold energy into infrared with increasing temperature due to electron–phonon interaction, the shift being strongly pronounced above RT (van Enckevort and Versteegen, 1992). But in our case the threshold of the FTPS signal for D_x level shifts just the opposite way—to a lower photon energy with *decreasing* temperature, in a whole temperature range. One possible explanation of this effect is the following: there is no sharp defect level at the threshold energy but rather a broad distribution of gap states in energy. This would explain various E_I for samples measured between 0.9 and 1.2 eV (Nesládek et al., 1998; Vanecek et al., 2002).

One of the explanations of the shift of D_x with temperature is that with decreasing temperature the occupation of these gap states is changing, the Fermi level and quasi-Fermi level for holes move with decreasing temperature towards the valence band edge, as discussed above. This means that within this broad distribution of levels less and less states are being occupied by electrons. Therefore, the threshold energy for the hole transitions to the valence band (electron transitions to the empty acceptor broad level) has to shift to lower photon energy with cooling, as it is experimentally observed.

In Figure 20 the FTPS signal, normalized by the pyrodetector signal is plotted for the sample S18 which is more resistant and less photosensitive. Here we cannot detect any trace of boron contamination and we can follow the

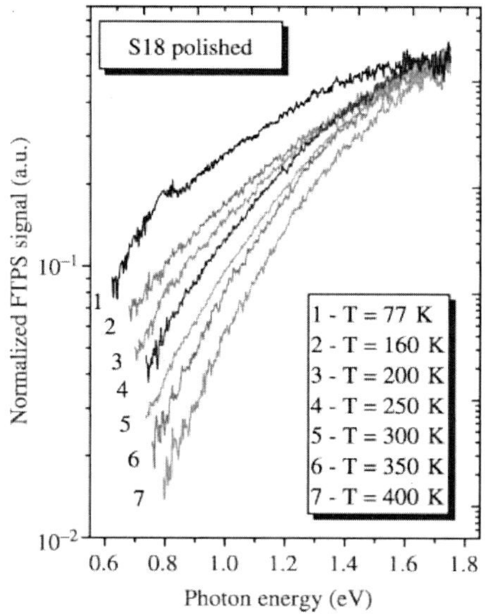

FIG. 20. The temperature dependence of the photocurrent spectra, measured by FTPS, for sample S18, containing N below 10^{16} cm^{-3} and no B.

development of the D_x level in a broad temperature range from 77 to 400 K. Normalized FTPS signals were matched at 1.7 eV. We can see again that the threshold of the FTPS signal, rising above the noise level, shift to a lower photon energy with decreasing temperature, in a whole temperature range.

At the end of this section, based on all the above experimental data, we would like to speculate what the origin of the D_x level is. As discussed above we see this defect in all polycrystalline diamond films also in some case we were able to detect this level even in homoepitaxial thin films. We can just suggest different possibilities. Since the D_x is strongly observed in polycrystalline diamond it could be related to defects on the grain boundaries such as those discussed in Section 2, measured by PDS. One can argue that the shape of the measured absorption is different, but we have to keep in mind that because PDS also measures spectrally independent absorption we cannot reject this possibility. The fact that D_x could be also related to crystal imperfection and eventually to presence of hydrogen (C dangling bond in close vicinity of H) could also be a plausible explanation based on the EPR measurements confirming a high concentration of a defect with $g = 2.0028$ related to hydrogen in a close vicinity of the C dangling bond (Talbot-Ponsonby, Newton, Baker, Scarsbrook, Sussmann and Whitehead, 1998). Because such dangling bonds can also exist in homoepitaxial CVD diamond on lattice defects, this could be a possible explanation. Could the D_x be a pure surface defect? Yes, it could be too. But we cannot explain why upon annealing to high temperatures we could see this defect always present (Nesládek et al., 1999a). Only annealing at very high temperatures at about 1100 °C destroyed completely the appearance of D_x defect, as well as the $g = 2.0028$ signal, pointing that it is not the case. But when hydrogenating the sample after such high-temperature annealing, in the plasma, the D_x defect appears again, so what is the origin, it is bulk or only surface-like? Well, we need more time to study it.

3.5. The Electronic Structure of B and P in CVD Diamond Films

With recent progress in CVD diamond technology it is possible to prepare B-doped layers with very low concentration of B-dopants. These layers, when measured in the 77–200 K range are typically very photosensitive (it means that they have low dark current and high photocurrent). When measuring with FTPS they give a very good interferogram—the necessary condition for a FTPS measurement. As grown or hydrogenated CVD polycrystalline layers have typical surface resistance around 10 kΩ per square. To study the fine electronic structure of boron we have carried out a set of measurements on natural crystals as well as CVD diamond plates containing various B-impurities starting from very pure samples, containing $10^{16}\ cm^{-3}$ B to samples containing $10^{18}\ cm^{-3}$ B (e.g., about 10 ppm). We were especially interested in the fine electronic structure of B impurity. It should be noted that the fine spectra of all boron excited states we can only obtain on low boron concentration samples of high crystallographic perfection, some of these measurements are currently in progress. The spectra for B-doped CVD polycrystalline diamond containing $10^{16}\ cm^{-3}$ impurities are still often broadened. Surprisingly we have got the finest spectra for a thin P-doped epitaxial layer, grown on a HPHT crystal. Figure 21 shows a high resolution FTPS spectrum of the boron impurity (and its one-phonon replica) from residual boron present in a HPHT crystal on which a P-doped diamond has been grown (Kravets et al., 2002).

It shows clearly that the FTPS technique is extremely suitable for a fine resolution studies of impurities. The peaks before the full ionization at 0.37 eV, e.g., the transition from the valence band maximum E_V to the empty B-ground state, are due to electron transitions from E_V to excited states of B (Fig. 13(a)), (see Serrano et al., 2000).

In Table II we compare fine B-spectra measured by FTPS with the work of Collins and Lightowlers (1968) and Anastassakis (1969). In a close inspection of the spectra we note that some of the sharp spectral lines are actually doublets with a 2 meV energy difference between the lines. Recently it has been suggested by a theoretical work by Serrano et al. (2000) that the spin–orbit splitting of the valence band in diamond can be reflected by the existence of doublets with an effective value of about 2 meV, induced by the valence band spin–orbital splitting. Thus the FTPS spectra can confirm this theoretical predication.

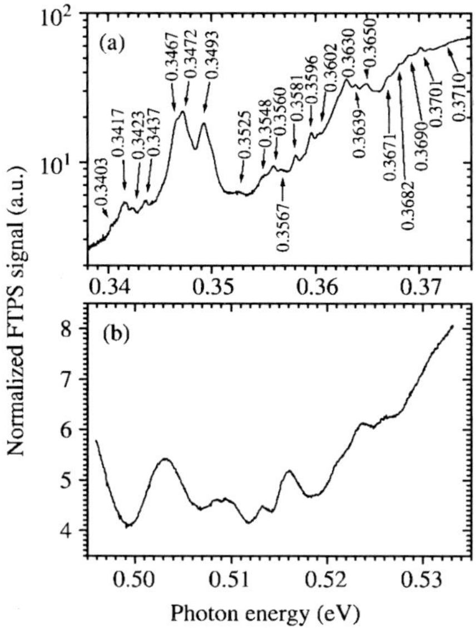

FIG. 21. Two high resolution FTPS fine spectra showing fine structures of the B-impurity on a (111) HPHT diamond substrate, on which a P-doped diamond has been grown later.

Another feature to discuss is a reverse process to the thermal excitation, so-called oscillatory photoconductivity, mentioned in previous sections and Figure 13(a). The free photocarriers, moving in the band under an electric field can thermalize fast by emitting a LO phonon, if their energy from the valence band maximum (or conduction band minimum) to the excited level of the impurity is a multiplication of the LO phonon value. In Figure 19 we show the equidistant minima in the photocondutivity spectra for the polycrystalline sample S20 with a period of the LO phonon, 165 meV, at the center of the Brillouin zone, e.g., at the valence band maximum. In reality the hole can be captured (i.e., electron emitted) by each of the excited states which are close enough to the E_V leading to a set of minima from which vice versa the position of the excited states as listed in Table 2 can be calculated, thus confirming the PTIS data.

Figure 22(a) shows the low resolution FTPS spectra for full energy range for the same sample as measured in Figure 21, the FTPS signal at 77 K, normalized by the pyrodetector signal is plotted for phosphorus-doped epitaxial layer on a Ib {111} oriented diamond substrate. Here we can detect both boron (below 0.52 eV) and phosphorus (starting around 0.52 eV) spectra (Gheeraert et al., 1999; Haenen, Meykens, Neslédek, Knuyt, Stals, Teraji, Koizumi and Gheeraert, 2001). They have quite different frequency dependence, as it is seen from this figure, the photocurrent modulation frequency being equal to two times the product of scanning velocity and wavenumber (Griffiths and deHaseth, 1986). Dependence of the photocurrent

TABLE II

Comparison between the PTIS Peak Energy Values Seen by FTPS and Earlier Work Done with PC and Absorption Experiments

E. Anastassakis	A. T. Collins, E. C. Lightowlers		K1	S20
	Absorption	PTIS		
0.3045				0.303
0.3375	0.3372			
0.3405	0.3402	0.3405	0.3403	
0.3415	0.3415	0.3416	0.3417	
0.3422	0.3423	0.3424	0.3423	
0.3435	0.3435	0.3438	0.3437	
	0.3444			
0.3452	0.3450	0.3452		
	0.3465	0.3465	0.3467	
0.3470	0.3471	0.3472	0.3472	0.347
	0.3480	0.3481		
0.3492	0.3492	0.3494	0.3493	
	0.3501	0.3502		
	0.3511			
0.3525	0.3524	0.3525	0.3525	
	0.3529			
0.3545	0.3547	0.3549	0.3548	
0.3558	0.3558	0.3559	0.3560	
0.3565	0.3567	0.3568	0.3567	
0.3577	0.3579	0.3581	0.3581	
	0.3587	0.3588		
0.3595	0.3593	0.3597	0.3596	
	0.3602	0.3603	0.3602	
0.3630	0.3627	0.3629	0.3630	
0.3638	0.3639	0.3639	0.3639	
0.3645				
	0.3650	0.3649	0.3650	
	0.3660	0.3659		
	0.3668	0.3669	0.3671	
		0.3675		
0.3680	0.3679	0.3680	0.3682	
		0.3685		
	0.3688	0.3689	0.369	
		0.3695		
	0.3700	0.3701	0.3701	
0.3705				
	0.3709	0.3709	0.3710	

signal on applied voltage is shown in Figure 22(b). The phosphorus spectrum (clearly seen just at low voltage) comes from the epilayer, doped with several hundreds ppm of phosphorus and the boron spectrum comes probably from the Ib HPHT (high-pressure high-temperature) substrate.

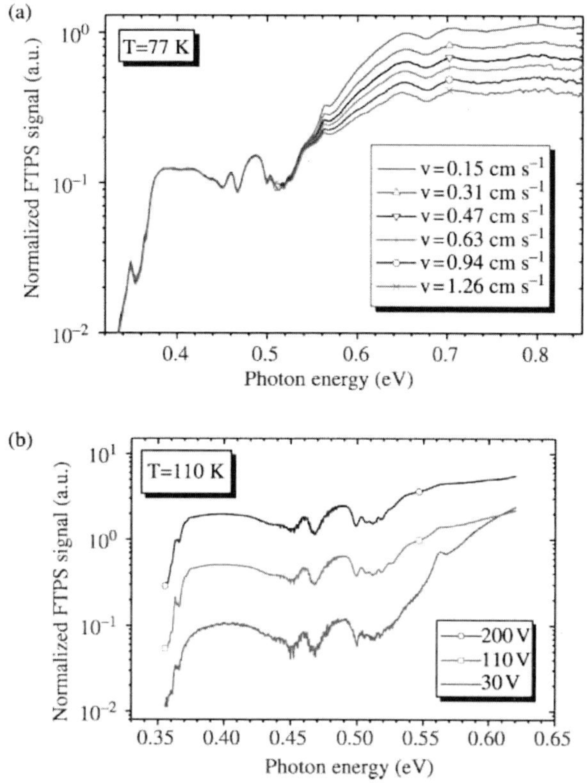

FIG. 22. (a) Normalized FTPS signal at 77 K of an epitaxial P-doped layer on a Ib {111} oriented substrate. The low resolution boron spectrum (0.30–0.52 eV) is independent of the measurement frequency (scan velocity v is the parameter) while the phosphorus part of spectra above 0.52 eV is frequency dependent. (b) Dependence of FTPS signal (measurement temperature 110 K) on applied voltage. For 30 V applied we are in the ohmic region, above 50 V, the current injection starts.

3.6. Electronic Structure of P in Homoepitaxial n-Type CVD Diamond Layers

In 1997, Satoshi Koizumi from NIMS in Japan was the first to successfully prepare phosphorus-doped CVD diamond films, using phosphine as a dopant source (Koizumi, Kamo, Sato, Ozaki and Inuzuka, 1997; Koizumi et al., 2001) by successfully growing epitaxial P-doped samples on (111) HPHT Sumitomo substrates, as discussed by Koizumi (2003). Though it was predicted that the P-incorporation in the diamond lattice is very low, nowadays samples with high mobility and high P-concentration can be prepared. Our first spectroscopic interest was to confirm the electronic and optical activity of P in the NIMS samples. In our first study in 1997, by using photocurrent, PTIS and oscillatory photoconductivity, we could unanimously confirm that P forms a level at about 0.6 eV from the conduction band minimum. In addition, we detected another defect, labeled X_{P2}, with $E_I = (0.81 \pm 0.03)$ eV (Nesládek, Meykens, Haenen, Navratil, Quaeyhaegens, Stals, Stesmans, Iakoubovskij, Adriaenssens, Rosa and Vanecek, 1999b) which was present in the first diamond samples, closely resembling the D_x level discussed earlier. It should be noted that based on cathodoluminescence studies, Sternschulte, Thonke, Sauer and Koizumi (1998) estimated the ionization energy of 630 ± 50 meV for the phosphorus donor, while Gheeraert et al. (1999) report an optical ionization energy of 600 ± 20 meV, based on infrared absorption measurements and theoretical calculations using an effective mass approximation. Our further interest was to clarify the P-spectroscopic data by studying the electronic structure of P-doped epitaxial layers (Haenen, Meykens, Nesládek, Knuyt, Quaeyhaegens, Stals, Koizumi and Gheeraert, 1999; Haenen et al., 2000a; Haenen, Meykens, Neslédek, Knuyt, Stals, Teraji and Koizumi, 2000b; Haenen et al., 2001). First experiments were done between liquid helium (LHeT) and RT. Recently, in close collaboration with Vanecek and co-workers we were able to study P-level also by the FTPS technique as shown above (Kravets, Poruba, Kromka, Nesládek, Rosa and Vanecek, 2003). In Figure 23, the measurements on the first P-doped samples from 1997 fitted with a photoionization cross-section for a deep impurity using the optical cross-section calculated using Eq. (11). Two defects—one at E_I about 0. 6 eV and the other one at about 0.81 eV—are clearly indicated. At higher energy a rise is attributed to the start of the onset of the N-impurity coming from the substrate absorption. The position of phosphorus agreed perfectly with the temperature dependence of the dark conductivity.

A set of samples prepared by MW-PECVD has been prepared with various phosphorus concentration of 100, 500 and 1000 ppm in the gas phase (Haenen et al., 2000b, 2001). In this optimized set of samples, we could not observe anymore the defects seen in the first samples at E_I of 0.81 eV. Figure 24 depicts the measured PC spectra.

The photocurrent spectra in Figure 25, showed a very good agreement with theoretical fitting of the P-photoionization cross-section. In our previous work (Haenen et al., 1999, 2000a; Nesládek, 1999b) discussed above, Eq. (11) yielded (0.56 ± 0.3) eV for the photoionization energy of X_{P1}, giving a good fit of the experimental curves. However, it cannot explain the 565 meV collection level (Figure 25). Therefore we suggest another model using Eq. (12), yielding $E_I = 0.60 \pm 0.03$ eV for P. The formula gives good results for the temperature range 4.2 K up to 170 K.

When looking to the spectra in Figure 25, there are a few features of interest. In the case of boron we can see a structure in the photocurrent spectra. Structure below the photoionization threshold E_I can be attributed to thermally/light induced transitions of electrons from the excited states of phosphorus to the conduction band minimum E_C (Fig. 13(a)). At energies higher than E_I the theoretical photoionization cross-section is modulated by emission of LO-phonons. Simultaneously, one or more LO phonons can be emitted. The first minimum is rather sharp, but the second minimum is broadened. In closer inspection this

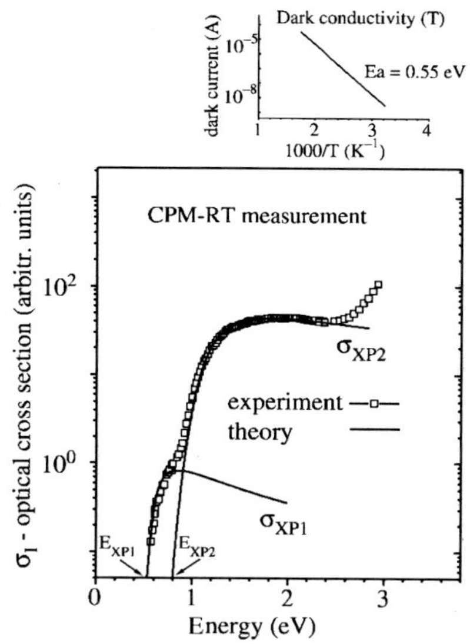

FIG. 23. The liquid nitrogen temperature photoionization cross-section of the P impurity at 0.6 eV (denoted as X_{P1}) and a second defect at 0.81 eV (denoted as X_{P2}) in a P-doped epitaxial sample.

minimum is actually a superposition of several minima. All minima separated by 155 meV can be grouped. This value corresponds to the LO-phonon located along the Δ-axes in the [100] direction of the diamond band structure as we have suggested in Haenen et al. (2000b, 2001) (Fig. 26). Thus the phonon-assisted oscillatory conductivity (OP) process can be used to calculate the position of excited states.

As discussed above for boron, the PAOP process takes place if the energy difference between the electron state in the conduction band (CB) and an excited

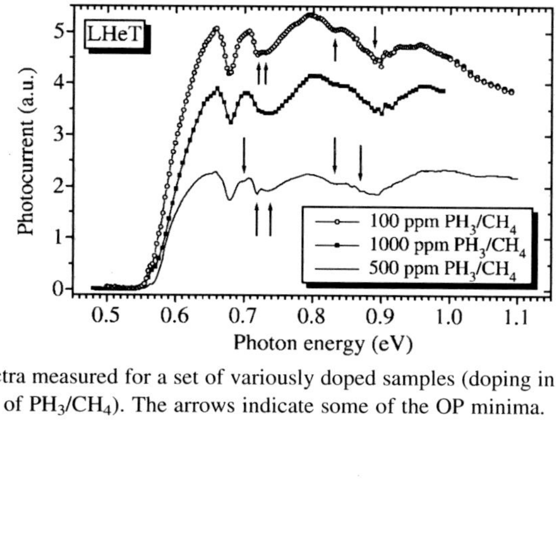

FIG. 24. PC spectra measured for a set of variously doped samples (doping in gas phase with 100, 500 and 1000 ppm of PH_3/CH_4). The arrows indicate some of the OP minima.

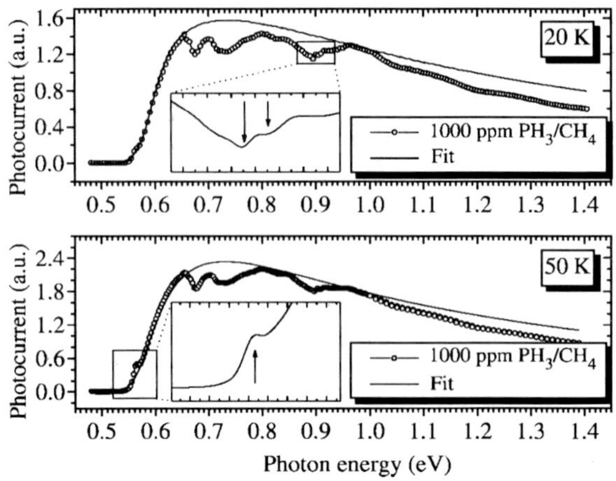

FIG. 25. The detailed PC fit using Eq. (12) showing also the oscillatory photoconductivity spectra for a 1000 ppm doped sample.

state of phosphorus, situated just under the CB, is exactly n-times the LO phonon energy, i.e., in analogy to the B transitions at the valence band. If one considers that the P-level—as it should be for a donor-like state–is close to the CB minimum, the value of the LO phonon energy at the CB minimum should give the energies of the excited states. Subtracting 155 meV n-times from the nth order minimum leads to 523($\pm$5), 565($\pm$5), 575($\pm$5) and 584($\pm$5) meV, for the four detected excited levels of phosphorus. Table III summarizes the energy values where a minimum in the photocurrent is detected, for 100, 500 and 1000 ppm samples. The most important difference with the 1000 ppm sample is the fact that the broad minimum at about 725 meV, reported in previous studies (Haenen et al.,

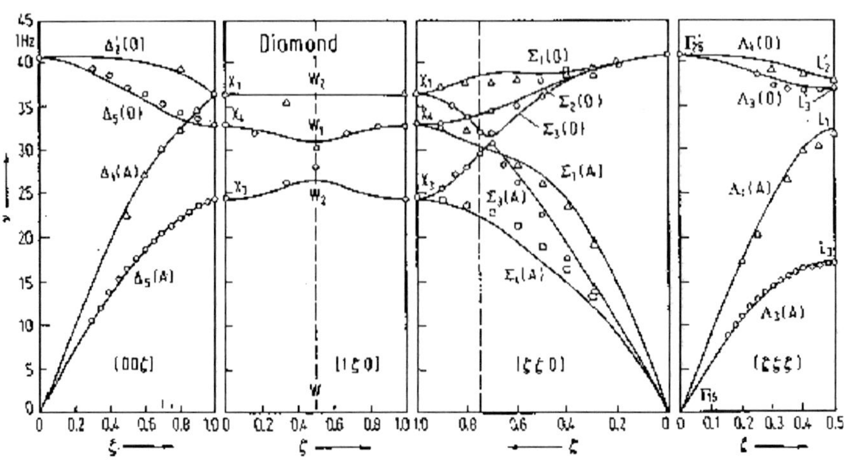

FIG. 26. The phonon dispersion curves in diamond, as can be found in Madelung (1996).

TABLE III

THE ENERGY OF THE EXCITED STATES OF P MEASURED BY PTIS AND COMPARED WITH THE VALUES CALCULATED FROM OSCILLATORY PHOTOCONDUCTIVITY

PH_3/CH_4 (ppm)	PTIS (meV)				PC (meV)—1 LO				PC (meV)—2 LO				910–915
	523	563	575	584	680	698	719	734	833	872	890	896	
1000		x			x			x	x			x	x
500		x	x	x	x	x	x	x	x	x	x	x	
100	x	x	x	x	x	x	x	x	x	x	x	x	x
					−155		−155	−155	−310	−310	−310	−310	
	523	563	575	584	525		564	579	523	562	580	586	
													−330
													580–585
	FTIR	FTIR		FTIR									
	$2P_0$	$2P_\pm$		$3P_\pm$									

1999, 2000a), is now clearly divided into two separate minima. Interesting to note is that now, the first two energy values agree perfectly with the two levels measured with FTIR by Gheeraert et al. (1999) and Gheeraert, Casanova, Koizumi, Teraji and Kanda (2001) and explained by the effective mass approximation. All minima found in the spectra are listed in the table, together with PTIS data, discussed further.

Table III shows a very good agreement of the PTIS peak position and the energy calculated by subtracting the LO phonon energies. When comparing with B in Figure 21, where a set of equidistant minima for quite a large number of phonons could be observed, the situation in phosphorus-doped samples the situation is different: only 2–3 sets of the equidistant minima can be observed. In addition the minima are broad and probably composed of various components which were not possible to distinguish for the samples measured. Thus further optimization towards lower doping levels is necessary to obtain fine spectra.

3.7. PTIS Spectra on P-doped Samples

As discussed above complementary to OP we could get the P-excited state energy from PTIS. In diamond, P forms rather deep dopant level and therefore to find PTIS peaks we need to go to temperatures where the thermal energy is comparable with the E-difference $\Delta = E_C - E_{IP}$ (photoionization of phosphorus level). But this is a difficult problem, when measuring at higher temperatures than LNT, the dark current stability is deteriorating, leading to worsened signal/noise ratio. So far we could only detect very clearly the transition at 565 meV for the samples prepared at NIMS, listed in Table III. Though the intensity of the 565 meV line was rather low we could also see peaks around 523 meV.

Recently n-type CVD diamond has been prepared in different groups. To get a high PC signal by increasing the optical absorption path, we have grown sample 15 μm thick. In Figure 27 we can see the P-photocurrent spectrum for the sample with 2×10^{19} cm^{-3} P and a mobility of 30 cm^2 V^{-1} s^{-1}, prepared by our group by the MW-PECVD method, using PH_3 doping in situ at conditions power 2000 W, 0.1% CH_4 in H_2, pressure 90 Torr and a substrate temperature of 900 °C. Surprisingly, even though the concentration of P was higher, we could see four OP minima and additionally the photocurrent was stable to almost RT. Thus for the first time we could try to measure PTIS at higher temperatures (Fig. 28). The measurement at higher temperatures is limited by a thermal broadening of the energy of the excited states. Whilst the structures at 523 meV were not possible to distinguish (we measure electric current as low as 10 fA at energy of 525 meV). We could see for the first time a structure appearing between 565 and 600 meV scale, though limited by the thermal broadening and the measurement noise. Clearly peaks at about 575 and 584 meV are present. The phase shift between the ac chopped light and the detected photocurrent plotted in Figure 27 reflect small changes of the carrier lifetime, associated with the transition from excited states

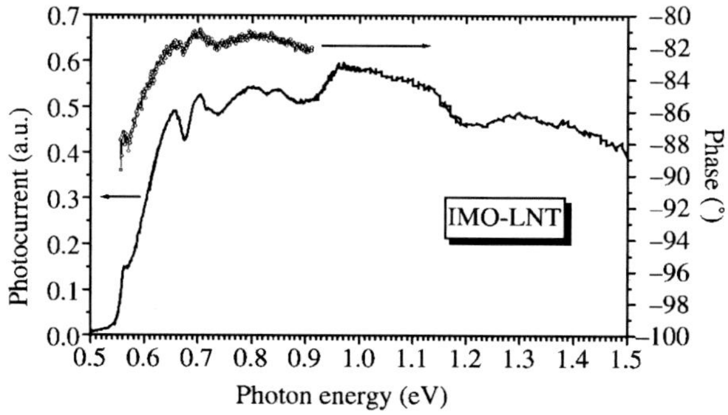

FIG. 27. The PC spectra, measured at 77.4 K, for a 15 μm thick sample prepared at IMO in Belgium from a 1000 PH_3/CH_4 mixture at a MW power of 2500 W and methane concentration of 0.1% containing 2×10^{19} cm^{-3} P impurities. Both the photocurrent and the phase shift reflecting changes in the carrier life-time are shown.

or capturing of the conduction band photoelectron by an excited state level. This confirms the PTIS and oscillatory conductivity detected excited states. Based on the currently available data we can draw a rough electronic structure of P. In Figure 29 a comparison of the work of Gheeraert et al. (2001), who used the effective mass approximation to calculate position of P excited states and our experimental data are plotted, giving good agreement for the main lines at 523 meV, 565 meV and 584 meV, taking into the account fine structure observed by PTIS in low concentration P-doped samples.

As we want to resolve a fine structure of P we need—similar to B—to measure samples with low P-concentration and high crystallographic quality. Figure 30 shows a fine FTPS spectrum of a sample prepared at our group from a mixture containing 1 ppm PH_3/CH_4, also confirming the 575 meV and 584 meV lines and

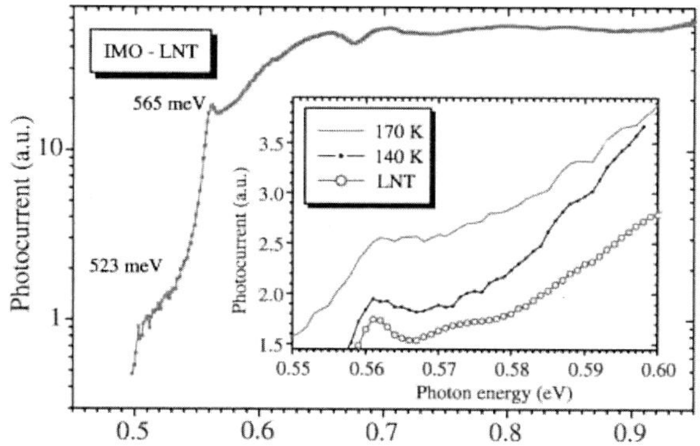

FIG. 28. The PTIS spectra measured for various temperatures, showing fine structure for a 15 μm thick sample prepared at IMO in Belgium from a 1000 PH_3/CH_4 mixture at MW power of 2500 W and methane concentration of 0.1% containing 2×10^{19} cm^{-3} P impurities.

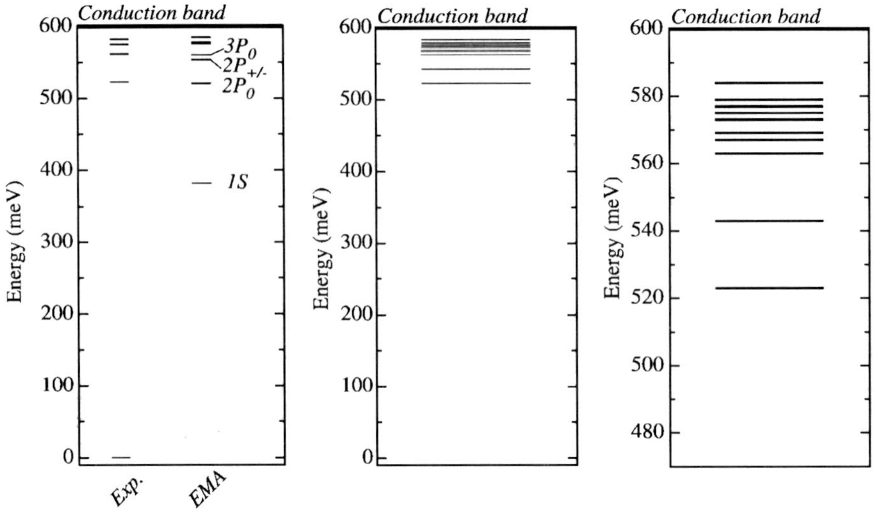

FIG. 29. A comparison between the work of Gheeraert et al. (2001), who used the effective mass approximation to calculate the position of the phosphorous excited states, and our experimental data.

additionally a peak at 590 meV. This sample showed a Hall mobility of 380 $cm^2\ V^{-1}\ s^{-1}$ and a P-concentration of $3 \times 10^{16}\ cm^{-3}$.

3.8. P-INCORPORATION IN POLYCRYSTALLINE CVD DIAMOND FILMS

It should be noted that all samples measured so far were single crystal homoepitaxial layers. The growth process has been optimized recently and the mobility and the crystalline purity has increased, but till now no incorporation of

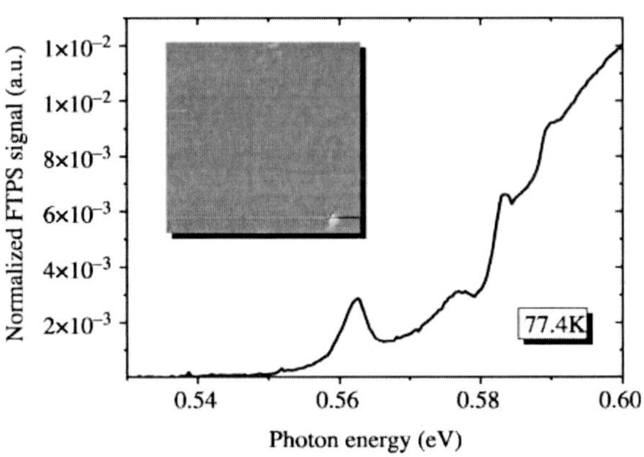

FIG. 30. A FTPS spectrum of an IMO grown sample prepared from a mixture containing 1 ppm PH_3/CH_4. The lines at 577 and 584 meV are clearly visible. The inset shows a SEM picture of the surface of such a CVD diamond layer.

P was possible besides in single crystal CVD diamond films. By optimization of the growth in (111) direction it was possible to prepare in our group polycrystalline diamond CVD films with an active P-incorporation. The samples were prepared using a 50–200 PH_3/CH_4 ratio, 0.125% CH_4 at temperatures of 880 °C. The growth rate at these temperatures and CH_4 and PH_3 concentrations was 1 μm h^{-1}. Generally, it is very difficult to observe free or bound excitons in polycrystalline material. Cathodluminescence (CL) spectra at LNT show P-bound excitons in our polycrystalline samples, independently confirming active incorporation of P. Figure 31 shows the photocurrent response of these layers together with P-bound exciton CL spectra measured on these materials. The Hall measurements confirmed n-type conductivity with mobility of 12 $cm^2\ V^{-1}\ s^{-1}$.

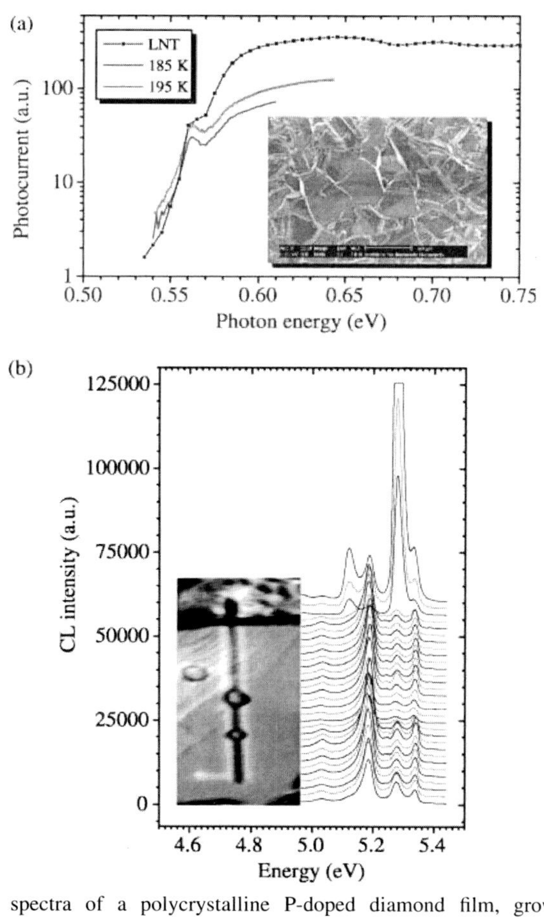

FIG. 31. (a) PC spectra of a polycrystalline P-doped diamond film, grown at IMO. The characteristic P-spectrum clearly indicates the incorporation of phosphorous in the film. The inset shows a SEM picture of the polycrystalline CVD diamond film. (b) The cathodluminescence scan along the P-containing grain in a polycrystalline P-doped sample showing the P- bound exciton in the grain and a free bound exciton at the adjacent grain just over the grain boundary.

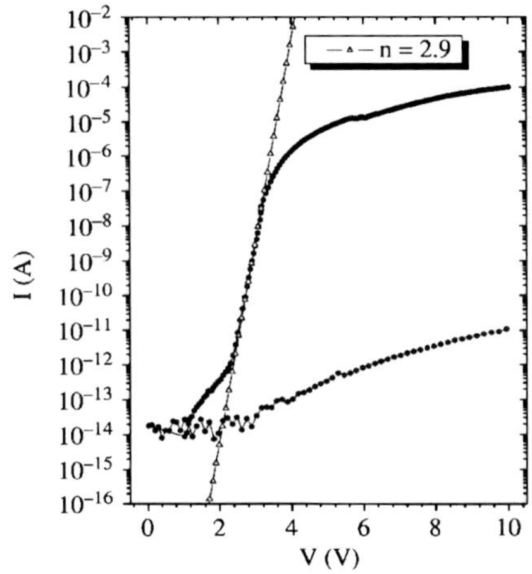

FIG. 32. The $I-V$ characteristics of the first polycrystalline pn junction, prepared by growing P-doped diamond on a B-doped 100 mm thick CVD diamond plate containing $2 \times 10^{18}\,cm^{-3}$ B. The calculated diode factor was about 2.9.

3.9. Concluding Remark: Further Progress and Way to Diamond Devices

The excellent work at NIMS, presented by Koizumi (2003) has shown how far the n-type diamond can be optimized to high degree, leading to a preparation of pn junctions, LEDs diodes and UV sensors on single crystal material. This development clearly shows the perspectives of CVD diamond for future high-temperature, high-frequency electronics. One of the future tasks will be the enlargement of growth area for homoepitaxial single crystal CVD diamond layers. One of the parallel ways, based on the latest possibility to prepare polycrystalline CVD diamond is to prepare bipolar devices based on this polycrystalline material which can be deposited on large areas. Of course, the main problem are the grain boundaries which can short circuit the junction or lead to high-tunneling currents. Thus till now the preparation of polycrystalline CVD diamond devices was thought to be impossible. As shown in Figure 31 a first bipolar device made on polycrystalline CVD diamond was recently prepared in our group. A 2 μm thick P-doped layer has been deposited on a polished B-doped sample with a carrier concentration of $2 \times 10^{18}\,cm^2\,V^{-1}\,s^{-1}$. The $I-V$ characteristics of this device is shown in Figure 32.

3.10. Conclusions

In this section we have explained the use of photocurrent techniques for monitoring optical transitions in nominally undoped, B- and P-doped CVD diamond layers. Photocurrent setups in three main configurations have been used: the CPM, PTIS and phonon-assisted Oscillatory Photoconductivity (OP) measurement and the novel FTPS. All these techniques allow to measure very low defect and impurity concentrations in CVD diamond thin films.

All undoped CVD diamond films show very specific characteristics in subgap photocurrent spectra. The photocurrent has ionization onset E_I at about 0.9 eV. This characteristic photocurrent is induced by a defect absorption which we called the D_x defect, rising up to the gap and found in all measured CVD diamond films. By photo-hall measurements we could prove that the level D_x is acceptor-like. Based on the D_x level and deep N-impurity we could explain the origin of photosensitivity changes for hydrogenated, oxidized and annealed CVD diamond samples. The photosensitivity is highest for hydrogenated samples with the Fermi level pinned by transfer doping by adsorbants at the hydrogenated diamond surface. Enhanced photoconductivity can be also reached by B-doping.

FTPS applied on CVD diamond for the first time enabled to monitor optical transition a fine electronic structure in B-doped CVD diamond. It has been shown that spin–orbital splitting of the valence band causes splitting of sharp electronic transitions, related to the excited states of B, to 2 meV equidistant doublets, related to the effective value of the spin–orbital splitting of the valence band.

CPM, PTIS and FTPS was successfully used to study the electronic structure of (111) epitaxial P-doped n-type CVD diamond films, prepared mainly at NIMS in Japan and at IMO in Belgium. It has been shown that the photoexcitation energy of P-ground state is 0.6 eV. The OP and PTIS allowed to monitor the optical transitions to and from excited states of P. The measured spectra confirmed the energy of excited states calculated by Gheeraert and co-workers (1999, 2001), using the effective mass approximation. Based on the comparison with experiment we have suggested the electronic structure of the P-dopant in diamond. With recent development of technology preparation of highest purity CVD diamond film allows lowering the active concentration of P-donors in diamond. Recently prepared P-doped samples show a fine structure of the rising edge of the photoconductivity which will allow, in future, more precise detection of the electronic structure of P. By a further technological development we were able for the first time to prepare P-doped polycrystalline CVD diamond and in addition to prepare first polycrystalline pn bipolar devices.

References

E. Anastassakis, Start effect on impurity levels in diamond, *Phys. Rev.* 186, 760–767 (1969).

E. Blank, Structural Imperfections in CVD Diamond Films in "Semiconductors and Semimetals" (C. E. Nebel and J. Ristein, Eds.), Vol. 76, pp. 49–145 (2003).

A. C. Boccara, D. Fournier, and J. Badoz, Thermooptical spectroscopy—detection by the mirage effect, *Appl. Phys. Lett.* 36, 130–132 (1980a).

A. C. Boccara, D. Fournier, W. B. Jackson, and N. M. Amer, Sensitive photothermal deflection technique for measuring absorption in optically thin media, *Opt. Lett.* 5(9), 377–379 (1980b).

K. W. Böer, "Survey of Semiconductor Physics: Electrons and Other Particles in Bulk Semiconductors". Van Nostrand Reinhold, New York, 1990.

R. H. Bube, "Photoelectronic Properties of Semiconductors". Cambridge University Press, Cambridge, 1992.

G. D. Cody, "Semiconductors and Semimetals" in (J. I. Pankove, Ed.), Vol. 21, pp. 11–82. Academic Press, Orlando, 1984.

A. T. Collins and E. C. Lightowlers, Photothermal ionization and photon-induced tunneling in the acceptor photoconductivity spectrum of semiconducting diamond, *Phys. Rev.* 171, 843–855 (1968).

G. Davies, A. Mainwood, C. Piccirillo, K. L. Lewis, T. P. Mollart, M. Nesládek, and Z. Remes, Why does diamond absorb infra-red radiation?, *Phys. Status Solidi a* 193(3), 442–447 (2002).

F. De Weerdt and I. N. Kupriyanov, Report on the influence of HPHT annealing on the 3107 cm^{-1} hydrogen related absorption peak in natural type Ia diamonds, *Diamond Relat. Mater.* 11(3–6), 714–715 (2002).

W. J. P. van Enckevort and E. H. Versteegen, Temperature dependence of optical absorption by the single substitutional nitrogen donor in diamond, *J. Phys.: Condens. Matter* 4(9), 2361–2373 (1992).

D. V. Fedoseev, I. G. Varshavskaya, A. V. Lavrentev, and B. V. Deryagin, Growth of diamond from methane above 1100 °C, *B. Acad. Sci. USSR Ch.* 26(4), 854–856 (1977).

J. E. Field (Ed.), "Properties of Natural and Synthetic Diamond", Academic Press, London, 1992.

D. Fournier, A. C. Boccara, and J. Badoz, "Digest of Topical Meeting on Photoacoustic Spectroscopy". Optical Society of America, Washington, DC, 1979.

O. Gaudin, M. D. Whitfield, J. S. Foord, and R. B. Jackman, Deep level transient spectroscopy of CVD diamond: the observation of defect states in hydrogenated films, *Diamond Relat. Mater.* 10(3–7), 614–617 (2001).

E. Gheeraert, N. Casanova, S. Koizumi, T. Teraji, and H. Kanda, Low temperature excitation spectrum of phosphorus in diamond, *Diamond Relat. Mater.* 10(3–7), 444–448 (2001).

E. Gheeraert, A. Deneuville, E. Bustarret, and F. Fontaine, Determination of weak optical absorption coefficients in polycrystalline diamond thin films by photothermal deflection spectroscopy, *Diamond Relat. Mater.* 4(5–6), 684–687 (1995).

E. Gheeraert, A. Deneuville, P. Gonon, M. Benabdesselam, and P. Iacconi, Thermally stimulated conductivity and luminescence in polycrystalline diamond films, *Phys. Status Solidi a* 172(1), 183–192 (1998).

E. Gheeraert, S. Kozumi, T. Teraji, H. Kanda, and M. Nesládek, Electronic states of boron and phosphorus in diamond, *Phys. Status Solidi (a)* 174(1), 39–51 (1999).

P. R. Griffiths and J. A. deHaseth, "Fourier Transform Infrared Spectroscopy". Wiley, New York, 1986.

D. M. Gruen, Ultrananocrystalline diamond in the laboratory and the cosmos, *MRS Bull.* 26(10), 771–776 (2001).

K. Haenen, K. Meykens, M. Nesládek, G. Knuyt, C. Quaeyhaegens, L. M. Stals, S. Koizumi, and E. Gheeraert, Low temperature photoconductivity detection of phosphorus in diamond, *Phys. Status Solidi a* 174(1), 53–58 (1999).

K. Haenen, K. Meykens, M. Nesládek, G. Knuyt, L. M. Stals, T. Teraji, and S. Koizumi, The electronic structure of phosphorus in n-type CVD diamond films: Revised, *Phys. Status Solidi (a)* 181(1), 11–16 (2000a).

K. Haenen, K. Meykens, M. Nesládek, G. Knuyt, L. M. Stals, T. Teraji, S. Koizumi, and E. Gheeraert, Temperature dependent spectroscopic study of the electronic structure of phosphorus in n-type diamond films, *Diamond Relat. Mater.* 9(3–6), 952–955 (2000b).

K. Haenen, K. Meykens, M. Nesládek, G. Knuyt, L. M. Stals, T. Teraji, S. Koizumi, and E. Gheeraert, Phonon-assisted electronic transitions in phosphorus-doped n-type chemical vapor deposition diamond films, *Diamond Relat. Mater.* 10(3–7), 439–443 (2001).

K. Hayashi, S. Yamanaka, H. Watanabe, T. Sekiguchi, H. Okushi, and K. Kajimura, Investigation of the effect of hydrogen on electrical and optical properties in chemical vapor deposited on homoepitaxial diamond films, *J. Appl. Phys.* 81(2), 744–753 (1997).

J. C. Inkson, Deep impurities in semiconductors. 2. The optical cross-section, *J. Phys. C: Solid State Phys.* 14(7), 1093–1101 (1981).

J. Isberg, J. Hammersberg, E. Johansson, T. Wikström, D. J. Twitchen, A. J. Whitehead, S. E. Coe, and G. A. Scarsbrook, High carrier mobility in single-crystal plasma-deposited diamond, *Science* 297, 1670–1672 (2002).

W. B. Jackson and N. M. Amer, Direct measurement of gap-state absorption in hydrogenated amorphous-silicon by photothermal deflection spectroscopy, *Phys. Rev. B* 25, 5559–5562 (1982) .

W. B. Jackson, N. M. Amer, D. K. Biegelsen, R. J. Nemanich, and J. C. Knights, Optical-absorption spectra of surface or interface states in hydrogenated amorphous-silicon, *Appl. Phys. Lett.* 42, 105–107 (1983).

W. B. Jackson, N. M. Amer, A. C. Boccara, and D. Fournier, Photothermal deflection spectroscopy and detection, *Appl. Optics* 20(8), 1333–1344 (1981).

I. Kiflawi, D. Fisher, H. Kanda, and G. Sittas, The creation of the 3107 cm^{-1} hydrogen absorption peak in synthetic diamond single crystals, *Diamond Relat. Mater.* 5(12), 1516–1518 (1996).

C. A. Klein, T. M. Hartnett, and C. J. Robinson, Critical-point phonon frequencies of diamond, *Phys. Rev. B* 45(22), 12854–12863 (1992).

D. S. Knight and W. B. White, Characterization of diamond films by Raman-spectroscopy, *J. Mater. Res.* 4(2), 385–393 (1989).

S. Koizumi, n-Type Diamond Growth in "Semiconductors and Semimetals" (C. E. Nebel and J. Ristein, Eds.), Vol. 76, pp. 239–259 (2003).

S. Koizumi, M. Kamo, Y. Sato, H. Ozaki, and T. Inuzuka, Growth and characterization of phosphorous doped {111} homoepitaxial diamond thin films, *J. Appl. Phys.* 71(8), 1065–1067 (1997).

S. Koizumi, K. Watanabe, M. Hasegawa, and H. Kanda, Ultraviolet emission from a diamond pn junction, *Science* 292, 1899–1901 (2001).

J. Koppitz, O. F. Schimer, and M. Seal, Pseudo-John–Teller optical-absorption of isolated nitrogen in diamond, *J. Phys. C: Solid State Phys.* 19(8), 1123–1133 (1986).

R. Kravets, V. Ogorodniks, A. Poruba, P. Moravec, M. Nesládek, J. Rosa, and M. Vanecek, Fourier-transform photocurrent spectroscopy of defects in CVD diamond layers, *Phys. Status Solidi (a)* 193(3), 502–507 (2002).

R. Kravets, A. Poruba, A. Kromka, M. Nesládek, J. Rosa, and M. Vanecek, Spectroscopy of dopants and defects in CVD diamond layers. Presented at "Surface and Bulk Defects in CVD Diamond Films, VIII, Limburgs Universitair Centrum, Diepenbeek-Hasselt, Belgium, 26–28 February 2003".

H. J. Looi, R. B. Jackman, and J. S. Foord, High carrier mobility in polycrystalline thin film diamond, *Appl. Phys. Lett.* 72(3), 353–355 (1998).

T. M. Lifshits, Photothermal ionisation spectroscopy of impurities in semiconductors, *Instrum. Exp. Tech.* 36(1), 1–39 (1993).

O. Madelung, (Ed.), "Semiconductors—Basic Data", Springer, Berlin, 1996.

F. Maier, M. Riedel, B. Mantel, J. Ristein, and L. Ley, Origin of surface conductivity in diamond, *Phys. Rev. Lett.* 85(16), 3472–3475 (2000).

K. Meykens, K. Haenen, M. Nesládek, L. M. Stals, C. S. J. Pickles, and R. S. Sussmann, Measurement and mapping of very low optical absorption of CVD diamond IR windows, *Diamond Relat. Mater.* 9(3–6), 1021–1025 (2000).

T. P. Mollart and K. L. Lewis, The infrared optical properties of CVD diamond at elevated temperatures, *Phys. Status Solidi (a)* 186(2), 309–318 (2001).

A. E. Mora, J. W. Steeds, and J. E. Butler, Relationship between grain boundaries and broad luminescence peaks in CVD diamond films. *Diamond Relat. Mater.* 12(3–7), 310–317 (2003).

0P. Muys, CVD diamond laser window applications. Presented at the "10th European Conference on Diamond, Diamond-like Materials, Carbon Nanotubes, Nitrides and Silicon Carbide, Hilton Atrium, Prague, Czech Republic, 12–17 September 1999". Unpublished.

C. E. Nebel, C. Sauerer, F. Ertl, M. Stutzmann, C. F. O. Graeff, P. Bergonzo, O. A. Williams, and R. B. Jackman, Hydrogen-induced transport properties of holes in diamond surface layers, *Appl. Phys. Lett.* 79(27), 4541–4543 (2001).

C. E. Nebel, Transport and Defect Properties of Intrinsic and Boron-Doped Diamond in "Semiconductors and Semimetals" (C. E. Nebel and J. Ristein, Eds.), Vol. 76, pp. 49–145 (2003).

M. Nesládek, K. Meykens, K. Haenen, J. Navratil, C. Quaeyhaegens, L. M. Stals, A. Stesmans, K. Iakoubovskij, G. J. Adriaenssens, J. Rosa, and M. Vanecek, Characteristic defects in CVD diamond: optical and electron paramagnetic resonance study, *Diamond Relat. Mater.* 8(8–9), 1480–1484 (1999b).

M. Nesládek, K. Meykens, K. Haenen, L. M. Stals, T. Teraji, and S. Koizumi, Low-temperature spectroscopic study of n-type diamond, *Phys. Rev. B* 59(23), 14852–14855 (1999a).

M. Nesládek, K. Meykens, L. M. Stals, M. Vanecek, and J. Rosa, Origin of characteristic subgap optical absorption in CVD diamond films, *Phys. Rev. B* 54(8), 5552–5561 (1996b).

M. Nesládek, L. M. Stals, A. Stesmans, K. Iakoubovskij, G. J. Adriaenssens, J. Rosa, and M. Vanecek, Dominant defect levels in diamond thin films: a photocurrent and electron paramagnetic resonance study, *Appl. Phys. Lett.* 72(25), 3306–3308 (1998).

M. Nesládek, M. Vanecek, J. Rosa, C. Quaeyhaegens, and L. M. Stals, Subgap optical absorption in CVD diamond films determined from photothermal deflection spectroscopy, *Diamond Relat. Mater.* 4(5–6), 697–701 (1995).

M. Nesládek, M. Vanecek, and L. M. Stals, Defect-induced optical absorption in CVD diamond films, *Phys. Status Solidi (a)* 154(1), 283–303 (1996a).

J. I. Pankove, "Optical Processes in Semiconductors", Dover Publications, New York, 1975.

C. Piccirillo, G. Davies, A. Mainwood, S. Scarle, C. M. Penchina, T. P. Mollart, K. L. Lewis, M. Nesládek, Z. Remes, and C. S. J. Pickles, Temperature dependence of intrinsic infrared absorption in natural and chemical-vapor deposited diamond, *J. Appl. Phys.* 92(2), 756–763 (2002).

C. S. J. Pickles, T. D. Madgwick, R. S. Sussman, and C. J. H. Wort, Optical performance of chemically vapour-deposited diamond at infrared wavelengths, *Diamond Relat. Mater.* 9(3–6), 916–920 (2000).

W. Piekarczyk, Diamond-vapour interface and processes proceeding on it during growth of diamond crystals I. Diamond (111) face, *J. Cryst. Growth* 119(3–4), 345–362 (1992).

Z. Remes, M. Nesládek, K. Meykens, C. S. J. Pickles, and R. S. Sussman, Photothermal deflection mapping of variations in the optical absorption in IR windows, *Phys. Status Solidi a* 181(1), 115–119 (2000).

Z. Remes, M. Nesládek, and C. S. J. Pickles, Local variations and temperature dependence of optical absorption coefficient in natural IIa type and CVD diamond optical windows, *Phys. Status Solidi a* 186(2), 297–301 (2001).

J. Robertson, π-bonded clusters in amorphous-carbon materials, *Philos. Mag. B* 66(2), 199–209 (1992).

J. Robertson, Electronic structure of diamond-like carbon, *Diamond Relat. Mater.* 6(2–4), 212–218 (1997).

E. Rohrer, C. F. O. Graeff, R. Janssen, C. E. Nebel, M. Stutzmann, H. Guttler, and R. Zachai, Nitrogen-related dopant and defect states in CVD diamond, *Phys. Rev. B* 54(11), 7874–7880 (1996).

E. Rohrer, C. E. Nebel, M. Stutzmann, A. Floter, R. Zachai, X. Jiang, and C. P. Klages, Photoconductivity of undoped, nitrogen- and boron-doped CVD- and synthetic diamond, *Diamond Relat. Mater.* 7(6), 879–883 (1998).

J. Rosa, J. Pangrac, M. Vanecek, V. Vorlicek, M. Nesládek, K. Meykens, C. Quaeyhaegens, and L. M. Stals, Simultaneous characterization of defect states in CVD diamond by PDS, EPR, Raman and photocurrent spectroscopies, *Diamond Relat. Mater.* 7(7), 1048–1053 (1998).

J. Rosa, M. Vanecek, M. Nesládek, and L. M. Stals, Photoionization cross-section of dominant defects in CVD diamond, *Diam. Relat. Mater.* 8(2–5), 721–724 (1999).

M. C. Rossi, S. Salvatori, F. Scotti, G. Conte, and E. Cappelli, Photocurrent and photoelectron yield spectroscopies of defect states in CVD diamond films, *Phys. Status Solidi (a)* 181(1), 29–35 (2000).

T. Ruf, M. Cardona, C. S. J. Pickles, and R. Sussmann, Temperature dependence of the refractive index of diamond up to 925 K, *Phys. Rev. B* 62(24), 16578–16581 (2000).

J. Serrano, M. Cardona, and J. Ruf, Spin–orbit splitting in diamond: excitons and acceptor related states, *Solid State Commun.* 113(7), 411–414 (2000).

R. E. Shroder, R. J. Nemanich, and J. T. Glass, Analysis of the composite structures in diamond thin films by Raman-spectroscopy, *Phys. Rev. B* 41(6), 3738–3745 (1990).

K. E. Spear and M. Frenklach, Mechanisms for CVD diamond growth in "Synthetic Diamond: Emerging CVD Science and Technology" (K. E. Spear and J. P. Dismukes, Eds.), pp. 243–304. Wiley, New York, 1994.

H. Sternschulte, K. Thonke, R. Sauer, and S. Koizumi, Optical evidence for 630-meV phosphorus donor in synthetic diamond, *Phys. Rev. B* 59(20), 12924–12927 (1998).

D. F. Talbot-Ponsonby, M. E. Newton, J. M. Baker, G. A. Scarsbrook, R. S. Sussmann, and A. J. Whitehead, EPR and optical studies on polycrystalline diamond films grown by chemical vapor deposition and annealed between 1100 and 1900 K, *Phys. Rev. B* 57(4), 2302–2309 (1998).

J. Tauc, in "Amorphous and Liquid Semiconductors" (J. Tauc, Ed.), Chapter 4, Plenum Press, New York, 1974.

M. E. Thomas, W. J. Tropf, and A. Szpak, Optical-properties of diamond, *Diamond Film. Technol.* 5(3), 159–180 (1995).

D. Tromson, P. Bergonzo, A. Brambilla, C. Mer, F. Foulon, and V. N. Amosov, Thermally stimulated investigations on diamond based radiation detectors, *Phys. Status Solidi a* 174(1), 155–164 (1999).

W. J. Tropf, M. E. Thomas, and M. J. Linevsky, Infrared refractive indices and thermo-optic coefficients for several materials in "Optical Diagnostic Methods for Inorganic Transmissive Materials" Proceedings of SPIE (R. U. Datla and L. M. Hanssen, Eds.), Vol. 3425, pp. 160–171. SPIE, Washington, 1998.

D. J. Twitchen, J. M. Baker, M. E. Newton, and K. Johnston, Identification of cobalt on a lattice site in diamond, *Phys. Rev. B* 61(1), 9–11 (2000).

M. Vanecek, J. Kocka, J. Stuchlik, Z. Kozisek, O. Stika, and A. Triska, Density of the gap states in undoped and doped glow-discharge a-Si–H, *Sol. Energy Mater.* 8(4), 411–423 (1983).

M. Vanecek, R. Kravets, A. Poruba, J. Rosa, M. Nesládek, and S. Koizumi, Fourier transform photocurrent spectroscopy of dopants and defects in CVD diamond. *Diamond Relat. Mater.* 12(3–7), 521–525 (2003).

M. Vanecek and A. Poruba, Fourier-transform photocurrent spectroscopy of microcrystalline silicon for solar cells, *Appl. Phys. Lett.* 80(5), 719–721 (2002).

V. Vorlicek, J. Rosa, M. Vanecek, M. Nesládek, and L. M. Stals, Quantitative study of Raman scattering and defect optical absorption in CVD diamond films, *Diamond Relat. Mater.* 6(5–7), 704–707 (1997).

U. Willamowski, D. Ristau, and E. Welsch, Measuring the absolute absorptance of optical laser components, *Appl. Optics* 37(36), 8362–8370 (1998).

B. L. Zimering and A. C. Boccara, Applications of a compact photothermal-deflection-based setup for trace-gas detection in real-time in situ environmental monitoring and chemical analysis, *Appl. Optics* 36(15), 3188–3194 (1997).

This page is intentionally left blank

Chapter 8

LUMINESCENCE FROM OPTICAL DEFECTS AND IMPURITIES IN CVD DIAMOND

Rolf Sauer

Abteilung Halbleiterphysik (Department of Semiconductor Physics), Universität Ulm, D-89069 Ulm, Germany

1. Introduction

The development of chemical vapor deposition (CVD) techniques enabling growth of thin films under low pressures and moderate temperatures (Eversole, 1962; Angus, Will, and Stanko, 1968; Deryaguin and Fedoseev, 1975; Sato, Kamo, Kanda, and Setaka, 1980; Spitsyn, Bouilov, and Deryaguin, 1981; Matsumoto, Sato, Kamo, and Setaka, 1982; Kamo, Sato, Matsumoto, and Setaka, 1983) has motivated interest in luminescence studies on such material. Relevant aspects include (i) a general characterization and quality control of CVD diamond, (ii) the question in how far defects known from natural or high-pressure high-temperature (HPHT) diamond also exist in CVD diamond, (iii) investigation of new defects unique to this material, and (iv) the ability to introduce shallow dopants and other impurities. Quality control refers to the fact that the recombination of excess electrons and holes excited across the bandgap by photons or electrons branches into non-radiative and radiative channels, and that luminescence near the bandgap will only be visible in films that are relatively free from non-radiative defects trapping charge carriers efficiently. It is expected that a novel growth technique will introduce defects unique to it, and that part of such defects will be optically active. Controlled doping of CVD diamond in the growth process not only supports identification of optical defects with chemical species but is also of great technological concern. In addition to boron, a natural acceptor and the only 'shallow' impurity in diamond, introduction and control of a donor with comparable or lower ionization energy is highly desirable for p- and n-type conductivity in electronic devices.

It has turned out that many of the optical defects known from natural or HPHT diamond also appear in CVD diamond. To mention a few examples, this applies to the 2.156 eV defect (a nitrogen–vacancy defect, NV^-), the 3.188 eV defect (a one nitrogen–carbon interstitial defect), or the 4.852 eV 5RL defect (involving a carbon $\langle 100 \rangle$ split interstitial) (Vavilov, Gippius, Zaitsev, Deryagin, Spitsyn and Aleksenko, 1980; Collins, Kamo and Sato, 1989a; Robins, Cook, Farabaugh and Feldman, 1989). Such data show that the CVD diamond films can contain

ISBN: 0-12-752185-2
ISSN: 0080-8784

nitrogen as a contamination and carbon interstitials and vacancies due to the imperfection of the growth or as a consequence of the polycrystalline structure of the layers. Many more optical defect spectra studied in natural or HPHT diamond are observed in CVD diamond; these are mostly not covered by this chapter as their mere appearance in CVD diamond does not contribute to an identification of the underlying defects. References which list optical defects in natural and high pressure-synthesized diamond discussing relevant properties and current problems are those by Davies (1977, 1994, 1999), Clark, Mitchell, and Parsons (1979), Walker (1979), Clark, Collins, and Woods (1992), Collins (1992, 1999) and Zaitsev (1998, 2001). The last mentioned reference seems to be the most comprehensive and actual compilation of data on the optical properties of diamond including data from CVD diamond. Doping studies of CVD diamond, e.g., with boron and phosphorus, have largely promoted identification of broad-band luminescence or have shown the potential of CVD growth methods of n-type doping diamond, with the concomitant appearance of novel sharp-line excitonic spectra. In these cases we feel it is necessary to compare in detail CVD diamond data with those obtained from natural and high-pressure synthetic diamond including free and bound exciton recombination radiation.

2. Intrinsic Luminescence

The optical gap of natural diamonds has been studied previously in absorption (Clark, Dean and Harris, 1964), and in luminescence and luminescence excitation measurements (Dean and Male, 1964). Dean, Lightowlers and Wight (1965) used natural semiconducting and deliberately boron-doped HPHT synthetic diamonds to study at low temperatures (~ 80 K) intrinsic free exciton (FE) and boron bound exciton (BE) luminescence. These fundamental investigations showed that TA-, TO- and LO-lattice phonons couple to the excitons to conserve the **k**-vector in the recombination transition between the hole in the valence band at $k = 0$ and the electron in one of the conduction band minima at $\vec{k} = 0.76\pi/a[100]$. The energies of the assisting phonons are (Dean et al., 1965)

$$\hbar\Omega(\mathrm{TA}) = (87 \pm 2)\ \mathrm{meV}$$
$$\hbar\Omega(\mathrm{TO}) = (141 \pm 1)\ \mathrm{meV}$$
$$\hbar\Omega(\mathrm{LO}) = (163 \pm 1)\ \mathrm{meV}.$$

The $\mathrm{FE}^{\mathrm{TO}}$ transition is the dominating recombination line. The TA- and LO-replicas are much weaker with relative intensities compared to $\mathrm{FE}^{\mathrm{TO}}$ of ≈ 9 and $\approx 5\%$, respectively, at ≈ 100 K and a spectral resolution of 3 meV (Dean et al., 1965). Multiple phonon replicas including a wavevector-conserving TO-phonon and one or more zone-center O^{Γ}-phonons ($\hbar\Omega = 163.9$ meV) are also observed. Comparing the threshold energy for light absorption by free excitons with that of free electrons and holes Dean and Male (1964) and Dean et al. (1965) determined the exciton binding energy to be $E_x = 70 - 80$ meV. For definiteness we shall refer in the following to a value of $E_x = 80$ meV. The FE lines show a doublet structure well discernable at temperatures around 100 K with two components spaced by ≈ 7 meV. It was argued (Dean et al., 1965) that the splitting reflects the structure of the valence band at $k = 0$, the lower-energy FE luminescence component involving a Γ_8^+ hole and the higher-energy FE luminescence component involving a Γ_7^+ hole whose energies are different by the spin–orbit interaction $\Delta_0 = 6$ meV (Rauch, 1963).

Recent high-resolution cathodoluminescence measurements on high-quality HPHT diamonds down to $T \approx 9$ K demonstrated that the FE recombination spectrum is much more complex consisting of up to four components in each phonon replica (Sauer, Sternschulte, Wahl, Thonke, and Anthony, 2000; Teofilov, Sauer, Thonke, Anthony and Kanda, 2001). Figure 1 exhibits FE cathodoluminescence spectra at selected temperatures and demonstrates that the lowest-energy line component (labeled 1) which is well discernable for $T \lesssim 30$ k appears significantly (= 3.3 meV) below the energy usually quoted as the FE low-energy threshold. This energy is important as the reference value from which BE localization energies E_{loc} are counted. The line component 1 in Figure 1 was also resolved at 18 K sample temperature by Sharp and Collins (1996), however, they did not report on the other fine structure in the FE emission. To keep consistent with previous literature data we shall refer to the onset of the dominating component 2 in defining the exciton threshold energies or optical gaps $E_{gx}^{TO} = 5.266$ eV, $E_{gx}^{TA} = 5.319$ eV, and $E_{gx}^{LO} = 5.245$ eV.

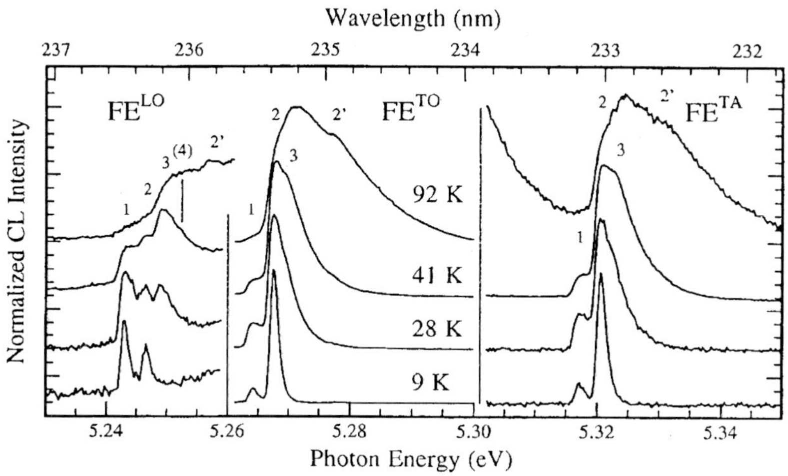

FIG. 1. High-resolution cathodoluminescence spectra of the intrinsic free exciton (FE) edge emission from an undoped HPHT diamond at selected temperatures. Displayed are the k-vector conserving LO-, TO-, and TA-phonon FE replicas. Identical fine structure is observed in all three spectra consisting of the line components 1, 2, 3, (4), and 2′. In standard measurements at $T \geq 50$ K and less resolution only the dominant components 3 and 2′ are usually observed, and the FE emission seems to be a doublet with 6–7 meV line spacing (from Teofilov et al., 2001).

Collins et al. (1989a) showed first that the same FE spectrum as in natural and HPHT crystals is also obtained from single particles of CVD diamond grown on a highly polished silicon substrate with widely spaced nucleation sites using methane–hydrogen gas mixtures in a microwave discharge (Fig. 2). In comparison with a natural semiconducting type IIb diamond, the spectrum is free from extrinsic luminescence, in particular there is no indication of transitions due to boron or nitrogen complexes. Also, the strength of the intrinsic luminescence suggests that there are essentially no deep non-radiative centers trapping the free excitons. Evidently, the CVD diamond particles can be grown with very low defect concentrations. This was found to be different for closed films of polycrystalline diamond grown identically which did not show any edge emission implying efficient trapping of excess electrons and holes by deep centers.

Similarly, Kawarada, Yokota and Hiraki (1990a) reported FE edge emission from undoped isolated crystalline particles grown by microwave plasma CVD on high-purity n-type silicon using CO (5%)/H_2 or CO (15%)/H_2 gas mixtures. They also observed weak FE intensities in the cathodoluminescence from grains just coalescing to a polycrystalline film. It is interesting that in monochromatic cathodoluminescence images of 40 μm-sized diamond particles recorded at the FE^{TO} emission energy at $\approx$5.27 eV free exciton emission was observed from (100) growth sectors of the particles but not from (111) growth sectors (Fig. 3). Collins (1992) compared this to the case of synthetic diamonds pointing out that there nitrogen and boron are taken up less readily in the (100) growth sectors than

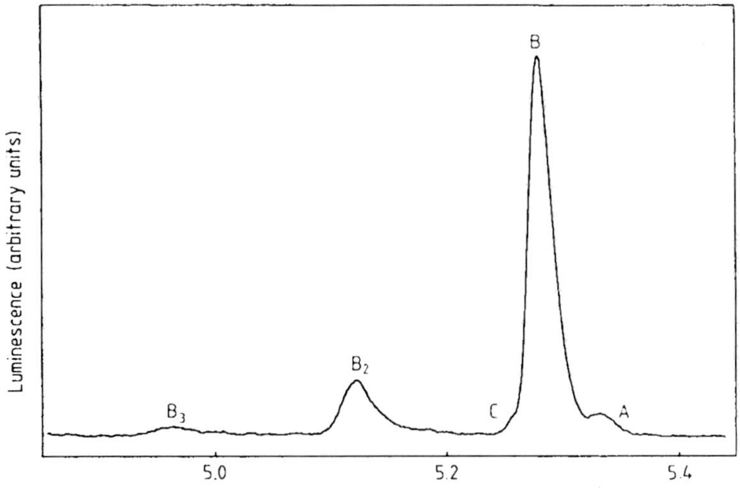

FIG. 2. Near-band-edge emission from a CVD single-crystal diamond at a crystal temperature of 77 K and electron beam excitation (45 kV). The notation follows that of Dean et al. (1965) where A, B, and C are the free exciton phonon replicas FE^{TA}, FE^{TO}, and FE^{LO}, respectively, of Figure 1, and B_2 and B_3 involve ($TO + O^{\Gamma}$) and ($TO + 2O^{\Gamma}$) phonons, respectively (from Collins et al., 1989a).

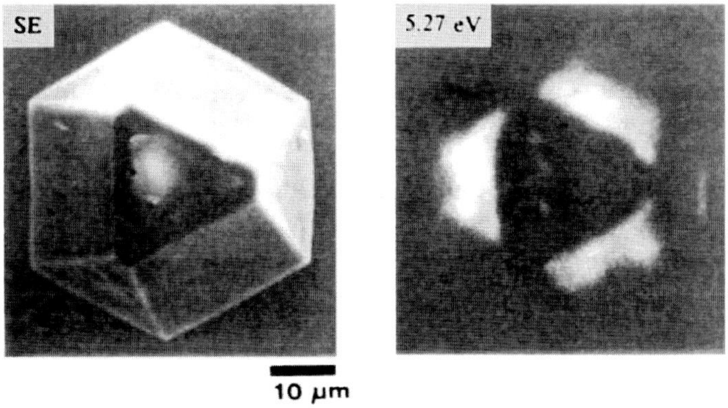

FIG. 3. *Left*: Secondary electron image of an undoped CVD diamond particle grown on Si by microwave plasma CVD using a CO (15%)/H_2 gas mixture. *Right*: Corresponding monochromatic cathodoluminescence image taken at a photon energy of 5.27 eV and a crystal temperature of 90 K (from Kawarada et al., 1991).

in the (111) growth sectors and that other defects such as nickel are present exclusively in the (111) sectors. The conclusion was that the (100) regions of the CVD particles also have a lower defect concentration. Edge emission was likewise observed from diamond single particles grown by hot-filament CVD on silicon wafers by Robins, Farabaugh and Feldman (1993).

In a comparative study using different growth methods, Kawarada and Yamaguchi (1993) observed FE edge emission from diamond films grown by microwave CVD on high-pressure synthetic Ib (001) diamonds, from single-crystal 10 μm sized particles grown by the combustion flame method with C_2H_2/O_2, and from polycrystalline diamond films grown by microwave CVD on Si substrates using CO/H_2 gas mixtures instead of CH_4/H_2 mixtures (Fig. 4). They argued that the former mixtures are much more effective than CH_4/H_2 mixtures in preventing the formation of non-diamond phases and non-radiative centers in polycrystalline microwave CVD films. Varying the gas mixture and using CO plus CO_2 in the H_2 carrier gas in the microwave-plasma CVD process Kawarada, Tsutsumi, Hirayama and Yamaguchi (1994) showed that diamond particles grown at conditions optimized with respect to the position of the silicon substrates in the plasma discharge exhibit dominant FE edge emission, exceeding in their samples by a factor of up to 26 the blue band A luminescence with peak at $\approx$2.9 eV ($\approx$430 nm) which is usually dominant.

Watanabe, Hayashi, Takenuchi, Yamanaka, Okushi, Kajimura and Sakiguchi (1998) used CH_4/H_2 gas mixtures in their microwave plasma CVD homoepitaxy on HPHT synthetic type Ib diamond substrates. Films grown at very low CH_4 concentrations ($\approx$0.5% CH_4 in H_2) exhibited strong FE radiation at room temperature. Temperature-controlled cathodoluminescence measurements yielded a thermal excitation energy of $E_x = (90 \pm 20)$ meV consistent with

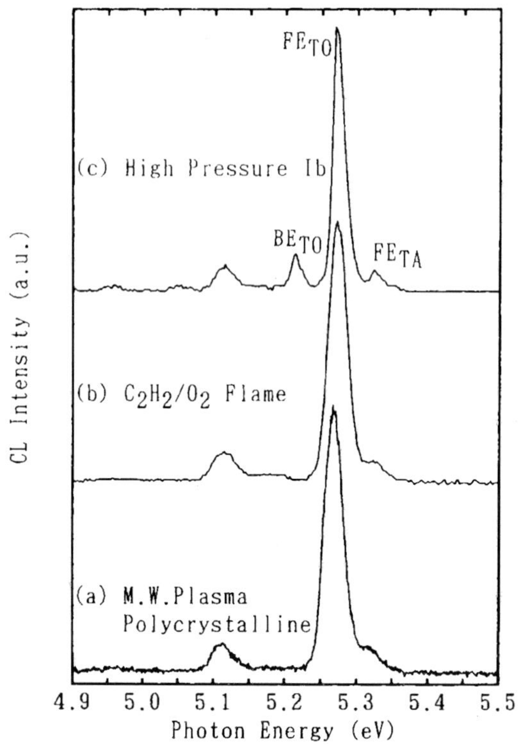

FIG. 4. Near-band-edge cathodoluminescence spectra of different diamond samples showing free exciton (FE) and boron bound exciton (BE) recombination: (a) polycrystalline film deposited by microwave plasma CVD with CO (5%)/H_2; (b) single-crystal particle (10 μm size) grown by the combustion flame method using C_2H_2/O_2; (c) HPHT synthetic type Ib diamond (from Kawarada and Yamaguchi, 1993).

the previous spectrally determined value of $\approx$80 meV from absorption measurements. Subsequently, Watanabe and Okushi (2000) studied homoepitaxial diamond films grown by the same method and reported non-linear FE luminescence intensities as a function of the exciting electron beam current. The penetration depth of the electron beam at 30 kV acceleration voltage amounted to 5.6 μm significantly exceeding the thickness of the grown CVD layer, $\approx$2 μm; as the substrates were known to show no edge emission, all FE recombination radiation must have come from the CVD layer. The superlinear intensity behavior was tentatively associated with excitonic molecules, electron–hole droplets, or Bose–Einstein condensation of the charge carrier system. Criticism on these interpretations implies that electron–hole drops have to be disregarded as this high-density condensed phase of the mobile charge carriers creates a broad-band emission with typical asymmetric line shape (Thonke, Schliesing, Teofilov, Zacharias, Sauer, Zaitsev, Kanda and Anthony, 2000).

Bose–Einstein condensation has occasionally been invoked in explaining high-excitation luminescence features in other semiconductors but could never be substantiated in the long run. Instead, an alternative interpretation might explain the measurements of Watanabe and Okushi: The data are consistent with an initial sublinear slope of the FE emission due to competing non-radiative defects trapping the excitons efficiently followed by a larger (linear) slope when these trap states become saturated at higher excitation levels.

3. Narrow-line Spectra from Optical Defects

Surveys on optical centers, mostly in natural or HPHT synthetic diamonds, were given, e.g., by Davies (1979) and Clark et al. (1992). Prominent photoluminescence (PL) and cathodoluminescence spectra, again with focus on natural and high-pressure synthetic diamonds but also including some data on CVD diamond, were summarized by Collins (1992) in a concise overview. Included there are major optical defects which are related to nitrogen (as the most common impurity in diamond), to the acceptor boron, to nickel (used as a catalyst in HPHT diamond growth), to silicon (found in many CVD diamond films), to radiation- and ion implantation-damage, and to dislocations. Most of these well-documented optical defects are not covered in the present section which aims at a compilation of centers characteristic of CVD grown films.

3.1. Unidentified Lines

Vavilov et al. (1980) first studied cathodoluminescence from epitaxial diamond films, grown by thermal decomposition of gaseous carbon compounds on (100) or (111) natural diamond crystals, and from polycrystalline films grown on heterosubstrates such as copper, tungsten, and silicon. They investigated undoped and nitrogen- or boron-doped films with thicknesses from 0.5 to 30 μm. Observed were in the undoped films on (100) diamond substrates sharp defect luminescence lines at 736 nm (1.68 eV), 575 nm (2.156 eV), 533 nm (2.326 eV), 389 nm (3.187 eV), and 441.5 nm (2.808 eV). Among these lines those at 1.68 and 2.33 eV were characteristic of the films, the others being known to emerge in diamonds containing nitrogen after radiation damage and annealing. These sharp lines were observed superimposed on a broad band at $\approx$440 nm ($\approx$2.8 eV), the blue band A. For films grown on (111) substrates, the same lines emerged except for the 2.156 eV system, and a second broad band at $\approx$540 nm ($\approx$2.3 eV) came up. The missing of the 2.156 eV nitrogen–vacancy line was consistent with the earlier observation of higher densities of non-radiative defects in (111) growth sectors competing with the optical defects in trapping excess carriers. Vavilov et al. noted that in their samples generally the intensities of the lines in the visible spectral range were significantly weaker than in natural diamonds, indicating a high density of non-radiative recombination centers. Also, with increasing film thicknesses, they observed a significant broadening of the 2.156 and 2.33 eV lines ascribed to local stress by structural defects, whereas the 3.187, 2.808, and 1.68 eV lines did not show much broadening. When the diamond films were grown on non-diamond substrates, the same narrow lines and broad bands were observed. Doping with nitrogen during the deposition of films on (100) substrates enhanced the 2.33 eV line intensity demonstrating that the optical defect contains nitrogen. Doping with boron led to a broad cathodoluminescence band at 580 nm (≈2.1 eV) to be discussed later in Section 4.

Collins, Kamo and Sato (1990a,b) and Collins, Lawson, Davies and Kanda (1990c) used methane–hydrogen gas mixtures in a microwave discharge to grow at substrate temperatures of 800–900 °C CVD diamond layers ≈20 μm thick on silicon substrates. In cathodoluminescence they observed new weak lines, not seen in natural or HPHT synthetic diamond (Fig. 5). For CH_4 concentrations above 0.3% the weak features became relatively stronger, and additional lines showed up. Bright emission came from the parts of the film with a rough surface, weak emission was observed from regions which were relatively glossy. These differences in surface texture were ascribed to temperature inhomogeneities of the substrate. The new lines found by Collins et al. are listed in Table I. In addition to these lines that are unique to CVD diamond, Collins et al. also detected the well-

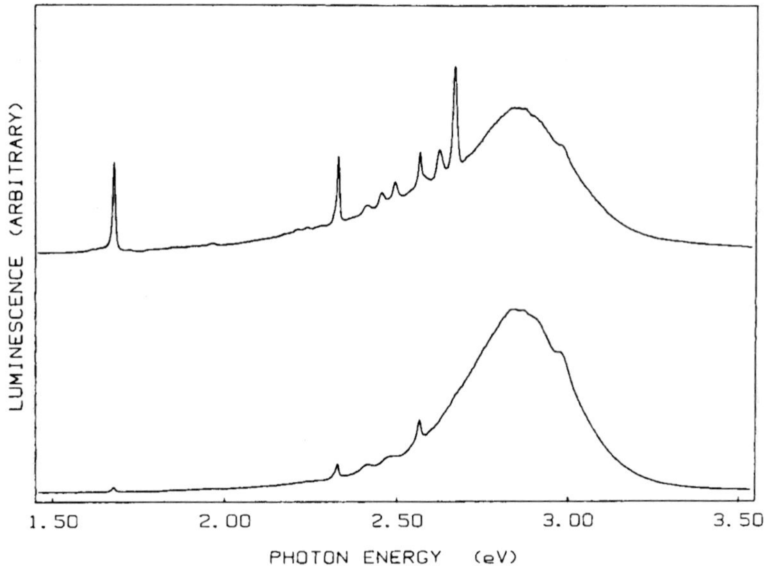

FIG. 5. Cathodoluminescence spectra at a sample temperature of 77 K, recorded under the same conditions for two different areas of a polycrystalline CVD diamond film separated by a few millimeters. The film was grown using 0.5% methane with a substrate temperature of 850 °C. The eight sharp lines in the upper spectrum are unique to CVD diamond. Their approximate positions are (from left to right) 1.68 eV (Si–V defect), 2.33, 2.41, 2.46, 2.50, 2.56, 2.64, and 2.67 eV (not identified, see text and Table I) (from Collins et al., 1990a).

TABLE I

NARROW LINE TRANSITIONS UNIQUE TO CVD DIAMOND

Vavilov et al. (1980)	Collins et al. (1989a,b)	Yacobi et al. (1991)	Ruan et al. (1991a,b)	Gheeraert et al. (1994)	Robins et al. (1993)	Dischler et al. (1994a,b)	($m =$)
1.68[a]	1.681[a]	1.680[a]	1.681[a]			1.681[a]	
			1.968			2.11, 2.07	(32, 50)
	2.200					2.24, 2.20	(13, 16)
2.33	2.322, 2.335	2.330	2.330, 2.320	2.33		2.37, 2.33, 2.28	(7, 8, 10)
		2.412		2.42		2.41	(6)
	2.466	2.480		2.47		2.46	(5)
	2.497					2.50	(4)
	2.567	2.567	2.567	2.57		2.57	(3)
	2.599						
	2.642		2.638(+)	2.63		2.62	(2)
	2.670	2.672				2.67	(1)
	2.756						
	2.786	(2.78)					
		(2.95)					
(2.80)	2.988	((2.987))		2.985			
	3.092						
	3.272						
			3.462*				
			3.57*				
			4.76		4.757		
					4.798		
			4.83		4.832		
					4.920		
					4.950		

Photon energies (in eV) of sharp luminescence lines characteristic of nominally undoped as-grown CVD diamond. All lines are observed in polycrystalline diamond films except for those reported by Robins et al. from single crystalline particles. Notation: (...) broadened lines, ((...)) shoulder, (+)TR12 defect line?

* After 1200 °C anneal.

[a]Silicon-related defect transition, see Section 3.3. The label m in the last entry is the shell number in the donor–aceptor pair model of Dischler et al. The values cited by Gheeraert et al. are not contained in the original references.

known no-phonon lines at 2.156 eV (from a nitrogen–vacancy defect), 3.188 eV (from a nitrogen–interstitial carbon defect) and 4.582 eV (5RL, incorporating a double split carbon interstitial) (Collins, 1993). Collins et al. noted, that some of the lines they observed were not detected by Vavilov et al. and vice versa. The appearance of lines not reported in natural or HPHT synthetic diamonds seemed to be quite natural regarding the very different conditions for the growth of such diamonds and CVD diamond.

Diamond films grown on boron-doped silicon substrates by plasma-assisted CVD were investigated in a cathodoluminescence study by Yacobi, Badzian and Badzian (1991). Growth temperatures were 910–975 °C at pressures of 70–80 Torr and the concentration of CH_4 in the H_2 gas stream amounted to 0.5 or 1%. The three samples studied were boron or silicon contaminated. Consistent with other studies, the luminescence revealed strong non-uniformities in the relative peak intensities in different regions of the samples, due to strong variations of the impurity and defect concentrations in the films. It was found that (i) increasing the substrate temperature above 950 °C and reducing the CH_4 concentration below 1% led to higher-quality diamond films, (ii) incorporation of silicon in diamond and the formation of Si-related defects led to a significant reduction of the overall CL intensities, and (iii) the relatively high luminescence intensity in high-quality samples is due to the reduced defect densities and not due to the boron. Raman and X-ray diffraction measurements indicated that (100) faces of the films grew much more perfect than (111) faces. The latter faces contained more planar defects (stacking faults and twins) that may have caused a significant reduction in CL intensity and could be responsible for the face-dependent CL variations. The Si contamination produces a strong peak at 738 nm ($\approx$ 1.68 eV) thought to be due to a vacancy center introduced by interstitial silicon atoms (Section 3.3). Extra peaks emerged at 483 nm (2.567 eV), 500 nm (2.480 eV), 514 nm (2.412 eV), and 532 nm (2.330 eV) on a strong broad-band background (Table I). The 2.33 eV line was thought to be associated with centers containing nitrogen atoms, supporting the same conclusion arrived at by Vavilov et al. (1980).

Ruan, Choyke and Partlow (1991a) deposited diamond films on silicon substrates from a microwave plasma at substrate temperatures of 900–1000 °C. The samples were then annealed between 500 and 1350 °C. The CL peak at 1.681 eV ascribed to a defect incorporating silicon was strong in all of the samples studied, and annealing temperatures of up to 1350 °C did not reduce its intensity. Apart from the known line at 2.156 eV and the 5RL spectrum they detected peaks at 1.968 eV, a doublet at 2.330 and 2.320 eV and, in one sample, a peak at 2.638 eV, similar in photon energy to the TR 12 peak but different in its annealing behavior (Table I). It was discussed that the doublet, which was resolved in this work, might produce the broader width of the 2.440 eV in other reports where it was not resolved. Two high-energy lines at 4.76 and 4.83 eV not previously documented in the literature were detected. There was no sufficient information on the lines for a detailed discussion. These peaks were destroyed after a 1000 °C anneal or were stable up to a 1200 °C anneal, respectively, and two new peaks at 3.462 and 3.57 eV appeared after a 1200 °C anneal. All these sharp lines were superimposed on the broad 2.8 eV blue band A luminescence.

Gheeraert, Fontaine, Deneuville, Khong and Collins (1994) characterized boron-implanted CVD-grown polycrystalline diamond films by electron paramagnetic resonance and cathodoluminescence. They showed CL spectra of as-grown films with a variety of sharp defect lines at approximate photon energies of 2.33, 2.42, 2.47, 2.57, 2.63, and 2.985 eV (Table I). These lines were cited to be specifically observed in virgin, good quality CVD films though their energies were not explicitly given in the works referred to by Gheeraert et al.

Nominally undoped and boron-doped crystalline diamond particles were grown by Robins et al. (1993) using the hot-filament CVD technique. In cathodoluminescence measurements from 80 to 280 K they studied in much detail the free exciton and boron bound exciton recombination spectra. In boron-doped and undoped particles, the exciton lines were much less intense in (111) crystal-growth sectors than in (100) sectors confirming equivalent data from other works. In addition, Robins et al. observed five near-band-edge lines at 4.757, 4.798, 4.832, 4.920, and 4.950 eV which they ascribed to excitons bound to lattice defects (possibly dislocations) which were not further identified.

Dischler, Rothemund, Maier, Wild, Biebl and Koidl (1994a) and Dischler, Rothemund, Wild, Locher, Biebl and Koidl (1994b) in cathodoluminescence measurements on diamond films prepared by microwave plasma-assisted CVD studied three samples in detail: (i) a boron-doped polycrystalline film grown on a silicon substrate at 820 °C with 1.5% CH_4 in the H_2 carrier gas incorporating nitrogen as evident from the nitrogen-defect lines observed at $\approx$2.16, $\approx$2.81, and $\approx$3.19 eV photon energy, (ii) a nominally undoped homoepitaxial film grown on a polished (100) natural diamond plate placed on a boron-doped silicon substrate which also contained nitrogen, and (iii) a polycrystalline diamond film grown on a silicon substrate at 980 °C; this sample was boron-contaminated and contained nitrogen and much silicon as concluded from the weakly detectable 2.16 eV line and the very intensive 1.68 eV line, respectively. The three samples were selected from more than 30 specimens prepared by plasma CVD. They all exhibited an extended set of 16 or more lines as listed in Table I, beginning on the high-energy side with a line at 2.67 eV. These lines were tentatively attributed to line-resolved donor–acceptor pair spectra following the classical work by Thomas, Gershenzon and Trumbore (1964). Dischler et al. noted that sets of lines very similar in energy to their had been seen in absorption in earlier work on synthetic diamond with aggregated nitrogen by Collins and Stanley (1985) and in annealed HPHT synthetic diamond containing nickel (Lawson and Kanda, 1993). It was also striking, that four lines in the set, at 2.67 eV (associated with the shell number $m = 1$), 2.50 eV ($m = 2$), 2.33 eV ($m = 8$), and 2.20 eV ($m = 16$), were much stronger in two of the three selected samples than the other lines. The authors could not explain this unusual behavior within their donor–acceptor recombination model. It is further remarkable that the four strong lines have energies identical or close to peaks observed independently by other groups (Table I). One might wonder whether all of these lines—in an alternative interpretation—could be due to individual defects, unrelated to donors or acceptors.

Comparing the lines listed in Table I and discussed above it is evident that many of them have been observed in independent work by several groups, and that in particular the lines at around 2.67, 2.57, 2.47, 2.42, and 2.33 eV are characteristic of CVD diamond irrelevant whether the films were grown by the microwave plasma technique, the hot-filament method, or by combustion flame deposition. Unfortunately, in all these cases an interpretation as to the chemical and structural identification of the optical defects is still missing.

The temperature dependence and decay times of the lines were studied by CL spectroscopy. Khong and Collins (1993) found that the measured decrease of their intensities at increasing temperatures could be fairly well described in an empirical model based on the capture and thermal detrapping of excitons at the optical defects. Values for the (de-)activation energies entering the model expressions were for all lines studied close to 200 meV: $E_{act} = 245$ meV (2.48 eV line), 185 meV (2.567 eV line), 200 meV (2.99 eV line), and 210 meV (2.33 eV line). Also the values of the 'effective density of band continuum states' into which the exciton ionization occurs were of similar size close to $900T^{3/2}$ for all lines. An additional fitting term had to be employed for the 2.33 eV line modeling the exciton capture and release by a competitive trap. These data, however, did not shed light on the nature of the optical defects. Luminescence decay time studies (Khong, Collins and Allers, 1994) showed that some of these lines unique to CVD diamond have relatively long life times (Table II).

The lines at 2.537 and 2.547 eV appear as near satellites to the 2.56 eV line in one sample. In another sample they can be detected due to 'some sharpening' during the decay while they are not observed in conventional luminescence scans. In comparison to these decay times there are various vibronic defects in diamond whose no-phonon lines exhibit decay times in the millisecond range, but others show a fast decay in the nanosecond range. In the latter group are the familiar centers at 2.985 eV (N3), 2.463 eV (H3), and 2.156 eV (nitrogen–vacancy complex) having lifetimes of 40, 16.7, and ~29 ns, respectively (Davies, 1979; Pereira, 1994).

Data obtained by Steeds, Charles, Davis, Gilmore, Hayes, Pickard and Butler (1999a), Steeds, Davis, Charles, Hayes and Butler (1999b) and Steeds, Charles, Davies and Griffin (2000) in studying electron-irradiated diamond by PL

TABLE II

LINES UNIQUE TO CVD DIAMOND: PHOTON ENERGIES AND DECAY TIMES

2.33 eV	2.48 eV	2.537 eV	2.547 eV	2.557 eV	2.567 eV	2.572 eV	2.99 eV
9.6 ns	1 μs	1.8 μs	1.3 μs	1.2 μs	1.5 μs	1.3 μs	63 ns

microscopy should be mentioned in the present context as CVD diamond films were also employed. In the new technique which these authors introduced the samples were first irradiated in a transmission electron microscope in a controlled way with electrons of sufficient energy to displace carbon atoms from their lattice sites. The irradiations could be performed over areas of 50–100 μm diameter from liquid helium temperature to 1000 °C, at variable doses, and acceleration voltages; subsequent annealing could be performed in situ. The samples were then transferred to a cryogenic stage on an optical microscope where they were studied by PL spectroscopy under UV or visible light illumination. Steeds et al. investigated diamond samples originating from different sources including type IIa and IIb samples, HPHT synthetic diamonds and large-grain polycrystalline CVD diamond layers produced by the techniques of microwave plasma-, hot filament-, or plasma torch-deposition. When different samples were irradiated at a high electron dose of 1.3×10^{20} cm^{-2} at room temperature many new PL lines were observed in the spectral region 500–600 nm (Fig. 6). A large number of

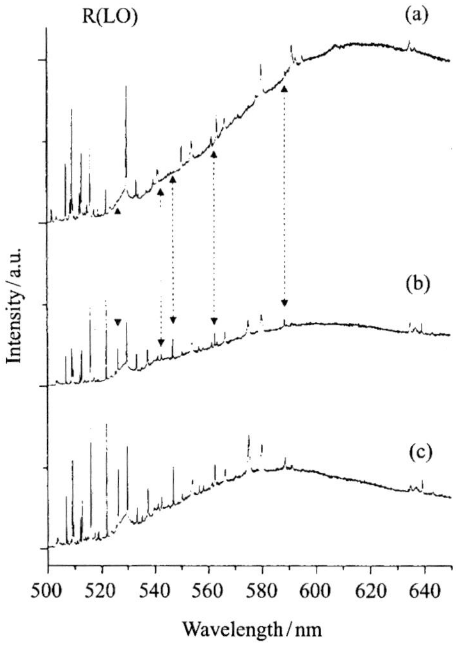

FIG. 6. Microscopic photoluminescence spectra (6 K sample temperature, 488 nm laser excitation) of three different kinds of diamond samples which were all irradiated at room temperature to 1.3×10^{20} electrons cm^{-2}. Remarkable is the similarity of the spectra: (a) natural IIa diamond with a very low level of nitrogen showing a completely featureless spectrum prior to irradiation, (b) low-nitrogen microwave plasma CVD diamond film, (c) higher-nitrogen microwave plasma CVD diamond film. The sharp lines at 526.3 nm (2.3558 eV), 542.6 nm (2.2850 eV), 547 nm (2.266 eV), and 588.6 nm (2.1064 eV) are possibly nitrogen-related. The peaks in the 630–650 nm region are believed to be boron-related (from Steeds et al., 1999a).

spectral lines in the same region had been previously reported in PL experiments on ion-implanted specimens (Gippius, Vavilov, Zaitsev and Zhakupbekov, 1983) and had been interpreted as related to self-interstitials (Zaitsev et al., 1996). This interpretation was also given in the lines observed by Steeds et al. Five of them, at 526.3 nm (2.356 eV), 542.6 nm (2.285 eV), 547 nm (2.267 eV), 562.5 nm (2.204 eV), and 588.6 nm (2.106 eV) did not show up in samples of very low nitrogen concentrations and are possibly due to self-interstitials trapped at nitrogen atoms. The other lines came up in a very similar manner in the various samples. A CVD sample exhibited originally, as-grown, the well-known no-phonon lines at 575 nm (2.156 eV) and 637 nm (1.946 eV), being related with a neutral nitrogen–vacancy complex $(NV)^0$ and a negatively charged nitrogen–vacancy complex $(NV)^-$, respectively. These optical centers were destroyed under electron irradiation at increasing doses hinting to the creation of interstitials above the displacement threshold which migrate to the $(NV)^0$ and $(NV)^-$ centers to eliminate them. Steeds et al. concluded from this finding and the similarities between the spectra in very different starting materials that 'most of the peaks are either related to isolated intrinsic defects, created by irradiation or to complexes formed between intrinsic defects and impurities that are common to natural diamond and microwave plasma-grown material'. This has bearing on one of the lines at 503.5 nm (2.462 eV), the 3H no-phonon transition, as studied intensely earlier (Walker, 1979). This spectrum appears weakly for irradiation at room temperature or liquid nitrogen temperature but comes up strongly for irradiation at liquid helium temperature (Fig. 7) (Steeds et al., 1999a). All of the other lines showed less variation in intensity as a function of the irradiation temperature. Further observations included the threefold splitting of the 552.5 nm (2.244 eV) local mode satellite of the no-phonon transition in a 50% ^{13}C isotope enriched HPHT sample. Taking also into consideration variances in the symmetry determination of the center in the literature Steeds et al. came to the important conclusion that the 3H center is the ⟨100⟩ double-split interstitial.

The peaks observed in the 630–650 nm region in Figure 6 were believed to be boron-related. In particular, there was an apparently unreported new PL system in CVD samples containing 10^{18}–10^{19} cm^{-3} average boron level. (Boron was very inhomogeneously distributed in the CVD diamonds having up to eight times higher concentrations in some grains than in the others as proven by secondary ion mass spectrometry (SIMS) images.) The new PL system consisted of a no-phonon line at 635.7 nm (1.950 eV) at cryogenic temperatures, often in conjunction with another line at 666.2 nm (1.861 eV), and was accompanied at 220 K by a weak transition at 622.6 nm (1.991 eV) and a strong transition at 649.7 nm (1.908 eV). The latter transitions were supposed to be anti-Stokes- and Stokes-satellites, respectively, of the no-phonon line. Furthermore, there was another weak line at 609.9 nm (2.033 eV). The new systems exhibited a complex relationship to the GR1 line at 741 nm (1.673 eV) emitted from the neutral vacancy V^0 (Collins, 1992, and references therein). Steeds et al. (2000) generally found that the 635.7 nm emission was quenched and replaced by the GR1

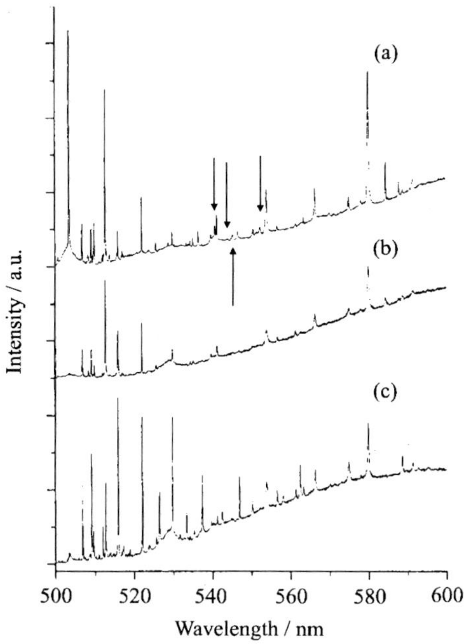

FIG. 7. Microscopic photoluminescence spectra (6 K sample temperature, 488 nm laser excitation) of a CVD diamond film containing initially a low concentration of (NV) centers, after electron irradiation at different temperatures: (a) irradiation at liquid helium temperature, (b) at liquid nitrogen temperature, (c) at room temperature. The intense peak in (a) at 503.5 nm (2.4625 eV) is the no-phonon emission of the 3H center, arrows indicate local vibrational modes of the center at 540.8 nm (2.2926 eV), 543.8 nm (2.280 eV), 544.9 nm (2.2754 eV), and 552.5 nm (2.2441 eV) (from Steeds et al., 1999a).

emission when the electron irradiation dose was increased. Vice versa, in a sample with comparable boron concentration which showed only GR1 emission after electron irradiation this luminescence was quenched under focused UV radiation and replaced by strong 635.7 nm emission. From these findings, Steeds et al. associated the 635.7 nm system with a positively charged vacancy explaining the last mentioned experiment by UV photo-ionization of the neutral vacancy and creation of a positive vacancy.

3.2. Oxygen-related Peaks

Ruan, Kobashi and Choyke (1992b) and Ruan, Choyke and Kobashi (1993) studied by cathodoluminescence measurements the effects of oxygen-doping together with boron-doping on films grown by microwave-assisted CVD at 800 °C on silicon substrates; 0.1% oxygen was added to the gas mixture (CH_4 (0.5%)/H_2) while different concentrations of boron were achieved by adding various amounts of B_2H_6 to the gas mixture. Boron-doping affected the relative

CL intensity of the blue band A peaking at ≈2.8 eV as will be discussed later in Section 4.1. Apart from the broad bands, the samples showed sharp CL lines (at 1.681, 1.967, 2.33, 2.41, 2.47, and 2.567 eV, Table I) when grown with or without oxygen, but in the oxygen-doped samples several peaks were significantly reduced in intensity, such as the 1.681 eV peak, or completely disappeared, such as the 1.967 eV peak (Ruan et al., 1993). New peaks were observed in the region between 450 and 490 nm. In this region, samples grown without oxygen exhibited a sharp line at 2.566 eV and a broad line at 2.67 (Fig. 8). Samples grown with 0.1% oxygen showed a new peak at 2.561 eV and six other new peaks, at 2.676, 2.681, 2.692, 2.696, 2.702, and 2.725 eV, which were in some way related to the 2.561 peak. The new lines needed oxygen doping to be detectable and were strong when in addition the samples were doped with a small concentration of boron. These findings and the fact that a certain amount of oxygen improved the film quality of CVD grown diamond led the authors to conclude that oxygen is incorporated into the diamond during the CVD

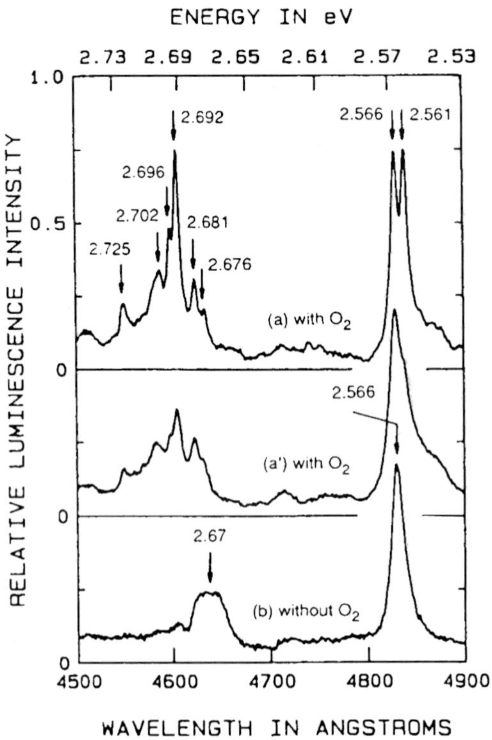

FIG. 8. Cathodoluminescence spectra of diamond films grown (a) with 0.1% oxygen and 0.002 ppm B_2H_6, (a′) with 0.1% oxygen, and (b) without oxygen. New peaks at 2.561 eV and from 2.676 to 2.725 eV appear in the samples grown with oxygen (from Ruan et al., 1993).

deposition and that the new lines are very likely due to centers associated with oxygen as an impurity in the diamond lattice.

3.3. 1.681 eV Silicon–Vacancy Defect

The 737.5 nm (1.681 eV) luminescence line has been observed in all kinds of CVD diamond irrespective whether grown in a microwave plasma (Collins, Kamo and Sato, 1989b, 1990a,b; Partlow, Ruan, Witkowski, Choyke and Knight, 1990; Yokota, Kawarada and Hiraki, 1990), by the hot filament technique (Vavilov et al., 1980; Robins et al., 1989), or in a combustion flame (Freitas, Butler, Bishop, Carrington and Strom, 1990a; Freitas, Butler and Strom, 1990b). The latter authors detected the spectrum when their polycrystalline films prepared by the hot filament technique or the oxygen–acetylene torch method were deposited either on silicon substrates or on nickel and molybdenum substrates. These data suggested that the optical center is universally existing in CVD diamond. In contrast, it is not detected in natural and undoped high-pressure synthetic diamond.

Vavilov et al. (1980) first reported the spectrum to appear in homoepitaxial diamond films, grown on (100) or (111) surfaces of natural crystals. Zaitsev, Vavilov and Gippius (1981) demonstrated that it could be produced by implantation of silicon atoms into natural diamonds, and from the observed quadratic dependence of the luminescence intensity on the implantation dose concluded that the optical center involves two interstitial silicon atoms. The correlation with Si in the implantation studies was later confirmed by Collins et al. (1990a) and Clark and Dickerson (1991a,b). Observation of the 1.681 eV line in absorption from a thin plasma grown polycrystalline film (Collins et al., 1990a) demonstrated that the defect had a rather high concentration.

In the early work there was some confusion of the 1.681 eV luminescence with the line at 1.673 eV (GR1) emitted by the neutral vacancy (Robins et al., 1989; Freitas et al., 1990a,b; Yokota et al., 1990). In some works, the 1.681 eV line was measured at halfwidth as large as 30 meV and could not be spectrally distinguished from the GR1 line; in other works, when the different spectral positions could be resolved, the energy difference of 7–8 meV was attributed to large strain in differently oriented grains of a polycrystalline film shifting the 1.681 eV line accordingly. However, it soon became clear that the 1.681 eV luminescence is definitely independent of the GR1 line. Collins et al. (1990a) argued that the line which they observed in their CVD films grown at temperatures above 850 °C was the 1.681 eV line since the vacancy anneals out at these temperatures in natural diamonds. Also, they saw no reason why the GR1 line should be shifted from its normal position of 1.673 eV while all other no-phonon lines remained at their standard energies. They also emphasized that random strain in the films may broaden the luminescence lines, but not shift them. Finally, they showed that in ^{29}Si implanted high-pressure synthetic diamonds both lines were separately visible. Their data seemed to indicate that apart from

Si also nitrogen is necessary in the samples for the creation of the 1.681 eV defect. This result was consistent with the observation of Yokota et al. (1990) that the 1.681 eV line first increased in intensity upon nitrogen doping the CVD diamond films, and decreased for higher nitrogen concentration in the gas mixture. Several other investigations corroborated the dependence of the 1.681 eV line on Si and N. When Sittas, Kanda, Kiflawi and Spear (1996) successfully doped HPHT synthetic diamond with Si using various catalysts they also found the 1.681 eV center only in specimens grown with nitrogen getters. However, micro-photoluminescence from selected regions of the samples showed that the presence of nitrogen excluded the presence of the Si center; possibly nitrogen competes more successfully in trapping the vacancies than Si. The final conclusion was that nitrogen is not involved in the optical center. In the original work of Zaitsev et al. (1981) there was also no correlation between the intensity of the line and the nitrogen content of the samples studied (Gippius, 1993a,b). It was found that in homoepitaxially grown CVD films having (001) and (111) surfaces the 1.681 eV line could be observed from the (111)-grown film but not from the (001)-grown film (Yacobi, Lebens, Vahala, Badzian and Badzian, 1993).

All these investigations made very plausible that the 1.681 eV optical defect incorporates silicon. Definite evidence for this conclusion came from a number of experiments three of which will be described in the following. Ruan, Choyke and Partlow (1991b) studied diamond films deposited by microwave CVD on silicon, on aluminum nitride (placed on top of a silicon wafer), and on tungsten. In this order, the 1.681 eV line was strong, weak, and absent suggesting that the defect contains silicon which had diffused-in from the wafers but did not originate from the quartz (SiO_2) walls of the CVD reactor. (These authors also compared the annealing behavior of the 1.681 eV center which was stable beyond 1350 °C annealing temperature with that of the 1.673 eV GR1 line which disappeared at 800 °C in natural diamond providing a clear distinction between the two lines. Therefore, the two lines must be emitted from different centers consistent with the conclusion of Collins et al. (1990a) mentioned earlier.) Robins, Farabough and Feldman (1991) performed cross-sectional cathodoluminescence measurements on diamond films deposited on a silicon substrate by hot-filament CVD. The 1.681 eV line intensity decreased with increasing distance from the interface between the film and the surface corroborating silicon being incorporated in the optical center (Fig. 9). Finally, Bergmann, Stoner, Turner, Glass and Nemanich (1993) studied the 1.681 eV line in diamond films deposited on silicon substrates by CVD as a function of the deposition time and the distance from the plasma during growth. As a function of time, the intensity first strongly increased up to a maximum at ≈ 7 h deposition time and then gradually decreased. Their interpretation was that initially the plasma etches the silicon substrate releasing silicon atoms into the gas phase from where they are incorporated in the initially growing isolated diamond grains. The process stops as the grains coalesce thus inhibiting the plasma to further etch the silicon wafer. Bergman et al. also found

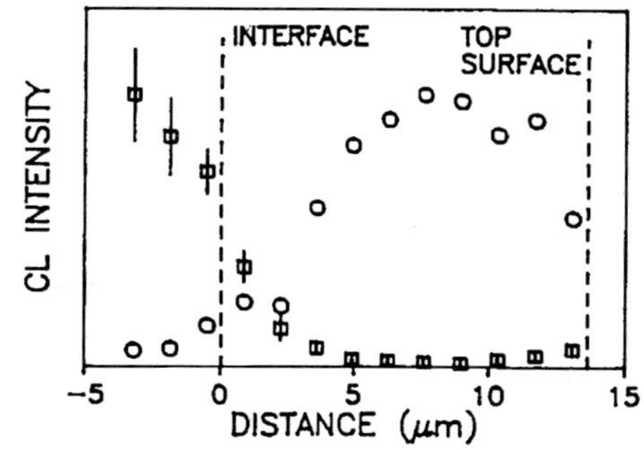

FIG. 9. Cathodoluminescence intensity ratio of the 1.681 eV line to the spectral background (squares), and intensity of background (circles), as a function of the distance from the film–substrate interface. The locations of the interface and the top surface of the film are indicated by dashed lines (from Robins et al., 1991).

that with increasing distance of the growing CVD film from the plasma the 1.681 eV line intensity was drastically reduced giving evidence that plasma interaction with the silicon substrate is responsible for the formation of the optical defect (Fig. 10).

Feng and Schwartz (1993) studied the PL intensity of the 1.681 eV line in free-standing diamond films which were grown by plasma-assisted CVD on silicon substrates using different methane concentrations in the gas mixture and which were then dissolved from the substrates. While the deposition rate increased for higher CH_4 concentrations in the gas up to 2% and the film quality deteriorated simultaneously due to a more defective microstructure, the line intensity dropped over nearly four orders of magnitude in the same range of gas composition.

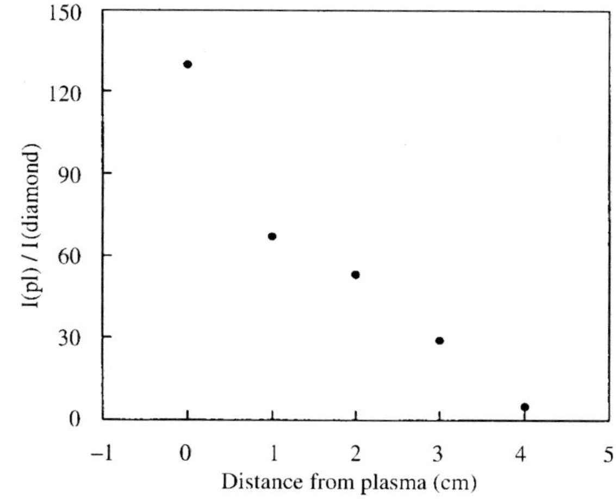

FIG. 10. The relative integrated photoluminescence intensity of the 1.681 eV line vs. distance from the plasma during growth (from Bergman et al., 1993).

The authors concluded that the center is characteristic of 'high-quality' diamond films.

The 1.681 eV center shows a number of sidebands down-shifted from the non-phonon (NP) line by energies of $\approx$42 meV (A), 64.6 meV (B), 125.3 meV (C), and 153.6 meV (D) (Fig. 11) (Feng and Schwartz, 1993). Other workers found slightly different energies and extra sidebands. Sternschulte, Thonke, Sauer, Münzinger and Michler (1994) reported a weak extra sideband shifted by 85 meV from the NP line, and a rather narrow band at a displacement energy of 166 meV, being clearly related with an O^{Γ} lattice phonon coupling to the optical center. Virtually the same vibronic structure was observed by Gorokhovsky, Turukhin, Alfano and Phillips (1995). The A sideband was ascribed to a local vibrational mode by Feng and Schwartz (1993) and Sternschulte et al. (1994). The latter group argued that a mode of similar energy is observed for a variety of other optical centers in diamonds, and that therefore the mode is not specific for the 1.681 eV center. It was also discussed that the energy of 42 meV could arise from a local softening of the crystal around the defect. Bilodeau, Doverspike, Strom, Freitas and Rameshan (1993) doubted the correlation of the 42 meV sideband (located at 1.639 eV absolute photon energy) with the 1.681 eV defect line since in their low-temperature calorimetric absorption spectroscopic (CAS) measurements the signal measuring the non-radiative portion of the recombination had the same intensity for excitation at 1.681 and 1.639 eV hinting to independent

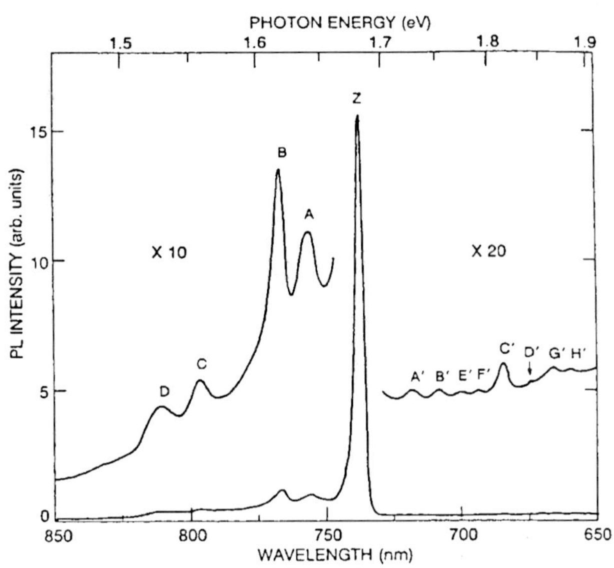

FIG. 11. Photoluminescence spectrum at 10 K sample temperature for a free-standing polycrystalline diamond film grown by microwave-plasma CVD with CH_4 (0.25%)/H_2. Z denotes the no-phonon line of the Si–V center at 1.681 eV (halfwidth 8.4 meV), A–D and A′–H′ are interpreted as vibronic satellites due to phonon emission and absorption, respectively (from Feng and Schwartz, 1993).

defects. The B sideband at 64.6 meV energy shift was ascribed to a TA lattice phonon by Feng and Schwartz (1993) while Sternschulte et al. (1994) associated it with a vibrational mode (LVM). Gorokhovsky et al. (1995) gave an explanation of the latter 64.6 meV line in terms of a Si–Si vibration of the defect implying a Si diatomic structure of the optical center. The 125.3 meV (C) and 153.6 meV (D) sidebands were ascribed to (perturbed) LA and LO lattice phonons, respectively, by Feng and Schwartz and Sternschulte et al. Other workers also observed the sideband structure without discussing its nature in detail (Clark and Dickerson, 1991a,b; Sittas et al., 1996). In their PL spectra, Feng and Schwartz also observed weak sidebands high-shifted from the parent NP line (in their work positioned at 1.684 eV) by 43.1 meV (A′), 67.4 meV (B′), 128.6 meV (C′), and 157.1 meV (D′) which they interpreted as anti-Stokes satellites corresponding to the Stokes-satellites A, B, C, and D. The relatively strong appearance of these lines, at a nominal sample temperature of 10 K, suggests that the true sample temperature might have been significantly higher or that these lines do not fully thermalize with the NP line. Additional lines above the NP transition, at 1.768_6 eV (E′), 1.784_9 eV (F′), 1.859_0 eV (G′), and 1.878_4 eV (H′), were discussed by Feng and Schwartz in terms of local modes or lattice phonon modes, but an alternative interpretation as electronic excited states of the optical center was left open.

The electronic structure of the 1.681 eV center was investigated using different measuring techniques. Iakoubovskii, Adriaenssens, Nesladek and Stals (1999) excited the 1.681 eV transition by 632.8 nm (He–Ne) or 514.5 nm (Kr) laser lines, and studied the quenching of the intensity when the sample was additionally illuminated by monochromatized light from a CW Xe lamp. Tuning this light they found quenching and enhancement thresholds at $E_C - 2.1$ eV and $E_V + 2.5$ eV, respectively. These values were associated with the decrease or increase, respectively, of the defect concentration in its ground state due to charging and de-charging of the defect by the quenching light. In their discussion, the authors associated the 1.681 defect with a negatively charged silicon–vacancy complex, $[Si-V]^-$, and placed its ground state at $E_C - 2.1$ eV while the neutral defect, $[Si-V]^0$, was located at $E_C - 3.0$ eV corresponding to $E_V + 2.5$ eV. A similar result was obtained by Iakoubovskii and Adriaenssens (2000a) in photoluminescence excitation (PLE) experiments. The PLE spectra, with excitation by monochromatized light from a 150 W Xe lamp and detection on the 1.681 eV line showed close lying excited states up to ≈ 1.9 eV which could also be detected in absorption measurements. Oscillatory structure in the PLE spectra from 2.0 to ≈ 2.5 eV was rather similar to that observed in the PLE spectra of the broad blue band A in the range of *intrinsic* absorption of natural and CVD diamonds which did not contain the 1.681 eV center (Fig. 12). Shifting the PLE energy scale so as to match the oscillatory structures in the two cases most satisfactorily led the authors to locate the ground state of the optical center at $E_C - 2.05$ eV consistent with the PL quenching data. This yields a dissociation energy of the upper luminescent defect state with respect to the bandgap of

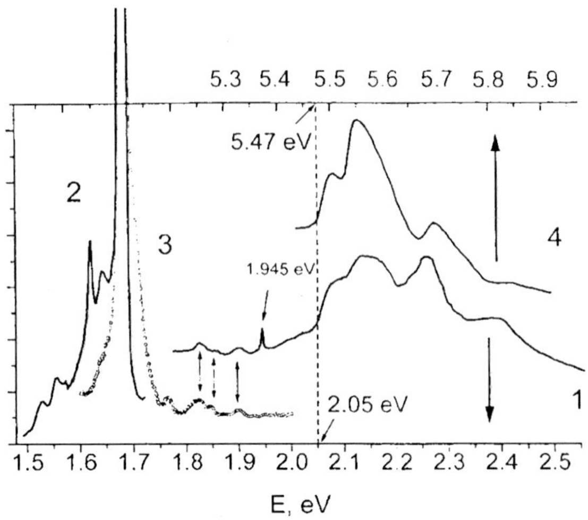

FIG. 12. Optical spectra at 77 K of the 1.681 eV Si–V defect in a CVD diamond film. Curve 1: photoluminescence excitation spectrum; curve 2: photoluminescence spectrum; curve 3: absorption spectrum; curve 4 (relating to the upper shifted energy scale): photoluminescence excitation spectrum of a natural diamond in the range of intrinsic absorption with detection on the broad blue band A taken for comparison from Dean and Male (1964) (from Iakoubovskii and Adriaenssens, 2000a).

≈370 meV which should be observable as a thermal de-activation energy of the PL intensity in temperature-controlled experiments. Actually, much lower de-activation energies were measured in a number of works, e.g., 70 meV (Feng and Schwartz, 1993) or 169 meV (Graham, Moustakas and Disko, 1991). This is no satisfactory agreement. It seems possible, however, that such values do not measure the excitation from the luminescent defect state into the band but to the higher lying electronic states at ≈1.83, ≈1.86, and ≈1.9 eV (Fig. 12) which then would constitute quite a good agreement with the latter de-activation energy. The electronic excited states of the defect observed by Iakoubovskii and Adriaenssens (2000a) are possibly the same states as detected by Feng and Schwartz (1993; their lines E′, F′, G′, and H′) though by these authors identified as vibrational states of the defect.

Rossi, Salvatori, Galluzzi, Montereali and Somma (1997) performed PLE measurements exciting the 1.681 eV center by a dye laser at selected photon energies between 1.9 and 2.0 eV and by an Ar-ion laser between 2.5 and 2.7 eV. They found a sharp resonance at ≈1.96 eV and a broad excitation band centered at ≈2.4 eV extending from 2.1 to 2.7 eV and higher energies. Their approach to an interpretation was to consider the optical defect as a nearly hydrogenic center, with the ground state of the 1.681 eV defect as $n = 1$, the luminescent upper state as $n = 2$, the resonance at 1.96 eV as $n = 3$, and the broad excitation band as a superposition of unresolved higher excited states plus a continuum above the ionization limit. The agreement of the hydrogen states with those experimentally observed was only very approximate. The defect ground state in this model lies at $E_C - 2.24$ eV which has to be compared with that obtained at $\approx E_C - 2.1$ eV by Iakoubovskii et al. (1999) and Iakoubovskii and Adriaenssens (2000a).

The decay kinetics of the 1.681 eV transition was studied by Khong et al. (1994) and Sternschulte et al. (1994). The first group investigated by electron beam excitation two samples from different sources which both exhibited first a slow decay ($\approx$3 ns decay time) followed by a faster decay ($\approx$1 ns decay time). This was suggested to be due to reservoir states feeding the 1.681 eV center initially whereas after depletion of the reservoir states the decay rate was determined by the intrinsic lifetime of the 1.681 eV center. Consistent with these data, Sternschulte et al. (1994) after laser pulse excitation found an exponential decay with 4 ns over three and a half decades of intensity for a homoepitaxial sample, and an exponential decay with 1.3 ns over nearly two decades for a polycrystalline diamond film. Both decay times were virtually constant between $\approx$5 and $\approx$100 K sample temperature and decreased at higher temperatures. It is evident that in the latter case additional recombination channels created by grain boundaries or other defect states in the polycrystalline material reduce the lifetime compared to the less perturbed case of crystalline diamond.

The 1.681 eV transition is often observed with large halfwidths (FWHM) of tens of millielectronvolts in diamond films on foreign substrates. When free-standing diamond films are used, the linewidths are reduced to a few millielectronvolts due to the strain release. For example, Ruan et al. (1991b) found 4–10 meV FWHM and Feng and Schwartz (1993) 8–9 meV FWHM. In a polycrystalline, free-standing as-grown CVD diamond film of 65 μm thickness Collins, Allers, Wort and Scarsbrook (1994) measured a linewidth of $\approx$0.6 meV allowing them to resolve a splitting of the 1.681 eV line into two components spaced by (0.8 ± 0.03) meV (Fig. 13). The doublet thermalized, and an Arrhenius fit to the data yielded a thermal activation energy of (0.76 ± 0.04) meV in excellent agreement with the spectroscopic splitting. A dramatically reduced halfwidth of $\approx$0.18 meV was obtained by Sternschulte et al. (1994) when the 1.681 eV defect was studied in a diamond layer deposited on a (100) oriented HPHT synthetic diamond crystal by plasma CVD. In this case, seven line components of the 1.681 meV transition could be resolved (Fig. 13). The two strongest lines labeled b and c are those observed by Collins et al. (1994). Sternschulte et al. (1994) obtained a thermal activation energy for the intensity ratio of the two main lines of (0.8 ± 0.04) meV confirming the data of Collins et al. They studied the complete seven-line spectrum in PL and PLE spectroscopy under $\langle 100 \rangle$ uniaxial stress at 10 and 13 K, respectively. The four strongest line components (a, b, c, and d) shifted linearly with the applied stress (Fig. 14) allowing to study their thermalization behavior. From these data a transition level scheme for the four strongest lines was derived (Fig. 14). They form a quartet of electronic transitions between the excited luminescent state and the ground state of the defect which are both split by the stress. There is

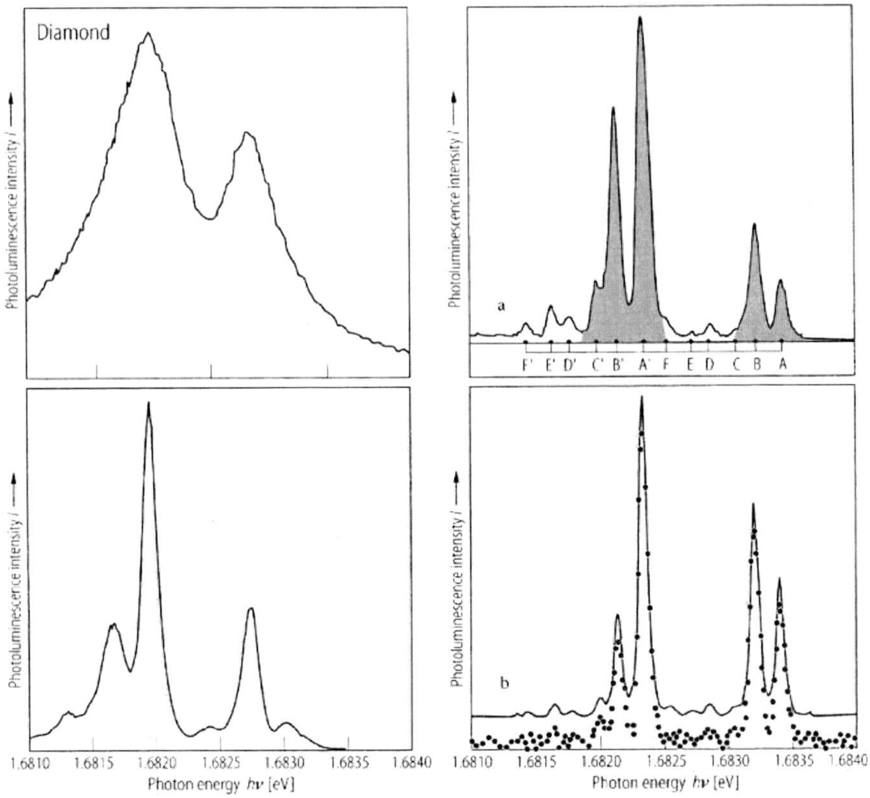

FIG. 13. Fine structure of the 1.681 eV Si–V defect no-phonon line. Halfwidths of the line for CVD diamond films on Si substrates are usually some millielectronvolts so covering the whole energy range of the panels. Reduction of linewidths: *Left top*, resolved doublet structure of the 1.681 eV line in a free-standing polycrystalline CVD diamond film (after Collins et al., 1994). *Left bottom*, resolved doublet with additional line components in a Si-doped homoepitaxial diamond film (after Sternschulte et al., 1994). Right (a), 12-line spectrum in a Si-doped HPHT diamond; (b) synthesis of the experimental spectrum by a superposition of 3 isotope-shifted partial spectra for the ^{28}Si, ^{29}Si, and ^{30}Si isotopes with relative intensities proportional to the natural isotope abundances. Each partial spectrum consists of four electronic transitions as shown shadowed in (a) for ^{28}Si (cf. recombination level scheme, Fig. 14) (after Clark et al., 1995) (Figure from Landolt-Börnstein, 2002).

a remaining splitting at external zero field creating the narrow quartet structure seen when no field is applied (Fig. 13). This zero-field splitting was interpreted as arising from an internal 'overpressure' in the defect of -0.6 GPa due to the large size of the Si defect in the narrow cage of the diamond lattice. The nature of the remaining three line components was not identified. This was accomplished by Clark, Kanda, Kiflawi and Sittas (1995) in a study of synthetic single crystal diamond doped with silicon. They observed, at further reduced linewidths of ≈ 0.1 meV, 12 line components of the 1.681 eV transition. The dominant four line components which Sternschulte et al. (1994) had shown to be correlated as

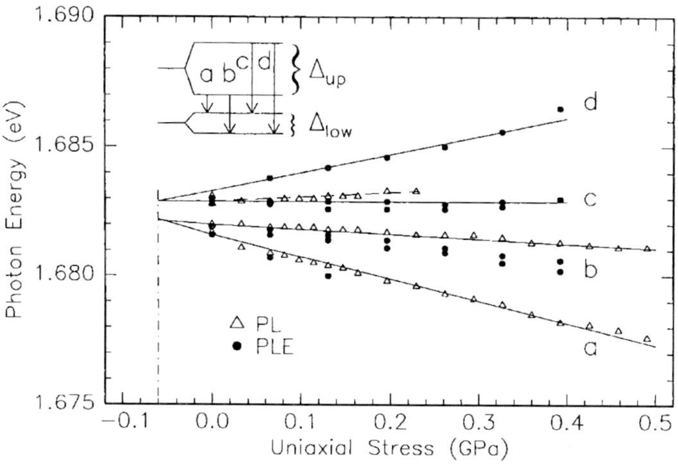

FIG. 14. Splitting of the 1.681 eV quartet line (cf. Fig. 13) under uniaxial stress applied along a ⟨001⟩ crystal direction. Triangles are from photoluminescence (PL) measurements and dots are from photoluminescence excitation (PLE) measurements. Straight lines are eye guides. Inset: recombination level scheme for the quartet spectrum consistent with the splitting data and the observed temperature dependence of the four lines in PL and PLE (from Sternschulte et al., 1994).

a quartet were interpreted as transitions at a ^{28}Si isotope in the defect. The complete 12-line spectrum could then very satisfactorily be interpreted on the assumption that ^{29}Si and ^{30}Si give rise to equivalent quartet patterns with intensities reflecting the natural abundances of the isotopes (^{28}Si: ^{29}Si: ^{30}Si as 92.3 : 4.7 : 3.0), however shifted by ≈ -0.354 meV for ^{29}Si and by twice this value for ^{30}Si (Fig. 13). Temperature-controlled PL studies enabled Clark et al. (1995) to deduce a level scheme identical to that derived earlier by Sternschulte et al. (1994) (Fig. 14).

Only few efforts were undertaken to determine the symmetry of the 1.681 eV defect. Standard procedures in determining the symmetry group of optical defects are splitting studies of the transitions in uniaxial stress fields or in homogeneous magnetic fields, accompanied by a comparison or fit of the observed splitting pattern to adequate theories. The uniaxial stress measurements performed by Sternschulte et al. (1994) were restricted to stress along a crystal direction of ⟨100⟩ type due to the geometrical constraints imposed by the film on the (100) HPHT diamond substrate. Sternschulte, Thonke, Gerster, Limmer, Sauer, Spitzer and Münzinger (1995) also performed Zeeman measurements with fields up to 14 Tesla along ⟨100⟩, ⟨111⟩, and ⟨111⟩ crystal directions (Fig. 15). Comparison of the observed number of Zeeman components with theory and the particularly 'simple' splitting pattern for $\vec{B} \parallel \langle 111 \rangle$ led the authors to conclude that the optical defect has symmetry close to tetragonal, however, slightly distorted towards monoclinic. This corresponds to a defect axis nearly along [001] but tilted in a

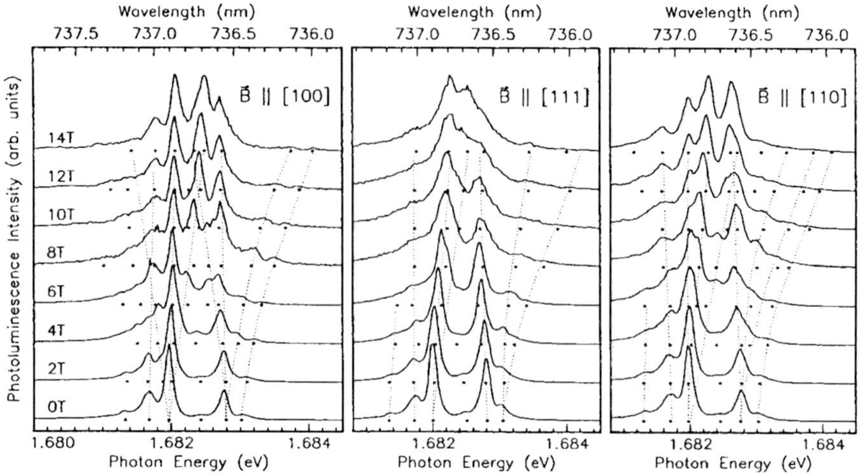

FIG. 15. Zeeman splitting of the 1.681 eV no-phonon quartet spectrum (cf. Fig. 13) for magnetic fields $\vec{B}$ along [100], [111], and [110] crystal directions in Faraday configuration. Spectra up to $B = 6$ Tesla recorded at ≈ 20 K sample temperature, for $B = 8-14$ Tesla at a nominal temperature of 5 K. Dots: line peak positions, dotted lines: eye guides (from Sternschulte et al., 1995).

(110) plane. A different approach was made by Brown and Rand (1995) to determine the site symmetry of the defect. They studied the polarization of the 1.681 eV line relative to the polarization of the electric field of the incident photons and compared the experiments with theoretical results expected for a number of assumed defect symmetries. Good agreement was found for a $\langle 100\rangle$ oriented defect with either C_2 monoclinic or D_2 orthorhombic symmetry, and the defect was then represented by a z dipole along the $\langle 110\rangle$ symmetry axis (Fig. 16). These data are at variance with a theoretical, ab initio cluster study by Goss, Jones, Breuer, Briddon and Öberg (1996). These authors found that for a [Si–V] complex—which the optical 1.681 eV defect was assumed to be in agreement with a large body of experimental work—the Si atom is unstable at its lattice site and spontaneously moves to the split-vacancy site yielding a defect of D_{3d} symmetry. The 1.681 eV line was ascribed to an internal transition between two doubly degenerate E levels, and the calculated transition energy was 1.86 eV in fair agreement with the experimental energy of 1.681 eV. Finally, the decay time was obtained theoretically as ≈ 3 ns in agreement with the experimental values of (1–4) ns obtained by Sternschulte et al. (1994).

3.4. HYDROGEN

Hydrogen is one of the main background impurities in natural diamond (Sellshop et al., 1980). Nonetheless, there is no luminescence spectrum which could be ascribed to hydrogen. Gippius, Zaitsev and Vavilov (1982), Gippius

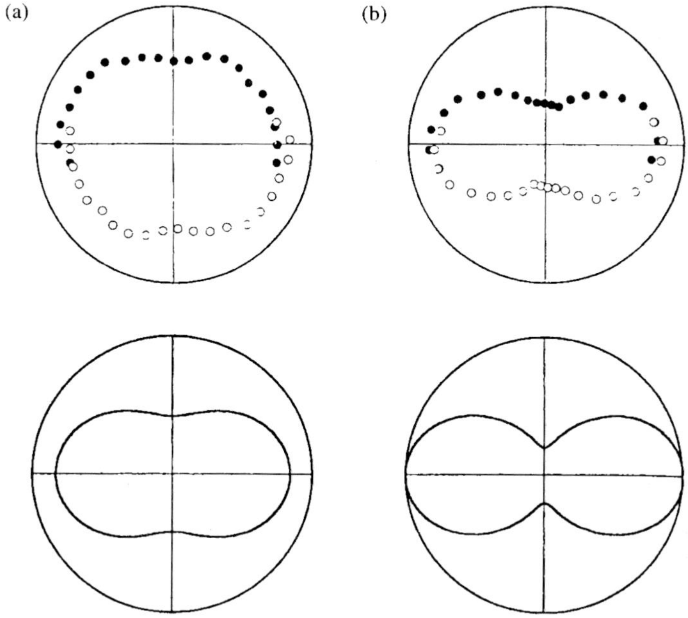

FIG. 16. *Top* (Experimental): Polar plots of the dependence of the 1.681 eV luminescence line intensity on the polarization of the incident exciting laser light (488 nm Ar ion laser). Analyzer fixed along (a) a ⟨100⟩ sample orientation and (b) a ⟨110⟩ sample orientation. Solid circles as measured are replicated by rotation through 180° to yield open circles. *Bottom* (Theoretical): Dependence of the polarized luminescence intensity as a function of sample rotation for a z dipole ($s \leftrightarrow p_z$ electronic transition). Analyzer aligned (a) parallel to a ⟨100⟩ crystallographic orientation and (b) parallel to a ⟨110⟩ orientation (after Brown and Rand, 1995).

et al. (1983) and Gippius (1993b) found that H ion implantation of natural diamonds and annealing above 600 °C caused a line at 545.8 nm (≈2.27 eV) to appear. The shift of the line to 545.2 nm (≈2.274 eV) upon deuterium implantation proved the correlation with hydrogen. As a function of annealing temperature the line intensity showed non-monotonous behavior between 600 and 1100 °C with several intensity maxima and minima. This indicated complicated formation and destruction kinetics of the optical defect which was not further identified.

In the CVD growth of polycrystalline and homoepitaxial diamond films hydrogen is used as a carrier gas, and the deposited films contain high concentrations of hydrogen up to 1 at.%. Therefore, evidence for hydrogen is expected either from vibrational transitions at low photon energies or from electronic transitions at higher energies.

Actually, hydrogen in CVD diamond has been observed by several authors using infrared spectroscopy (McNamara, Levy, Gleason and Robinson, 1991; Janssen, van Enckevort, Vollenberg and Giling, 1992). Dischler, Wild,

Müller-Sebert and Koidl (1993) observed a number of C–H stretching absorption bands in the region 2800–3000 cm^{-1} ($\approx$0.347–0.372 eV) when they studied by infrared absorption free-standing polycrystalline diamond films grown by hot-filament or plasma-assisted CVD on silicon substrates which were chemically dissolved prior to the measurements. They showed that hydrogen is incorporated in non-crystalline and defective regions of the polycrystalline films, such as grain boundaries and dislocations. Similar results were obtained in EPR and optical absorbance studies by Talbot-Ponsonby, Newton, Baker, Scarsbrook, Sussmann and Whitehead (1998) on polycrystalline diamond films. An earlier EPR study by Zhou, Watkins, McNamara, Rutledge, Messmer and Chawla (1996) had already given evidence for two defects in which a hydrogen atom has entered a stretched bond at a grain boundary or other extended misfits region in the CVD films. Finally, again in an infrared spectroscopic absorption study, Fuchs, Wild, Schwarz and Koidl (1995) observed hydrogen-related electronic transitions in homoepitaxial diamond layers around 7300 cm^{-1} ($\approx$0.905 eV) with spectral positions sensitive to isotopic replacement of H by D and of ^{12}C by ^{13}C as well. These electronic transitions were not reported to appear in luminescence.

Hydrogen although not involved in sharp line luminescence spectra influences strongly a broad luminescence band in the green (2.3 eV) spectral region. These data are reported in Section 4.1.

3.5. Metals

McCauley and Vohra (1994) reported a PL line at 1.6 eV in diamond films deposited on silicon substrates by hot-filament CVD using a gas mixture containing only methane, hydrogen, and oxygen as precursors. The line is broad (FWHM $\approx$ 100 meV) and superimposed on an even broader band (FWHM $\approx$ 460 meV) centered at $\approx$1.9 eV which was not further discussed. As potential causes for this emission the authors considered the incorporation of tungsten from the filament during deposition and the formation of an interfacial SiC layer between the silicon substrate and the polycrystalline diamond film. Neither of the two possibilities could be substantiated or ruled out by further analysis of the samples. The authors discussed that the optical defect possibly involves the combination of a lattice defect and an impurity atom (incorporated into the diamond lattice during deposition) and found it likely that boron or nitrogen would be involved.

Another PL spectrum was observed by Lal, Dallas, Yi, Gangopadhyay, Holtz and Anderson (1996) and Lal, Lin, Gangopadhyay, Hague and Naseem (1997) to emerge in polycrystalline diamond films grown by arc-jet CVD on molybdenum substrates. The spectrum has a general appearance typical for vibronic defects with weak lattice coupling. The sharp main line at 1.735 eV ($T = 20$ K) is accompanied by multiple phonon sidebands with 24 meV phonon energy. Sharp lines at higher photon energy were shown to be related with a ground-state splitting of the defect. The intensity of the 1.735 eV line is nearly constant from low temperatures to 100 K, and decreases then exponentially with a de-activation energy of ≈ 37 meV ascribed to the activation of competing non-radiative recombination processes. The decay of the luminescence was found to be very fast, below < 100 ps, limited by the instrumental resolution. The authors ascribed the optical defect to tungsten incorporated into the diamond films during deposition due to erosion of the filament whereas substrate impurities seemed to have no influence on the spectrum. The vibronic spectrum was superimposed on a broad band (FWHM $\approx$ 500 meV) centered near 1.80 eV. This band and the band underlying the 1.6 eV emission first described in this section could have the same origin as the red broad bands in the 1.7–1.9 eV region discussed in Section 4.2.

4. Broad-band Luminescence

4.1. Blue and Green Band A Luminescence ($\approx$ 2.8 and $\approx$ 2.3 eV)

Band A is a prominent radiative recombination feature which can be excited in virtually all diamonds by energetic particles (α, electrons) or photons (UV, X-rays) (Walker, 1979). Dean (1965) observed this strong emission in the luminescence spectra of natural and p-doped high-pressure synthetic diamonds and found it could appear centered at nearly any energy between ≈ 2.0 and ≈ 3.2 eV, with halfwidths of ≈ 500–1000 meV. The spectral appearance depended on the samples being non-semiconducting or semiconducting, on the excitation by X-rays, light, or electrons, on the excitation strengths and on the measurement temperature. The results indicated there were two distinct luminescence band at ≈ 2.2 and 2.8 eV. The green band at 2.2 eV appeared strong in the high-pressure synthetic samples whereas the blue band at 2.8 eV was intense in natural diamonds. The bands shifted to higher energy when the excitation power was increased, and, despite complex behavior in detail, there was a trend that the luminescence decay times were slower at low photon energies and faster at higher photon energies. Based on these findings Dean suggested that both bands originated from electron–hole recombination processes at the same donor–acceptor pairs (DAP): The pairs are closer in the natural samples due to their growth on a geological time scale yielding the blue luminescence band, and are more distant ('dilute') in the synthetic samples yielding the green band. The donor was identified with nitrogen and the acceptor with aluminum. It was demonstrated later that the acceptor determining the p-type semiconducting properties of diamond is boron (Section 5). Electron–hole recombination from individual pairs should cause discrete line spectra due to the Coulomb interaction of the ionized impurities after the recombination. Such lines were not observed in the band A luminescence, but it was argued that random strain in the diamonds would broaden the discrete lines into bands.

Dean's DAP model stood for many years but it became obvious that it was inconsistent with many of the subsequent experimental findings. We follow Collins (1992) in summarizing arguments against the model. (i) The low energy edge of the green band corresponds to a donor with an ionization energy of 4 eV as, e.g., formed by aggregated nitrogen atoms. Isolated nitrogen impurities, however, give rise to a donor state 1.7 eV below the conduction band which should result in a DAP band in the ultraviolet region. Such a band is not observed. (ii) Blue band A luminescence, ascribed in the DAP model to close pairs formed preferentially during the long growth time of natural diamond is also observed from the small {100} growth sectors of high-pressure synthetic diamond in which the nitrogen content is low. (iii) Furthermore, the model does not account for the differences between natural type Ia and type IIb diamonds which can luminesce with well-distinguishable spectral features. Other problems with the model arose when cathodoluminescence experiments on band A were combined with panchromatic and monochromatic microimaging of the luminescence. In a number of works (Kiflawi and Lang, 1974; Hanley, Kiflawi and Lang, 1977; Pennycook, Brown and Craven, 1980; Pirouz, Cockayne, Sumida, Hirsch and Lang, 1983; Yamamoto, Spence and Fathy, 1984) it was demonstrated that the 2.8 eV blue band A emission originates primarily from dislocations. Yamamoto et al. in particular found that all luminescent dislocations were dissociated and that straight dislocations exhibited the largest luminescence intensities. The suggestion was that donor and acceptor states formed in the strain field of the dislocation core were the sources of electron–hole recombination in the 2.8 eV photon region. (It is interesting to note in passing that in silicon such a model was firmly shown to apply: Widely split dislocation partials prepared in a two-step deformation process exhibit sets of narrow line spectra which are closely related in energy and annealing behavior to the discrete-valued dissociation widths and originate from the recombination of electrons and holes localized on opposite partials of a dislocation (Sauer, Kisielowski-Kemmerich and Alexander, 1986, 1987)).

Blue band A emission also occurs in CVD single-crystal diamond particles and in as-grown polycrystalline CVD films very similar to that observed from natural IIa diamonds (Figs. 5 and 17) (Collins et al., 1989a, 1990b). Extended studies on polycrystalline diamond films demonstrated that the green band A is associated with boron. While the recombination mechanism remained unexplained at first, later investigations of synthetic diamond crystals showed clearly that this band originates from DAP recombination with boron as the acceptor so finally verifying, at least partially, Dean's original model of the band A emission. Some of these data are discussed in the following in more detail.

Vavilov et al. (1980) reported that lightly boron-doped epitaxial films on (111) natural diamond crystals exhibited a broad band without narrow point-defect lines. The band increased in intensity with higher boron concentrations and peaked in the region of 2.1 eV before it decreased in intensity and shifted to $\approx$2.7 eV for heavy doping where the films appeared blue. Kawarada, Nishimura,

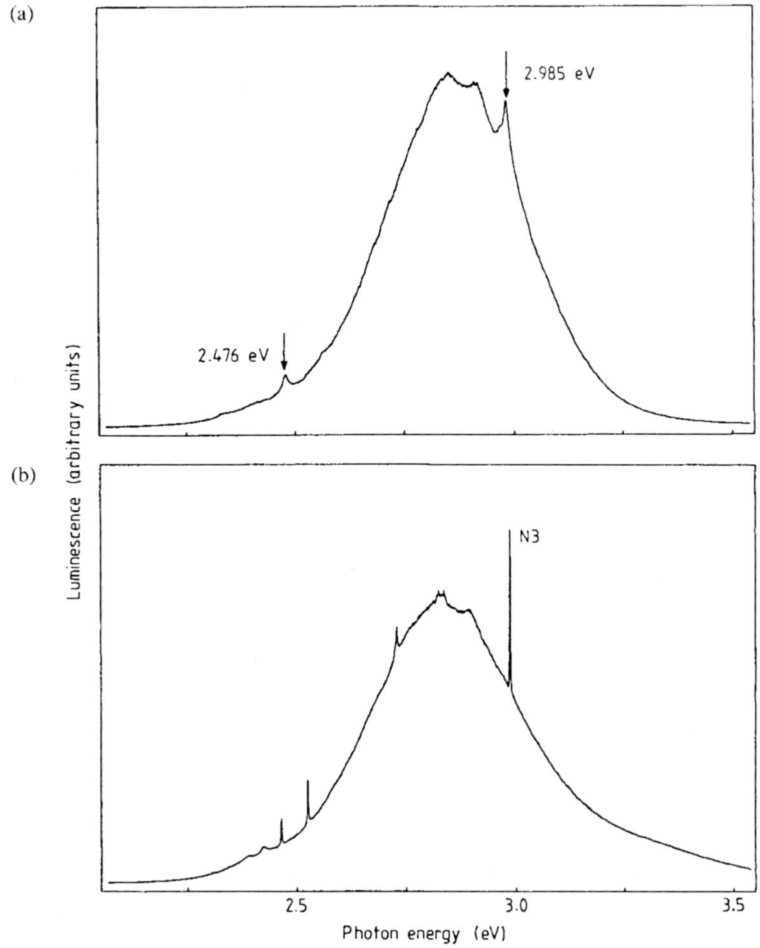

FIG. 17. Blue band A cathodoluminescence spectra taken at 77 K sample temperature. (a) CVD single-crystal diamond grown with CH_4 (1%)/H_2, electron beam energy 15 kV. (b) Type IIa natural diamond, electron beam energy 45 kV. Indicated are two point defect lines at 2.985 eV (N3) and 2.476 eV (from Collins et al., 1989a).

Ito, Suzuki, Mar, Yokota and Hiraki (1988) initially observed in many films grown by plasma-assisted CVD on Si (100) substrates either green luminescence at around 2.5 eV or blue luminescence without being able to correlate this difference to relevant growth parameters. (Similarly stated Burns, Cvetkovic, Dodge, Evans, Rooney, Spear and Welbourn (1990) in their CL study of the sector dependence of the blue (2.9 eV) and green (2.3–2.6 eV) bands in synthetic diamonds grown by the temperature-gradient technique that there seemed to be no obvious pattern to the occurrence of these blue and green bands with respect to donor or acceptor concentrations and felt it was difficult to comment on the applicability of Dean's model.) However, Kawarada et al. (1990a) in a subsequent paper studied CVD diamond films grown on Si substrates which were intentionally boron-doped and compared these films with (nominally) undoped films (Fig. 18). In undoped CVD diamond they found the prominent CL peak at 2.8–2.9 eV to occur. The luminescence characteristics such as the peak position, the halfwidths of the spectra and the temperature dependence of the intensities were found almost equal to those of the band A luminescence of natural type IIa crystals where the emission came from dislocation regions. In the boron-doped CVD diamond, this luminescence was much stronger at the optimum doping level than in the undoped samples, and a band at 2.3–2.4 eV was additionally seen. The authors concluded from the temperature dependence and the excitation dependence of the two bands that the higher energy band (≈2.9 eV) was the same as that observed in undoped CVD diamond ascribed there to dislocations, and that the lower energy band (≈2.3 eV), clearly related to boron, was identical to the band seen in natural type IIb semiconducting diamond. That the 2.9 eV band in CVD diamond layers is also directly correlated with dislocations as it is in natural and high-pressure synthetic crystals was shown by Graham et al. (1991) in a cathodoluminescence and transmission electron microscopy imaging study. The

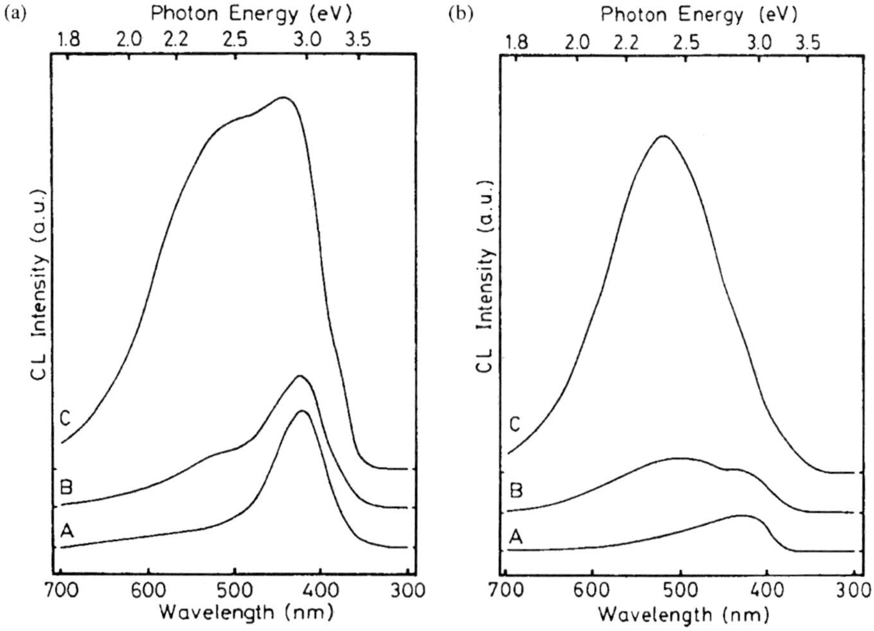

FIG. 18. High- and low-excitation cathodoluminescence spectra of undoped (curve A), and boron-doped (curves B and C) CVD diamond films grown using a CO (5%)/H_2 gas mixture. Boron-doping in curve B corresponds to a flux ratio $[B]/[C] = 200$ ppm during growth, that in curve C to 1000 ppm. *Left*: High excitation with 10 μA mm^{-2} electron current density; *Right*: low excitation with 0.2 μA mm^{-2} electron current density (from Kawarada et al., 1990b).

origin of the blue and green band A luminescence was further investigated in a thorough CL study of undoped and deliberately boron-doped CVD diamond layers grown on (111) silicon substrates by Ruan et al. (1992a). They observed the 2.3 eV green band to emerge and to grow in intensity at increasing boron doping levels. The band was superimposed on the blue 2.8 eV band seen at constant intensity (Fig. 19, right). Summing up all their data Ruan et al. came to the unambiguous conclusion that in their CVD samples the 2.8 eV band was due to dislocations whereas the 2.3 eV band was related to boron. These correlations appear now to be firmly established and generally accepted though the detailed nature of the recombination process was for both bands not identified at that time.

Further progress towards the identification of the green band was advanced by Klein, Crossfield, Freitas and Collins (1995). They studied by steady-state and time-resolved photoluminescence and cathodoluminescence measurements synthetic type IIb boron-doped semiconducting diamond samples which

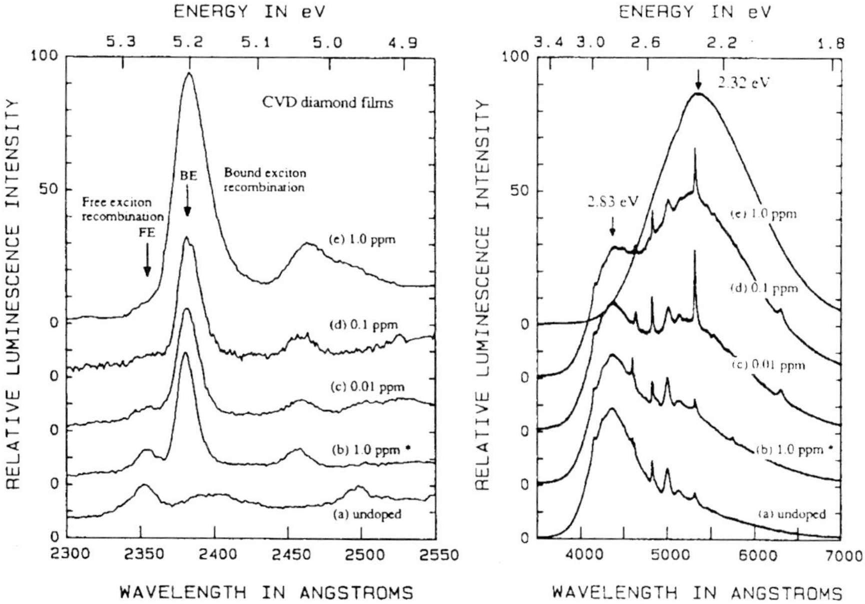

FIG. 19. Systematic variation of the boron doping of diamond films grown by microwave plasma-assisted CVD on ⟨111⟩ p-type silicon substrates. The concentration of diborane in the CH_4 (0.5%)/H_2 reaction gas mixture is indicated on the cathodoluminescence spectra shown. The notation in (b), 1.0 ppm*, indicates that 0.1% oxygen was used for the deposition of this sample which reduced the effective boron concentration of the film by a factor of ≈ 100 (cf. Fig. 21). *Left*: Increase of the boron bound exciton (BE^{TO}) recombination line in competition with the intrinsic free exciton (FE^{TO}) recombination line for increasing boron doping levels. *Right*: Increase of the 2.32 eV green band A luminescence relative to the 2.83 eV blue band A luminescence in the same samples as shown left (from Ruan et al., 1992a).

exhibited the blue 2.8 eV band, the green 2.3 eV band, and a red band centered at 1.84 eV. Clues to an identification of the green and red bands came from three essential experiments: (i) Gated CL spectra of the green band at 77 K showed that the maximum of the band shifted to lower photon energies at increasing delay times consisted with DAP recombination; (ii) the CL intensities at three representative emission energies across the green band decayed non-exponentially following precisely theoretical decay curves obtained from numerical calculations in the DAP recombination scheme; and (iii) the recombination rate of the red PL band excited at 355 nm was shown to be sum of the recombination rates of the red and green bands excited at 532 nm. An earlier study of the same group (Freitas, Doverspike, Klein, Khong and Collins, 1994a) had given strong evidence that the red band itself is due to DAP recombination (Section 4.2). Combining these results, Klein et al. (1995) concluded that the green band and the red band are due to pair emission from two different donors to the neutral boron acceptor which are in direct competition. The lower edge of the red DAP band was at ≈ 1.5 eV leading to a donor binding energy of $E_d \approx 3.6$ eV which is close to that of nitrogen A aggregates. For the green band, the low-edge emission occurs at ≈ 1.8–1.9 eV resulting in an approximate donor binding energy of $E_d \approx 3.2$–3.3 eV which could not be assigned to a specific donor. These results finally revived and confirmed Dean's original DAP model for the 2.3 eV green luminescence band.

Other assignments of the 2.3 eV luminescence band than discussed above have been given possibly implying that such bands can exist with similar spectral appearance but of different origin. Hayashi, Watanabe, Yamanaka, Okushi, Kajimura and Sekiguchi (1996), Hayashi, Sekiguchi and Okushi (1997a) and Hayashi, Watanabe, Yamanaka, Sekiguchi, Okushi and Kajimura (1997b) reported on a green band centered close to 540 nm (2.3 eV) with halfwidth of ≈ 150 nm (570 meV). In their study, diamond films were deposited by microwave CVD on high-pressure synthetic Ib diamond (001) substrates to thicknesses of ≈ 2.4 μm. After deposition the samples were hydrogenated by keeping them in the hydrogen plasma for 15 min without CH_4 gas. When the hydrogenation was performed at 800 °C, these undoped layers exhibited a broad CL spectrum composed of two components with maxima at ≈ 2.3 and ≈ 2.0 eV at room temperature. At cryogenic temperature (20 K) the 2.3 eV green band alone was observed. Oxidization of the films made the 2.3 eV band disappear (while the 2.0 eV band remained) but it could be restored by re-hydrogenation. Under these experimental conditions the samples had low, almost temperature-independent resistivities around 10^5 Ω cm (Fig. 20). When the samples were hydrogenated at 500 or 300 °C, the 2.3 eV luminescence was essentially absent and the resistivities were high, up to some 10^8 Ω cm, and strongly temperature-dependent. However, low-energy electron beam irradiation at ≈ 10 kV decreased the resistivities by up to three decades and simultaneously generated the 2.3 eV band (Fig. 20). The authors concluded that the 2.3 eV band is directly related to the hydrogenation of the diamond films. Based on experiments with variable

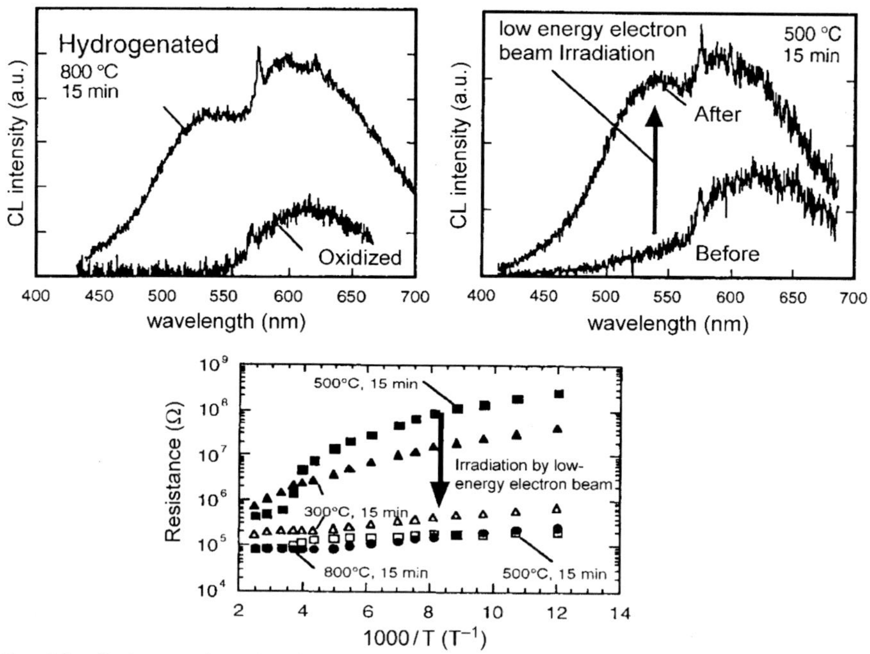

FIG. 20. Hydrogen-related broad-band cathodoluminescence in microwave-plasma CVD diamond films grown on synthetic Ib (001) diamond substrates using a CH_4 (0.5%)/H_2 gas mixture. *Top left*: CL spectrum after oxidation by an H_2SO_4 and HNO_3 solution at 200 °C for 15 min and thermal annealing at 400 °C in N_2 ambient for 30 min. Appearance of a broad band centered at 2.3 eV after subsequent exposure to a hydrogen (deuterium) plasma at 800 °C for 15 min. *Top right*: CL spectra observed for hydrogen plasma treatment at 500 °C/15 min ('before') and after low energy ($\approx$ 10 keV) electron beam irradiation ('after'). *Bottom left*: Sheet resistances of the diamond films hydrogenated at 300, 500, and 800 °C as a function of reciprocal temperature (from Hayashi et al., 1997a).

electron-beam acceleration voltages and calculated penetration depths, they suggested that the low resistivities originate from a subsurface region of the homoepitaxial films in which also the luminescent hydrogen states are located.

The author believes it could be worth to discuss these data on the assumption that basically the 2.3 eV band is associated with boron as documented in the works cited earlier in this chapter. Then, the influence of oxygen in quenching the 2.3 eV band is understandable in the light of the data reported by Ruan et al. (1992b). They found that oxygen incorporation into their CVD diamond films decreased the actual boron concentration dramatically. This was concluded from two observations (Fig. 21): (i) The boron BE line observed at the standard energy position with moderately broad linewidth in films containing oxygen shifted to lower energy and broadened significantly at constant boron concentrations when no oxygen was incorporated. Concomitantly, the free exciton (FE) line disappeared (Fig. 21, left). This was explained by an effective boron

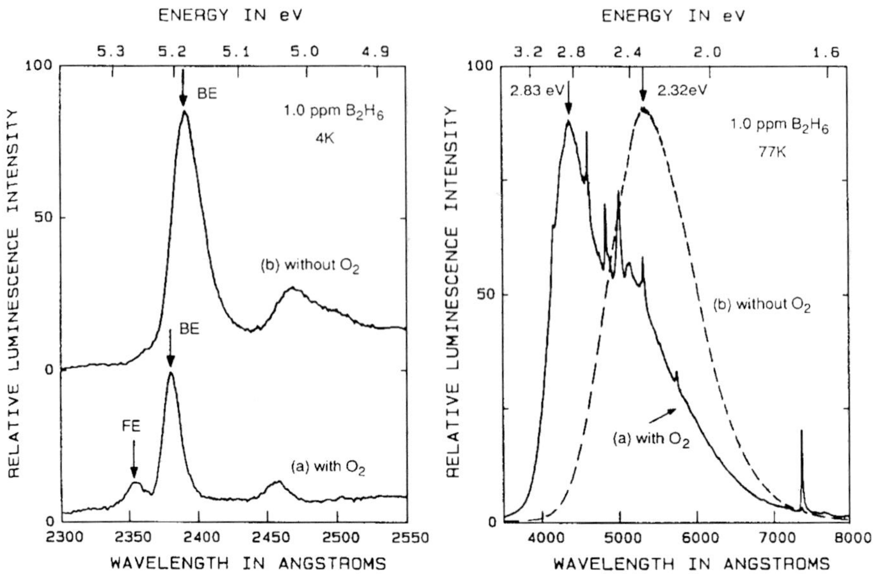

FIG. 21. Effect of oxygen on boron doping of CVD diamond films. *Left*: Cathodoluminescence spectra in the near-band-edge region showing boron bound exciton (BE^{TO}) and free exciton (FE^{TO}) recombination lines in diamond films doped with 1.0 ppm B_2H_6 in the reaction gas mixture. The shift in energy and the increase of the halfwidth of the BE line demonstrates that the sample of curve (b) without oxygen has a much higher effective boron concentration than the sample of curve (a) with 0.1% oxygen added. *Right*: Cathodoluminescence spectra of the same samples in the visible spectral region. The band peaking at 2.32 eV is related to the presence of boron in the diamond but the effective boron concentration is strongly reduced by the oxygen doping (from Ruan et al., 1992b).

concentration governed by the amount of incorporated oxygen. Following the data of Sternschulte, Albrecht, Thonke and Sauer (1996a) and Sternschulte, Albrecht, Thonke, Sauer, Grießer and Grasserbauer (1996b) discussed in Section 5.1, these effects are characteristic of increasing boron concentrations exceeding $\approx 10^{18}$ cm^{-3}, and their physical reasons are the exchange interactions between the charge carriers at high boron concentrations when the BE wavefunctions located at different acceptor sites overlap. (ii) The 2.3 eV band was essentially reduced in intensity with oxygen incorporation compared to samples where no oxygen was introduced (Fig. 21, right). With respect to the data of Hayashi et al. (1997a) it is furthermore tempting to associate the differences in their luminescence spectra and in their resistivity values for hydrogenation at 800 °C and at 500/300 °C, respectively, to boron–hydrogen interactions as studied by Chevallier, Ballutaud, Theys, Jomard, Deneuville, Gheeraert and Pruvost (1999). This group found that H–B pairs form in boron-doped crystalline epitaxial diamond layers grown by microwave plasma CVD on high-pressure synthetic diamond substrates. The pairs are stable up to 550 °C where they start to dissociate. (This extended earlier work by Landstrass and Ravi (1989) showing that hydrogen passivates electrically active defects in diamond.) It appears possible to the author that the electron beam irradiating the diamond films of Hayashi et al. (1997a,b) at the low hydrogenation temperatures of 300 and 500 °C transfer enough energy to the films to dissociate H–B pairs so activating the boron acceptors and causing the 2.3 eV band to appear. A more critical discussion of such interrelations and mutual influences among boron, oxygen, and hydrogen must be conducted elsewhere.

Another study of the 2.3 eV green band luminescence should be mentioned. Iakoubovski and Adriaenssens (2000b) performed PL and PLE measurements on the band using high-pressure synthetic Ib diamonds and undoped, free-standing diamond films obtained by microwave plasma CVD on silicon substrates after removal of the substrates by chemical etching (Fig. 22). For the CVD films, they found two components in the PLE spectra, a symmetric band centered at 2.95 eV and a step-like feature exhibiting a relatively sharp threshold at 3.25 eV. Both components were temperature- and sample-dependent. The second PLE component of the 2.3 eV band disappeared after annealing of the sample at 400 °C for 30 min in vacuum but could be restored by exposing the sample to a humid atmosphere for about a day. The annealing and restoration cycles could be repeated many times. In contrast, the PLE spectrum of the green 2.3 eV band from a high-pressure synthetic Ib diamond showed merely a step-like threshold at 3.2 eV—this was similar to the CVD film—but the PLE spectrum did not change upon annealing. The 2.3 eV band in the CVD film when excited in the region of the first PLE band (at 2.7–3.1 eV) correlated with the content of amorphous carbon phases in the film as monitored by confocal micro-Raman spectroscopy. This is consistent with earlier studies by Bergman et al. (1993) and Bergmann,

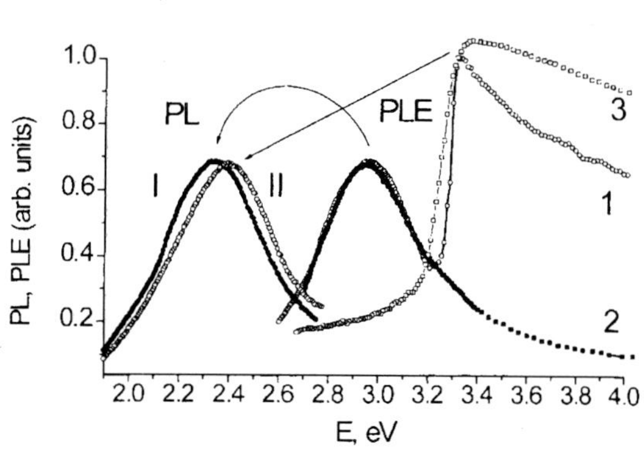

FIG. 22. PL and PLE spectra at 7 K sample temperature of the broad green 2.3 eV band in an undoped CVD film and in a HPHT synthetic type Ib diamond crystal. Curves I and II: PL spectra excited at any energy in the range 2.7–3.1 eV in the CVD film (curve I) and in the range 3.3–4.9 eV in the CVD film or in the Ib diamond (curve II). Curves 1 and 2: PLE spectra of the CVD film before and after annealing (400 °C, 30 min), respectively. Curve 3: PLE spectrum of the Ib diamond (from Iakoubovski and Adriaenssens, 2000b).

McClure, Glass and Nemanich (1994) on microwave CVD diamond films grown on Si (100) substrates. [There, an almost linear relationship was established between the PL intensity of the broad 2.1 eV band and the Raman intensity of the disordered sp^2-bonded graphitic carbon phases.] In summary, Iakoubovskii and Adriaenssens were led to conclude that their 2.3 eV luminescence band was composed of two components which are nearly identical in line shape but have different origin; the first component was ascribed to recombination in regions of amorphous carbon, the second component was associated with hydrogenation and a low-resistivity p-type surface region consistent with the conclusions reached by Hayashi et al. (1997a,b) as discussed above. Iakoubovskii and Adriaenssens also reasoned, based on the characteristic form of the PLE spectrum, that for the second, hydrogen-related component of the 2.3 eV band the most logical model was a DA pair recombination process without specifying the donor and acceptor species involved. This, in turn, would concur with the conclusions of Klein et al. (1995).

4.2. Red Broad Luminescence Bands (1.95–1.9, 1.84, 1.7, and 1.5 eV)

A number of red broad luminescence bands have been observed in natural and high-pressure synthetic diamonds as well as in CVD diamond films. Roughly, the bands fall into the regions 1.84–1.95, 1.7, and 1.5 eV. Robins et al. (1989) found in their films grown by the hot-filament CVD technique on silicon substrates between 600 and 250 °C, and in isolated diamond particles grown at temperatures of 800 °C a broad band after electron beam excitation centered at $\approx$ 1.9 eV with a halfwidth of $\approx$ 400 meV. Since the 2.156 eV defect line weakly appeared on the high-energy side of the broad band the authors identified the spectrum as the vibronic Stokes sideband of this nitrogen–vacancy point defect assuming a strong phonon coupling of the defect and additional strain-induced broadening of the phonon modes. A similar luminescence band with maximum at $\approx$ 2.1 eV was observed by Partlow et al. (1990) in microwave plasma CVD diamond films ('coatings') superposed on the 2.156 eV defect line and its local modes. It was not discussed whether in this case the broad band was part of the vibronic spectrum of the 2.156 eV defect or an independent feature. Clark and Dickerson (1992) studied free-standing microwave plasma CVD diamond films grown originally on silicon substrates after chemical dissolution of the substrates. Their PL spectra exhibited a broad band at 1.95 eV with halfwidth of $\approx$ 700 meV and above whose excitation behavior and temperature dependence were similar to those found for donor–acceptor recombination in amorphous $Si_{0.3}C_{0.7}$: H. This comparison seemed to imply tacitly that the 1.95 eV band was considered to be due to DAP recombination. A band at 1.9 eV was also observed by Deneuville, Gonon, Gheeraert, Collins and Khong (1993) in diamond layers grown by microwave plasma-assisted CVD on roughened silicon substrates. It increased in intensity relative to the 2.8 eV band which was also observed in the samples as the deposition temperature decreased and appeared in almost all samples grown from gases not containing oxygen. In this study there was no indication of the 2.156 eV vibronic spectrum, therefore the authors concluded that either their diamond films were highly strained following the reasoning of Robins et al. (1989), or that the luminescence was not associated with the vibronic band.

A broad red band, with maximum emission intensity around 1.84 eV, also appeared in high-pressure synthetic diamonds studied by Freitas, Strom and Collins (1993). The red band was superimposed on a weaker green band and shifted to higher energies by about 70 meV at very strong excitation. Under these conditions, the green band became dominant. The red band was discussed in terms of DAP recombination, where boron was suggested as the acceptor ($E_a = 370$ meV) and a nitrogen complex center or an unknown impurity as the donor with an ionization energy of $E_d = 3.59$ eV. This interpretation was substantiated in subsequent work (Freitas et al., 1994b) by steady-state and time-resolved PL studies with UV or visible light excitation. In that work, a synthetic type IIb boron-doped diamond showed the 1.84 eV band dominantly. Again, the band shifted to higher energies by ≈ 45 meV when the excitation power was increased by approximately three orders of magnitude. The luminescence decay times were significantly faster for the highest photon energies in the band than for lower energies. At 650 nm emission wavelength the band showed an initial fast non-exponential decay followed by a much slower power law decay proportional to $t^{-0.7}$. All these data conspired to suggest a DAP recombination model for the red band. Freitas et al. (1994b) identified the acceptor with boron ($E_a = 370$ meV) from which a donor ionization energy $E_d = 3.63$ eV followed consistent with a nitrogen complex. However, it was admitted this was not a clear identification of the donor.

A subsequent investigation (Klein et al., 1995) already discussed in Section 4.1 extended this work. Boron-doped high-pressure synthetic diamonds exhibiting the red 1.84 eV luminescence band together with the green 2.3 eV band were again studied by steady-state and time-resolved photoluminescence and cathodoluminescence measurements. The data demonstrated that both bands originated from DAP recombinations and competed for the same acceptor boron whereas the donors were different, with an ionization energy of $E_d \approx 3.6$ eV for the red band and $E_d \approx 3.2$–3.3 eV for the green band.

The same group (Freitas, Klein and Collins, 1994a) observed in CVD flame-grown diamond films which were nominally undoped, nitrogen- or boron-doped, and nitrogen plus boron co-doped, a PL band at 1.95 eV. The band although spectrally close to the afore discussed 1.84 eV band was definitely of different origin as it did not exhibit any evidence for DAP characteristics in steady-state and pulsed-excitation measurements. The 1.95 eV band appeared to be nitrogen-related but was clearly distinct from the 2.156 eV N–V center and its vibronic band.

Cao, Driessen, Bauhuis, Giling and Alkemade (1995) studied by PL and CL homoepitaxial CVD diamond films co-doped with phosphorus and nitrogen, which were grown on natural diamond substrates. They observed a PL band more than 300 meV broad which was composed of two overlapping components with maxima at ≈1.5 and ≈1.7 eV. These two bands were also found in diamond films grown by the flame technique (which also emitted the characteristic 1.95 and 2.15 eV nitrogen-related defect lines) and in a (100) substrate of type IIa diamond (known to contain nitrogen as well). Hence, the bands seemed to be related to the incorporation of nitrogen. The authors had indications from selective excitation PL experiments that the 1.7 eV band is a DAP recombination process. Finally, te Nijenhuis, Olsthoorn, van Enckevort and Giling (1997) reported a broad CL band at 1.9 eV with halfwidth up to 400 meV in phosphorus-doped homoepitaxial and polycrystalline diamond films deposited by hot-filament CVD. The band could only be observed for above-gap excitation and did not show up in PL measurements with green Ar ion laser excitation. From this observation it was argued that the band is not of vibronic origin since in that case absorption of the 2.54 eV laser light by the Anti-Stokes vibronic component would have excited the band. The authors tentatively related the band to electron–hole recombination related with phosphorus or a phosphorus–vacancy complex.

4.3. Ultraviolet Bands (4.6, 4.0, and 3.75–3.5 eV)

There are several reports on a broad band centered around 4.6 eV observed in cathodoluminescence studies on p-type semiconducting diamond specimens (Freitas, Klein and Collins, 1994c; Lawson et al., 1994; Prins, 1994a). This band only appeared when boron acceptors were present. Lawson et al. (1994) studied this band and related lower-energy bands at 3.5 and 2.3 eV (Fig. 23) in detail. The samples investigated were boron-doped high-pressure synthetic diamonds grown by the temperature-gradient method using either nickel or cobalt solvent catalysts, and microwave-plasma CVD films homoepitaxially deposited on type Ib synthetic diamond substrates which were also doped with boron. In both types of specimens the 4.6 eV band was only observed when boron-doping had been performed, hence directly establishing the incorporation of boron in the luminescent states. However, this band was more common for CVD samples than for those produced by the high-pressure technique since in the latter case the band could only be seen when nitrogen was gettered, otherwise it was quenched. This behavior was different than that for the 2.3 eV band: As discussed earlier, the 2.3 eV is related to boron but it can emerge in nitrogen-containing samples showing little sensitivity to the concentration of atomically dispersed nitrogen. Lawson et al. observed in boron-doped high-pressure synthetic diamonds also a band at 3.12 eV which could be even dominating the spectrum depending on the sample regions investigated. Spatially resolved CL images taken from a cross section of nitrogen-gettered high-pressure diamond demonstrated that the 4.6 eV band and the 2.3 eV band are emitted from complementary regions of the crystal (Fig. 24). The 4.6 eV band was mainly observed from boron-rich

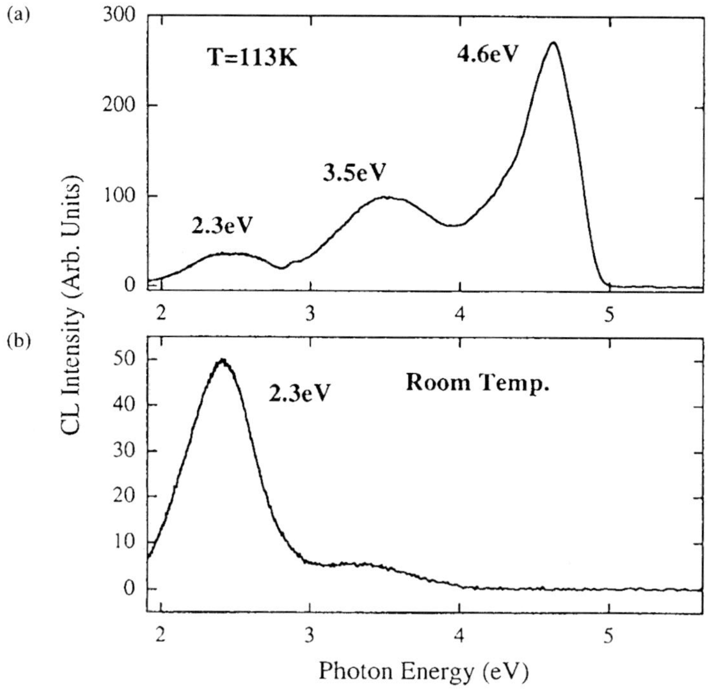

FIG. 23. Cathodoluminescence spectra from a boron-doped CVD diamond film (a) at 113 K and (b) at room temperature. The spectrum in (a) is very similar to that from a nitrogen-gettered HPHT boron-doped diamond. At room temperature the 4.6 eV band has completely disappeared (from Lawson et al., 1995).

regions in which the 2.3 eV band was not present whereas the 3.12 eV band appeared confined to regions that contain negligible amounts of uncompensated boron. The 3.5 eV band was correlated in intensity with the 4.6 eV band and had a similar temperature dependence which in detail was not investigated due to its overlap with the neighboring bands at 2.3 and 4.6 eV hampering such studies. The 4.6 eV band had the maximum intensity at ≈ 130 K and was completely quenched for temperatures above ≈ 200 K. The dominant cathodoluminescence feature at room temperature was the 2.3 eV band. The temperature dependence of the 4.6 eV was measured over the range 77–300 K and its characteristic behavior was explained in terms of a competition between exciton capture and release at unidentified traps with an exciton binding energy of (104 ± 5) meV.

The 4.6 eV band was also observed by Prins (1994a, 1998a,b) in his implantation studies of natural type IIa diamonds employing the cold implantation rapid annealing (CIRA) technique. Before implantation the diamonds showed blue band A 2.9 eV luminescence. After boron implantation and annealing up to 1600 °C, either in a single implantation step or with repeated

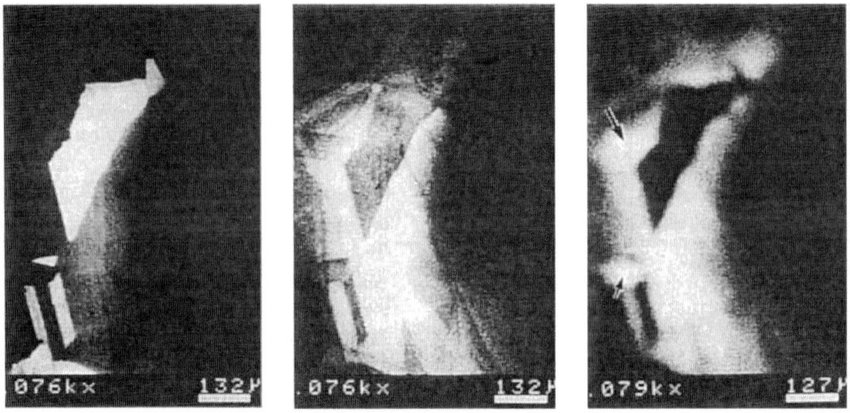

(a) (b) (c)

FIG. 24. Spatially resolved cathodoluminescence (CL) images taken from a cross section of a nitrogen-gettered HPHT diamond. Image (a) shows the CL at 2.3 eV, image (b) that at 3.12 eV, and image (c) that at 4.6 eV. Complementary luminescence regions are easily discernable. The arrows in (c) illustrate minor differences between the CL distribution of the 3.12 and 4.6 eV bands (from Lawson et al., 1995).

boron implantation CIRA steps, the 4.6 eV band emerged. It was associated with a second band centered at 3.5 eV similar to the findings of Lawson et al. (1995). Also, after boron implantation and annealing, the green band appeared (specified by Prins to be centered at 2.4 eV). When type IIa diamonds were implanted by carbon ions (below the amorphization threshold as was the case in the aforementioned CIRA experiments), the blue 2.9 eV luminescence was found to increase and a UV band at 4 eV developed after annealing at 1200 °C which disappeared for annealing at 1600 °C (Prins, 1996) (Fig. 25). This band could not only be observed in cathodoluminescence but also in electroluminescence (Prins, 1994b, 1998a). From his observations Prins suggested that the 4 eV band may originate from electron capture at 'vacloids', a random arrangement of vacancies that produce a donor state $\approx$4 eV below the conduction band. On the other hand, the blue 2.9 eV band may be caused by hole capture transitions from the valence band to deep-lying acceptor states $\approx$2.9 eV above the valence band made up by vacancy clusters in a more periodic arrangement labeled 'vacancy crystallites'. Briefly discussed was the possible relationship of such vacancy clusters with the established finding of the 2.9 eV band as being emitted from straight dislocation states. The idea that some of the vacancies in a cluster collapse to form dislocation loops was discarded since such loops would be too small to be compatible with the required straightness of the dislocations. However, in an alternative model Prins suggested that vacancies within the core of dislocations could form 'decorated' dislocations, and that luminescence from such dislocations would be most intense if the vacancies were periodically arranged in 'vacancy crystallites', and if the dislocation structure would be

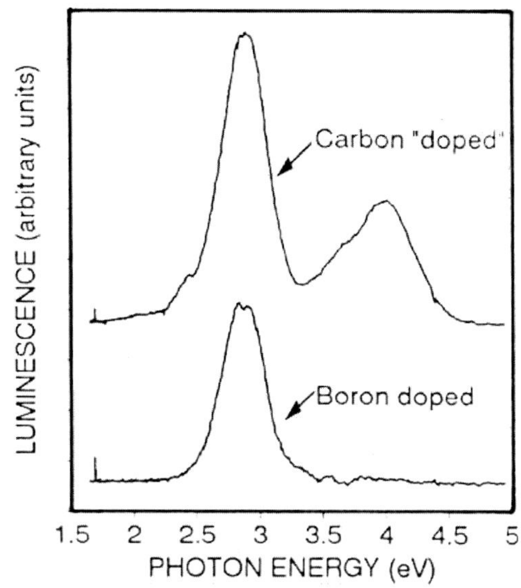

FIG. 25. Comparison of the cathodoluminescence spectra from CIRA carbon and boron implanted type IIa diamonds after two implantation–annealing steps. Both diamonds contain the same amount and density of deep lying centers N_d but the boron implanted diamond also contains boron acceptors N_a with $N_a > N_d$ (from Prins, 1994b).

straight without kinks and bends. Finally, Prins suggested that the four bands he studied are pairwise correlated, the 2.9 eV band with the 3.5 eV band, and the 4.0 eV band with the 4.6 eV band; their identical spectral spacings of 0.6 eV should then be due to the Coulomb energy of negatively charged boron acceptors effective in the 3.5 eV and in the 4.6 eV bands.

The 4 eV band as reported by Prins (1994a,b, 1996, 1998a,b) may have been observed earlier by Collins et al. (1989a). In the latter cathodoluminescence study on microwave CVD diamond films a very broad (halfwidth $\approx$ 0.7 eV) intense band centered at $\approx$3.9 eV was observed. However, the authors described it as a vibronic spectrum associated with the 5RL defect spectrum whose no-phonon line at 4.582 eV and sharp local mode replicas were superimposed on the broad band with low intensities.

Other UV bands were mentioned earlier in conjunction with oxygen. Mori, Eimori, Kozuka, Yokota, Moon, Ma and Hiraki (1992) observed a strong CL band centered at 3.75 eV (halfwidth $\approx$ 1 eV) accompanied by an overlapping much weaker and apparently narrower band with maximum at 4.64 eV. The bands appeared after oxygen was diffused into CVD diamond layers, grown either on high-pressure synthetic (111) diamonds or on p^+Si (111) wafers, by boiling the samples in a saturated solution of CrO_3 in H_2SO_4 at 200 °C, rinsing in H_2O_2 and NH_4OH, and then exposing the specimens to a hydrogen microwave plasma. Rutherford backscattering showed that an oxygen concentration of $\approx 10^{20}\,cm^{-3}$ was homogeneously distributed in the films to a depth of 300 nm or more whereas S or Si remained at the surface. Damage could not be detected by Raman spectroscopy and the hydrogen plasma did not affect the optical spectra. Hence, the two CL peaks at 3.75 and 4.64 eV excited by the electron beam to a depth comparable with the oxygen diffusion depth, were associated with the presence of oxygen.

5. Acceptor- and Donor-related Luminescence: Bound Excitons and Line-resolved DAP Recombination

A basic investigation on the radiative recombination of excitons bound to natural acceptors is the paper by Dean et al. (1965). An earlier paper by Dean and Jones (1964) only briefly reported on two close lying sharp emission lines ascribed to the unionized acceptor center thought to be substitutional aluminum. Dean et al. also referred to aluminum doping in studying p-type semiconducting natural diamond specimens. It was shown later (Chrenko, 1971, 1973; Collins and Williams, 1971; Lightowlers and Collins, 1976) by various techniques such as Hall measurements, nuclear activation analysis, and capacitance–voltage characteristics that boron is the dominant acceptor in semiconducting diamond. In the following we will first briefly review BE luminescence related to (neutral) boron acceptors as observed in natural or high-pressure synthetic diamonds. We will then discuss such spectra appearing in boron-doped CVD grown diamond particles and films. Finally, we will focus on bound exciton and line-resolved DAP luminescence spectra related with the donor phosphorus which are observed in CVD diamond films doped with phosphorus in the growth. Excitons localized at shallow impurities will be generally labeled BE or, with respect to neutral donors D^0 and acceptors A^0, D^0X, and A^0X, respectively.

5.1. Boron Bound Exciton Spectra in Natural and HPHT Diamond

A cathodoluminescence spectrum at 95 K sample temperature dominated by boron bound exciton emission is displayed in Figure 26. Lines belonging to this spectrum are labeled D, and appear as no-phonon (NP) transitions and transitions assisted by 'momentum' or wavevector conserving phonons (Table III) corresponding to $\vec{k}_0 = 0.76\pi/a[100]$, the location of the conduction band minimum in k-space (Dean et al., 1965). NP transitions become allowed when the presence of the acceptors perturbs the translational invariance of the lattice and relaxes the k-conservation rule. More quantitatively, the spreading of the BE wavefunction in k-space, including components at $k = 0$ necessary to yield NP transitions, is given by the Fourier transform of the BE wavefunction in real space whose extension depends on the exciton localization energy. The more strongly the excitons are localized the more are the NP transitions expected to dominate over the phonon-assisted transitions. An excellent experimental example for this trend represents silicon with BE spectra originating from the acceptors B, Al, Ga,

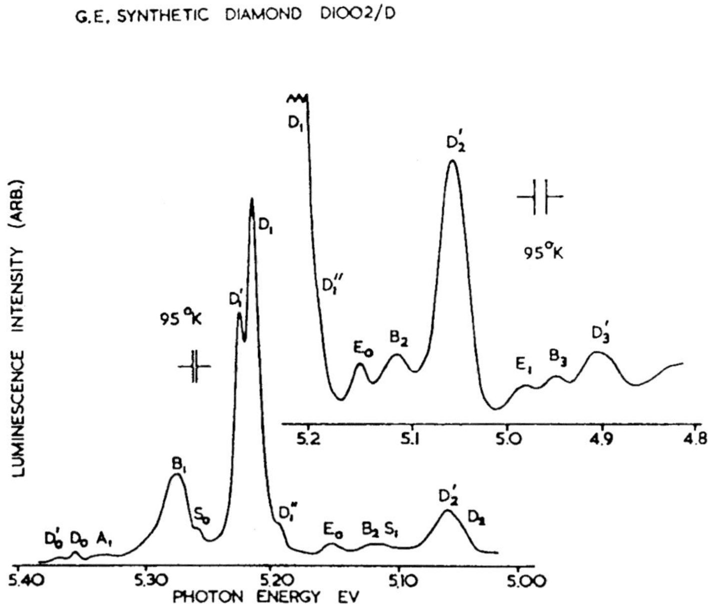

FIG. 26. Edge recombination radiation spectrum from a relatively strain-free natural p-type semiconducting diamond at 95 K. The line labeling is arbitrary and translates to the notation used here as (D'_O, D_O) : boron bound exciton, BE^{NP} doublet, (D'_1, D_1) : BE^{TO} doublet, D''_1 : BE^{LO}, (D'_2, D_2) : $BE^{TO+O^{\Gamma}}$ doublet; A_1: free exciton, FE^{TA}, B_1: FE^{TO}. Resolution, 3 meV (from Dean et al., 1965).

and In, and from the donors P, As, Sb, and Bi highlighting this relationship (Dean, Haynes and Flood, 1967).

The BE transitions are split into doublets with a spacing of ≈ 12 meV of the two components. The doublet thermalizes demonstrating that the splitting is in the upper exciton states of the transitions. Dean et al. (1965) tentatively ascribed the splitting to

TABLE III

BORON BOUND EXCITON TRANSITION ENERGIES $h\nu$ AT 100 K

Line label[a]	$h\nu$ (eV)[a]	Assisting phonon	Phonon energy (meV)
D'_0	5.368	NP	–
D_0	5.356		
D'_1	5.227	TO	141
D_1	5.215		
–	–	LO	163
D''_1	5.193		
D'_2	5.060	TO + O^{Γ}	167 (O^{Γ})
D_2	5.048		

[a]Notation and energy values after Dean et al. (1965).

the spin–orbit interaction of the two holes in the BE state: To yield a large pairing energy (causing a large exciton binding to the acceptor) both holes should come from the same band, either the Γ_8 or the Γ_7 valence band separated by the spin–orbit interaction energy Δ_0, and therefore, the splitting should be roughly twice the value of Δ_0 which was experimentally determined to be 6 meV (Rauch, 1963).

The exciton localization energy E_{loc} at the boron acceptors is spectroscopically given by the spacing of the lower-energy BE doublet-line and the low-energy threshold of the FE transition, equal to the excitonic gap E_{gx}. For the TO-phonon transition $E_{gx} = 5.266$ eV so that $E_{loc}(B) = 51$ meV using 5.215 eV for $BE^{TO}(D_1)$ from Table III.

Recent low-temperature, high-resolution cathodoluminescence studies on high-quality boron-doped HPHT crystals revealed more fine structure in the BE spectrum (Sharp, Collins, Davies and Joyce, 1997; Sternschulte, Wahl, Thonke, Sauer, Ruf, Cardona and Anthony, 1997b). Each of the two doublet lines discussed above is split into four components in an identical manner (Fig. 27). The spectral positions of these line components for the NP transition are listed in Table IV. From temperature-controlled measurements the level scheme in Figure 28 was derived (Sternschulte et al., 1997b) showing (i) the acceptor ground state, A^0, is

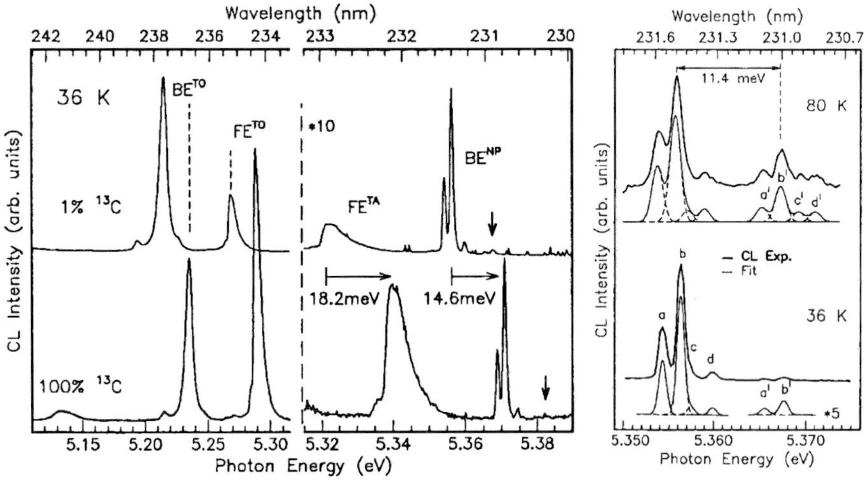

FIG. 27. *Left*: Cathodoluminescence spectra at 36 K sample temperature of two HPHT diamonds with 1 and 100% ^{13}C isotope contents. Upper and lower spectra exhibit an isotopic shift of 14.6 meV for the bound exciton (BE^{NP}) line and 18.2 meV for the free exciton (FE^{TA}) line not discussed in the text (like other isotopic shifts seen in the spectra. Isotope shifts are specifically reported and discussed for the energy gap of ^{13}Diamond by Collins et al. (1990c), for the FE by Sharp and Collins (1996) and for the boron BE by Sternschulte et al. (1997b)). Arrows mark the bound exciton component D_1' of Figure 26 which has disappeared in these spectra due to thermalization, and the very narrow doublet 11.4 meV lower in energy is the component D_1 of Figure 26. *Right*: Complete bound exciton recombination spectra in high resolution with eight fine splitting components a, b, c, d (corresponding to D_1 in Figure 26) and a′, b′, c′, d′ (corresponding to D_1' in Figure 26) (from Sternschulte et al., 1997b).

TABLE IV

FINE SPLITTING OF BORON BOUND EXCITON NO-PHONON TRANSITION

Line component		$h\nu$ (eV)	Line component		$h\nu$ (eV)
D_0	a	5.3538	D'_0	a′	5.3652
	b	5.3558		b′	5.3672
	c	5.3572		c′	5.3692
	d	5.3591		d′	5.3711

split by 2.0 meV, (ii) there are two main exciton states, A^0X and $(A^0X)'$ spaced by 11.4 meV (this value corresponds to the older value of 12 meV of Dean et al. (1965), (iii) both A^0X states are identically split by 3.3 meV. The 2.0 meV A^0 ground state splitting is in perfect agreement with the value obtained from other investigations on boron-doped diamond such as infrared acceptor excitation spectroscopy or Raman scattering (see references in Sternschulte et al., 1997b). The 11.4 meV splitting could be interpreted as the spin–orbit interaction energy of a (bound) hole if a spin–orbit splitting of 13 meV is adopted for free holes as obtained theoretically by Willatzen, Cardona and Christensen (1994). The latter value is often considered more reliable than the previous experimental value of 6 meV (Rauch, 1963). The extra splitting of 3.3 meV in the A^0X exciton states remained unidentified. It could be due to various effects such as jj-coupling of the holes and the electrons in the exciton, to valley–orbit interactions of the bound electron owing to the many-valley nature of the diamond band structure, and to crystal field effects.

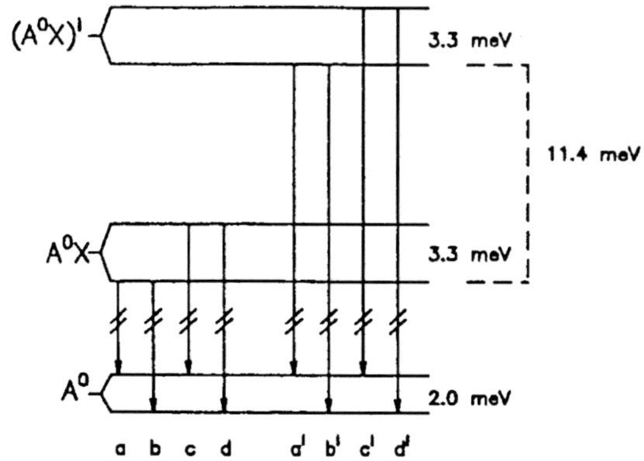

FIG. 28. Recombination level scheme of the boron bound exciton. The transition initial exciton states are denoted A°X and A°X′ (exciton bound to neutral acceptor), the transition final states of the neutral boron acceptor are labeled A° (from Sternschulte et al., 1997b).

5.2. Boron-bound Exciton Luminescence in CVD Diamond

Polycrystalline CVD diamond layers with grains sufficiently large and free from defects also showed bound exciton cathodoluminescence spectra upon boron doping (Kawarada et al., 1990a; Kawarada, Yokota, Mori, Nishimura and Hiraki, 1990b; Kawarada, Yokota, Matsuyama, Sogi and Hiraki, 1991; Kawarada, Matsuyama, Yokota, Sogi, Yamaguchi and Hiraki, 1993). The layers were grown on n-type (100) Si wafers at temperatures between 850 and 900 °C by microwave plasma CVD using CO (5%)/H_2 and CO (15%)/H_2 gas mixtures. Boron was added as B_2H_6 gas to the H_2 carrier gas at boron/carbon flux ratios of 50–1000 ppm. The layers were 4 μm thick with grain sizes of 2 μm. The edge recombination spectra showed FE and BE transitions whose relative intensities depended on the boron/carbon flux ratios. Figure 29 displays a series of cathodoluminescence spectra with boron concentrations in the range of 1.5×10^{17}–3×10^{18} cm^{-3}. An undoped layer shows only FE recombination, but at increased doping levels the BE lines appear and become successively stronger. The spectra in Figure 29 were recorded using an instrumental resolution

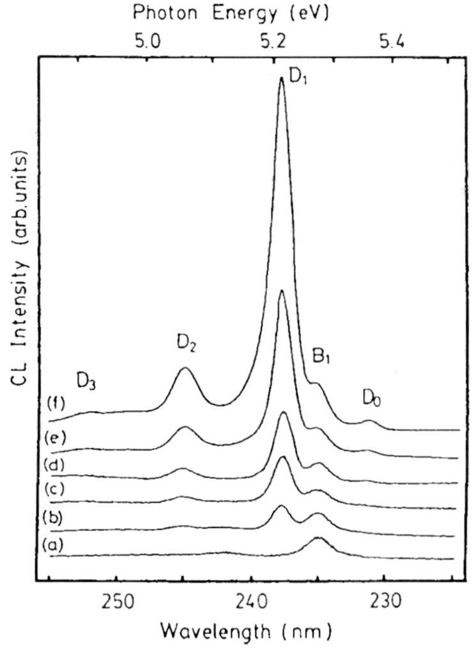

Fig. 29. Near-band-edge cathodoluminescence spectra from boron-doped CVD polycrystalline diamond films at 80 K sample temperature. Gas mixture CO (5%)/H_2, film thicknesses 4 μm. Boron concentrations (a) undoped, (b) 1.5×10^{17} cm^{-3}, (c) 3×10^{17} cm^{-3}, (d) 7.5×10^{17} cm^{-3}, (e) 1.5×10^{18} cm^{-3}, and (f) 3×10^{18} cm^{-3}. Resolution, 0.9 nm (19.6 meV). Line labels: B_1: free exciton emission, FE^{TO}; D_0: boron bound exciton emission, BE^{NP}; D_1: BE^{TO}; D_2: $BE^{TO+O^{\Gamma}}$; D_3: $BE^{TO+2O^{\Gamma}}$ (from Kawarada et al., 1993).

of ≈19 meV. At an improved resolution of ≈9.5 meV, BE linewidths (FWHM) of ≈24 meV were measured whereas the FE line widths amounted to ≈29 meV (at 80 K). These values are much larger than in HPHT synthetic diamonds where BE linewidths of approximately 1 meV were obtained (cf. Fig. 27). The large linewidths in the CVD layers probably indicate residual strain in the grains of the polycrystalline diamond layers. However, BE spectra measured in boron-doped diamond films which were grown by CVD homoepitaxially on high-pressure synthetic diamond substrates did not show narrower linewidths (Kawarada and Yamaguchi, 1993).

A quantitative evaluation of exciton spectra as in Figure 29 yielded an initial linear increase of the BE/FE intensity ratios up to some 10^{18} cm^{-3} boron concentration followed by saturation at higher doping levels exceeding $\approx 5 \times 10^{18}$ cm^{-3} (Fig. 30). The linear part of the diagram can be easily understood provided there are no efficient recombination channels other than free and bound exciton recombination. Then, the ratio of BE-to-FE densities follows a Boltzmann law $\exp(-E_{loc}/kT)$ with a pre-factor containing the degeneracies of BE and FE states given by the boron doping level and the effective exciton density of states. Rate equations can be used to model the BE and FE densities and their luminescence intensities as a function of excitation including parameters such as capture and release probabilities. Such studies have been performed experimentally and in modeling minutely for donor and acceptor doped silicon by Tajima (1978) showing they can serve to determine donor or acceptor concentrations quantitatively. Kawarada et al. (1993) suggested that their BE/FE intensity ratio

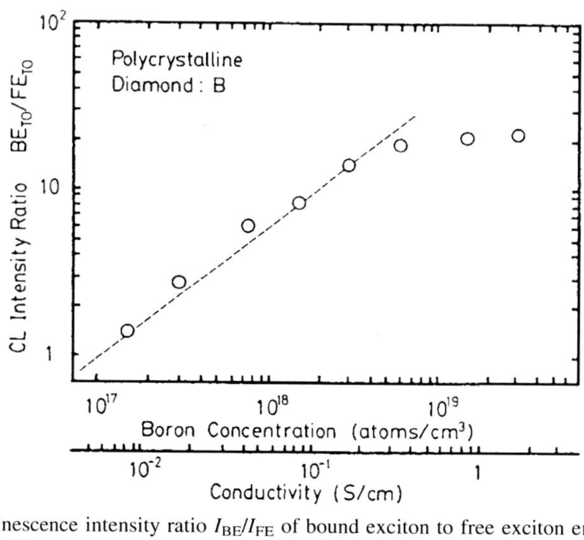

FIG. 30. Luminescence intensity ratio I_{BE}/I_{FE} of bound exciton to free exciton emission in boron doped polycrystalline diamond films vs. boron concentration and sample conductivity. For the intensity ratio and the original spectra data in Figure 29, compare also the equivalent data of Ruan et al. (1992a) in Figure 19 (from Kawarada et al., 1993).

curves (Fig. 30) can be used in a similar way to estimate boron concentrations. In the 'saturation' regime, for boron concentrations above $\approx 10^{19}$ cm^{-3}, an additional peak emerged at ≈ 5.15 eV accompanied by a second peak at ≈ 4.99 eV ascribed to a phonon replica of the first peak. Kawarada et al. (1993) argued that in this regime the nature of the recombination process changed and ascribed these lines to the recombination of an excess electron with a hole on the acceptor (e, A^0 process). Energetically, this process is consistent with the observed photon energy, $h\nu = E_g - E_a = (5.49 - 0.37)$ eV $= 5.12$ eV; however, the interpretation was not substantiated by further data.

The potential of the microwave and hot-filament CVD techniques to vary the doping level relatively delete easily over a large range of boron concentrations has been utilized to study the bound exciton spectrum from low concentrations at 10^{14} cm^{-3} to very high concentrations at around 10^{21} cm^{-3} (Fig. 31, Sternschulte et al., 1996a). For total boron concentrations up to [B] $\leq 10^{20}$ cm^{-3} as determined

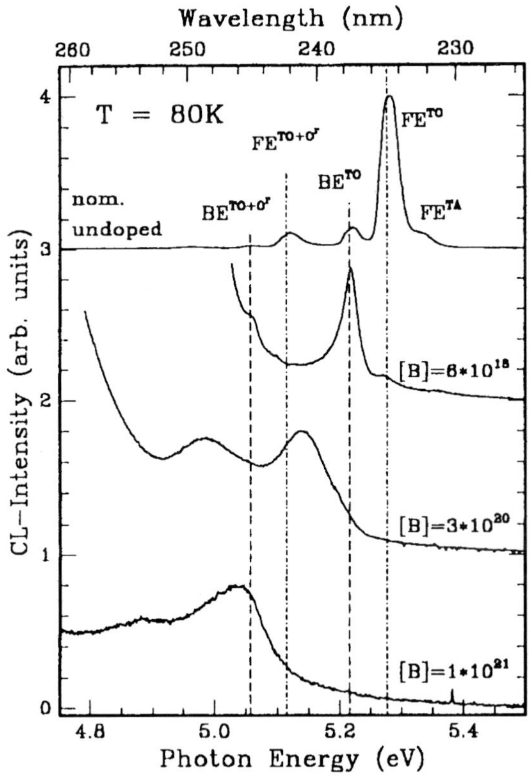

FIG. 31. Near-band-edge cathodoluminescence spectra of four boron-doped CVD diamond films: Shift of the bound exciton lines (BE^{TO} and $BE^{TO+O^{\Gamma}}$) with increasing boron concentration. The rise of the intensity on the low energy side is towards the broad emission band due to boron peaking at 4.6 eV (cf. Figure 23) (from Sternschulte et al., 1996a).

by SIMS the BE spectrum (BE^{TO} line) appeared essentially at a constant energy position $h\nu \approx 5.22$ eV, apart from some scatter (Fig. 32). Concomitantly, the linewidths increased monotonously from ≈ 30 to ≈ 60 meV (Fig. 32). For $[B] > 10^{20}$ cm^{-3} a rapid shift to lower photon energies was observed, with absolute line positions down to $h\nu \approx 5.0$ eV while the linewidths further increased monotonously when the scatter was averaged out. Line shape modeling made sure that even in this case the observed luminescence was due to bound exciton recombination. The boron concentrations were rescaled to electrically active doping levels using Hall measurements. The dependence of the BE photon energies on the effective doping levels (Fig. 32) was then modeled based on theoretical calculations by Mahan (1980). This work deals with donor bound excitons in Si and Ge taking into account the electron–electron exchange energy, the hole–electron, and the donor–electron interactions as a function of the donor concentration. These energy terms were reformulated so as to apply to the actual situation with acceptor bound excitons in diamond, and were calculated choosing the parameters $\varepsilon = 5.7$, $m_h^* = 0.87m_0$, an acceptor Bohr-radius of $a_0^* = 3.7$ Å, and a boron ionization energy of $E_a = 370$ meV. Out of two leading terms the hole–hole exchange energy alone, $\delta(h\nu) \approx -25$ meV$(p_0/10^{18}\ \text{cm}^3)^{1/3}$, obtained with no adjustable parameters, was sufficient to describe the experimentally observed shift of the BE transition energy. For $p_0 = 9 \times 10^{20}$ cm^{-3} the impurity-Mott transition at $E_g - E_i - \hbar\Omega(\text{TO}) \approx 4.98$ eV was reached defined as $p_0^{-1/3} = 4a_0^*$ (Mott, 1956). For higher densities, the acceptor potential is completely screened, the high-doped sample becomes metallic, and exciton luminescence will no longer be observable. Experimentally, the lowest-energy BE luminescence is visible at $h\nu \approx 5.03$ eV rather close to the energy of the Mott transition. It must be stated that all values involving the Bohr radius a_0^* are to be considered with caution: a_0^* is derived from the effective-mass-theory using the experimental ionization energy $E_a = 370$ meV, and the value obtained is close to the diamond lattice constant $a_0 = 3.57$ Å in contradiction to the conditions of the theory requiring $a_0^* >> a_0$.

5.3. Phosphorus-related Excitonic Spectra

The only natural donor in diamond is nitrogen with an ionization energy of $E_d = 1.7$ eV (Farrer, 1969; Collins, 1990). Nitrogen A aggregates give rise to an even deeper donor level ≈ 4 eV below the conduction band (Collins and Lightowlers, 1979). Isolated substitutional nitrogen does not bind excitons giving rise to BE luminescence in the near band-edge region. Instead, nitrogen complexes with other defects, as e.g. vacancies, to emit numerous vibronic spectra at photon energies far below the band-edge (Davies and Manson, 1994, and references therein).

Electronic needs for efficient n-type doping of diamond have motivated intense efforts in the past years to find atomic species that can be technologically handled and incorporated into diamond as a shallow donor. From the analogy with silicon,

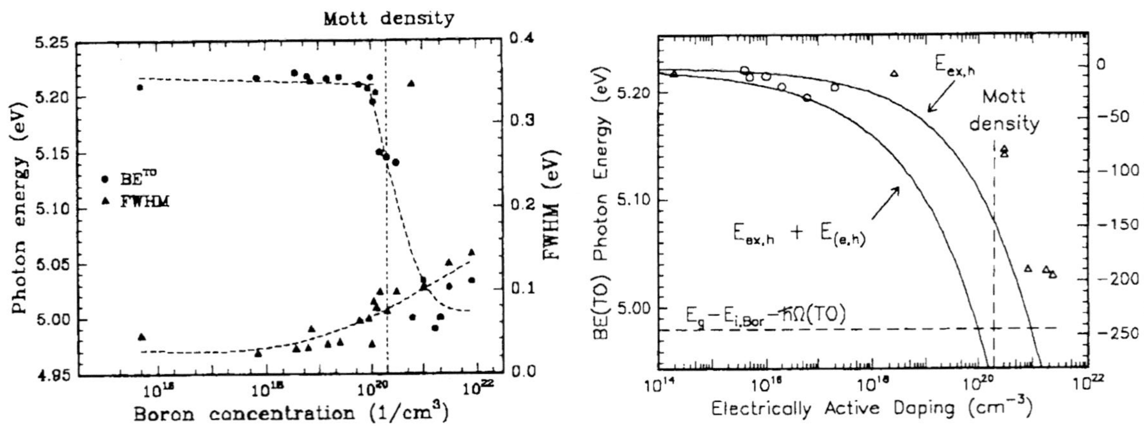

FIG. 32. *Left*: Photon energies of the boron bound exciton (BE^{TO}) transition and related linewidths (FWHM) vs. total boron concentrations. The vertical dotted line indicates the Mott density following a theoretical estimate and Hall measurements. Dashed lines are eye guides. *Right*: Photon energies of the boron bound exciton (BE^{TO}) transition vs. electrically active hole concentrations. Full lines correspond to two models for the bound exciton energy shift including either the hole-exchange energy E_{exch} in the bound exciton complex alone (upper curve), or E_{exch} plus the electron–hole interaction $E_{e,h}$ (lower curve). The horizontal broken line at $E_g - E_i - \hbar\Omega(TO) = 4.98$ eV energetically defines the model Mott transition at 10^{20}–10^{21} cm^{-3} boron concentration (from Sternschulte et al., 1996a).

group V and VI elements would be naively expected to be donors in diamond. First principles calculations suggest Li and Na to be interstitial donors with $E_d = 100$ and 300 meV, respectively, and P to be a substitutional donor with E_d between 200 and 1090 meV (Jackson, Pederson and Harrison, 1990; Kajihara, Antonelli and Bernholc, 1990; Kajihara, Antonelli, Bernholc and Car, 1991). Li is estimated to be a fast diffusing element and therefore unstable even at room temperature for longer storage times. In all cases, extremely low solid solubilities are theoretically predicted. Therefore, the probability is low that these elements are incorporated in the lattice under quasi-equilibrium conditions such as diffusion or during growth. More recent first principles calculations predict that P introduces a donor state with $\approx$400 meV below the conduction band (Saada, Adler and Kalish, 2000). Initial experimental data with S doping (Sakaguchi, Nishitani-Gamo, Kikuchi, Yasu, Haneda, Suzuki and Ando, 1999) motivated theoretical investigations on S as a donor in diamond. Widely scattered ionization energies were obtained, with values of 150 meV for neutral S and 500 meV for ionized S^+ (Saada et al., 2000), 375 meV (Zhou, Yokoi, Tamura, Takami, Kubo, Miyamato, Gamo and Ando, 2001), and 1.2 eV for a trigonally distorted substitutional S donor and 1.1 eV for a S donor distorted to orthorhombic symmetry (Miyazaki and Okushi, 2001). First principles calculations by Wang and Zunger (2002) suggested that phosphorus in comparison to sulfur gives rise to a shallower level, $E_d(P) = 370$ meV vs. $E_d(S) = 770$ meV, and has a higher bulk solid solubility.

Experimentally, attempts to n-dope diamond by implantation and annealing failed, although this technique was successfully employed in p-doping diamond with boron (Braunstein and Kalish, 1983; Prins, 1988; Kalish, Uzan-Saguy, Samoiloff, Locher and Koidl, 1994); even boron bound exciton luminescence was obtained in boron implanted and annealed synthetic type IIa diamond samples (Sternschulte et al., 1997a). The technique is basically problematic since diamond is not the stable phase of carbon. Damage-induced states and graphitic bonds due to the implantation can give rise to electrical conductivity. The conductivity may persist at high temperatures owing to incomplete annealing-out of the damage but does not represent the donor-to-band activated conductivity with as high mobilities as possible. A survey of ion implantation in diamond is given by Prins (1992). Li ion implantation of natural type IIa diamonds (Prawer, Uzan-Saguy, Braunstein and Kalish, 1993) and synthetic diamonds (Job, Werner, Denisenko, Zaitsev and Fahrner, 1996) followed by annealing led to n-type conductivity with largely varying thermal activation energies in the range from $\approx$80 to $\approx$270 meV. Control implantations with $^{12}C^+$ ions in the former case showed no remaining conductivity after annealing. Hence, it was concluded that the conductivity was rather due to the presence of Li than to implantation-induced damage. However, the observed temperature dependence was typical for variable-range hopping and definitely different from conductivity in the conduction band exponentially activated from donor states. Other attempts to n-dope CVD diamond films with Li and Na by different techniques similarly did not lead to fully conclusive results (Prawer et al.,

1993; Borst and Weis, 1995; Popovici and Prelas, 1995, Popovici, Melnikov, Varichenko, Khasawinah, Sung, Prelas, Denisenko, Penina, Martinovich, Drozdova, Zaitsev, Fahrner, Farmer, White and Chamberlain, 1996; Sternschulte et al., 2000). A fuller account of the activites on the search for donors in diamond was recently given by Kalish (2001).

Phosphorus doping experiments include (i) synthesis of hot-filament CVD films on Si substrates (Okano, Kiyota, Iwasaki, Nakamura, Akiba, Kurosu, Iida and Nakamura, 1990; Bohr, Haubner and Lux, 1995), (ii) growth of homoepitaxial films by microwave-assisted CVD (Koizumi, Kamo, Sato, Ozaki and Inuzuka, 1997; Kalish, Reznik, Uzan-Saguy, Koizumi and Kamo, 1998; Koizumi, Kamo, Sato, Mita, Sawabe, Reznik, Uzan-Saguy and Kalish, 1998) with P_2O_5 or PH_3 as the doping source, (iii) growth of HPHT-diamonds using P as a catalyst (Akaishi, Kanda and Yamaoka, 1993), and (iii) P ion implantation of CVD films (Schreck, 1998) and of natural IIa-type diamonds (Sandhu, Kao, Swanson and Chu, 1990; Kalish et al., 1998). Significant progress was finally made with P-doped diamond films grown by microwave-assisted CVD on synthetic high-pressure diamonds (Koizumi et al., 1997, 1998; Koizumi, 1999). Resistivities $\rho \approx 50\ \Omega$ cm and electron concentrations $n \approx 10^{16}\ \mathrm{cm}^{-3}$ at temperatures in the region of 500 °C were obtained. As a function of temperature, ρ and n followed Arrhenius laws with activation energies between 430 and 500 meV; these values were directly interpreted as P donor ionization energies E_d implying compensation by non-doping impurities effective over the whole range of temperatures because for non-compensated doping half of the donor ionization energy, $E_d/2$, would have been obtained. Recent improvements of this P doping technique (Koizumi, Teraji and Kanda, 2000) have raised the initial low mobilities of order 20–30 $\mathrm{cm}^2\ (\mathrm{V\ s})^{-1}$ to values above 200 $\mathrm{cm}^2\ (\mathrm{V\ s})^{-1}$ at room temperature. Also, electron concentrations of order $10^{16}\ \mathrm{cm}^{-3}$ at 500 K (227 °C) were achieved. Activation energies of 600 meV were found for the temperature dependence of the electron concentration independent of the P doping level exceeding significantly the activation energies measured earlier. All these data were interpreted in the way that P is incorporated in the CVD growth as a donor with ionization energy E_d at around 600 meV.

Confirmatory evidence for P being a donor in diamond came from three kinds of optical experiments, photocurrent spectroscopies, infrared donor excitation spectroscopy, and luminescence experiments. Haenen, Meykens, Nesládek, Knuyt, Quaeyhaegens, Stals, Koizumi and Gheeraert (1999a), Haenen, Meykens, Nesládek, Stals, Teraji, Koizumi and Gheeraert (1999b), Nesládek, Meykens, Haenen, Stals, Teraji and Koizumi (1999), and Haenen, Meykens, Nesládek, Knuyt, Stals, Teraji, Koizumi and Gheeraert (2000) found in P-doped, n-conducting CVD films thresholds and dips in their photoconductivity measurements as a function of the incident photon energy which after evaluation of the data resulted in two defect optical photoionization energies with values of (0.56 ± 0.03) eV and (0.81 ± 0.03) eV. The first value was in very good agreement with the activation energy for the electron concentration as obtained from temperature dependent Hall measurements and associated with the donor ionization energy E_d. Infrared excitation spectroscopy on P-doped CVD diamond films yielded consistently up to three absorption lines in the range 500–600 meV which were interpreted as dipole transitions from the $1s(A_1)$-donor ground state to electronic excited $2p_0$, $2p_\pm$, and $3p_\pm$ donor states. The energy spectrum of the excited p states is expected to be well described by the effective-mass-approximation yielding accurate energies relative to the conduction band. Indeed, an excellent experimental-to-theoretical agreement was found for these states after the transverse and longitudinal electron masses reported in the literature were slightly modified. Addition of the optical transition energies to these EMT-energy values yielded the donor ground-state energy E_d below the conduction band. While values close to 600 meV (Gheeraert, Koizumi, Teraji, Kanda and Nesládek, 1999) and slightly higher (610 meV) were first reported (Gheeraert, Koizumi, Teraji and Kanda, 2000a; Gheeraert, Koizumi, Teraji, Kanda and Nesládek, 2000b) the most recent value was $E_d = 604$ meV (Gheeraert, Casanova, Koizumi, Teraji and Kanda, 2001).

Luminescence spectroscopy has demonstrated in two different ways the role of P as a donor in polycrystalline and homoepitaxial diamond layers: P gives rise to a shallow bound exciton spectrum characteristic of shallow impurities and, in samples co-doped with boron, to line-resolved DAP spectra.

First indications for a P-related BE spectrum came from samples which were grown by the hot-filament CVD technique on boron-doped Si substrates and were doped by adding PH_3 to the CH_4/H_2 gas. These samples exhibited a new CL spectrum with two lines close to 5.16 and 4.99 eV which were attributed to TO- and (TO + O^Γ)-phonon assisted recombination transitions from an isolated phosphorus impurity or from a phosphorus-related complex (Sternschulte et al., 1996b; Sternschulte, Thonke, Sauer and Koizumi, 1999b). Generally, the large photon energy close to the band-edge suggested a bound exciton recombination process, and in this case the exciton localization energy would be $E_{loc} = 106$ meV, equal to the shift of the 5.16 eV TO-phonon line from the excitonic gap, $E_{gx}^{TO} =$ 5.266 eV. This value was larger than that for the boron BE, $E_{loc}(B) = 51$ meV, suggesting an impurity binding potential of phosphorus larger than 370 meV as for boron. However, theory at that time had predicted a shallower phosphorus donor level (Kajihara et al., 1990), and electrical measurements on phosphorus-implanted type IIa diamonds had shown n-type conductivity with only ≈ 200 meV activation energy (Prins, 1995). Compared to these data, the identification of the optical center with isolated phosphorus was not straightforward favoring a phosphorus-related complex.

Following the early attempts, phosphorus doping of homoepitaxially grown CVD diamond films showed n-type conductivity in Hall effect measurements with significantly larger thermal activation energies in the range of 500–600 meV (Koizumi, 1999; Koizumi et al., 2000). Such samples were also studied in cathodoluminescence spectroscopy (Sternschulte, Thonke and Sauer, 1999a; Sternschulte et al., 1999b) and yielded a slightly structured bound exciton

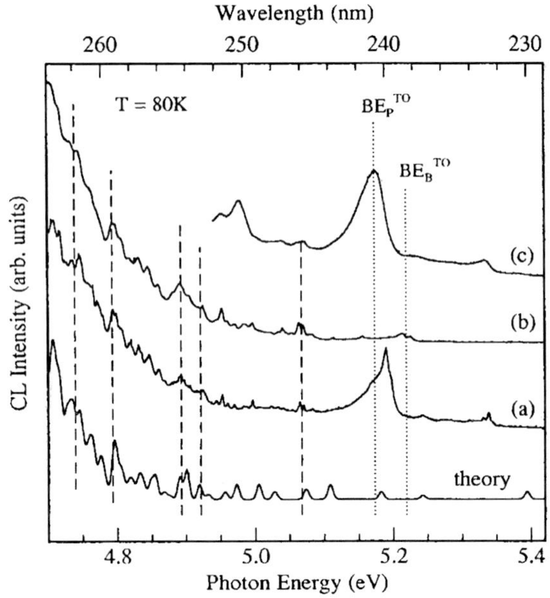

FIG. 33. Near-band-edge cathodoluminescence spectra of phosphorus-doped diamonds: (a) microwave-assisted CVD film, (b) HPHT synthetic crystal, and (c) hot-filament CVD film. Peaks close to 5.19 and 5.34 eV are phosphorus bound exciton (BE_P^{TO} and BE_P^{NP}) recombination lines. The multiple-line spectra at lower energies are resolved DAP spectra modeled by a theoretical spectrum assuming $\varepsilon_0 = 5.7$, diamond lattice constant $a_0 = 3.568$ Å and boron as the participating acceptor with an ionization energy of $E_a = 370$ meV. In the theoretical spectrum, the doublet lines at $= 4.9$ eV correspond to shell numbers $m = 14$ and 15 of the DAP and the dominant line at 4.8 eV corresponds to $m = 26$ or $m = 13$ (type I) (from Sternschulte et al., 1999b).

spectrum with peaks at ≈ 5.19 eV (BE^{TO}) and ≈ 5.34 eV (BE^{NP}) reasonably close to the lines in the hot filament specimens (Fig. 33). In addition, these samples exhibited complex narrow-line spectra in the wide spectral range from ≈ 5.07 to ≈ 4.6 eV which were identified as line-resolved DAP spectra.

DA pair spectra originate from the radiative recombination of an electron at a donor with a hole at an acceptor leaving both impurities ionized after the recombination. Typical discrete-line DAP spectra were first observed in GaP with various donors (S, Te, Se) on P sites and group IV acceptors (Si) and group II acceptors (Zn, Cd) on P and Ga sites, respectively (Hopfield, Thomas and Gershenzon, 1963; Thomas et al., 1964). In elemental crystals, donors and acceptors can occupy both sublattices of the crystal. Thus, when discussing DAP spectra in diamond Si may be a more interesting semiconductor for comparison. Actually, many years after various investigations in III-V-compounds, line-resolved DA pair spectra were also observed in Si owing to (B, In) acceptors and P donors (Ziemelis, Parsons and Voos, 1979; Ziemelis and Parsons, 1981; Schmid, Nieper and Weber, 1983).

The photon energy emitted in the radiative electron–hole recombination process is given by (Hopfield et al., 1963; Thomas et al., 1964)

$$h\nu = E_g - E_d - E_a + \frac{e^2}{4\pi\varepsilon_0\varepsilon r_{DA}} + E_{correction} \tag{1}$$

where E_g is the bandgap, E_d and E_a are the donor and acceptor ionization energies, respectively, and r_{DA} is the DA pair separation. The last term is a correction formulated either as a Van der Waals interaction energy proportional to r_{DA}^{-6} (Hopfield et al., 1963; Thomas et al., 1964), or as a Coulomb integral including four Coulomb interaction terms between the two mobile and the two locally fixed charges (Williams, 1968). The correction term is essential only for very close pairs.

For donors and acceptors on substitutional lattice sites, r_{DA} takes discrete values which in the notation of Thomas et al. (1964) can be expressed as

$$r_{DA}(m) = a_0\sqrt{(m - b)/2} \tag{2}$$

Here, $m = 1, 2, \ldots$ is the shell number, a_0 the lattice constant, and $b = 0$ when the donors and acceptors are located on the same sublattice (type I) or $b = 5/8$ when they occupy sites on different sublattices (type II). Some of the recombination lines may be split due to various effects, such as strain on the bound hole states, jj-momentum coupling of electrons and holes, or valley–orbit splitting of the bound electron states in a many-valley indirect band structure semiconductor. Other reasons for a fine structure may be the crystallographic inequivalence of DA pairs as discussed by Thomas et al. (1964). When the assumption is made that the transition matrix elements do not drastically vary in a limited range of DA distances, and donors and acceptors are randomly distributed in the lattice, the intensities of the DA lines will approximately reflect the statistical probability of a particular pair occurring. The probabilities are listed for various lattices including diamond and zinc blende lattices up to shell numbers of $m = 50$ (type II) and $m = 53$ (type I) (Wiley and Seman, 1970).

The diamond films studied by Sternschulte et al. (1999a,b) in CL (Fig. 33) were deliberately P doped and showed n-type conductivity. There was no other intentional doping. An acceptor necessary to create the observed characteristic DAP spectra can be reasonably assumed only to be boron. Unintentional boron doping is very plausible, either from a contamination of the CVD apparatus when boron was used in earlier experiments, or from in-diffusion from a boron-doped silicon substrate. Hence, it was assumed that boron is the acceptor in the DAP line spectra. There was additional evidence for boron incorporation in the films as the 4.6 eV broad band was strongly observed which is characteristic of boron (Section 4.3). Sternschulte et al. (1999a,b) generated theoretical DAP line spectra using the statistical weights for the intensities and assuming only Coulomb interactions between the DA pairs. After Gaussian line broadening of the discrete transitions this theoretical spectrum was compared to the experimental DA spectrum by shifting the energy scale so as to yield by visual inspection the best possible agreement (Fig. 33). The correlation of experimental line energies $h\nu$ and shell numbers m (equivalent to DA spacings r_{DA}) so obtained was plotted as a graph $h\nu(1/r_{DA})$. In this representation the data points were with little scatter on a straight line whose extrapolation intersected the energy scale for $1/r_{DA} \rightarrow 0$ at $h\nu = E_g - E_a - E_d = 4.50$ eV. With $E_g = 5.5$ eV and $E_a = 0.37$ eV, the ionization energy of the acceptor boron, a donor ionization energy $E_d = 630$ meV was obtained. A second possible though less probable correlation between the experimental and the theoretical spectrum yielded a larger value of E_d which was finally taken into account by an accuracy limit of $E_d = (630 \pm 50)$ meV for the phosphorus donor.

The procedure of plotting DAP luminescence line energies $h\nu$ vs. $1/r_{DA}$ to finally determine $E_g - E_a - E_d$ has become customary over many years in numerous studies. However, the information contained in such a plot is no more than that obtained originally in shifting theoretical against experimental spectra for the best possible agreement. The most reliable correlation between experimental and theoretical spectra can be made for DA pairs which are distant enough so that the Coulomb term is sufficient to describe their interaction. Then, one can directly take individual lines or line groups which are reasonably well modeled by the theoretical DAP spectrum after the visual match, and calculate $(E_g - E_a - E_d)$ from their experimental photon energies and the associated theoretical values of m or r_{DA}. For the correlation shown in Figure 33, considered to represent the best experimental-to-theoretical agreement we may take the asymmetric line whose peak is experimentally positioned at 4.795 eV. This is the $m = 13$ (type I) transition (which can be easily seen for vanishing theoretical line widths). With $a_0 = 3.568$ Å for the diamond lattice constant, $r_{DA} = 0.9097$ nm for this line, and $\varepsilon = 5.7$ the Coulomb energy becomes 278 meV. Taking the low-temperature band-gap energy as $E_g = 5.49$ eV (which seems to be the more accurate value than 5.50 eV as employed by Sternschulte et al.) and $E_a(\mathrm{B}) = 0.37$ eV, a donor ionization energy of $E_d = (5.49\ \mathrm{eV} - 0.37\ \mathrm{eV} - 4.795\ \mathrm{eV} + 278\ \mathrm{meV}) = 603$ meV is obtained. This is nearly identical to the latest value of 604 meV deduced from infrared donor excitation spectroscopy. Within an error limit of a few meV this value now appears to be the most reliable ionization energy of the donor phosphorus.

We now return to phosphorus bound excitons. The results by Sternschulte et al. (1999a,b) were confirmed by Nakazawa, Tanabe, Tachiki, Kawarada and Koizumi (2001) who observed in CL experiments on phosphorus-doped thin diamond films grown by microwave plasma-assisted CVD on high-pressure synthetic type Ib (111) diamond a luminescence spectrum with lines at 5.19 and 5.02 eV which they ascribed to $\mathrm{BE}^{\mathrm{TO}}$ and $\mathrm{BE}^{\mathrm{TO+O}^{\Gamma}}$ exciton transitions, respectively, at phosphorus donors (Fig. 34). These line energies differ from those of Sternschulte et al. (Fig. 32) by ≈ 10 meV. As a new feature Nakazawa et al. observed an additional line at 5.32 eV which they identified as the $\mathrm{BE}^{\mathrm{NP}}$ transition. Also, the $\mathrm{FE}^{\mathrm{TO}}$ transition was visible close to 5.27 eV indicating a very

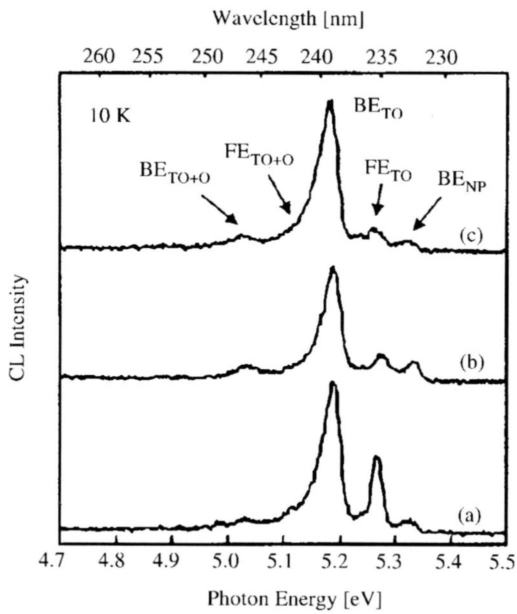

FIG. 34. Near-band-edge cathodoluminescence spectra of phosphorus-doped CVD diamond films at 10 K sample temperature. Phosphorus-to-carbon flux ratios in the growth process are (a) 200 ppm, (b) 500 ppm, and (c) 1000 ppm. The emission of excitons bound to neutral phosphorus donors is observed in no-phonon (NP) and **k**-vector conserving phonon replicas: BE^{NP} at $h\nu = 5.32$ eV, BE^{TO} at $h\nu = 5.19$ eV, and $BE^{TO+O^{\Gamma}}$ at $h\nu = 5.02$. Linewidth (FWHM) of $BE^{TO} \approx 44$ meV (from Nakazawa et al., 2001).

good quality of the film. The measurements showed an increase of the FE luminescence at elevated temperatures when the luminescence of BEs decreased owing to thermal dissociation. This interplay of free and bound excitons could be fairly well modeled using simple rate equations for the exciton trapping at and release from the phosphorus donors.

Improved quality of phosphorus-doped CVD diamond films in a metal-chamber microwave plasma CVD system (Koizumi et al., 2000) has recently led to BE cathodoluminescence spectra with very high intensities and narrow line widths down to 2 meV allowing for the resolution and identification of many more spectral details than before (Teofilov, Sauer, Thonke and Koizumi, 2003) (Fig. 35). Two diamond films were studied which are $\approx 1\ \mu$m thick and were grown by microwave CVD on synthetic Ib (111) diamond substrates. One film was phosphorus doped to $7–9 \times 10^{18}$ cm^{-3} according to temperature-dependent Hall measurements and had an electron mobility of ≈ 80 cm^2 V s^{-1} at room temperature. This film showed very strong line-resolved DAP spectra confirming exactly the DAP structures reported by Sternschulte et al. (1999a,b) (Fig. 33). It also exhibited a well-resolved phosphorus BE spectrum including four

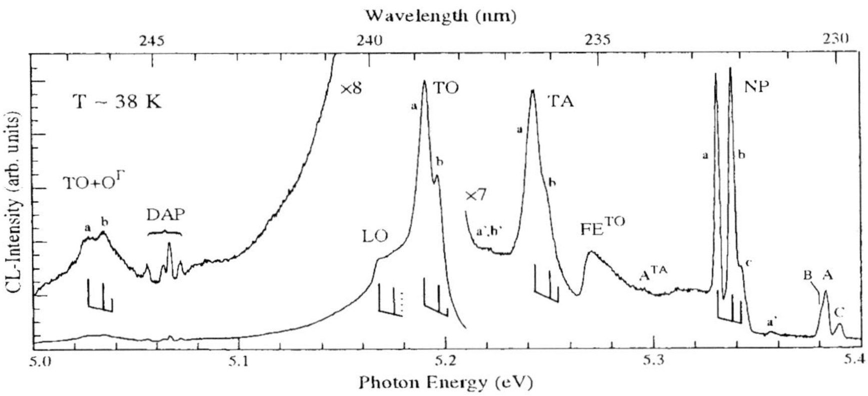

FIG. 35. Cathodoluminescence spectrum of a CVD diamond film grown by microwave-assisted CVD and doped by the addition of phosphine to some 10^{17} cm^{-3} with phosphorus. The phosphorus bound exciton spectrum includes no-phonon (NP) and k-vector conserving TA-, TO-, LO-, and (TO + O$^\Gamma$)-phonon transitions. Due to the narrow halfwidths down to 2.0 eV FWHM in the NP-transition a threefold fine structure is resolved which is also visible in the various phonon replicas. The triplet emission at high photon energies at around 5.38 eV is tentatively ascribed to a phosphorus–boron isoelectronic center (see text) (from Teofilov et al., 2003).

wavevector-conserving phonon replicas. We discuss in the following the second sample where this BE spectrum was much stronger. The second sample was lightly doped to 4×10^{17} cm^{-3} and had an electron mobility of 240 cm^2 (V s)$^{-1}$ at room temperature. It exhibited again the extended, complex-structured DAP spectrum though at relatively low intensities, and had an extremely well-developed BE spectrum which even at very high resolution showed only little noise (Fig. 35). The NP line of the BE spectrum consists of three components a, b, c visible at the measurement temperature of $\approx$40 K. This triplet is replicated by TA-, TO-, and (TO + O$^\Gamma$)-phonon satellites. Even in the weak LO-phonon replica two components of the triplet can be well recognized. In addition to these BE components luminescence from free excitons (FE^{TO}) is visible underlining the superior quality of the diamond film. When the NP triplet is studied in greater detail and at various temperatures the three components exhibit thermally-activated behavior, and two more components a$'$ and b$'$ come up (Table V). Comparing the line spacings in this fine structure it is suggestive to assume that the excited exciton lines a$'$, b$'$ are replicas of the triplet a, b, c at 25 meV higher energy where the faint component c$'$ is not visible. Splitting of donor bound excitons into multiple states is not basically new. It has been observed for the P-site donors S, Se, and Te in GaP (Dean, 1967) and for the Ga-site donor Sn in GaP (Dean, Faulkner and Kimura, 1970). Multiple exciton states have also been observed in the case of the boron bound exciton in high-quality HPHT diamonds where the NP transition is split into eight components (Section 5.1, Figs. 27 and 28). Apparently, valley–orbit effects of the localized electrons and jj-momentum coupling of the three mobile particles in the bound exciton,

TABLE V

PHOTON ENERGIES OF PHOSPHORUS-RELATED EXCITON SPECTRA

Line label	Assignment of assisting phonon	$h\nu$ (eV)	Spacing (meV) from line a	Phonon energy (meV)
Phosphorus bound exciton				
a	NP	5.3310	–	
b		5.3380	7.0	
c		5.3420	11.0	
a′		5.356_4	25.4	
b′		≈5.364	≈33.0	
a	TA	5.2430	–	88.0
b		≈5.250	≈7.0	
a	TO	5.1910	–	140.0
b		5.197_0	7.0	
a′, b′ unresolved		≈5.22	≈29.0	
a	LO	5.168_6	–	162.4
b		5.175_5	≈7.0	
a	TO + O^{Γ}	5.0268	–	304.2 (O^{Γ} : 164.2)
b		5.0340	≈7.0	
Line label	**Assignment of assisting phonon**	**$h\nu$ (eV)**	**Spacing (meV) from line B**	**Phonon energy (meV)**
Phosphorus–boron isoelectronic bound exciton				
B	NP	5.380_7	–	
A		5.383_2	2.5	
C		5.390_0	9.3	
A	TA	5.295_7		87.5

D^0X complex play an important role. It will be a challenging task for the future to unravel the exact nature of these excited bound exciton states.

After it is now well documented that phosphorus can be incorporated into diamond as a donor giving rise to a bound exciton luminescence spectrum with many fine characteristic details similar to other, well-studied semiconductors one may correlate the exciton localization energy E_{loc} with the donor ionization energy E_{d} and compare the donor (phosphorus) case with the acceptor (boron) case. The correlation was discussed by Sternschulte et al. (1999a) and Nakazawa et al. (2001). The most recently available data deliver

$$\frac{E_{\mathrm{loc}}}{E_{\mathrm{i}}} = \begin{cases} \dfrac{75\ \mathrm{meV}}{604\ \mathrm{meV}} = 0.124\ \text{(phosphorus)} \\[2ex] \dfrac{51\ \mathrm{meV}}{370\ \mathrm{meV}} = 0.138\ \text{(boron)} \end{cases}$$

where E_{i} stands for E_{d} (donor) or E_{a} (acceptor) and the FE threshold energy was taken as $E_{\mathrm{gx}}^{\mathrm{TO}} = 5.266$ eV. The two ratios are rather similar, hence one may write on average $E_{\mathrm{loc}} \approx 0.13E_{\mathrm{i}}$. This is an analogue of the empirical relation in silicon valid for many shallow donors and acceptors, $E_{\mathrm{loc}} \approx 0.1E_{\mathrm{i}}$, often addressed to as 'Haynes' rule' (Haynes, 1960). Although the general validity of this relation in diamond is not proved by just two examples, it may serve as a tentative guide in further potential cases of 'shallow' Coulomb impurities and their associated excitonic spectra. In diamond the intensity ratio $I^{\mathrm{NP}}/I^{\mathrm{TO}}$ of the BE lines is significantly larger for P than for B, with approximate values of ≈ 0.03 (Fig. 35) and ≈ 0.002 (Fig. 27), respectively. This is what one expects from the larger ionization energy of P rendering more allowed $\Delta k = 0$ direct transitions in comparison to those assisted by wavevector-conserving phonons.

An intriguing novel luminescence feature emerges at photon energies around 5.38 eV (Fig. 35) significantly higher in energy than the phosphorus $\mathrm{BE}^{\mathrm{NP}}$ lines. This spectrum consists of a doublet line whose lower-energy component is further split into two transitions which are not completely resolved (Table V). The narrow linewidths suggest that the triplet is due to an exciton localized at an (unknown) center X. The exciton localization energy taken from the shift of the peak at ≈ 5.38 eV from the optical gap, at $E_{\mathrm{gx}}^{\mathrm{TO}} + \hbar\Omega(\mathrm{TO}) = 5.407$ eV, amounts to $E_{\mathrm{loc}}(\mathrm{X}) \approx 27$ meV. This energy is extremely low and significantly less than $E_{\mathrm{loc}}(\mathrm{B}) = 51$ meV for excitons bound to boron, the shallowest known impurity in diamond. At increasing temperatures, the high-energy component of the triplet is thermally activated relative to the lower doublet lines while the triplet intensity as a whole drops with a de-activation energy close to the value of $E_{\mathrm{loc}}(\mathrm{X})$. This clearly confirms that the triplet originates from an exciton at a very shallow trap X. A number of emission line spectra with similarly high energies and narrow halfwidths have previously been observed in relatively defect-free natural diamonds by Dean and Jones (1964). These are line pairs (C′: 5.375 eV, D′: 5.361 eV), (A: 5.325 eV, B: 5.276 eV), and (X: 5.267 eV, Y: 5.259 eV) interpreted as originating from excitons bound to impurity or defect centers which were not identified. However, the C′ and D′ lines were only detected for type IIb specimens and could be related with neutral aluminum acceptors (nowadays: boron). Another prominent high energy narrow emission line pair is the N9 doublet with positions at 5.254 and 5.263 eV, and a very weak third line at 5.280 eV (Dean and Male, 1964). The N9 spectrum was also studied earlier by Halperin and Nahum (1961) and Nahum and Halperin (1962). Later intensive investigations of the spectrum (Crowther and Dean, 1967; Denham, Lightowlers and Dean, 1967) advanced some evidence that the binding center of the exciton is a nearest-neighbor nitrogen–aluminum (nowadays: boron) pair thought to be common in the natural diamonds used. The latter model and the association of some of the lines with boron is a very interesting issue when compared to the argumentation of Teofilov et al. (2003) to follow.

In the case of the A, B, C triplet (Fig. 35) the binding center X cannot be a Coulomb impurity such as a donor or an acceptor for two reasons: (i) There is no donor or acceptor expected to be shallower than boron; if X were Coulombic and Haynes' rule is assumed to apply, X would have an approximate ionization energy of $E_i(X) \approx (1/0.13)E_{loc} \approx 208$ meV. (ii) If X were a donor or acceptor that shallow, the perturbation of the translational invariance of the lattice would be very weak. Then the luminescence spectrum must show TA- and TO-phonon replicas much stronger than the NP transition in comparison to phosphorus and boron. Experimentally, the expected TO-replica would then fall on the phosphorus BE^{TA} line in Figure 35 and no conclusion can be made; however, the expected strong TA-replica would lie close to 5.30 eV where no line appears. Teofilov et al. (2003) concluded furthermore the novel spectrum is neither vibronic (since no local modes are observed) nor is X a structural complex defect incorporating vacancies, interstitials, and chemical species. Excitons at such defects would be expected to be much more strongly localized. The highest photon energy for a sufficiently well-studied structural defect in diamond is that for the 5RL center, incorporating a (001) double split carbon interstitial (Collins, 1993), at 4.582 eV, yielding $E_{loc} \approx 818$ meV. While all these explanations for the binding center can be discarded with some certainty the novel triplet structure is fully consistent with what one expects for an exciton bound to an axial isoelectronic trap. A well-documented and illustrative example of this kind of localized exciton represent the A, B, C lines in Al-doped Si at 1.122 eV (Weber, Schmid and Sauer, 1980) where the isoelectronic trap incorporates close Al and N atoms (Alt and Tapfer, 1985; Modavis and Hall, J. Appl. Phys. 67, 545 (1990)). The trap itself does not introduce an extra change into the lattice, and the excess electron–hole pair localized at the trap couples to yield exciton states with total angular momenta $J = 2$ and $J = 1$ which are split in the $\langle 111 \rangle$ axial strain field of the Al–N pair into states $|J, M_J\rangle = |2, \pm 2\rangle$, $|2, \pm 1\rangle$, $|2, 0\rangle$, $|1, \pm 1\rangle$, and $|1, 0\rangle$. Transitions from the first state at the lowest energy are dipole forbidden but become allowed under the perturbation of a magnetic field yielding the B line. Dipole-allowed transition from the second state give rise to the dominant A line at higher energy, and line C observed at increasing temperatures originates from higher energy-exciton states which cannot be spectrally resolved.

This isoelectronic trap model can account for the present structure of the 5.38 eV triplet labeled A, B, C in Figure 35. The triplet has been observed in five phosphorus-doped CVD diamond films which exhibit the phosphorus BE spectrum and the DAP lines as well, indicating that both, P and B, are incorporated. Hence, it is suggestive to assume that the isoelectronic trap X consists of these two species on nearest-neighbor sites or in close proximity. The weak appearance of the dipole forbidden B line must then be related with a relaxation of the dipole selection rule due to some additional perturbation of the axial center. Isoelectronically bound excitons have been studied excessively in GaP, and various examples show the trend that the exciton localization energy E_{loc} is small when the trap induces little lattice distortion as is the case for trap atoms neighboring the host lattice atoms in the periodic table; in contrast, E_{loc} can be very large for elements distorting the lattice heavily. To exemplify this we cite the following cases studied previously in detail N_P: $E_{loc} = 11.5$ meV (Thomas and Hopfield, 1966), Bi_P: $E_{loc} = 98.5$ meV (Dean, Cuthbert and Lynch, 1969), $Li_{Ga}-Li_{Ga}-O_P$: $E_{loc} = 238.1$ meV (Dean, 1971), and $Cd_{Ga}-O_P$: $E_{loc} = 420.2 \times$ meV (Dean, Henry and Frosch, 1968). Following these trends, a P–B isoelectronic trap in diamond is expected to have a very small exciton localization energy consistent with the experimental value of $E_{loc}(X) \approx 27$meV. Summing up the arguments given by Teofilov et al., nearest-neighbor or close P–B pairs could account for the 5.38 eV triplet spectrum in Figure 35.

5.4. Other Donor Candidates

The presently available experimental results on sulfur-doped CVD diamond films are ambiguous. Sakaguchi et al. (1999) and Nishitani-Gamo, Yasu, Xiao, Kikuchi, Ushizawa, Sakaguchi, Suzuki and Ando (2000) doped films grown homoepitaxially on (100) synthetic diamonds by adding hydrogen sulfide to the CH_4/H_2 gas mixture in the microwave-assisted plasma deposition process. They reported n-type conductivity with a carrier concentration of $> 10^{16}$ cm^{-3} and mobility up to ≈ 600 cm^2 (V s)$^{-1}$ at room temperature. The carrier concentration was thermally activated with an energy of 0.38 eV interpreted as the ionization energy of a sulfur donor. This interpretation was doubted by Kalish, Reznik, Uzan-Saguy and Cyterman (2000) who studied three of the samples grown by Sakaguchi et al. and found p-type conduction in Hall effect measurements with activation energies, carrier concentrations, and mobilities very close to those of boron-doped p-type diamond. In addition, SIMS measurements indicated high enough boron concentrations to explain the observed p-type features by unintentional boron doping. One of these samples was investigated by the group of the author, and again contradictory results were obtained: The sample showed n-type conductivity at room temperature but CL measurements exhibited exclusively the boron BE spectrum with no indication of any other luminescence features which could be ascribed to sulfur (Teofilov, Aleksov, Thonke and Sauer, 2000). Similarly negative results were obtained in CL investigations on sulfur-doped CVD diamond films prepared by Sternschulte, Schreck and Stritzker (2003).

Lithium-doped CVD diamond layers as obtained from various laboratories were studied by CL measurements in the group of the author. In a particularly close cooperation with the Institute of Physics, University of Augsburg, many lithium-doped diamond films were investigated grown on synthetic type Ib (100) single crystal diamonds by addition of lithium compounds to the CH_4/H_2 microwave plasma (Sternschulte, Schreck, Stritzker, Bergmaier and Dollinger, 2000). In no case were luminescence signals detected which could be correlated with lithium in an unambiguous way. This applies also to the spectra displayed in the paper by Sternschulte, Thonke and Sauer (1999a) where a Li-doped polycrystalline diamond film exhibited a new transition a $\approx$5.12 eV which might be related with the Li-doping; this was not substantiated by further data.

References

M. Akaishi, H. Kanda, and S. Yamaoka, Phosphorus: an elemental catalyst for diamond synthesis and growth, *Science* 259, 1592 (1993).

H. Ch. Alt and L. Tapfer, Photoluminescence observation of the A, B, C exciton in nitrogen implanted silicon, *J. Electron. Mater.* 14a, 833 (1985).

J. C. Angus, H. A. Will, and W. S. Stanko, Growth of diamond seed crystals by vapor deposition, *J. Appl. Phys.* 39, 2915 (1968).

L. Bergmann, M. T. McClure, J. T. Glass, and R. J. Nemanich, Recombintion of the broadband and 1.681 eV optical centers in diamond films, *Mater. Res. Soc. Symp. Proc.* 339, 663 (1994).

L. Bergmann, B. R. Stoner, K. F. Turner, J. T. Glass, and R. J. Nemanich, Microphotoluminescence and Raman scattering study of defect formation in diamond films, *J. Appl. Phys.* 73, 3951 (1993).

T. G. Bilodeau, K. Doverspike, U. Strom, J. A. Freitas, and R. Rameshan, Calorimetric absorption spectroscopy and photoluminescence study of defects in diamond, *Diamond Relat. Mater.* 2, 699 (1993).

S. Bohr, R. Haubner, and B. Lux, Influence of phosphorus addition on diamond CVD, *Diamond Relat. Mater.* 4, 133 (1995).

T. H. Borst and O. Weis, Electrical characterization of homoepitaxial diamond films doped with B, P, Li and Na during crystal growth, *Diamond Relat. Mater.* 4, 948 (1995).

G. Braunstein and R. Kalish, Effective p-type doping of diamond by boron ion implantation, *J. Appl. Phys.* 54, 2106 (1983).

S. W. Brown and S. C. Rand, Site symmetry analysis of the 738 nm defect in diamond, *J. Appl. Phys.* 78, 4069 (1995).

R. C. Burns, V. Cvetkovic, C. N. Dodge, D. J. F. Evans, M. T. Rooney, P. M. Spear, and C. M. Welbourn, Growth-sector dependence of optical features in large synthetic diamonds, *J. Cryst. Growth* 104, 257 (1990).

G. Z. Cao, F. A. J. M. Driessen, G. J. Bauhuis, L. J. Giling, and P. F. A. Alkemade, Homoepitaxial diamond films codoped with phosphorus and nitrogen by chemical-vapor deposition, *J. Appl. Phys.* 78, 3125 (1995).

J. Chevallier, D. Ballutaud, B. Theys, F. Jomard, A. Deneuville, E. Gheeraert, and F. Pruvost, Hydrogen in monocrystalline CVD boron doped diamond, *Phys. Status Solidi (a)* 174, 73 (1999).

R. M. Chrenko, Boron content and profiles in large laboratory diamonds, *Nature (Physical Science)* 229, 165 (1971).

R. M. Chrenko, Boron, the dominant acceptor in semiconducting diamond, *Phys. Rev. B* 7, 4560 (1973).

C. D. Clark, A. T. Collins, and G. S. Woods, Absorption and luminescence spectroscopy in "The Properties of Natural and Synthetic Diamond" (J. E. Field, Ed.), p. 35. Academic Press, London, 1992.

C. D. Clark, P. J. Dean, and P. V. Harris, Intrinsic edge absorption in diamond, *Proc. R. Soc.* A277, 312 (1964).

C. D. Clark and C. B. Dickerson, The 1.681 eV centre in polycrystalline diamond, in "Proc. 1st Europ. Conf. Diamond and Diamond-Like Carb. Coat. (Crans-Montana, Switzerland)", p. 336. Elsevier Science, Amsterdam, 1991a.

C. D. Clark and C. B. Dickerson, The 1.681 eV centre in polycrystalline diamond, *Surf. Coat. Technol.* 47, 336 (1991b).

C. D. Clark and C. B. Dickerson, Raman and photoluminescence spectra of as-grown diamond CVD films, *J. Phys.: Condens. Matter* 4, 869 (1992).

C. D. Clark, H. Kanda, I. Kiflawi, and G. Sittas, Silicon defects in diamond, *Phys. Rev. B* 51, 16681 (1995).

C. D. Clark, E. W. J. Mitchell, and B. J. Parsons, Colour centres and optical properties in "The Properties of Diamond" (J. E. Field, Ed.), p. 23. Academic Press, London, 1979.

A. T. Collins, The electronic and optical properties of diamond; do they favour device applications? *Mater. Res. Soc. Symp. Proc.* 162, 3 (1990).

A. T. Collins, The characterization of point defects in diamond by luminescence spectroscopy, *Diamond Relat. Mater.* 1, 457 (1992).

A. T. Collins, Intrinsic and extrinsic absorption and luminescence in diamond, *Physica B* 185, 284 (1993).

A. T. Collins, Things we still don't know about optical centers in diamond, *Diamond Relat. Mater.* 8, 1455 (1999).

A. T. Collins, L. Allers, C. J. H. Wort, and G. A. Scarsbrook, The annealing of radiation damage in De Beers colourless CVD diamond, *Diamond Relat. Mater.* 3, 932 (1994).

A. T. Collins, M. Kamo, and Y. Sato, Intrinsic and extrinsic cathodoluminescence from single-crystal diamonds grown by chemical vapour deposition, *J. Phys: Condens. Matter* 1, 4029 (1989a).

A. T. Collins, M. Kamo, and Y. Sato, Optical centres related to nitrogen, vacancies and interstitials in polycrystalline diamond films grown by plasma-assisted chemical vapour deposition, *J. Phys. D: Appl. Phys.* 22, 1402 (1989b).

A. T. Collins, M. Kamo, and Y. Sato, A spectroscopic study of optical centers in diamond grown by microwave-assisted chemical vapor deposition, *J. Mater. Res.* 5, 2507 (1990a).

A. T. Collins, M. Kamo, and Y. Sato, Characterisation of thin films and single-crystal CVD diamond by absorption and luminescence spectroscopy, *Mater. Res. Soc. Symp. Proc.* 162, 225 (1990b).

A. T. Collins, S. C. Lawson, G. Davies, and H. Kanda, Indirect energy gap of ^{13}C diamond, *Phys. Rev. Lett.* 65, 891 (1990c).

A. T. Collins and E. C. Lightowlers, Electrical properties in "The Properties of Diamond" (J. E. Field, Ed.), p. 79. Academic Press, New York, 1979.

A. T. Collins and M. Stanley, Absorption and luminescence studies of synthetic diamond in which the nitrogen has been aggregated, *J. Phys. D: Appl. Phys.* 18, 2537 (1985).

A. T. Collins and A. W. S. Williams, The nature of the acceptor centre in semiconducting diamond, *J. Phys. C: Solid State Phys.* 4, 1789 (1971).

P. A. Crowther and P. J. Dean, Phonon interactions, piezo-optical properties and the inter-relationship of the N3 and N9 absorption–emission systems in diamond, *J. Phys. Chem. Solids* 28, 1115 (1967).

G. Davies, The Optical Properties of Diamond, in "Chemistry and physics of carbon" (P. L. Walker and P. A. Thrower, Eds.), Vol. 13, p. 1. Marcel Dekker, New York, 1977.

G. Davies, Cathodoluminescence in "The Properties of Diamond" (J. E. Field, Ed.), p. 165. Academic Press, New York, 1979.

G. Davies, (Ed.), "Properties and Growth of Diamond", EMIS Datareviews Series No. 9 INSPEC, IEE, London, 1994.

G. Davies, Current problems in diamond: towards a quantitative understanding, *Physica B* 273/274, 15 (1999).

G. Davies and N. B. Manson, Optical properties of vacancy–nitrogen aggregates in diamond in "Properties and Growth of Diamond" EMIS Datareviews Series No. 9 (G. Davies, Ed.), p. 173. INSPEC, IEE, London, 1994.

P. J. Dean, Bound excitons and donor–acceptor pairs in natural and synthetic diamond, *Phys. Rev.* 139, A588 (1965).

P. J. Dean, Absorption and luminescence of excitons at neutral donors in gallium phosphide, *Phys. Rev.* 157, 655 (1967).

P. J. Dean, Isoelectronic trap Li–Li–O in GaP, *Phys. Rev. B* 4, 2596 (1971).

P. J. Dean, J. D. Cuthbert, and R. T. Lynch, Interimpurity recombinations involving the isoelectronic trap bismuth in gallium phosphide, *Phys. Rev.* 179, 754 (1969).

P. J. Dean, R. A. Faulkner, and S. Kimura, Optical properties of the donor tin in gallium phosphide, *Phys. Rev. B* 2, 4062 (1970).

P. J. Dean, J. R. Haynes, and W. F. Flood, New radiative recombination processes involving neutral donors and acceptors in silicon and germanium, *Phys. Rev.* 161, 711 (1967).

P. J. Dean, C. H. Henry, and C. J. Frosch, *Phys. Rev.* 168, 812 (1968).

P. J. Dean and I. H. Jones, Recombination radiation from diamond, *Phys. Rev.* 133, A1698 (1964).

P. J. Dean, E. C. Lightowlers, and D. R. Wight, Intrinsic and extrinsic recombination radiation from natural and synthetic aluminum-doped diamond, *Phys. Rev.* 140, A352 (1965).

P. J. Dean and J. C. Male, Luminescence excitation spectra and recombination radiation of diamond in the fundamental absorption region, *Proc. R. Soc.* A277, 330 (1964).

A. Deneuville, P. Gonon, E. Gheeraert, A. T. Collins, and Y. L. Khong, The influence of oxygen, in gas mixtures and various substrate positions, on the broad cathodoluminescence bands of MPCVD diamond films, *Diamond Relat. Mater.* 2, 737 (1993).

P. Denham, E. C. Lightowlers, and P. J. Dean, Ultraviolet intrinsic and extrinsic photoconductivity of natural diamond, *Phys. Rev.* 161, 762 (1967).

B. V. Deryaguin and D. V. Fedoseev, The synthesis of diamond at low pressure, *Sci. Am.* 233, 102 (1975).

B. V. Derjaguin, B. V. Spitsyn, A. E. Gorodetsky, A. P. Zakharov, L. L. Bouilov, and A. E. Aleksenko, Structure of autoepitaxial diamond films, *J. Cryst. Growth* 31, 44 (1975).

B. Dischler, W. Rothemund, K. Maier, C. Wild, H. Biebl, and P. Koidl, Diamond luminescence: resolved donor–acceptor pair recombination lines, *Diamond Relat. Mater.* 3, 825 (1994a).

B. Dischler, W. Rothemund, C. Wild, R. Locher, H. Biebl, and P. Koidl, Resolved donor–acceptor pair-recombination lines in diamond luminescence, *Phys. Rev. B* 49, 1685 (1994b).

B. Dischler, C. Wild, W. Müller-Sebert, and P. Koidl, Hydrogen in polycrystalline diamond: an infrared analysis, *Physica B* 185, 217 (1993).

W. G. Eversole. Synthesis of diamond by deposition on seed crystals, US Patent No. 3030187, 1962.

R. G. Farrer, On the substitutional nitrogen donor in diamond, *Solid State Commun.* 7, 685 (1969).

T. Feng and B. D. Schwartz, Characteristics and origin of the 1.681 eV luminescence center in chemical-vapor-deposited diamond films, *J. Appl. Phys.* 73, 1415 (1993).

J. A. Freitas, J. E. Butler, S. G. Bishop, W. A. Carrington, and U. Strom, Photoluminescence spectroscopy of diamond films, *Mater. Res. Soc. Symp. Proc.* 162, 237 (1990a).

J. A. Freitas, J. E. Butler, and U. Strom, Photoluminescence studies of polycrystalline diamond films, *J. Mater. Res.* 5, 2502 (1990b).

J. A. Freitas, K. Doverspike, P. B. Klein, Y. L. Khong, and A. T. Collins, Luminescence studies of nitrogen- and boron-doped diamond films, *Diamond Relat. Mater.* 3, 821 (1994a).

J. A. Freitas, P. B. Klein, and A. T. Collins, Evidence of donor–acceptor pair recombination from a new emission band in semiconducting diamond, *Appl. Phys. Lett.* 64, 2136 (1994b).

J. A. Freitas, P. B. Klein, A. T. Collins, Radiative recombination processes in semiconducting diamond, in "Advances in New Diamond Science and Technology" (S. Saito, N. Fujimori, O. Fukunaya, M. Kamo, K. Kobashi, and M. Yoshikawa, Eds.), p. 321. MYU, Tokyo, 1994c.

J. A. Freitas, U. Strom, and A. T. Collins, Photoluminescence and cathodoluminescence studies of semiconducting diamond, *Diamond Relat. Mater.* 2, 87 (1993).

F. Fuchs, C. Wild, K. Schwarz, and P. Koidl, Hydrogen-related IR absorption in chemical vapour deposited diamond, *Diamond Relat. Mater.* 4, 652 (1995).

E. Gheeraert, N. Casanova, S. Koizumi, T. Teraji, and H. Kanda, Low temperature excitation spectrum of phosphorus in diamond, *Diamond Relat. Mater.* 10, 444 (2001).

E. Gheeraert, F. Fontaine, A. Deneuville, Y. L. Khong, and A. T. Collins, Characterization of defects in boron implanted chemically vapour deposited diamond films by electron parametric resonance and cathodoluminescence, *Diamond Relat. Mater.* 3, 737 (1994).

E. Gheeraert, S. Koizumi, T. Teraji, H. Kanda, and M. Nesládek, Electronic states of boron and phosphorus in diamond, *Phys. Status Solidi (a)* 174, 39 (1999).

E. Gheeraert, S. Koizumi, T. Teraji, and H. Kanda, Electronic transitions of electrons bound to phosphorus donors in diamond, *Solid State Commun.* 113, 577 (2000a).

E. Gheeraert, S. Koizumi, T. Teraji, H. Kanda, and M. Nesládek, Electronic states of phosphorus in diamond, *Diamond Relat. Mater.* 9, 948 (2000b).

A. A. Gippius, Private communication (1993a).

A. A. Gippius, Luminescent characterization of radiation damage and impurities in ion-implanted natural diamond, *Diamond Relat. Mater.* 2, 640 (1993b).

A. A. Gippius, V. S. Vavilov, A. M. Zaitsev, and B. S. Zhakupbekov, Defects production and interaction in ion-implanted diamond, *Physica B* 116, 187 (1983).

A. A. Gippius, A. M. Zaitsev, and V. S. Vavilov, Formation, annealing, and interaction of defects in ion-implanted layers of natural diamond, *Sov. Phys. Semicond.* 16, 256 (1982) ([Fiz. Techn. Poluprov. 16, 404 (1982)]).

A. A. Gorokhovsky, A. V. Turukhin, R. R. Alfano, and W. Phillips, Photoluminescence vibrational structure of Si center in chemical-vapor deposited diamond, *Appl. Phys. Lett.* 66, 43 (1995).

J. P. Goss, R. Jones, S. J. Breuer, P. R. Briddon, and S. Öberg, The twelve-line 1.682 eV luminescence center in diamond and the vacancy–silicon complex, *Phys. Rev. Lett.* 77, 3041 (1996).

R. J. Graham, F. D. Moustakas, and M. M. Disko, Cathodoluminescence imaging of defects and impurities in diamond films grown by chemical vapor deposition, *J. Appl. Phys.* 69, 3212 (1991).

K. Haenen, K. Meykens, M. Nesládek, G. Knuyt, C. Quaeyhaegens, L. Stals, S. Koizumi, and E. Gheeraert, Low temperature photoconductivity detection of phosphorus in diamond, *Phys. Status Solidi (a)* 174, 53 (1999a).

K. Haenen, K. Meykens, M. Nesládek, G. Knuyt, L. Stals, T. Teraji, S. Koizumi, and E. Gheeraert, Temperature dependent spectroscopic study of the electronic structure of phosphorus in n-type CVD diamond films, *Diamond Relat. Mater.* 9, 952 (2000).

K. Haenen, K. Meykens, M. Nesládek, L. M. Stals, T. Teraji, S. Koizumi, and E. Gheeraert, Study of the electronic structure of the phosphorus level in n-type CVD diamond, *Phys. Status Solidi (a)* 174, R1 (1999b).

A. Halperin and J. Nahum, Some optical and electrical properties of semiconducting diamonds, *J. Phys. Chem. Solids* 18, 297 (1961).

P. L. Hanley, I. Kiflawi, and A. R. Lang, On topographically identifiable sources of cathodoluminescence in natural diamonds, *Phil. Trans. R. Soc. A* 284, 329 (1977).

K. Hayashi, T. Sekiguchi, and H. Okushi, Formation and relaxation of hydrogen-related defects in the subsurface region of diamond films, *Mater. Sci. Forum* 258–263, 745 (1997a).

K. Hayashi, H. Watanabe, S. Yamanaka, H. Okushi, K. Kajimura, and T. Sekiguchi, Direct observation of hydrogen-related luminescent states in subsurface region of homoepitaxial diamond films, *Appl. Phys. Lett.* 69, 1122 (1996).

K. Hayashi, H. Watanabe, S. Yamanaka, T. Sekiguchi, H. Okushi, and K. Kajimura, Hydrogen-related gap states in the near surface of chemical vapor deposited homoepitaxial diamond films, *Diamond Relat. Mater.* 6, 303 (1997b).

J. R. Haynes, Experimental proof of the existence of a new electronic complex in silicon, *Phys. Rev. Lett.* 4, 361 (1960).

J. J. Hopfield, D. G. Thomas, and M. Gershenzon, Pair spectra in GaP, *Phys. Rev. Lett.* 10, 162 (1963).

K. Iakoubovskii and G. J. Adriaenssens, Optical detection of defect centers in CVD diamond, *Diamond Relat. Mater.* 9, 1349 (2000a).

K. Iakoubovski and G. J. Adriaenssens, Characterization of the broad green band luminescence in CVD and synthetic Ib diamond, *Diamond Relat. Mater.* 9, 1017 (2000b).

K. Iakoubovskii, G. J. Adriaenssens, M. Nesladek, and L. M. Stals, Photoluminescence excitation and quenching spectra in CVD diamond films, *Diamond Relat. Mater.* 8, 717 (1999).

K. Jackson, M. R. Pederson, and J. G. Harrison, Donor levels and impurity-atom relaxation in nitrogen- and phosphorus-doped diamond, *Phys. Rev. B* 41, 12641 (1990).

G. Janssen, W. J. P. van Enckevort, W. Vollenberg, and L. J. Giling, Characterization of single-crystal diamond grown by chemical vapour deposition processes, *Diamond Relat. Mater.* 1, 789 (1992).

R. Job, M. Werner, A. Denisenko, A. Zaitsev, and W. R. Fahrner, Electrical properties of lithium-implanted layers on synthetic diamond, *Diamond Relat. Mater.* 5, 757 (1996).

S. A. Kajihara, A. Antonelli, and J. Bernholc, N-type doping and diffusion of impurities in diamond, *Mater. Res. Soc. Symp. Proc.* 162, 315 (1990).

S. A. Kajihara, A. Antonelli, J. Bernholc, and R. Car, Nitrogen and potential n-type dopants in diamond, *Phys. Rev. Lett.* 66, 2010 (1991).

R. Kalish, The search for donors in diamond, *Diamond Relat. Mater.* 10, 1749 (2001).

R. Kalish, A. Reznik, C. Uzan-Saguy, and C. Cyterman, Is sulfur a donor in diamond? *Appl. Phys. Lett.* 76, 757 (2000).

R. Kalish, A. Reznik, C. Uzan-Saguy, S. Koizumi, and M. Kamo, n-type diamond, in "Advanced Materials '98, The 5th NIRIM Internat. Symp. Advanced Materials (ISAM '98)", Tsukuba, Ibaraki, Japan (M. Kamo, et al., Eds.), p. 105. National Institute for Research in Inorganic Materials, 1998.

R. Kalish, C. Uzan-Saguy, A. Samoiloff, R. Locher, and P. Koidl, Doping of polycrystalline diamond by boron ion implantation, *Appl. Phys. Lett.* 64, 2532 (1994).

M. Kamo, Y. Sato, S. Matsumoto, and N. Setaka, Diamond synthesis from gas phase in microwave plasma, *J. Cryst. Growth* 62, 642 (1983).

H. Kawarada, H. Matsuyama, Y. Yokota, T. Sogi, A. Yamaguchi, and H. Hiraki, Excitonic recombination radiation in undoped and boron-doped chemical-vapor-deposited diamonds, *Phys. Rev. B* 47, 3633 (1993).

H. Kawarada, K. Nishimura, T. Ito, J. Suzuki, K. S. Mar, Y. Yokota, and A. Hiraki, Blue and green cathodoluminescence of synthesized diamond films formed by plasma-assisted chemical vapour deposition, *Jpn. J. Appl. Phys.* 27, L683 (1988).

H. Kawarada, T. Tsutsumi, H. Hirayama, and A. Yamaguchi, Dominant free-exciton recombination radiation in chemical vapor deposited diamonds, *Appl. Phys. Lett.* 64, 451 (1994).

H. Kawarada and A. Yamaguchi, Excitonic recombination radiation as characterization of diamonds using cathodoluminescence, *Diamond Relat. Mater.* 2, 100 (1993).

H. Kawarada, Y. Yokota, and H. Hiraki, Intrinsic and extrinsic recombination radiation from undoped and boron-doped diamonds formed by plasma chemical vapor deposition, *Appl. Phys. Lett.* 57, 1889 (1990a).

H. Kawarada, Y. Yokota, Y. Mori, K. Nishimura, and A. Hiraki, Cathodoluminescence and electroluminescence of undoped and boron-doped diamond formed by plasma chemical vapor deposition, *J. Appl. Phys.* 67, 983 (1990b).

H. Kawarada, Y. Yokota, H. Matsuyama, T. Sogi, A. Hiraki, Free Exciton and Bound Exciton Recombination Radiation in Undoped and Boron-doped CVD Diamonds, in "Proc. 2nd Internat.

Symp. Diamond Materials" (A. J. Purdes, J. C. Angus, R. F. Davis, B. M. Meyerson, K. E. Spear, and M. Yoder, Eds.), Vol. 91, p. 420. The Electrochemical Society, New Jersey, 1991.

Y. L. Khong and A. T. Collins, Temperature dependence of cathodoluminescence from CVD diamond, *Diamond Relat. Mater.* 2, 1 (1993).

Y. L. Khong, A. T. Collins, and L. Allers, Luminescence decay time studies and time-resolved cathodoluminescence spectroscopy of CVD diamond, *Diamond Relat. Mater.* 3, 1023 (1994).

I. Kiflawi and A. R. Lang, Linearly polarized luminescence from linear defects in natural and synthetic diamond, *Philos. Mag.* 30, 219 (1974).

P. B. Klein, M. D. Crossfield, J. A. Freitas, and A. T. Collins, Donor–acceptor pair recombination in synthetic type-IIb semiconducting diamond, *Phys. Rev. B* 51, 9634 (1995).

S. Koizumi, Growth and characterization of phosphorus doped n-type diamond thin films, *Phys. Status Solidi (a)* 172, 71 (1999).

S. Koizumi, M. Kamo, Y. Sato, S. Mita, A. Sawabe, A. Reznik, C. Uzan-Saguy, and R. Kalish, Growth and characterization of phosphorus doped n-type diamond thin films, *Diamond Relat. Mater.* 7, 540 (1998).

S. Koizumi, M. Kamo, Y. Sato, H. Ozaki, and T. Inuzuka, Growth and characterization of phosphorus doped {111} homoepitaxial diamond thin films, *Appl. Phys. Lett.* 71, 1065 (1997).

S. Koizumi, T. Teraji, and H. Kanda, Phosphorus-doped chemical vapor deposition of diamond, *Diamond Relat. Mater.* 9, 935 (2000).

S. Lal, T. Dallas, S. Yi, S. Gangopadhyay, M. Holtz, and F. G. Anderson, Defect photoluminescence in polycrystalline diamond films grown by arc-jet chemical-vapor deposition, *Phys. Rev. B* 54, 13428 (1996).

S. Lal, S. Lin, S. Gangopadhyay, M. S. Hague, and H. H. Naseem, Effect of diamond film quality on tungsten related photoluminescence peaks, *Appl. Phys. Lett.* 71, 3640 (1997).

Landolt-Börnstein: Numerical Data and Functional Relationships in Science and Technology, Vol. 41, Semiconductors (O. Madelung et al., Eds.), Subvolume A2: Impurities and Defects in Group IV Elements, IV–IV and III–V Compounds (M. Schulz, Ed.), p. 4, Springer, Berlin, 2002.

M. I. Landstrass and K. V. Ravi, Hydrogen passivation of electrically active defects in diamond, *Appl. Phys. Lett.* 55, 1391 (1989).

S. C. Lawson and H. Kanda, Nickel in diamond: an annealing study, *Diamond Relat. Mater.* 2, 130 (1993).

S. C. Lawson, H. Kanda, H. Kiyota, T. Tsutsumi, and H. Kawarada, Cathodoluminescence from high-pressure synthetic and chemical-vapor-deposited diamond, *J. Appl. Phys.* 77, 1729 (1995).

S. C. Lawson, H. Kanda, H. Kiyota, T. Tsutsumi, H. Kawarada, New cathodoluminescence from boron-doped diamond, in "Advances in New Diamond Science and Technology" (S. Saito, N. Fujimori, O. Fukunaya, M. Kamo, K. Kobashi, and M. Yoshikawa, Eds.), p. 315. MYU, Tokyo, 1994.

E. C. Lightowlers and A. T. Collins, Determination of boron in natural semiconducting diamond by prompt particle nuclear microanalysis and Schottky barrier differential-capacitance measurements, *J. Phys. D: Appl. Phys.* 9, 951 (1976).

G. D. Mahan, Energy gap in Si and Ge: impurity dependence, *J. Appl. Phys.* 51, 2634 (1980).

S. Matsumoto, Y. Sato, M. Kamo, and N. Setaka, Vapor deposition of diamond particles from methane, *Jpn. Appl. Phys.* 21, L183 (1982).

T. S. McCauley and Y. K. Vohra, Defect center in diamond thin films observed by micro-Raman and photoluminescence studies, *Phys. Rev. B* 49, 5046 (1994).

K. M. McNamara, D. H. Levy, K. K. Gleason, and C. J. Robinson, Nuclear magnetic resonance and infrared absorption studies of hydrogen incorporation in polycrystalline diamond, *Appl. Phys. Lett.* 60, 580 (1991).

T. Miyazaki and H. Okushi, A theoretical study of a sulfur impurity in diamond, *Diamond Relat. Mater.* 10, 449 (2001).

R. A. Modavis and D. G. Hall, Aluminum-nitrogen isoelectronic trap in silicon, *J. Appl. Phys.* 67, 545 (1990).

Y. Mori, N. Eimori, H. Kozuka, Y. Yokota, J. Moon, J. S. Ma, and A. Hiraki, Oxygen diffusion into diamond induced by hydrogen microwave plasma, *Appl. Phys. Lett.* 60, 47 (1992).

N. F. Mott, On the transition to metallic conduction in semiconductors, *Can. J. Phys.* 34, 1356 (1956).

J. Nahum and A. Halperin, Excitation spectra and temperature dependence of luminescence and photoconductivity of diamond, *J. Phys. Chem. Solids* 23, 345 (1962).

K. Nakazawa, K. Tanabe, M. Tachiki, H. Kawarada, and S. Koizumi, Excitonic recombination radiation in phosphorus-doped CVD diamonds, *Phys. Rev. B* 64, 235203 (2001).

M. Nesládek, K. Meykens, K. Haenen, L. M. Stals, T. Teraji, and S. Koizumi, Low-temperature spectroscopic study of n-type diamond, *Phys. Rev. B* 59, 14852 (1999).

J. te Nijenhuis, S. M. Olsthoorn, W. J. P. van Enckevort, and L. J. Giling, Red luminescence in phosphorus-doped chemically vapor deposited diamond, *J. Appl. Phys.* 82, 419 (1997).

M. Nishitani-Gamo, E. Yasu, C. Xiao, Y. Kikuchi, K. Ushizawa, I. Sakaguchi, T. Suzuki, and T. Ando, Sulfur-doped homoepitaxial (001) diamond with n-type semiconductive properties, *Diamond Relat. Mater.* 9, 941 (2000).

K. Okano, H. Kiyota, T. Iwasaki, Y. Nakamura, Y. Akiba, T. Kurosu, M. Iida, and T. Nakamura, Synthesis of n-type semiconducting diamond film using diphosphorus pentaoxide as the doping source, *Appl. Phys. A* 51, 344 (1990).

W. D. Partlow, J. Ruan, R. E. Witkowski, W. J. Choyke, and D. S. Knight, Cryogenic cathodoluminescence of plasma-deposited polycrystalline diamond coatings, *J. Appl. Phys.* 67, 7019 (1990).

S. J. Pennycook, L. M. Brown, and A. J. Craven, Observation of cathodoluminescence at single dislocations by STEM, *Philos. Mag.* 41, 589 (1980).

E. Pereira, Tabulation of luminescence decay times in diamond in "Properties and Growth of Diamond" EMIS Datareviews Series No. 9 (G. Davies, Ed.), p. 228. INSPEC, IEE, London, 1994.

P. Pirouz, D. J. H. Cockayne, N. Sumida, P. Hirsch, and A. R. Lang, *Proc. R. Soc. A* 396, 241 (1983).

G. Popovici, A. A. Melnikov, V. S. Varichenko, S. Khasawinah, T. Sung, M. A. Prelas, A. B. Denisenko, N. M. Penina, V. A. Martinovich, E. N. Drozdova, A. M. Zaitsev, W. R. Fahrner, J. W. Farmer, H. White, and J. Chamberlain, Properties of Li doped diamond films, obtained by transmutation of ^{10}B into ^{7}Li, *Diamond Relat. Mater.* 5, 761 (1996).

G. Popovici and M. A. Prelas, Prospective n-type impurities and methods of diamond doping, *Diamond Relat. Mater.* 4, 1305 (1995).

S. Prawer, C. Uzan-Saguy, G. Braunstein, and R. Kalish, Can n-type doping of diamond be achieved by Li or Na ion implantation? *Appl. Phys. Lett.* 63, 2502 (1993).

J. F. Prins, Activation of boron-dopant atoms in ion-implanted diamonds, *Phys. Rev. B* 38, 5576 (1988).

J. F. Prins, Modification, doping and devices in ion implanted diamond in "The Properties of Natural and Synthetic Diamond" (J. E. Field, Ed.), p. 301. Academic Press, London, 1992.

J. F. Prins, Optical studies on ion-beam-doped diamond layers, in "Advances in New Diamond Science and Technology" (S. Saito, N. Fujimori, O. Fukunaya, M. Kamo, K. Kobashi, and M. Yoshikowa, Eds.), p. 443. MYU, Tokyo, 1994a.

J. F. Prins, Cathodoluminescence and electroluminescence in ion implanted type II diamonds, *Diamond Relat. Mater.* 3, 922 (1994b).

J. F. Prins, Ion-implanted n-type diamond: electrical evidence, *Diamond Relat. Mater.* 4, 580 (1995).

J. F. Prins, Increased band A cathodoluminescence after carbon ion implantation and annealing of diamond, *Diamond Relat. Mater.* 5, 907 (1996).

J. F. Prins, Ion implantation of diamond below the amorphization threshold: defects, impurities, and their interactions, in "The 5th NIRIM International Symposium on Advanced Materials (ISAM '98)", Tsukuba, Ibaraki, Japan (M. Kamo, et al., Eds.), p. 93. National Institute for Research in Inorganic Materials, 1998a.

J. F. Prins, Ultraviolet cathodoluminescence from diamond layers after doping by means of boron-ion implantation, *Appl. Phys. Lett.* 73, 2308 (1998b).

C. J. Rauch, "Proc. Int. Conf. Phys. Semicond., (Exeter) 1962", p. 276. Institute of Physics and the Physical Society, London, 1963.

L. H. Robins, L. P. Cook, E. N. Farabaugh, and A. Feldman, Cathodoluminescence of defects in diamond films and particles grown by hot-filament chemical-vapor deposition, *Phys. Rev. B* 39, 13367 (1989).

L. H. Robins, E. N. Farabough, and A. Feldman, Spatially and spectrally resolved cathodoluminescence measurements of CVD-grown diamond particles and films in "Proc. 2nd Int. Symp. on Diamond Materials" (A. J. Purdes, J. C. Angus, R. F. Davis, B. M. Meyerson, K. E. Spear, and M. Yoder, Eds.), Proc. Vol. 91, p. 427. The Electrochem. Society, Pennington, NJ, 1991.

L. H. Robins, E. N. Farabaugh, and A. Feldman, Cathodoluminescence spectroscopy of free and bound excitons in chemical-vapor-deposited diamond, *Phys. Rev. B* 48, 14167 (1993).

M. C. Rossi, S. Salvatori, F. Galluzzi, R. M. Montereali, and F. Somma, Emission and excitation spectra of silicon-related luminescent centers in CVD-grown diamond films, *Diamond Relat. Mater.* 6, 1564 (1997).

J. Ruan, W. J. Choyke, and K. Kobashi, Oxygen-related centers in chemical vapor deposition of diamond, *Appl. Phys. Lett.* 62, 1379 (1993).

J. Ruan, W. J. Choyke, and W. D. Partlow, Cathodoluminescence and annealing study of plasma-deposited polycrystalline diamond films, *J. Appl. Phys.* 69, 6632 (1991a).

J. Ruan, W. J. Choyke, and W. D. Partlow, Si impurity in chemical vapor deposited diamond films, *Appl. Phys. Lett.* 58, 295 (1991b).

J. Ruan, K. Kobashi, and W. J. Choyke, On the "band-A" emission and boron related luminescence in diamond, *Appl. Phys. Lett.* 60, 3138 (1992a).

J. Ruan, K. Kobashi, and W. J. Choyke, Effect of oxygen on boron doping in chemical vapor deposition of diamond as deduced from cathodoluminescence studies, *Appl. Phys. Lett.* 60, 1884 (1992b).

D. Saada, J. Adler, and R. Kalish, Sulfur: a potential donor in diamond, *Appl. Phys. Lett.* 77, 878 (2000).

I. Sakaguchi, M. Nishitani-Gamo, Y. Kikuchi, E. Yasu, H. Haneda, T. Suzuki, and T. Ando, Sulfur: a donor dopant for n-type diamond semiconductors, *Phys. Rev. B* 60, R2139 (1999).

G. S. Sandhu, C. T. Kao, M. L. Swanson, and W. K. Chu, Doping of diamond by co-implantation with dopant atoms and carbon, *Mater. Res. Soc. Symp. Proc.* 162, 321 (1990).

Y. Sato, M. Kamo, H. Kanda, and N. Setaka, *J. Surf. Sci. Soc. Jpn* 1, 60 (1980).

R. Sauer, Ch. Kisielowski-Kemmerich, and H. Alexander, Dissociation-width dependent radiative recombination of electrons and holes at widely split dislocations in silicon, *Phys. Rev. Lett.* 57, 1472 (1986).

R. Sauer, Ch. Kisielowski-Kemmerich, and H. Alexander, Photoluminescence of straight dislocations in silicon, *Izv. Akad. Nauk, SSSR Ser. Fiz.* 51, 638 (1987) ([Bull. Acad. Sci. USSR, Phys. Ser. 51, 13 (1987)]).

R. Sauer, H. Sternschulte, S. Wahl, K. Thonke, and T. R. Anthony, Revised fine splitting of excitons in diamond, *Phys. Rev. Lett.* 84, 4172 (2000).

W. Schmid, U. Nieper, and J. Weber, Donor–acceptor pair spectra in Si:In LPE-layers, *Solid State Commun.* 45, 1007 (1983).

M. Schreck, Private communication (1998).

J. P. F. Sellshop, C. C. P. Madiba, and H. J. Annegarn, Light volatiles in diamond: physical interpretation and genetic significance, *Nuclear Instr. and Methods* 168, 529 (1980).

S. J. Sharp and A. T. Collins, A study of free- and bound-exciton recombination in device quality diamond, *Mater. Res. Soc. Symp. Proc.* 416, 125 (1996).

S. J. Sharp, A. T. Collins, G. Davies, and G. S. Joyce, Higher resolution studies of shallow bound exciton luminescence in diamond, *J. Phys: Condens. Matter* 9, L451 (1997).

G. Sittas, H. Kanda, I. Kiflawi, and P. M. Spear, Growth and characterization of Si-doped diamond single crystals grown by the HPHT method, *Diamond Relat. Mater.* 5, 866 (1996).

B. V. Spitsyn, L. L. Bouilov, and B. V. Deryaguin, Vapor growth of diamond on diamond and other surfaces, *J. Cryst. Growth* 52, 219 (1981).

J. W. Steeds, S. J. Charles, J. Davies, and I. Griffin, Photoluminescence microscopy of TEM irradiated diamond, *Diamond Relat. Mater.* 9, 397 (2000).

J. W. Steeds, S. Charles, T. J. Davis, A. Gilmore, J. Hayes, D. Pickard, and J. E. Butler, Creation and mobility of self-interstitials in diamond by use of a transmission electron microscope and their subsequent study by photoluminescence microscopy, *Diamond Relat. Mater.* 8, 94 (1999a).

J. W. Steeds, T. J. Davis, S. J. Charles, J. M. Hayes, and J. E. Butler, 3H luminescence in electron-irradiated diamond samples and its relationship to self-interstitials, *Diamond Relat. Mater.* 8, 1847 (1999b).

H. Sternschulte, T. Albrecht, K. Thonke, and R. Sauer, Band-gap narrowing and Mott transition in heavily boron-doped diamond: optical investigation in "The Physics of Semiconductors" (M. Scheffler and R. Zimmermann, Eds.), p. 169. World Scientific, Singapore, 1996a.

H. Sternschulte, T. Albrecht, K. Thonke, R. Sauer, M. Grießer, and M. Grasserbauer, Cathodoluminescence studies of bound excitons and near band gap emission lines in boron- and phosphorus-doped CVD-diamond, *Mater. Res. Soc. Symp. Proc.* 423, 693 (1996b).

H. Sternschulte, M. Schreck, and B. Stritzker, Growth and properties of CVD diamond films grown under H_2S conditions, *Diamond Relat. Mater.* 12, 318 (2003).

H. Sternschulte, M. Schreck, B. Stritzker, A. Bergmaier, and G. Dollinger, Lithium addition during CVD diamond growth: influence on the optical emission of the plasma and properties of the films, *Diamond Relat. Mater.* 9, 1046 (2000).

H. Sternschulte, K. Thonke, J. Gerster, W. Limmer, R. Sauer, J. Spitzer, and P. C. Münzinger, Uniaxial stress and Zeeman splitting of the 1.681 eV optical center in a homoepitaxial CVD diamond film, *Diamond Relat. Mater.* 4, 1189 (1995).

H. Sternschulte, K. Thonke, and R. Sauer, Near-band-edge luminescence studies of diamond during CVD growth or by ion implantation, *Phys. Status Solidi (a)* 172, 37 (1999a).

H. Sternschulte, K. Thonke, R. Sauer, and S. Koizumi, Optical evidence for 630-meV phosphorus donor in synthetic diamond, *Phys. Rev. B* 59, 12924 (1999b).

H. Sternschulte, K. Thonke, R. Sauer, P. C. Münzinger, and P. Michler, 1.681-eV luminescence center in chemical-vapor-deposited homoepitaxial diamond films, *Phys. Rev. B* 50, 14554 (1994).

H. Sternschulte, S. Wahl, K. Thonke, R. Sauer, M. Dalmer, C. Ronning, and H. Hofsäss, Observation of boron bound excitons in boron-implanted and annealed natural IIa diamonds, *Appl. Phys. Lett.* 71, 2668 (1997a).

H. Sternschulte, S. Wahl, K. Thonke, R. Sauer, T. Ruf, M. Cardona, and T. R. Anthony, Fine structure of the boron bound exciton in diamond, *Mater. Sci. Forum* 258–263, 757 (1997b).

M. Tajima, Determination of boron and phosphorus concentrations in silicon by photoluminescence analysis, *Appl. Phys. Lett.* 32, 719 (1978).

D. F. Talbot-Ponsonby, M. E. Newton, J. M. Baker, G. A. Scarsbrook, R. S. Sussmann, and A. J. Whitehead, EPR and optical studies on polycrystalline diamond films grown by chemical vapor deposition and annealed between 1100 and 1900 K, *Phys. Rev. B* 57, 2302 (1998).

N. Teofilov, A. Aleksov, K. Thonke, and R. Sauer. Reported on the Final Meeting of the D-A-CH joint research program of Germany, Austria, and Switzerland on the synthesis of superhard materials, March 2000, Neuhofen/Ybbs, Austria.

N. Teofilov, R. Sauer, K. Thonke, T. R. Anthony, and H. Kanda, Fine structure of excitons in diamond: a cathodoluminescence study, "Proc. 25th Int. Conf. Physics Semicond., Osaka 2000" in "Springer Proceedings in Physics Vol. 87" (N. Miura and T. Ando, Eds.), p. 95. Springer, Berlin, 2001.

N. Teofilov, R. Sauer, K. Thonke, and S. Koizumi, Bound exciton luminescence related to phosphorus donors in CVD diamond. Physica B (2003, in press).

D. G. Thomas, M. Gershenzon, and F. A. Trumbore, Pair spectra and 'edge' emission in gallium phosphide, *Phys. Rev.* 133, A269 (1964).

D. G. Thomas and J. J. Hopfield, Isoelectronic traps due to nitrogen in gallium phosphide, *Phys. Rev.* 150, 680 (1966).

K. Thonke, R. Schliesing, N. Teofilov, H. Zacharias, R. Sauer, A. M. Zaitsev, H. Kanda, and T. R. Anthony, Electron-hole drops in synthetic diamond, *Diamond Relat. Mater.* 9, 428 (2000).

V. S. Vavilov, A. A. Gippius, A. M. Zaitsev, B. V. Deryagin, B. V. Spitsyn, and A. E. Aleksenko, Investigation of the cathodoluminescence of epitaxial diamond films, *Fiz. Tekh. Polyprovodn.* 14, 1811 (1980) ([Sov. Phys. Semicond. 14, 1078 (1980)]).

J. Walker, Optical absorption and luminescence in diamond, *Rep. Prog. Phys.* 42, 1605 (1979).

L. G. Wang and A. Zunger, Phosphorus and sulphur doping of diamond, *Phys. Rev. B* 66, 161202 (R) (2002).

H. Watanabe, K. Hayashi, D. Takenuchi, S. Yamanaka, H. Okushi, K. Kajimura, and T. Sakiguchi, Strong excitonic recombination radiation from homoepitaxial diamond thin films at room temperature, *Appl. Phys. Lett.* 73, 981 (1998).

H. Watanabe and H. Okushi, Nonlinear effects (in the) excitonic emission from high quality homoepitaxial diamond films, *Jpn. J. Appl. Phys.* 39, L835 (2000).

J. Weber, W. Schmid, and R. Sauer, Localized exciton bound to an isoelectronic trap in silicon, *Phys. Rev. B* 21, 2401 (1980).

J. D. Wiley and J. A. Seman, The enumeration of neighbours on cubic and hexagonal-based lattices, *Bell Syst. Tech. J.*, 50, 355 (1970).

M. Willatzen, M. Cardona, and N. E. Christensen, Linear muffin-tin-orbital and k·p calculations of effective masses and band structure of semiconducting diamond, *Phys. Rev. B* 50, 18054 (1994).

F. Williams, Donor–acceptor pairs in semiconductors, *Phys. Status Solidi* 25, 493 (1968).

B. G. Yacobi, A. R. Badzian, and T. Badzian, Cathodoluminescence study of diamond films grown by microwave plasma assisted chemical vapor deposition, *J. Appl. Phys.* 69, 1643 (1991).

B. G. Yacobi, J. Lebens, K. J. Vahala, A. R. Badzian, and T. Badzian, Preferential incorporation of defects in monocrystalline diamond films, *Diamond Relat. Mater.* 2, 92 (1993).

N. Yamamoto, J. C. H. Spence, and D. Fathy, Cathodoluminescence and polarization studies from individual dislocations in diamond, *Philos. Mag. B* 49, 609 (1984).

Y. Yokota, H. Kawarada, and A. Hiraki, Effect of doping with nitrogen and boron on cathodoluminescence of CVD-diamond, *Mater. Res. Soc. Symp. Proc.* 162, 231 (1990).

A. M. Zaitsev, Characterization techniques in "Handbook of Industrial Diamond and Diamond Films" (M. A. Prelas, G. Popovici, and L. K. Bigelow, Eds.), p. 227. Dekker, New York, 1998.

A. M. Zaitsev, "Optical Properties of Diamond" A Data Handbook. Springer, Heidelberg, 2001.

A. M. Zaitsev, V. S. Vavilov, and A. A. Gippius, Cathodoluminescence of diamond associated with silicon impurity, *Sov. Phys. Lebedev, Inst. Rep.* 10, 15 (1981).

A. M. Zaitsev, A. A. Melnikov, A. V. Denisenko, V. S. Varichenko, R. Job, and W. R. Fahrner, Luminescence characterization and application of diamond, *Mater. Res. Soc. Symp. Proc.* 416, 113 (1996).

H. Zhou, G. D. Watkins, K. M. McNamara Rutledge, R. P. Messmer, and S. Chawla, Hydrogen-related defects in polycrystalline CVD diamond, *Phys. Rev. B* 54, 7881 (1996).

H. Zhou, Y. Yokoi, H. Tamura, S. Takami, M. Kubo, A. Miyamato, M. Gamo, and T. Ando, Quantum chemical calculations of sulfur doping reactions in diamond CVD, *Jpn. J. Appl. Phys.* 40, 2830 (2001).

U. O. Ziemelis and R. R. Parsons, Sharp line donor–acceptor pair luminescence in silicon, *Can. J. Phys.* 59, 784 (1981).

U. O. Ziemelis, R. R. Parsons, and M. Voos, The puzzle of double-doped Si(B,In): sharp line series in near-band-edge photoluminescence, *Solid State Commun.* 32, 445 (1979).

Index

D

P